ShockWave Science and Technology Reference Library

The new Springer collection, Shock Wave Science and Technology Reference Library, conceived in the style of the famous Handbuch der Physik has as its principal motivation to assemble authoritative, state-of-the-art, archival reference articles by leading scientists and engineers in the field of shock wave research and its applications. A numbered and bounded collection, this reference library will consist of specifically commissioned volumes with internationally renowned experts as editors and contributing authors. Each volume consists of a small collection of extensive, topical and independent surveys and reviews. Typical articles start at an elementary level that is accessible to non-specialists and beginners. The main part of the articles deals with the most recent advances in the field with focus on experiment, instrumentation, theory, and modeling. Finally, prospects and opportunities for new developments are examined. Last but not least, the authors offer expert advice and cautions that are valuable for both the novice and the well-seasoned specialist.

ShockWave Science and Technology Reference Library

Collection Editors

Hans Grönig

Hans Grönig is Professor emeritus at the Shock Wave Laboratory of RWTH Aachen University, Germany. He obtained his Dr. rer. nat. degree in Mechanical Engineering and then worked as postdoctoral fellow at GALCIT, Pasadena, for one year. For more than 50 years he has been engaged in many aspects of mainly experimental shock wave research including hypersonics, gaseous and dust detonations. For about 10 years he was Editor-in-Chief of the journal Shock Waves.

Yasuyuki Horie

Professor Yasuyuki (Yuki) Horie is internationally recognized for his contributions in high-pressure shock compression of solids and energetic materials modeling. He is a co-chief editor of the Springer series on Shock Wave and High Pressure Phenomena and the Shock Wave Science and Technology Reference Library, and a Liaison editor of the journal Shock Waves. He is a Fellow of the American Physical Society, and Secretary of the International Institute of Shock Wave Research. His current interests include fundamental understanding of (a) the impact sensitivity of energetic solids and its relation to microstructure attributes such as particle size distribution and interface morphology, and (b) heterogeneous and nonequilibrium effects in shock compression of solids at the mesoscale.

Kazuyoshi Takayama

Professor Kazuyoshi Takayama obtained his doctoral degree from Tohoku University in 1970 and was then appointed lecturer at the Institute of High Speed Mechanics, Tohoku University, promoted to associate professor in 1975 and to professor in 1986. He was appointed director of the Shock Wave Research Center at the Institute of High Speed Mechanics in 1988. The Institute of High Speed Mechanics was restructured as the Institute of Fluid Science in 1989. He retired in 2004 and became emeritus professor of Tohoku University. In 1990 he launched Shock Waves, an international journal, taking on the role of managing editor and in 2002 became editorin-chief. He was elected president of the Japan Society for Aeronautical and Space Sciences for one year in 2000 and was chairman of the Japanese Society of Shock Wave Research in 2000. He was appointed president of the International ShockWave Institute in 2005. His research interests range from fundamental shock wave studies to the interdisciplinary application of shock wave research.

F. Zhang (Ed.)

Shock Wave Science and Technology Reference Library, Vol. 6

Detonation Dynamics

With 264 Figures and 20 Tables

Fan Zhang
Defence Research and Development Canada - Suffield
PO Box 4000, Station Main
Medicine Hat, Alberta, T1A 8K6
Canada
fan.zhang@drdc-rddc.gc.ca

Fan Zhang

Fan Zhang is a Senior Scientist and the Head of the Advanced Energetics Group at Defence Research and Development Canada – Suffield and an adjunct Professor at the University of Waterloo in the Department of Mechanical Engineering. He specializes in shock waves, detonations and explosions, more specifically in multiphase reactive flow, heterogeneous explosives and high energy density systems. He obtained a Doctoral degree in Science in 1989 from the Aachen University of Technology (RWTH), Germany, and received a Borchers Medal, a Friedrich-Wilhelm Prize and several defence community awards. He is the author or co-author of more than 200 refereed journal and proceedings papers including book chapters and journal special issues. He has served for a number of international defence technical panels and academic committees.

ISBN 978-3-642-22966-4 e-ISBN 978-3-642-22967-1
DOI 10.1007/978-3-642-22967-1
Springer Heidelberg Dordrecht London New York

Library of Congress Control Number: 2011941778

Printed on acid-free paper

Springer is part of Springer Science+Business Media (www.springer.com)

Preface

As a volume in the *Shock Wave Science and Technology Reference Library*, this book is primarily concerned with the fundamental theory of detonation physics in gaseous and condensed-phase reactive media.

A detonation is a chemical reaction transport process accompanied by high-speed energy release. It comprises a shock wave standing upstream of the chemical reaction, where both propagate as a tightly coupled complex with supersonic velocity (2–10 km/s). After the passage of the detonation wave, the reactive medium is transformed into high temperature and pressure gaseous products, thereby providing enormous thermodynamic work potential for commercial and military applications. The detonation process involves complex interactions between reactive chemical dynamics and fluid dynamics, accompanied by intricate effects of heat, light, electricity and magnetism. These characteristics make detonation dynamics an important field of applied physics, spanning numerous theoretical and applied research topics. Outstanding summaries of the fundamental physics and theory of detonation can be found in a number of classic books (e.g., Zeldovich, Ya.B., Kompaneets, A.S.: Theory of Detonation, Academic Press, New York, 1960; Soloukhin, R.J.: Shock Waves and Detonations in Gases, Mono Book Corp., Baltimore, 1966; Fickett, W., Davis, W.C.: Detonation, University of California Press, Berkeley, 1979). A recent book "The Detonation Phenomenon" by J.H.S. Lee (Cambridge University Press, Cambridge, 2008) offers a very informative reading on the up-to-date phenomenology of gaseous detonation physics. The current book, however, tries to address the recent developments in the theoretical foundation of detonation physics, in both gaseous and condensed-phase energetic materials. The book contains seven chapters which were written by a number of subject experts. The chapters are thematically interrelated in a systematic descriptive approach, where each chapter is self-contained. It offers a timely reference in theoretical detonation physics for graduate students as well as professional scientists and engineers.

The first chapter, by Sorin Bastea and Laurence E. Fried, describes the equilibrium Chapman–Jouguet (CJ) detonation theory. The CJ theory simplifies the detonation wave to a strong discontinuity after which the flow is sonic with chemical reaction complete at an equilibrium detonation products state. With a brief introduction to the classic CJ theory, the chapter focuses on modern challenging topics of equilibrium detonation with emphasis on condensed-phase energetic materials. These subjects include the equations of state and high pressure chemistry and physics of detonation products such as dissociation, ionization and phase separation. Conditions and limitations are discussed for applying the theory to practical explosives with finite size and heterogeneity. This chapter offers a solid foundation for the application of modern equilibrium detonation theory to increase predictability in computing macroscopic average detonation states for a wide variety of both existing and novel energetic materials.

The simplicity of the CJ theory prevents one from gaining insights into the variety of detonation wave structures responsible for detonation propagation, initiation and failure. The second chapter, by Andrew Higgins, introduces the Zeldovich, von Neumann and Döring (ZND) model that emerged in the 1940s to offer a steady planar reaction zone structure of detonation. The chapter then provides a state-of-the-art description of generalized, steady, quasi-one-dimensional detonation theory, in which the rate of energy loss competes with the rate of exothermic chemical energy release within the reaction zone, thus resulting in the flow passing a sonic point at a chemically non-equilibrated state. The energy loss inherent in the materials or its boundaries has been systematically depicted in the chapter through endothermic chemical reaction, friction, heat loss and flow divergence. The common characteristics and differing varieties of steady detonation zone structures are discussed under various energy loss mechanisms through the coupling of the chemical reaction and fluid dynamics. This chapter provides a one-dimensional average framework upon which the detonation instability and multidimensional structures are built. The generalized ZND detonation theory also offers a foundation to predict detonation performance of nonideal energetic materials.

Numerous experiments since 1959 have demonstrated that the dynamic trajectory of a detonation wave front generally manifests itself as an unsteady, non-planar cellular structure in both homogeneous and a number of heterogeneous materials. The steady one-dimensional ZND detonation structure is theoretically unstable and will be transformed into an ordered unsteady oscillating structure under given conditions. The third chapter, by Hoi Dick Ng and Fan Zhang, describes the fundamental theory of detonation instability and its recent development. The kernel of unstable detonation theory lies within a nonlinear dynamics consideration, in which the oscillating instability modes generally transit through a period bifurcation or trifurcation sequence to chaos. After an introduction to classic linear stability analysis, the chapter presents in detail this latest nonlinear detonation instability theory in one-dimensional space. This theory helps to reveal the hidden nature of the reactive fluid dynamics equations, where the coupled chemistry dominates the control parameters for instability transition. The effect of coupled chemistry on the detonation instability is further addressed through different reaction kinetic models. Recent two-dimensional numerical simulations also confirm the instability hierarchy of detonation cellular structure in the same bifurcation sequence, with additional multidimensional effects including transverse waves and turbulent mixing to help us move towards a quantitative description of cellular detonation instability. Finally, nonideal detonation instability is analyzed, where the dissipation resulting from energy loss sources create an additional class of stability control parameters that lead to detonation velocity reduction and instability.

The fourth chapter, by Anatoly Vasil'ev, provides a fundamental description of dynamic parameters of cellular detonation in gaseous mixtures with an emphasis on the theoretical analysis, mostly from the author's own distinguished studies, accompanied with numerous experimental data. The dynamic parameters described in this chapter include detonation cell size, critical energy for direct initiation of detonation and critical opening size for detonation diffraction re-initiation. These subjects are followed by an introduction to advanced schemes for detonation initiation and critical geometry scales of limiting detonation waves. The theoretical models capture the essential detonation phenomenon of transverse wave collisions as local explosions to form detonation cells; their agreement with a large quantity of experimental data demonstrates an adequate functional relation between critical initiation energy and cell size. This chapter lays out a theoretical foundation for the calculation of dynamic parameters of the regular cellular detonation, as well as the irregular cellular detonation with a dominant mean cell size.

The fifth chapter, by Daniel Desbordes and Henri-Noël Presles, offers an informative review of recent advances in the area of gaseous detonation cellular structure. The detonation cellular structure has been characterized by a main cell size that is deduced from a cell size spectrum and may contain substructures, depending on the degree of detonation instability. The authors of this chapter, along with their coworkers, have newly discovered a two-cell structure where a large cellular structure is superimposed on that of smaller cells, thus opening an avenue which extends existing detonation physics and theory. This chapter begins with an overview of the experimental observation of various detonation wave structures including the latest finding of two-cell structure. It then presents a detonation energy release theory, in which the traditional cellular structure dominated by one main cell size, is replicated by one-step chemical heat release with one peak of thermicity, while the two-cell detonation structure is a result of two-step heat release with two maxima in thermicity within the ZND detonation reaction zone. Between the two models, there is a transition regime where the two-step heat release only exhibits one peak in thermicity. This theory is applied in direct numerical simulations to predict various cellular detonation structures in different mixtures, and in an attempt to explain disputed critical problems including detonation diffraction re-initiation and the limiting detonation in tubes. The multi-cell detonation wave might display a more comprehensive structure for mixtures of multiple fuels with different chemical kinetic length scales as the reacting flow moves towards the sonic point. Hence, the theory invoked in this chapter, if valid, might be extended to a multi-scale heat release law with multiple maxima in thermicity to predict multi-cell detonation structure. While the detonation propagation, transmission and instabilities have been associated with the role and type of transverse waves, detonation chemistry is no doubt dominant in controlling the dynamics of cell structure.

The description in Chaps. 2–5 primarily uses gas as a prototypical material, because the inherent low density permits sufficient observation resolution. It is expected that the range of applicability of the main principles described in these chapters would extend to detonation in generic homogeneous energetic materials including liquid and solid, as well as in a number of heterogeneous materials (see Heterogeneous Detonation, Volume 4 of *Shock Wave Science and Technology Reference Library*).

Due to high material density and detonation pressure of the order of tens of gigapascals, the condensed-phase detonation structure has been a subject of debate between nonequilibrium thermal initiation behind the shock front and nonequilibrium mechanical initiation within the shock front. Experimental support to answer these questions must be derived under extreme conditions of pressure from observations within the shock front, thereby requiring measurement resolution on the tens of femtosecond time scale. The sixth chapter by Craig M. Tarver presents an overview on the development and challenge towards a condensed matter detonation theory. The author proposes a nonequilibrium ZND model based on the timescale analysis of various nonequilibrium processes. While much more research will be required to fully understand the nonequilibrium processes coupled with the high pressure chemistry to establish a condensed matter detonation theory, the chapter also offers a simple phenomenological reaction model of ignition and growth, and its use in practical detonation modeling in comparison with a number of experiments.

The last chapter, by John B. Bdzil and D. Scott Stewart, systematically describes the fundamentals of detonation shock dynamics (DSD) as a technique to efficiently compute engineering-scale effects of condensed-phase detonation. A detonation wave front for condensed-phase energetic materials, from a macroscopic average view point, generally experiences curvature in the vicinity of an inert lateral boundary. The detonation velocity is influenced by this front curvature and can experience a strong deficit due to divergent flow. In the DSD theory, the relationship between the normal detonation velocity and the front curvature therefore plays a key role. The authors of this chapter have laid the main foundation for the rigorous development of modern DSD theory. They extended Whitham's geometrical inert shock dynamics to a detonation shock evolution through an asymptotic theory applied to the quasi-one-dimensional ZND detonation model. This chapter offers a rigorous derivation and description of the theory in a gradual and logical fashion. It clearly demonstrates the superiority of this theory over the classic Huygens formulation where the detonation velocity is assumed to be constant. It also shows great potential for high simulation efficiency in engineering-scale explosive applications when the DSD theory is coupled with the numerical solution of the flow field behind the detonation. This precludes the necessity to apply expensive, multidimensional simulations with sufficient resolution to capture the correct solution within the 10^{-5} m or less thickness of the detonation reaction zone.

The editor is indebted to all of the authors for their willingness to prepare and make available their timely and authoritative materials to a wide audience. Sincere gratitude is directed to Hans Grönig, Yasuyuki Horie and Kazuyoshi Takayama for their strong support. Particular thanks go to Andrew Higgins and Hoi Dick Ng for their helpful commentary and useful suggestions during the final edition of this book to maintain consistency between the closely related chapters. Andrew Higgins also spent many hours carefully proofreading the proofs. François-Xavier Jetté and Patrick Lee Batchelor are thanked for their proof read of the draft and Akio Yoshinaka for the proof read of selected parts. The editor also wishes to thank the dedicated work of Springer-Verlag in their final process of editing and publication in order to ensure a good appearance of this book.

Medicine Hat, Canada *Fan Zhang*

Contents

Contributors

Sorin Bastea
Lawrence Livermore National Laboratory
Physical and Life Sciences
L-350, 7000 East Ave.
Livermore, CA 94550 USA
sbastea@llnl.gov

Laurence E. Fried
Lawrence Livermore National Laboratory
Physical and Life Sciences
L-282, 7000 East Ave.
Livermore, CA 94550 USA
lfried@llnl.gov

Andrew Higgins
McGill University
Mechanical Engineering
817 Sherbrooke St. W.,
Montreal, Quebec,
H3A 2K6 Canada
andrew.higgins@mcgill.ca

Hoi Dick Ng
Concordia University
Mechanical & Industrial Engineering
1455 de Maisonneuve Blvd. W.,
Montreal, Quebec,
H3G 1M8 Canada
hoing@encs.concordia.ca

Fan Zhang
Defence R&D Canada - Suffield
PO Box 4000, Stn. Main
Medicine Hat, AB T1A 8K6 Canada
fan.zhang@drdc-rddc.gc.ca

Anatoly A. Vasil'ev
Lavrentyev Institute of Hydrodynamics, Novosibirsk,
630090 Russia
gasdet@hydro.nsc.ru

Daniel Desbordes
Ecole Nationale Supérieure de Mécanique et d'Aérotechniques (ENSMA), Laboratoire de Combustion et de Détonique(LCD)
86961 Futuroscope-Chasseneuil,
France
desbordes@lcd.ensma.fr

Henri-Noël Presles
Ecole Nationale Supérieure de Mécanique et d'Aérotechniques (ENSMA), Laboratoire de Combustion et de Détonique(LCD)
86961 Futuroscope-Chasseneuil,
France
presles@lcd.ensma.fr

Craig M. Tarver
Lawrence Livermore National Laboratory
Energetic Materials Center
Livermore, CA 94550 USA
tarver1@llnl.gov

John B. Bdzil
Los Alamos National Laboratory
Los Alamos, NM USA
University of Illinois at Urbana-Champaign
Mechanical Sciences & Engineering
1206 West Green Street, Urbana, IL 61801 USA
jbbdzil@gmail.com

D. Scott Stewart
University of Illinois at Urbana-Champaign
Mechanical Sciences & Engineering
1206 West Green Street, Urbana, IL 61801 USA
dss@illinois.edu

1

Chemical Equilibrium Detonation

Sorin Bastea and Laurence E. Fried

1.1 Introduction

Energetic materials are unique for having a strong exothermic reactivity, which has made them desirable for both military and commercial applications. The fundamental principles outlined in this chapter pertain to the study of detonation in both gas-phase and condensed-phase energetic materials, but our main focus will be on the condensed ones, particularly on high explosives (HEs). They share many properties with other classes of condensed energetic compounds such as propellants and pyrotechnics, but a detailed understanding of detonation is especially important for numerous HE applications. The usage and study of HE materials goes back more than a century, but many questions remain to be answered, e.g., on their reaction pathways at high pressures and temperatures, chemical properties, etc.

Knowledge of the chemical reaction mechanisms of condensed energetic materials at high densities and temperatures is essential for understanding events that occur at the reactive front under combustion or detonation conditions. Under shock conditions, for example, energetic materials undergo rapid heating to a few thousand degrees and are subjected to a compression of hundreds of kilobars, resulting in an approximate volume reduction of 30% [84]. Complex chemical reactions are thus initiated, in turn releasing large amounts of energy to sustain the detonation process. Detailed investigations into the reactive process have been undertaken over the past 2 decades [49]. However, the sub-microsecond timescales of explosive reactions, in addition to the highly exothermic conditions of an explosion, make experimental study of the decomposition pathways exceedingly difficult.

Chemical equilibrium thermodynamic modeling is a relatively well-established approach to understanding reactivity in energetic materials, although there are still many challenges in creating models that adequately describe the growing knowledge of chemical processes at extreme conditions. The chemical equilibrium approach uses a simplified thermodynamic picture of the reaction process, leading to a convenient and predictive model of

F. Zhang (ed.), *Shock Wave Science and Technology Reference Library, Vol. 6:* Detonation Dynamics, DOI 10.1007/978-3-642-22967-1_1,

detonation and other decomposition processes. Chemical equilibrium codes are often used in the design of new materials, both at the level of synthesis chemistry and formulation.

One of the attractive features of chemical equilibrium thermodynamic modeling is that it requires limited information regarding the unreacted energetic material: elemental composition, density, and heat of formation of the material are the only properties needed. Since elemental composition is known once the material is specified, only density and heat of formation need to be measured or predicted. While predicting these properties from first principles is a significant theoretical challenge [56, 57], much work in energetic material development involves combining well-characterized materials where the density and heat of formation are known. In such cases chemical equilibrium calculations can be an invaluable aid to the experimentalist.

1.2 Chemical Equilibrium States

The energy content of an HE material often determines its practical utility. Accurate estimates of the energy content are essential in the design of new materials [26] and for understanding quantitative detonation tests [66]. The useful energy content is determined by the anticipated release mechanism. Because many detonation events of condensed-phase explosives typically occur on a microsecond time frame, chemical reactions significantly faster than this may be assumed to induce instantaneous chemical equilibrium. It is generally believed that reactions involving the production of small gaseous molecules (CO_2, H_2O, etc.) at temperatures over 2,000 K are fast enough to justify treating such molecules as being in chemical equilibrium for most energetic materials. This belief is based partly on success in modeling a wide range of materials with the assumption of chemical equilibrium [13, 19, 20, 53, 71].

Unfortunately, direct measurements of chemical species involved in the detonation of a solid or liquid HE material are difficult to perform. Blais et al. [8] have measured some of the species produced in detonating nitromethane using a special mass spectroscopic apparatus. These measurements pointed to the importance of condensation reactions in detonation. The authors estimate that the length of the hydrodynamic reaction zone of detonating base-sensitized liquid nitromethane is 50 μm with a reaction time of 7 ns. Typical explosive experiments are performed on parts with dimensions on the order of 1–10 cm. In this case, hydrodynamic confinement is expected to last for roughly 10 μs, before the expansion reaches the center of the charge, based on a high-pressure sound speed of several mm/μs. Thus, chemical equilibrium is expected to be a valid assumption for nitromethane, based on the timescale separation between the 7 ns reaction zone and the 10 μs timescale of confinement. Chemical equilibrium can be reached rapidly under high temperature (up to 6,000 K) conditions produced by detonating energetic materials [47]. The formation of solids, however, such as carbon, or the combustion of metallic fuels, such as

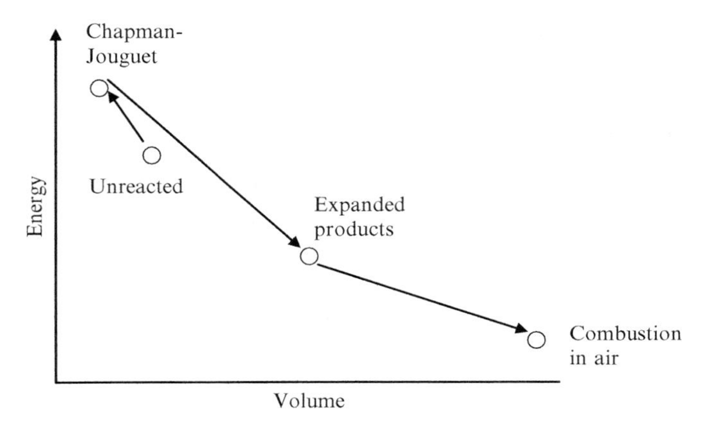

Fig. 1.1. A thermodynamic picture of detonation: the unreacted material is compressed by the shock front and reaches the Chapman–Jouguet point. From there adiabatic expansion occurs, leading to a high volume state. Finally, detonation products may mix in air and combust, in a process called afterburning

Al, is believed to yield significantly longer timescales of reaction [17]. In this case detonation is nonideal and chemical equilibrium is a rough, although still valuable, approximation to the state of matter of a detonating material.

Thermodynamic paths are a useful concept for understanding energy release mechanisms. Detonation can be thought of as a path that transforms the unreacted explosive into stable product molecules at the Chapman–Jouguet (CJ) state [22,82], which is a steady, chemical equilibrium shock state that conserves mass, momentum, and energy (see Fig. 1.1). Similarly, the deflagration of a propellant converts the unreacted material into product molecules at constant enthalpy and pressure. The nature of the CJ state and other special thermodynamic states important to energetic materials is determined by the equation of state of the stable reaction products, as described in more detail below.

A purely thermodynamic treatment of detonation ignores the important question of reaction timescales. The finite timescale of reaction can lead to strong deviations in detonation velocities from values based on the CJ theory [40]. For molecular explosives, this difficulty is mitigated by the observation that the thermodynamic limit is reached for sufficiently large charges, since for a large charge expansion waves take longer to interrupt the detonation process. Unfortunately, the decomposition kinetics of even simple molecules under high-pressure conditions is not well understood.

High-pressure experiments promise to provide insight into chemical reactivity under extreme conditions. For instance, chemical equilibrium analysis of shocked hydrocarbons predicts the formation of condensed diamond and molecular hydrogen [51, 74]. This prediction has been validated by static compression experiments demonstrating the formation of diamond in compressed methane [7]. Similar mechanisms are at work when detonating energetic materials form condensed carbon [71].

Most thermochemical calculations treat solid products such as graphite, diamond, and Al_2O_3 as being distinct phases from the gaseous detonation products. The free energy of the entire system is determined by summing the free energy of each phase. It is typically difficult to find chemical equilibrium for a multiphase system, and therefore research into algorithms for chemical and phase equilibrium remains an active field [65]. Very often condensed phases formed in detonations occur in the form of nanosized particles [32], which have properties substantially different from those of the bulk phase [63, 73]. The accurate treatment of nanoparticle formation in detonation is a current challenge in thermochemical modeling [74].

We begin our discussion by examining thermodynamic path theory as applied to high explosive detonation. This is a current research topic because high explosives produce detonation products at extreme pressures and temperatures: up to 40 GPa and 6,000 K. These conditions make it very difficult to probe chemical speciation. Nonetheless, shock experiments on a wide range of materials have generated sufficient information to allow reliable thermodynamic modeling to proceed.

The main assumption of the CJ detonation theory [22] is that the performance of an explosive is determined by the CJ state, and its connected expansion region as illustrated in Fig. 1.1. Since detonation processes are extremely rapid, there is insufficient time for thermal conduction during expansion. This implies that the expansion from the CJ state lies on an adiabat: $\mathrm{d}E = -p\mathrm{d}V$. Here, E is the energy per unit mass, p is the pressure, and V is the volume per unit mass. The expansion of the detonation products releases energy in the form of work and heat. Subsequent turbulent mixing of the fuel-rich detonation products in air surrounding the energetic material leads to combustion processes that release more energy.

Thermochemical codes use thermodynamics to calculate states illustrated in Figs. 1.1 and 1.2, and thus predict explosive detonation performance. Since a detonation wave is essentially a self-propagating shock wave whose energy is supplied by the chemical reaction of the original explosive, the fundamental equations that describe the steady propagation of a shock wave in a material are introduced first. These equations are best derived by considering the classical example of a fluid in a semi-infinite cylinder which is being

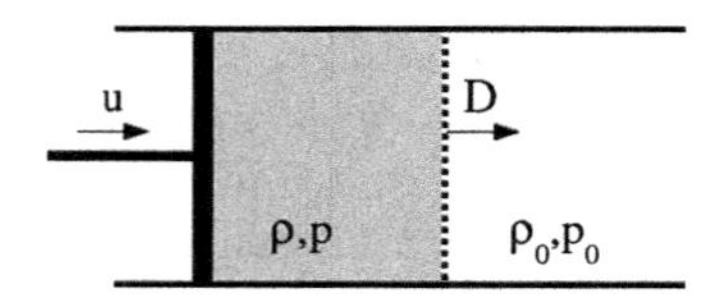

Fig. 1.2. Schematic diagram of a shock wave generated into a cylinder-enclosed fluid by a moving piston

compressed by a piston moving with constant velocity u. We further assume that the compression is sufficiently fast so that dissipation processes associated with heat conduction and viscosity can be neglected. If the fluid is initially at rest with constant mass density ρ_0 and pressure p_0, the steady state flow solution that satisfies the boundary conditions (both next to the piston and far away in front of it where the gas is still undisturbed) corresponds to a moving, infinitely thin surface of discontinuity commonly referred to as a shock wave.

The shock wave separates two uniform flow regions, one in front of it extending to infinity, where the fluid is at rest with the initial thermodynamic conditions ρ_0 and p_0, and another behind it, bounded by the piston, which contains compressed fluid with density ρ and pressure p moving with velocity u; the shock wave itself propagates with a constant velocity D, which is greater than u. The shock wave properties can be determined by explicitly considering the conservation of mass, momentum, and energy across the shock front. This is best accomplished in a moving frame with velocity D, in which the discontinuity is stationary. In this frame of reference the undisturbed fluid with density ρ_0 moves towards the surface of discontinuity with velocity $-D$, and the compressed fluid moves away from it with velocity $-(D-u)$. By equating the mass fluxes on both sides of the surface we obtain

$$\rho_0 D = \rho(D-u) - \text{Mass conservation.} \tag{1.1}$$

Moreover, from Newton's second law the change of momentum flux across the surface is equal to the difference in pressure:

$$\rho(D-u)^2 - \rho_0 D^2 = p_0 - p - \text{Momentum conservation.} \tag{1.2}$$

Here, p is understood to be the diagonal component of the stress tensor in the direction of the shock front. For a material with no strength, p is the same as the pressure. Furthermore, the change of energy (internal and kinetic) flux is equal to the work done per unit area and per unit time by the pressure forces:

$$\begin{aligned} &\rho(D-u)\left(E + \frac{(D-u)^2}{2}\right) - \rho_0 D\left(E_0 + \frac{D^2}{2}\right) \\ &\quad = p_0 D - p(D-u) - \text{Energy conservation,} \end{aligned} \tag{1.3}$$

where E and E_0 are internal energies per unit mass. If the initial state of the fluid (ρ_0, p_0, E_0) and the thermodynamic function $E(\rho,\ p)$ of the shocked fluid are known, the above three equations can be used to determine the state of the fluid behind the shock (ρ, p, E) and its velocity u for any given shock velocity D. (Note that $E(\rho,\ p)$ can be obtained for example from the knowledge of the equations of state (EOS) $p = p(\rho, T)$ and $E = E(\rho, T)$ by substituting the temperature). Eliminating the fluid velocity u from the momentum and energy equations by using the mass conservation equation, the following equations are obtained, respectively:

$$p - p_0 = \rho_0^2 D^2 (V_0 - V) \tag{1.4}$$

$$E - E_0 = \frac{1}{2}(p + p_0)(V_0 - V), \tag{1.5}$$

where $V = 1/\rho$ and $V_0 = 1/\rho_0$ are the specific volumes. The first of these two equations, expressing the linear dependence of the shock pressure p on the shock specific volume V at a given shock velocity D, is traditionally called the Rayleigh line. The second equation expresses the energy E of the shocked fluid as a function of its pressure p and specific volume V and it is called the Rankine–Hugoniot or simply the Hugoniot relation. Together with the thermodynamic function $E = E(V, p)$ this equation yields the shock pressure p as a function of shock specific volume V, typically denoted as the Hugoniot curve for initial conditions $(V_0,\ p_0,\ E_0)$. The Hugoniot curve encompasses all the thermodynamic states that can be reached by shocking a material starting from its $(V_0,\ p_0,\ E_0)$ state. Because mass and momentum must be conserved at all times, as embedded in the Rayleigh line, it is clear that for any given shock velocity D the shock pressure p and specific volume V are found at the intersection of the Rayleigh line with the Hugoniot curve. Since in most cases the Hugoniot curve is a convex function of V in the typical domain of interest $[0, 0.5V_0]$, its intersection with the Rayleigh line will yield at most two points, one possibly at $(V_0,\ p_0)$. This assumption will be made in the following, but it is worth noting that phase and/or chemical transformations can lead to more complicated Hugoniot curves that do not satisfy this condition.

It should be noted that not all intersections of the Rayleigh line with the Hugoniot curve necessarily correspond to stable, steadily propagating shock waves. Two fundamental criteria of stability for a shock wave are $D > c_0$ and $|D - u| < c$, where c_0 and c are the adiabatic sound speeds for the undisturbed and shocked fluid, respectively. They are best understood in the frame of reference moving with the shock, and state that disturbances from behind the shock front should not be able to propagate ahead of it, and thus widen it and lead to its decay (first condition), while behind the shock front sound waves from the piston should be able to reach it to deliver the energy necessary to sustain its propagation (second condition). Additional shock wave stability criteria are related to the resilience of the surface of discontinuity against deviations from planarity and translate into mathematical conditions that involve the properties of the Hugoniot curve; for a detailed discussion of these issues see for example [68] and references therein.

As previously noted a detonation wave can be regarded as a self-sustaining shock wave, in which the compression due to the detonation products produced behind the shock front plays the role of the piston. Consequently, the above mass, momentum, and energy equations apply to detonation waves as well. The simplest detonation theory, proposed by Chapman and Jouguet [22], assumes that the transformation of the explosive into chemical products

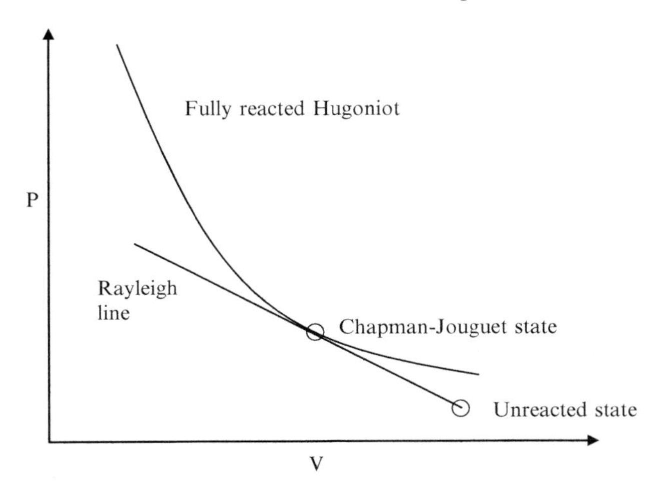

Fig. 1.3. Allowed thermodynamic states in detonation are constrained to the shock Hugoniot. Steady state shock waves follow the Rayleigh line

occurs across an infinitely thin surface and that the chemical reactions reach completion instantaneously. As a result, the detonation products behind the shock front are in thermodynamic equilibrium and their properties can be calculated using standard thermodynamics. The relevant Hugoniot curve for detonation problems is therefore the one corresponding to full chemical equilibrium, often referred to as the fully reacted Hugoniot. Note that this curve does not pass through the initial conditions $(V_0,\ p_0)$, since an explosive material is in a metastable thermodynamic state, with a significant amount of energy locked into chemical bonds.

The CJ theory assumes that a stable detonation occurs when the Rayleigh line is tangent to the shock Hugoniot, as shown in Fig. 1.3. At the tangent point, the flow velocity relative to the shock front equals the local equilibrium sound speed, i.e., it is a sonic point. Since any other intersections of the Rayleigh line with the shock Hugoniot are above the tangent line, the CJ detonation wave is the slowest shock wave that transforms the explosive into fully equilibrated reaction products.

This point of tangency can be determined, assuming that the equation of state $P = P(V,T)$ and $E = E(V,T)$ of the products is known. In most thermochemical calculations P and E are generated from the Helmholtz free energy $F(V,T) = E(V,T) - TS(V,T)$. $S(V,T)$ is the entropy per unit mass. $F(V,T)$ is referred to as a complete equation of state, because all thermodynamic properties can be derived through the appropriate derivatives. For example,

$$
\begin{aligned}
P(V,T) &= -\left[\frac{\partial F(V,T)}{\partial V}\right]_T , \\
E(V,T) &= F(V,T) - T\left[\frac{\partial F(V,T)}{\partial T}\right]_V .
\end{aligned} \tag{1.6}
$$

Given the equation of state, the Hugoniot can be found at a specified volume by iterating over temperatures until (1.5) is satisfied. Given the Rayleigh line and the Hugoniot curve, it is possible to calculate D as follows. Assume that the initial conditions (ρ_0, p_0, E_0) are known. We take a trial detonation velocity D'. If $D' > D$, there will be two intersections between the Rayleigh line and the Hugoniot (see Fig. 1.3). If $D' < D$, there will be no intersections. Therefore by iterating over D' it is possible to determine D to within arbitrary precision.

A brief discussion of the equations solved in a thermochemical calculation is provided in [77], with a more detailed discussion to be found in the book of Smith and Missen [64]. A thermochemical code solves thermodynamic equations between product species to find chemical equilibrium. For instance, in a system comprised solely of condensed carbon, gaseous CO, and gaseous CO_2, a thermochemical code can find the equilibrium of the reaction $C + CO_2 = 2CO$ at a specified pressure and temperature. Chapman–Jouguet (CJ) theory says that the detonation point is a state in thermodynamic and chemical equilibrium, so a thermochemical calculation can predict the thermodynamic and chemical properties of this state. These properties and CJ detonation theory yield the detonation velocity and other performance indicators. Thermochemical codes can also calculate thermodynamic states where the pressure and temperature are not explicitly indicated. For instance, the volume and the entropy can be specified instead of the pressure and temperature. These types of calculations implicitly define pressure and temperature, i.e.,

$$\begin{aligned} V(P,T) &= V_0, \\ S(P,T) &= S_0. \end{aligned} \tag{1.7}$$

States at constant entropy play for example an important role when undertaking an adiabatic expansion from the CJ state. To see why this is so, note that for an adiabatic expansion, $\mathrm{d}E = -p\mathrm{d}V$. Recall that $T\mathrm{d}S = \mathrm{d}E + p\mathrm{d}V - \sum \mu_i \mathrm{d}N_i$. Here, μ_i is the chemical potential of the ith species and N_i is the number of molecules of the ith species per unit mass. If the system is at chemical equilibrium, then S is maximized at fixed V and E, implying that $\sum \mu_i \mathrm{d}N_i = 0$ for changes in the concentrations that conserve the elemental abundances. Combining these results, it is found that $\mathrm{d}S = 0$ for adiabatic expansion under chemical equilibrium. Portions of the adiabatic expansion are sometimes done under the assumption that the concentrations are fixed. This is a simple way to model "frozen" systems that are too cold to maintain chemical equilibrium. In this case $\mathrm{d}S = 0$, since $\mathrm{d}N_i = 0$. The result is that thermochemical calculations usually treat adiabatic expansions from the Chapman–Jouguet state as being at constant entropy, whether the chemical concentrations change with time or not. This is somewhat paradoxical, since the conversion of the solid explosive to the high-temperature detonation products obviously involves a large change in entropy. The resolution to the paradox is that the entropy is fixed after the high explosive has fully reacted at the CJ state.

Thermodynamic equilibrium can be found by balancing chemical potentials. The above carbon example leads for example to the equation $\mu_{\mathrm{C}} + \mu_{\mathrm{CO}_2} = 2\,\mu_{\mathrm{CO}}$. The chemical potentials of condensed species are functions of pressure and temperature only, while the potentials of gaseous species depend on concentrations:

$$\begin{aligned} \mu_{\mathrm{C}} &= \mu_{\mathrm{C}}\,(P,T) \\ \mu_{\mathrm{CO}_2} &= \mu_{\mathrm{CO}_2}\,(P,T,N_{\mathrm{CO}_2},\,N_{\mathrm{CO}}) \\ \mu_{\mathrm{CO}} &= \mu_{\mathrm{CO}}\,(P,T,N_{\mathrm{CO}_2},\,N_{\mathrm{CO}})\,. \end{aligned} \tag{1.8}$$

Here, N_{CO_2} is the number of molecules of CO_2, and N_{CO} is the number of molecules of CO. The chemical potential is the Gibbs free energy difference ($G = F + pV$) obtained from adding one more molecule into the system:

$$\mu_{\mathrm{CO}} = G\,(P,T,N_{\mathrm{CO}}+1,N_{\mathrm{CO}_2}) - G\,(P,T,N_{\mathrm{CO}},N_{\mathrm{CO}_2})\,. \tag{1.9}$$

Balancing the chemical potentials is the same as minimizing G. In fact many thermochemical codes numerically minimize G instead of explicitly treating chemical potentials.

In many thermochemical codes there are no chemical reactions specified (for an exception to this see [52]). How then, it might be asked, does the code get the chemical reactions for balancing chemical potentials? The reactions used do not matter, as long as there are enough of them to change concentrations arbitrarily; the chemical equilibrium state will be the same no matter what reactions are used. The Cheetah thermochemical code [25, 27], for example, specifies a decomposition reaction for each product molecule. This reaction is an internal construct of the code and does not have to occur physically. The product molecule (called a "constituent") decomposes into a small number of "component" molecules. The number of components is equal to the number of elements in the problem. For example Cheetah might choose C and CO as components. The decomposition reactions would then be

$$\begin{aligned} \mathrm{C} &\Leftrightarrow \mathrm{C} \\ \mathrm{CO} &\Leftrightarrow \mathrm{CO} \\ \mathrm{CO_2} &\Leftrightarrow 2\,\mathrm{CO} - \mathrm{C}. \end{aligned} \tag{1.10}$$

Since decomposition reactions in Cheetah are just a convenient mathematical device, they can have negative stoichiometric coefficients. The first two reactions are trivial and therefore have no effect on the chemical equilibrium.

The equations to be solved in the thermochemical calculation are

$$\begin{aligned} &\mu_{\mathrm{CO}_2}\,(P,T,N_{\mathrm{CO}},\,N_{\mathrm{CO}_2}) = 2\,\mu_{\mathrm{CO}}\,(P,T,N_{\mathrm{CO}},N_{\mathrm{CO}_2}) - \mu_{\mathrm{C}}\,(P,T) \\ &N_{\mathrm{CO}} + N_{\mathrm{CO}_2} + N_{\mathrm{C}} = W_{\mathrm{C}} \\ &N_{\mathrm{CO}} + 2\;N_{\mathrm{CO}_2} = W_{\mathrm{O}}. \end{aligned} \tag{1.11}$$

Here W_C and W_O are the number of atoms of the elements C and O in the system, respectively.

The above equations can be summarized as follows: balance the chemical potentials while keeping the total moles of each element fixed. A thermochemical code can therefore be regarded to a large extent as a complicated algorithmic device for balancing chemical potentials.

1.3 Equations of State

As noted above the detonation products consist of fluid and condensed phases in chemical equilibrium. One of the most difficult parts of the thermochemical calculation is describing the EOS of the fluid components accurately. Because of its simplicity, the Becker–Kistiakowski–Wilson (BKW) [42] EOS has been used in many practical applications involving energetic materials. The BKW EOS specifies the pressure as a function of mole numbers n_i:

$$\begin{aligned}
\frac{pV}{nRT} &= 1 + x\ \mathrm{e}^{\beta x}, \\
x &= \frac{k}{V(T+\theta)^{\alpha}}, \\
k &= \kappa \sum_i n_i k_i, \\
n &= \sum_i n_i.
\end{aligned} \tag{1.12}$$

There are several features of the BKW equation of state to note. The ideal gas equation of state is recovered when $x = 0$. This occurs for sufficiently large T, sufficiently large V, or for sufficiently small values of k. k has the interpretation of an excluded volume, with the parameters k_i being proportional to the molecular volume. The parameters α, β, κ, and θ are adjustable. This gives the equation of state considerable flexibility, but it comes at the cost of a clear physical interpretation. Numerous parameter sets have been proposed for the BKW EOS [23,34,37,44]. Souers and Kury [66] have critically reviewed these choices by comparing their predictions to a database of detonation tests. They concluded that the BKW EOS does not model the detonation of a copper-lined cylindrical charge adequately. The reason for this is likely to be that the BKW equation of state yields $pV/nRT > 1$. This implies that attractions between molecules, which can lead to states with $pV/nRT < 1$, are neglected in the BKW EOS. This limits the accuracy of the equation of state when dealing with expanded gases where attractions are important.

It has long been recognized that the validity of the BKW EOS is questionable [13]. Better EOS accuracy is particularly important when designing new materials that may have unusual elemental compositions. Efforts to develop better EOSs have been based largely on the concept of model potentials. The fundamental idea is that product molecules can be assumed to interact via idealized spherical pair potentials. Although a spherical pair potential may seem to be a poor approximation for many molecules, it has long been recognized that the equation of state of many nonspherical bodies can be well modeled by a reference spherical system [78]. Statistical mechanics is then employed to calculate the EOS of the interacting mixture of effective spherical particles. Most often, the *exponential*-6 (exp-6) potential, $V(r)$, is used for the pair interactions:

$$V(r) = \frac{\varepsilon}{\alpha - 6}\left[6\exp\left(\alpha - \frac{\alpha r}{r_m}\right) - \alpha\left(\frac{r_m}{r}\right)^6\right]. \quad (1.13)$$

Here, r is the distance between particles, r_m is the position of the minimum of the potential well, ε is the well depth, and α is the softness of the potential well. Both attractions and repulsions between molecules are included in the exp-6 potential.

The Jacobs–Cowperthwaite–Zwissler (JCZ3) EOS was the first successful model based on a pair potential that was applied to detonation [18]. This EOS was based on fitting Monte Carlo simulation data to an analytic functional form. Ross, Ree, and others successfully applied a soft-sphere EOS based on perturbation theory to detonation and shock problems [14, 52, 59, 71, 72]. Computational cost can be a significant hindrance with EOS based on fluid perturbation theory. Byers Brown [11] developed an analytic representation of the exp-6 EOS using Chebyshev polynomials. Fried and Howard [24] have used a combination of integral equation theory [69] and Monte Carlo simulations to generate a highly accurate EOS for the exp-6 fluid.

Despite its usefulness the exp-6 model is not well suited for molecules with large dipole moments. To account for this Ree [53] used a temperature-dependent well depth $\varepsilon(T)$ in the exp-6 potential to model polar fluids and fluid phase separations. Jones et al. [41] have applied thermodynamic perturbation theory to polar detonation product molecules.

Efforts have been made to develop EOS for detonation products based on direct Monte Carlo simulations instead of analytical approaches [9,10,62]. This approach is promising given recent increases in computational capabilities. One of the advantages of direct simulation is the ability to go beyond van der Waals 1-fluid (vdW1) theory, which approximately maps the equation of state of a mixture onto that of a single component fluid [43]; vdW1 coupled with thermochemical modeling remains however the method of choice for complex chemical equilibrium calculations. Following Ree [52] we define:

$$\begin{aligned} a &= \sum_{i,j} x_i x_j \varepsilon_{ij} r_{m,ij}^3, \\ b &= \sum_{ij} x_i x_j r_{m,ij}^3, \\ c &= \sum_{i,j} x_i x_j \alpha_{ij} \varepsilon_{ij} r_{m,ij}^3, \end{aligned} \quad (1.14)$$

where x_i is the mole fraction of species i. Then for a fluid mixture of molecules interacting through exp-6 potentials the vdW1 theory yields the effective potential corresponding to a single component fluid through ${r_m}^3 = b$, $\varepsilon = a/b$, and $\alpha = c/a$. The mixture is thus mapped into a simple single component effective fluid, with nearly identical equation of state properties if the size asymmetry between the component molecules of the mixture is not too large.

In most cases, interactions between unlike molecules (treated as single spherical sites) are obtained using the Lorentz–Berthelot combination rules [54]. The rules are used to determine the interactions between unlike molecules from those of like molecules. They specify the interactions between unlike molecules to be the arithmetic or geometric averages of single molecule pairwise interactions. The combination rules most commonly used for exp-6 potentials are

$$\begin{aligned} r_{m,ij} &= \frac{(r_{m,ii} + r_{m,jj})}{2}, \\ \varepsilon_{ij} &= \sqrt{\varepsilon_{ii}\varepsilon_{jj}}, \\ \alpha_{ij} &= \sqrt{\alpha_{ii}\alpha_{jj}}. \end{aligned} \tag{1.15}$$

In these equations, the parameters with repeated indices (e.g., ε_{ii}) pertain to interactions between like molecules, whereas the parameters with distinct indices (e.g., ε_{ij}) pertain to interactions between unlike molecules. It appears that these rules work well in practice, although they have not been extensively tested at high densities through experiments or simulations. The success of the Lorentz–Berthelot rules in many situations dramatically simplifies the parameterization of thermochemical codes. Without the Lorentz–Berthelot rules, $3N(N-1)/2$ parameters would be required to model an N-component exp-6 mixture. With the use of the combination rules, this number is reduced to $3N$.

The exp-6 potential has proved successful in modeling chemical equilibrium at the high pressures and temperatures characteristic of detonation. However, to calibrate the parameters for such models, it is necessary to have experimental data for product molecules and mixtures of molecular species at high temperature and pressure. Static compression, Hugoniot, and sound-speed measurements provide important data for these models [80, 82].

In addition to the intermolecular potential contribution to the Helmholtz free energy of the system there is also an intramolecular portion. The thermochemical code Cheetah uses for example a polyatomic model to account for the intramolecular part, including electronic, vibrational, and rotational states. Such a model can be expressed conveniently in terms of the heat of formation, standard entropy, and constant-pressure heat capacity of each species at standard thermodynamic conditions.

We now consider how the EOS outlined above predicts the detonation behavior of condensed explosives. The overdriven shock Hugoniot of an explosive is an appropriate EOS test, since it accesses a wide range of high pressures.

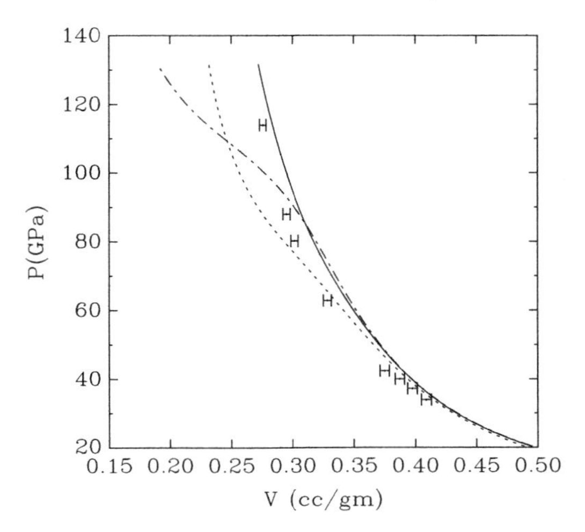

Fig. 1.4. The shock Hugoniot of PETN as calculated with exp-6 (*solid line*), the JCZS library (*dotted line*), and the BKWC library(*dot-dashed line*) versus experiment (*error bars*)

Overdriven states lie on the shock Hugoniot at pressures above the CJ point (see Fig. 1.3). The Hugoniot of pentaerythritol tetranitrate (PETN) is shown in Fig. 1.4. Fried et al. [27] have calculated the Hugoniot with the exp-6 model and also with the JCZS [38] product library. Figure 1.4 shows that the exp-6 model is within 1% of the measured data for pressures up to 120 GPa (1.2 Mbar). The JCZS model is accurate to within 1% up to a pressure of 90 GPa, but shows a disagreement with experiment at 120 GPa. The BKWC model also deviates from experiment above 90 GPa. Since the exp-6 model is not calibrated to condensed explosives, such agreement is a strong indication of the validity of the chemical equilibrium approximation to detonation.

Although the exp-6 modeling of fluid detonation products is generally successful, there remain a number of outstanding EOS issues that still need to be addressed, in particular the treatment of electrostatic interactions for polar and charged (ionic) molecular products. In addition, effects such as dielectric screening and charge-induced dipoles may also need to be considered, as well as the possible occurrence of non-molecular phases under high pressure and temperature conditions. It should also be noted that molecular shape is neglected in exp-6 models. While the small size of most detonation product molecules limits the importance of shape, lower temperature conditions could yield long-chain molecules, where molecular shape becomes more important.

The possibility of having ionic species in the detonation products is a complication that cannot be modeled using the exp-6 representation alone. Recent results on the superionic behavior of water at high pressures [29, 31] provide compelling evidence for a high-pressure ionic dissociation scenario. These results suggest, for example, that polar and ionic species interactions may account for approximately 10% of the Chapman–Jouguet (CJ) pressure of PETN. In addition, thermochemical calculations of high explosive formulations rich in highly electronegative elements, such as F and Cl, typically have substantially higher errors than calculations performed on formulations containing only the elements H, C, N, and O. The difficulty in modeling the CJ states of these formulations may be due to the neglect of ionic species.

Bastea et al. [6] have extended the exp-6 free energy approach to include the explicit thermodynamic contributions arising from dipolar and ionic interactions to the equation of state. The main task of their theory involves calculating the Helmholtz free energy (per particle) of the detonation products, f, when polar and charged species are present. The theory starts with a mixture of molecular species whose short range interactions are well described by isotropic, exp-6 potentials. This includes, for example, all molecules commonly encountered as detonation products, such as N_2, H_2O, CO_2, CO, NH_3, and CH_4. As documented previously [50], a one-fluid representation of this system, where the different exp-6 interactions between species are replaced by a single interaction depending on both individual interactions and mixture composition, is a very good approximation when the molecular diameter varies by less than a factor of approximately 1.5 between different components of the mixture. This theory also dramatically reduces the complexity of the thermochemical calculation.

Bastea et al. therefore chose the equivalent one component fluid to be the reference system for the perturbation approach outlined below. If the mixture components possess no charge or permanent dipole moments, the calculation of the corresponding free energy per particle, designated as f_0, suffices to yield the mixture thermodynamics and all desired detonation properties. This has been the physical model previously used in many thermochemical codes for the calculation of high explosives behavior.

It is worth noting that at high detonation pressures and temperatures the behavior of the exp-6 fluid so introduced is dominated by short range repulsions and is similar to that of a hard sphere fluid. In fact, the variational theory treatment [58] of the exp-6 thermodynamics employs a reference hard sphere system with an effective, optimal diameter σ_{eff} that depends on density and temperature. Bastea et al. pursued this connection to the hard sphere fluid by considering first a fluid of equisized hard spheres of diameter σ with dipole moments μ. For this simple model of a polar fluid, Stell et al. [60, 67] had previously suggested a Padé approximation approach for calculating the free energy f_d:

$$f_d = f_0 + \Delta f_d,$$
$$\Delta f_d = \frac{f_2}{1 - f_3/f_2}, \tag{1.16}$$

where f_0 corresponds to the simple hard sphere fluid and f_2 and f_3 are terms (second and third order, respectively) of the perturbation expansion in the dipole–dipole interaction ($\sim\mu^2$) such that

$$f_d = f_0 + f_2 + f_3 + \cdots . \tag{1.17}$$

The first-order term f_1 can be shown to be identically zero, while f_2 and f_3 have been calculated explicitly [67]. The resulting thermodynamics can be written in scaled variables as

$$\begin{aligned} \Delta f_d &= \Delta f_d\left(\rho^*, \beta_d^*\right), \\ \rho^* &= \rho\sigma^3, \\ \beta_d^* &= \frac{\mu^2}{k_\mathrm{B} T \sigma^3}, \end{aligned} \tag{1.18}$$

where ρ is the (number) density and T is the temperature. The same Padé approximation also holds for a mixture of identical hard spheres with different dipole moments μ_i [33, 70]. It should be noted that within this approximation it is easy to show that the mixture thermodynamics is equivalent to that of a simple hard sphere polar fluid with an effective dipole moment μ given by

$$\mu^2 = \sum_i x_i \mu_i^2, \tag{1.19}$$

where $x_i = \rho_i/\rho$ is the concentration of particles with dipole moment μ_i.

The above combination rule (1.19) is also adopted for the general case of exp-6 mixtures that include polar species. Moreover, in this case the calculation of the polar free energy contribution Δf_d uses the effective hard sphere diameter σ_{eff} of the variational theory.

A comparison of this procedure with molecular dynamics (MD) simulation results for an exp-6 model of polar water is shown in Figs. 1.5 and 1.6. Also

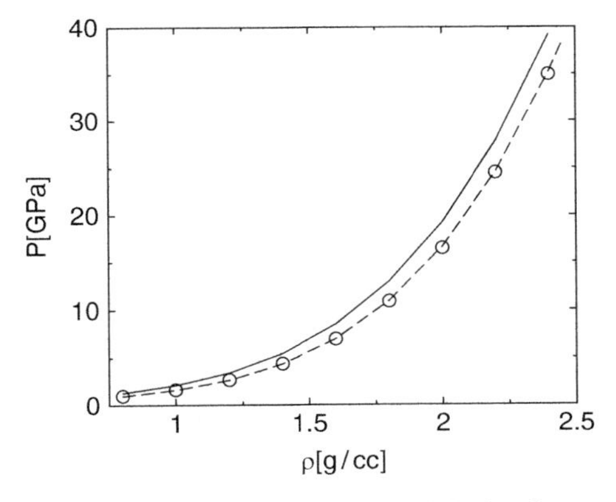

Fig. 1.5. Comparison of pressure results for a model of polar water at $T = 2{,}000$ K: MD simulations (*symbols*), newly developed theory for polar fluids (*dashed line*), and exp-6 calculations alone (*solid line*)

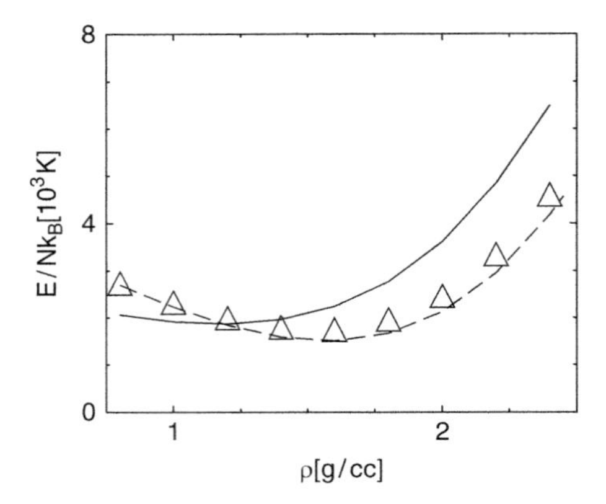

Fig. 1.6. Same as Fig. 1.5 for energy per particle

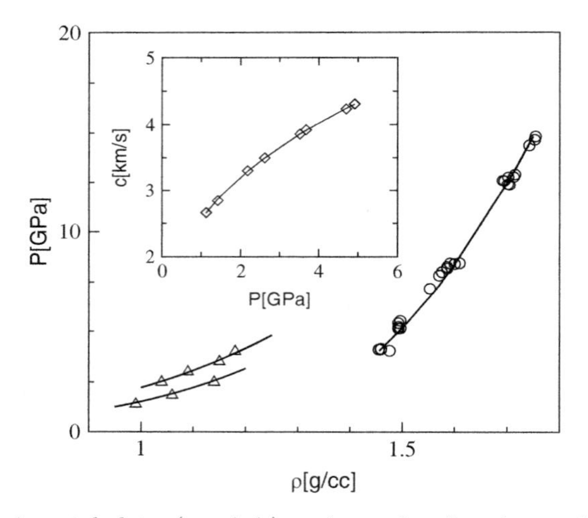

Fig. 1.7. Experimental data (*symbols*) and exp-6 polar thermodynamics results (*lines*) for dense, supercritical water. *Triangles*: $T = 983\,\mathrm{K}$ and $T = 1{,}373\,\mathrm{K}$ isotherms; *circles*: shock Hugoniot; *inset*: sound speed at $T = 673\,\mathrm{K}$

shown are the results of exp-6 thermodynamics alone. For both the pressure and energy the agreement is very good and the dipole moment contribution is sizeable.

The thermodynamic theory for exp-6 mixtures of polar molecules is now implemented in the thermochemical code Cheetah [5, 6]. In this case the major polar detonation products H_2O, NH_3, CO, and HF are considered. The optimal exp-6 parameters and dipole moment values for these species were determined by fitting to a variety of available experimental data. It is found, for example, that a dipole moment of 2.2 D for water reproduces very

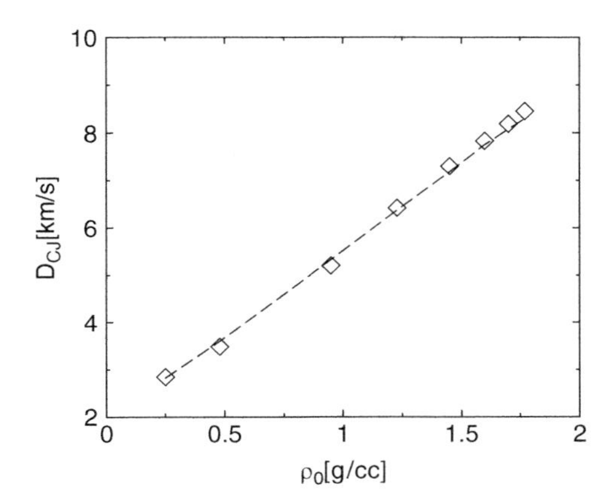

Fig. 1.8. PETN detonation velocity as a function of initial density; experiments (*symbols*) and CHEETAH calculation (*line*)

well all available experiments. Incidentally, this value is in very good agreement with values typically used to model supercritical water [35].

A comparison of the Cheetah polar water model predictions with both high pressure Hugoniot data [48] and low-density (steam at 800 K) experimental data [76] is presented in Fig. 1.7. The agreement is very good for both cases.

The newly developed equation of state has been applied to the calculation of detonation properties. In this context, one stringent test of any equation of state is the prediction of detonation velocities as a function of initial densities, and PETN has been chosen for this purpose. The Cheetah results are shown in Fig. 1.8 along with the experimental data [39] with a very good agreement being obtained.

1.4 Fluid Phase Separations

The detonation of modern condensed high explosives yields complex multiphase, multicomponent mixtures which are generally understood to reach chemical equilibrium very rapidly behind the shock front. Although some carbon rich explosives, most notably TNT and TATB, also produce significant amounts of condensed carbon, the major detonation products are usually small molecules mixed in a supercritical fluid phase. This dense, hot fluid typically consists of significant amounts of N_2, H_2O, CO_2, CO, NH_3, H_2, etc. The thermodynamic properties of this mixture play a crucial role in determining the detonation velocity of explosives as well as the work that they can deliver upon subsequent expansion, and therefore their prediction is a major task of explosives modeling [4, 6, 22, 27, 53, 83].

Thermochemical calculations based on the exp-6 potential were performed by Ree [53], who showed that they can reproduce the experimental data available for some common high explosives, including PETN. He also found that analogous calculations were less successful for HMX-based formulations like PBX 9404, although the composition of the detonation products at the CJ point was to a large extent similar. The theoretical predictions also seriously overestimated the experimental shock Hugoniot of PBX 9404, which is significantly more compressible than predicted. Given that the detonation temperature of PBX 9404 is lower than that of PETN while its detonation pressure is higher, Ree conjectured that the discrepancy was due to the occurrence of a supercritical phase separation of nitrogen and water at temperatures as high as 4,000 K. This was later introduced in the calculations via an empirical correction of the nitrogen–water interaction, and agreement with experiments was to a reasonable extent restored. The interaction correction was based on the premise that the phase separation was due to significant positive nonadditivity of the pair interactions, i.e.,

$$r^0_{\mathrm{N_2-H_2O}} > \frac{r^0_{\mathrm{N_2}} + r^0_{\mathrm{H_2O}}}{2}. \tag{1.20}$$

Thus it was assumed that N_2–H_2O phase segregation is largely a temperature-independent effect, similar to the one encountered in nonadditive hard sphere mixtures.

Supercritical phase separations are a common occurrence for a wide range of fluid mixtures, and such a phase transition is well documented for nitrogen-water systems at temperatures as high as about 800 K and pressures up to approximately 2 GPa [61]. Unfortunately they remain quite difficult to predict theoretically, largely because such phase transition lines result from an equality of chemical potentials and are therefore quite sensitive to any uncertainties in the interactions. The unlike-pair potentials can play an important role in determining phase separation, but they are difficult to determine with any degree of certainty. Theoretical efforts are further hampered by a lack of equation of state experimental data on mixtures at high pressures, which could be used to provide direct constraints on these interactions.

The like-pair potentials, however, also play crucial roles in determining phase boundaries and need to be well understood and modeled if reasonable estimates are to be made. The exp-6 potential has long been proven to enable accurate calculations of the high pressure-high temperature thermodynamics of simple molecules and remains the best representation of intermolecular interactions at these conditions. Polar molecules on the other hand have always been challenging to model within the constraints of a simple exp-6 potential. Indeed, for his water modeling Ree [53] adopted an empirical adjustment in which the strength of the interaction was assumed to be temperature dependent. Unfortunately such approximations further limit the reliability of phase boundary calculations.

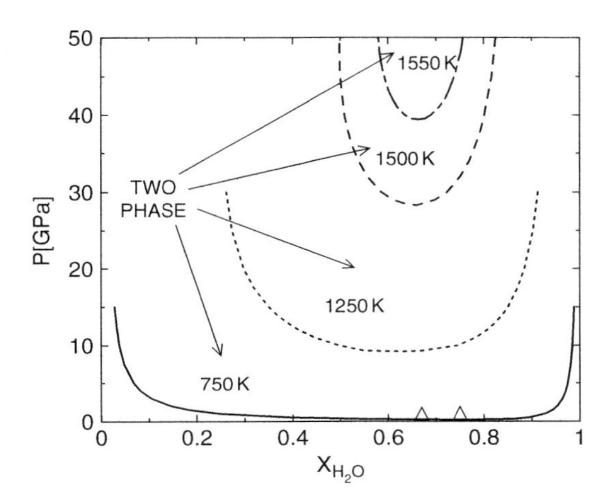

Fig. 1.9. Phase coexistence diagram for N_2–H_2O mixtures at temperatures between 750 and 1,550 K calculated using the exp-6 polar mixture theory (*lines*); *triangles* denote the experimental data of [16]

Nitrogen–water mixtures are typical of detonation product mixtures having polar (H_2O) and nonpolar (N_2) species. Bastea and Fried [5] employed the exp-6 polar theory described above to examine phase segregation in nitrogen–water mixtures. The nitrogen–water interaction was assumed to follow from standard combination rules, i.e., without arbitrary additivity corrections such as that introduced in [53] to induce the occurrence of phase segregation.

It is found that phase coexistence is indeed predicted by the exp-6 polar thermodynamic theory. The results of the calculations—solubility curves for the N_2–H_2O system as a function of temperature—are summarized in Fig. 1.9. Although both these calculations and those of [53] indicate the existence of phase separation, both the origin and the predicted features of this transformation are different. As opposed to [53], here the phase transition is largely an energy not entropy driven occurrence, due to the energy-lowering effect of dipole–dipole interactions. In fact, slightly better agreement with the low-temperature experimental results could be obtained with a very small *negative, not positive* N_2–H_2O nonadditivity.

Even more importantly, the phase diagram differs in some major respects from that of [53]. There the nitrogen–water coexistence region remains essentially identical up to temperatures as high as 4,000 K, and it only moves to higher pressures as the temperature is increased. On the other hand, as shown in Fig. 1.9, the present calculations predict that the miscibility gap narrows strongly at higher temperatures, with the net result that above approximately 1,600 K the two fluids become fully miscible at all pressures. This behavior is in agreement with the theoretical calculations reported in [15] and, moreover, the results also agree better with the experimental data [16].

Nevertheless, due to the usage of simple effective classical potentials in a wide range of pressures and temperatures, it cannot be established conclusively that phase segregation of nitrogen and water at high pressures and temperatures does indeed have the features suggested by the present results. However, it should be noted that this thermodynamic modeling is the most physical to date, and it does not employ any arbitrary assumptions. Furthermore, the simulations of [46] also suggest that, just as in the above calculations, the dipole moment of water and not the unlike-pair nonadditivity is in fact the major driving force for phase separation. Experimental results at higher pressures and temperatures would be very useful in fully elucidating this issue, particularly because the relevant thermodynamic regime is now within the reach of new high pressure–high temperature techniques.

Next to consider are the consequences of these calculations for the modeling of high explosives. To this end the CJ points and expansion isentropes of three common explosives, PETN ($C_5H_8N_4O_{12}$), HMX ($C_4H_8N_8O_8$), and RDX ($C_3H_6N_6O_6$), are determined. The results are shown in Fig. 1.10 together with the phase coexistence boundary for a typical nitrogen–water relative concentration encountered in detonation products. They indicate that both the CJ points and the major parts of the isentropic expansion paths are well outside the nitrogen–water immiscibility region. Phase separation appears therefore to be potentially relevant only for strongly expanded states, or situations where strong confinement is maintained to very low temperatures.

Phase separation as a possible high pressure–high temperature effect was originally proposed due to the difficulty in modeling detonation product mixtures with large water and nitrogen content. Although the previous analysis strongly suggests that this effect is not operational at detonation conditions,

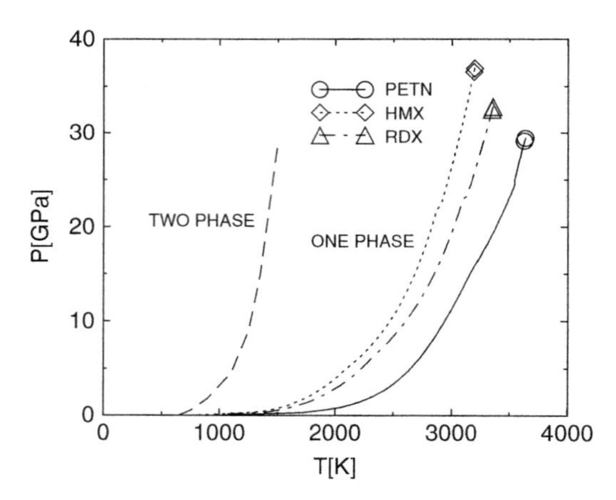

Fig. 1.10. Chapman–Jouguet expansion isentropes of PETN, RDX, and HMX and the coexistence boundary of the 33% N_2 + 67% H_2O water mixture

difficulties remain in accurately predicting the behavior of water-rich mixtures at high pressures and temperatures. We consider likely explanations for this behavior below.

1.5 Ionic Dissociation at High Pressures and Temperatures

Quantum molecular dynamics (QMD) simulations of hydrogen bonding liquids, e.g., H_2O, HCl, HF, etc., [8] indicate that at extreme conditions these systems exhibit a behavior that is different from what is traditionally assumed for molecular fluids. For such systems the breaking of molecular bonds that occurs upon the application of high pressures and temperatures, e.g., tens of GPa and thousands of K, leads to the formation of small products, in particular hydrogen (H), that carry an effective electrical charge. As a result, the process can be understood as ionic dissociation, although the resulting ionic products are better treated as radicals rather than ions in solution [79]. This has important consequences for the equation of state of these fluids, as well as their conduction properties [30, 79]. Direct QMD simulations of PETN [79] at CJ conditions in fact indicate that the ionic dissociation of water is also likely to appear at detonation conditions, which is in agreement with experimental results on the electrical conductivity of detonation products [36].

We have previously proposed and tested an explicitly ionic thermodynamics to model ionic mixtures that treats the charge–charge interactions at the perturbation level [6]. This thermodynamic concept is now largely integrated in chemical equilibrium calculations and it is tested here first on water.

To this end the shock Hugoniot of water is calculated, which has been measured with good accuracy experimentally [48], and for which detailed QMD

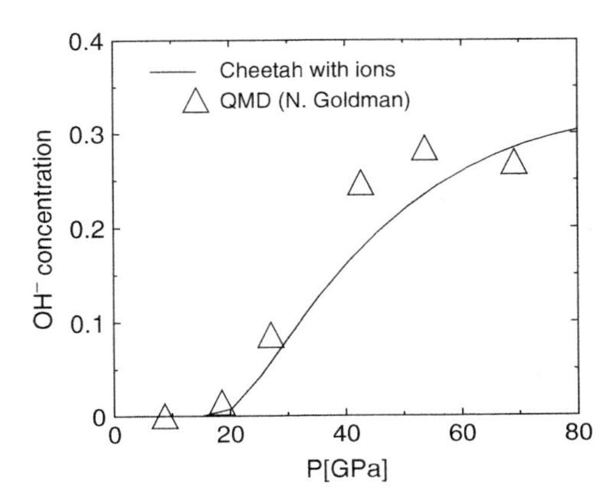

Fig. 1.11. Molar fraction of OH^- ions along the shock Hugoniot of water: QMD simulations of [30] (*triangles*) and exp-6 polar ionic thermodynamics (*solid line*)

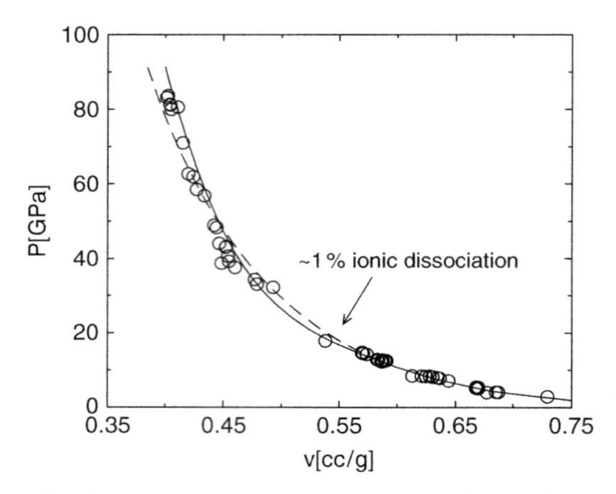

Fig. 1.12. Water shock Hugoniot: experimental data (*circles*), exp6-polar thermodynamics without ions (*dashed line*) and exp-6 polar ionic thermodynamics (*solid line*)

simulations are also available [30]. The theoretical chemical equilibrium calculations include H_2O modeled using the exp-6 polar thermodynamics, as well as OH^- and H^+ radicals. The results are shown in Figs. 1.11 and 1.12. It is found that the H_2O concentration decreases, while the concentration of ions steadily increases at high pressures. This behavior is in qualitative and reasonably good quantitative agreement with the QMD simulations—see Fig. 1.11. Moreover, the introduction of ions leads to improved modeling of the shock Hugoniot—see Fig. 1.12, compared with calculations assuming no such breakup. The subtle "softening" of the Hugoniot curve around 15 GPa thus appears to be due to the inception of ionic dissociation as observed in ab initio simulations and also suggested by conductivity measurements.

Advances continue in the treatment of detonation in mixtures that includes explicit polar and ionic contributions. The new thermodynamic formalism places the modeling of polar species on a solid footing, opens the possibility of realistic multiple fluid phase chemical equilibrium calculations (polar–nonpolar phase segregation), extends the validity domain of the exp-6 library [27], and opens the possibility of applications in a wider regime of pressures and temperatures. Predictions of high explosive detonation based on the new approach yield excellent results. A similar theory for ionic species modeling [6] agrees very well with MD simulations. Nevertheless, high explosive chemical equilibrium calculations that include standard ionization are beyond the current abilities of the Cheetah code, due to the presence of multiple minima in the free energy surface. Such calculations will require additional algorithmic developments. However, the possibility of partial ionization, suggested by first principles simulations of water and PETN, appears to be a very promising avenue for improving the accuracy of detonation predictions in the chemical equilibrium framework, if electrostatic interactions are accounted for using the conceptual thermodynamic structure outlined above.

1.6 Nonideal Explosives

Despite the many successes in the thermochemical modeling of energetic materials, several significant limitations still exist. One such limitation is that real systems do not always reach chemical equilibrium during the relatively short (ns–μs) timescales of detonation. When this occurs, quantities such as the energy of detonation and the detonation velocity can be predicted by a thermochemical calculation to be 10% or higher than the actual experimental values.

Thermochemical codes have been developed as tools for predicting the detonation performance of standard and insensitive plastic bonded high explosives. More recently, interest has expanded to the modeling of a wide range of energetic materials, including newly synthesized compounds, explosive mixtures, propellants, rocket motors, and metal-loaded explosives. Thermochemical calculations assume that the material is not initially undergoing any chemical or thermodynamic transformations and is to a large extent homogenous. The detonation behavior of compounds or mixtures that react or transform over time can be predicted by a thermochemical code only if the precise thermodynamic state immediately preceding detonation is well understood and quantified, and if the material does not develop large scale inhomogeneities during its previous chemical and/or physical evolution.

Thermochemical equilibrium (standard Chapman–Jouguet) calculations are customarily believed to apply with the highest accuracy to ideal explosives, i.e., reactive materials that reach chemical equilibrium "quickly" behind the shock front. A more precise applicability criterion, however, is that the size of the charge should be much larger than the reaction zone of the material and/or heavy confinement be present, thereby limiting the boundary energy losses that are not accounted for in thermochemical equilibrium calculations [83], which diminish performance in practice. It is well known that such losses lead to lower detonation velocities as the charge size is decreased and ultimately detonation failure at a small enough size, i.e., critical diameter [21].

Since practical considerations typically limit the range of charge diameters studied experimentally, the value of the detonation velocity corresponding to a very large charge can only be extrapolated from the experimental data, based on reasonably well-understood and tested assumptions about the dependence of the detonation velocity D on charge radius R. For many explosives this extrapolation is trivial since their reaction zones are so small that they reach the ideal, "infinite" charge velocity at configurations commonly studied experimentally, e.g., 1 in. diameter steel cylinders. On the other hand some materials exhibit significant dependence of the detonation velocity on the charge diameter at common experimental sizes, and a more thorough analysis of the data is necessary to deduce the asymptotic, large size behavior, which a thermochemical code predicts; such materials are termed nonideal explosives. It is worth emphasizing, however, that all explosives show strong size dependence for diameters close to the critical one. In fact, theoretical considerations [21] predict that the rate of change of the detonation velocity as a function of charge diameter is infinite at the critical value, and experimental results largely confirm this scenario for a variety of compounds. Thus, broadly speaking, if chemical and physical transformations occurring during detonation have a finite relaxation path for energy release, the difference between ideal and nonideal explosives is only one of scale, and most energetic materials can be expected to behave "ideally" at sizes much larger than the critical diameter. Consequently, thermochemical equilibrium detonation calculations should apply to both ideal and nonideal formulations, as long as they are understood to predict the behavior of large and/or strongly confined charges. This conclusion is confirmed by Cheetah and experimental comparisons for a number of well known nonideal explosives (see Table 1.1). We note that the stand-alone Cheetah code can also perform detonation kinetics (Wood–Kirkwood [28] calculations), thus enabling the study of performance dependence on explosive size.

Apart from chemical nonequilibrium effects the detonation of a solid particulate two-phase mixture can exhibit other nonequilibrium processes of mass, momentum, and energy transfer between the two phases due to finite sizes of solid particles. A full equilibrium state includes chemical, mechanical (pressure and particle velocity), and thermal equilibrium between the phases. In general, the nonequilibrium momentum and heat transfer depend on the physical properties of the particles and do not have the same relaxation length scales as that of the mass transfer or chemical nonequilibrium processes.

Table 1.1. Experimental detonation velocities versus Cheetah predictions

Explosive	Density [g/cc]	$D_{\mathrm{experiment}}$ [km s^{-1}]	D_{Cheetah} [km s^{-1}]	Difference [%]
Ammonium perchlorate (AP)	1.26	4.95[a]	4.83	2.4
Ammonium nitrate (AN)	0.83	3.95[a]	3.92	0.8
ANFO (95.5% AN + 5.5% fuel oil)	0.87	5.03[a]	4.86	3.4
Black powder	0.7	1.3[b]	1.3	0
Hydrogen peroxide (90.5%)	1.39	6.29[a,b]	6.19	1.6
TATB	1.85	7.78[a]	8.01	3.0
TNT	1.57	6.84[a]	6.86	0.3

Experimental data are from P.C. Souers, private communication
[a]Experimental value is extrapolated to infinite radius
[b]Experimental value is based on a heavily confined experiment

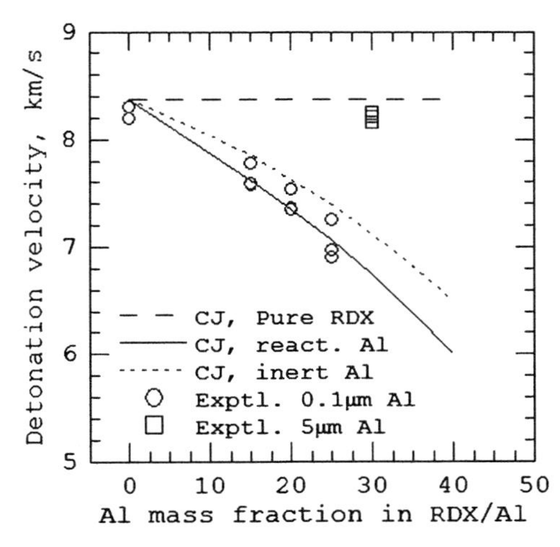

Fig. 1.13. Comparison of experimental detonation velocities with the equilibrium CJ theory for aluminum particle-RDX mixtures at a 1.66 g cm^{-3} initial density [85]

Since the CJ theory assumes a unique state of full-equilibrium detonation products at the equilibrium sonic locus, it cannot predict the detonation velocity precisely for finite-sized particulate mixtures with large interphase momentum and heat transfer length scales.

Figure 1.13 illustrates the predictability of the CJ equilibrium theory by comparing the theoretical predictions from the Cheetah code with the experimental results obtained in various aluminum particle-RDX ($C_3H_6N_6O_6$) mixtures at a common initial mixture density $\rho = 1.66\,\mathrm{g\,cm^{-3}}$. For sufficiently small particles (e.g., 0.1 µm), an increase in aluminum mass fraction results in a decrease in detonation velocity. The experimental detonation velocities are in agreement with the theoretical prediction, regardless of the reactive or chemically frozen nature of the particles. This fact indicates the significance of the momentum and heat transferred to the particles during the process toward the mixture equilibrium state as the flow approaches the sonic locus. The momentum and heat transferred are responsible for the velocity deficit with respect to pure RDX detonation. In contrast, for sufficiently large particles (e.g., 5 µm), the experimental velocity is much higher than the full equilibrium prediction and close to that of pure RDX, thus suggesting a nearly frozen transfer of momentum and heat between the two phases within the detonation zone. Most probably the experimental detonation velocity is therefore a strong function of particle size and mass fraction and ranges between the full equilibrium value and the phase-frozen limit.

We mention in closing that chemical kinetics modeling is a more detailed way to treat detonation. There are several well-developed chemical kinetic mechanisms for highly studied materials such as hexahydro-1,3,5-trinitro-1,3,5-s-triazine (RDX) and 1,3,5,7-tetranitro-1,3,5,7-tetraazacyclooctane

(HMX) [12]. Unfortunately, reliable chemical kinetic mechanisms are generally not available for high-pressure conditions. Some researchers have applied simplified chemical kinetics to detonation processes [40]. The primary difficulty in high-pressure chemical kinetic models is a lack of experimental data on speciation. First principles quantum simulations [45,55], have the potential to provide chemical kinetic information for fast processes and thus fill this knowledge gap. This information could then conceivably be applied to longer timescales and lower temperatures using high-pressure chemical kinetics.

1.7 Concluding Remarks

The ability to model chemical reaction processes in condensed-phase energetic materials at the extreme conditions typical for detonation is progressing with contributions from theory, simulations, and experiments. Chemical equilibrium modeling is a useful methodology that can rapidly predict the performance of energetic materials on the basis of only a few material properties. Traditional chemical equilibrium ideas are however challenged by developments in high-pressure chemistry. It is now clear that ionic dissociation plays a more significant role under detonation conditions than previously thought. At the same time, it is argued that some processes thought to be important for detonation, such as supercritical phase separation, likely do not occur during typical detonations. As the understanding of the important physical processes occurring during a detonation improves, the reliability of chemical equilibrium calculations will likely also improve. Chemical equilibrium modeling is limited to predicting results in the limit of large charge size and a large degree of homogeneity. Where this is not sufficient, chemical equilibrium can be supplemented by chemical kinetic or kinetic/hydrodynamic treatments [75]. Such treatments often require a knowledge of the transport properties of detonation products. Fortunately these properties can be predicted [1–3] in a wide range of thermodynamic conditions and can be easily integrated in chemical equilibrium calculations.

Acknowledgments

The authors are grateful for the contributions of many collaborators to the work described here. Nir Goldman, Evan Reed, M. Riad Manaa, and Christine Wu played a central role in the QMD simulations. Fan Zhang provided helpful comments and additions to this chapter. I.-F. Will Kuo, W. Michael Howard, Kurt R. Glaesemann, P. Clark Souers, and Peter Vitello contributed to many of the thermochemical simulation techniques discussed here. J. Zaug performed relevant high-pressure experimental work. S.B. would like to thank Francis Ree for introducing him to the subject of detonation modeling using chemical equilibrium. This work was performed under the auspices of the U.S. Department of Energy by Lawrence Livermore National Laboratory under Contract No. DE-AC52-07NA27344.

References

1. Bastea, S.: Transport properties of fluid mixtures at high pressures and temperatures. Application to the detonation products of HMX: In: Proceedings of the 12th International Detonation Symposium, San Diego, CA (2002)
2. Bastea, S.: Transport properties of dense fluid argon. Phys. Rev. E **68**, 031204 (2003)
3. Bastea, S.: Transport in a highly asymmetric binary fluid mixture. Phys. Rev. E **75**, 031201 (2007)
4. Bastea, S., Fried, L.E.: Exp-6 polar thermodynamics of dense supercritical water. J. Chem. Phys. **128**, 174502 (2008)
5. Bastea, S., Fried, L.E.: Major effects in the thermodynamics of detonation products: Phase segregation versus ionic dissociation. In: Proceedings of the 14th Symposium (International) on Detonation. Office of Naval Research, Coeur d'Alene, ID (2010)
6. Bastea, S., Glaesemann, K., Fried, L.E.: Equation of state for high explosives detonation products with explicit polar and ionic species. In: Proceedings of the 13th Symposium (International) on Detonation. Office of Naval Research, Norfolk, VA (2006)
7. Benedetti, L.R., Nguyen, J.H., Caldwell, W.A., Liu, H., Kruger, M., Jeanloz, R.: Dissociation of CH_4 at high pressures and temperatures: Diamond formation in giant planet interiors? Science **286**, 100–102 (1999)
8. Blais, N.C., Engelke, R., Sheffield, S.A.: Mass spectroscopic study of the chemical reaction zone in detonating liquid nitromethane. J. Phys. Chem. A **101**, 8285–8295 (1997)
9. Brennan, J.K., Rice, B.M.: Molecular simulation of shocked materials using the reactive Monte Carlo method. Phys. Rev. E **66**, 021105 (2002)
10. Brennan, J.K., Lisal, M., Gubbins, K.E., Rice, B.M.: Reaction ensemble molecular dynamics: Direct simulation of the dynamic equilibrium properties of chemically reacting mixtures. Phys. Rev. E **70**, 061103 (2004)
11. Byers Brown, W.: Analytical representation of the excess thermodynamic equation of state for classical fluid mixtures of molecules interacting with alpha-exponential-six pair potentials up to high densities. J. Chem. Phys. **87**, 566 (1987)
12. Chakraborty, D., Muller, R.P., Dasgupta, S., Goddard III, W.A.: Mechanism for unimolecular decomposition of HMX (1,3,5,7-tetranitro-1,3,5,7-tetrazocine), an ab initio study. J. Phys. Chem. A **105**, 1302 (2001)
13. Charlet, F., Turkel, M.L., Danel, J.F., Kazandjian, L.: Evaluation of various theoretical equations of state used in calculation of detonation properties. J. Appl. Phys. **84**, 4227–4238 (1998)
14. Chirat, R., Pittion-Rossillon, G.: A new equation of state for detonation products. J. Chem. Phys. **74**, 4634 (1981)
15. Churakov, S.V., Gottschalk, M.: Perturbation theory based equation of state for polar molecular fluids: II. Fluid mixtures. Geochem. Cosmochim. Acta **67**, 2415 (2003)
16. Costantino, M.S., Rice, S.F.R.: Supercritical phase separation in water-nitrogen mixtures. J. Phys. Chem. **95**, 9034 (1991)
17. Cowperthwaite, M.: In: Short, J.M. (ed.) Tenth International Detonation Symposium, pp. 656–664. Office of Naval Research, Boston (1993)

18. Cowperthwaite, M., Zwisler, W.H.: The JCZ equations of state for detonation products and their incorporation into the TIGER code. In: Sixth Detonation Symposium, p. 162 (1976)
19. Cowperthwaite, M., Zwisler, W.H.: Thermodynamics of nonideal heterogeneous systems and detonation products containing condensed Al_2O_3, Al, and C. J. Phys. Chem. **86**, 813–817 (1982)
20. Davis, W.C., Fauquignon, C.: Classical theory of detonation. J. De Physique IV **5**, 3–21 (1995)
21. Eyring, H., Powell, R.E., Duffey, G.H., Parlin, R.B.: The stability of detonation. Chem. Rev. **45**, 69 (1948)
22. Fickett, W., Davis, W.C.: Detonation. University of California Press, Berkeley (1979)
23. Finger, M., Lee, E.L., Helm, F.H., Hayes, B., Hornig, H.C., McGuire, R.R., Kahara, M., Guidry, M.: The effect of elemental composition on the detonation behavior of explosives. In: Sixth Symposium (International) on Detonation, pp. 710–722. Office of Naval Research, Coronado (1976)
24. Fried, L.E., Howard, W.M.: An accurate equation of state for the exponential-6 fluid applied to dense supercritical nitrogen. J. Chem. Phys. **109**, 7338–7348 (1998)
25. Fried, L.E., Souers, P.C.: BKWC: An empirical BKW parametrization based on cylinder test data. Propellants Explos. Pyrotech. **21**, 215–223 (1996)
26. Fried, L.E., Manaa, M.R., Pagoria, P.F., Simpson, R.L.: Design and synthesis of energetic materials. Annu. Rev. Mater. Res. **31**, 291 (2001)
27. Fried, L.E., Howard, W.M., Souers, P.C.: Exp6: A new equation of state library for high pressure thermochemistry. In: Twelfth Symposium (International) on Detonation, pp. 567–575. Office of Naval Research, San Diego (2002)
28. Glaesemann, K.R., Fried, L.E.: Improved Wood-Kirkwood detonation kinetics. Theor. Chem. Acc. **120**, 37–43 (2008)
29. Goldman, N., Fried, L.E., Kuo, I.F.W., Mundy, C.J.: Bonding in the superionic phase of water. Phys. Rev. Lett. **94**, 217801 (2005)
30. Goldman, N., Reed, E.J., Kuo, I.-F. W., Fried, L.E., Mundy, C.J., Curioni, A.: Ab initio simulation of the equation of state and kinetics of shocked water. J. Chem. Phys. **130**, 124517 (2009)
31. Goncharov, A.F., Goldman, N., Fried, L.E., Crowhurst, J.C., Kuo, I.-F. W., Mundy, C.J., Zaug J.M.: Dynamic ionization of water under extreme conditions. Phys. Rev. Lett. **94**, 125508 (2005)
32. Greiner, N.R., Phillips, D.S., Johnson, J.D., Volk, F.: Diamonds in detonation soot. Nature **333**, 440 (1988)
33. Gubbins, K.E., Twu, C.H.: Thermodynamics of polyatomic fluid mixtures – I. Chem. Eng. Sci. **33**, 863 (1977)
34. Gubin, S.A., Odintsov, V.V., Pepekin, V.I.: BKW-RR EOS. Sov. J. Chem. Phys. **3**, 1152 (1985)
35. Guillot, B.: A reappraisal of what we have learned during three decades of computer simulation. J. Mol. Liq. **101**, 219–260 (2002)
36. Hayes, B.: On electrical conductivity in detonation products. In: Fourth Detonation Symposium (International) pp. 595 (1965)
37. Hobbs, M.L., Baer, M.R.: Calibrating the BKW-EOS with a large product species data base and measured C-J properties. In: Tenth International Detonation Symposium, pp. 409–418. Boston (1993)

38. Hobbs, M.L., Baer, M.R., McGee, B.C.: JCZS: An intermolecular potential database for performing accurate detonation and expansion calculations. Propellants Explos. Pyrotech. **24**, 269–279 (1999)
39. Hornig, H.C., Lee, E.L., Finger, M., Kurrle, J.E.: Equation of state of detonation products. In: Proceedings of the 5th Symposium (International) on Detonation. Office of Naval Research, Boston (1970)
40. Howard, W.M., Fried, L.E., Souers, P.C.: Kinetic modeling of non-ideal explosives with CHEETAH. In: Eleventh International Symposium on Detonation, pp. 998–1006. Snowmass, CO (1998)
41. Jones, H.D.: Theoretical equation of state for water at high pressures. In: Furnish, M.D., Thadhani, N.N., Horie, Y. (eds.) Shock Compression of Condensed Matter, 2001, pp. 103–106. AIP, Atlanta (2001)
42. Kistiakowski, G.B., Wilson, E.B.: Report on the prediction of the detonation velocities of solid explosives. Office of Scientific Research and Development, Report OSRD-69 (1941)
43. Leland, T.W., Rowlinson, J.S., Sather, G.A.: Statistical thermodynamics of mixtures of molecules of different sizes. Trans. Faraday Soc., **64**, 1447–1460 (1968)
44. Mader, C.L.: Numerical Modeling of Detonations. University of California Press, Berkeley (1979)
45. Maillet, J.B., Bourasseau, E.: Ab initio simulations of thermodynamic and chemical properties of detonation product mixtures. J. Chem. Phys. **131**, 084107 (2009)
46. Maiti, A., Gee, R.H., Bastea, S., Fried, L.E.: Phase separation in N_2–H_2O mixture: Molecular dynamics simulations using atomistic force fields. J. Chem. Phys. **126**, 044510 (2007)
47. Manaa, M.R., Fried, L.E., Melius, C.F., Elstner, M., Frauenheim, T.: Decomposition of HMX at extreme conditions: A molecular dynamics simulation. J. Phys. Chem. A **106**, 9024 (2002)
48. Marsh, S.P.: LASL Shock Hugoniot Data. University of California Press, Berkeley (1980)
49. Nomura, K., Kalia, R.K., Nakano, A., Vashishta, P., van Duin, A.C.T., Goddard, W.A.: Dynamic transition in the structure of an energetic crystal during chemical reactions at shock front prior to detonation. Phys. Rev. Lett. **99**, 148303 (2007)
50. Ree, F.H.: Simple mixing rule for mixtures with exp-6 interactions. J. Chem. Phys. **78**, 409 (1983)
51. Ree, F.H.: Systematics of high-pressure and high-temperature behavior of hydrocarbons. J. Chem. Phys. **70**, 974–983 (1979)
52. Ree, F.H.: A statistical mechanical theory of chemically reacting multiple phase mixtures: Application to the detonation properties of PETN. J. Chem. Phys. **81**, 1251 (1984)
53. Ree, F.H.: Supercritical fluid phase separations – implications for detonation properties of condensed explosives. J. Chem. Phys. **84**, 5845–5856 (1986)
54. Reed, T.M., Gubbins, K.E.: Statistical Mechanics. McGraw-Hill, New York (1973)
55. Rice, B.M., Byrd, E.F.C.: Ab initio study of compressed 1,3,5,7-tetranitro-1,3,5,7-tetraazacylooctane (HMX), cyclotrimethylenetrinitramine (RDX), 2,4,6,8,10,12-hexanitrohexaazaisowurzitane (CL-20), 2,4,6-trinitro-1,3,5-benzenetriamine (TATB) and pentaerythritol tetanitrate (PETN). J. Phys. Chem. C **111**, 2787–2796 (2007)

56. Rice, B.M., Byrd, E.F.C.: A comparison of methods to predict solid phase heats of formation of molecular energetic salts. J. Phys. Chem. A **113**, 345–352 (2009)
57. Rice, B.M., Hare, J.J., Byrd, E.F.C.: Accurate predictions of crystal densities using quantum mechanical molecular volumes. J. Phys. Chem. A **111**, 10874–10879 (2007)
58. Ross, M.: A high density fluid-perturbation theory based on an inverse 12th power hard-sphere reference system. J. Chem. Phys. **71**, 1567 (1979)
59. Ross, M., Ree, F.H.: Repulsive forces of simple molecules and mixtures at high density and temperature. J. Chem. Phys. **73**, 6146–6152 (1980)
60. Rushbrooke, G.S., Stell, G., Hoye, J.S.: Theory of polar liquids I. Dipolar hard spheres. Mol. Phys. **26**, 1199 (1973)
61. Schouten, J.A.: What is different in mixtures? From critical point to high pressures. Int. J. Thermophys. **22**, 23 (2000)
62. Shaw, M.S.: Monte Carlo simulation of equilibrium chemical-composition of molecular fluid mixtures in the N_{atoms} PT ensemble. J. Chem. Phys. **94**, 7550–7553 (1991)
63. Shaw, M.S., Johnson, J.D.: Carbon clustering in detonations. J. Appl. Phys. **62**, 2080–2085 (1987)
64. Smith, W.R., Missen, R.W.: Chemical Reaction Equilibrium Analysis: Theory and Algorithms. Wiley, New York (1982)
65. Sofyan, Y., Ghajar, A.J., Gasem, K.A.M.: Multiphase equilibrium calculations using Gibbs minimization techniques. Ind. Eng. Chem. Res. **42**, 3786–3801 (2003)
66. Souers, P.C., Kury, J.W.: Comparison of cylinder data and code calculations for homogeneous explosives. Propellants Explos. Pyrotech. **18**, 175 (1993)
67. Stell, G., Rasaiah, J.C., Narang, H.: Thermodynamic perturbation theory for simple polar fluids, 1. Mol. Phys. **23**, 393 (1972)
68. Swan, G.W., Fowles, G.R.: Shock wave stability. Phys. Fluids **18**, 28–35 (1975)
69. Talbot, J., Lebowitz, J.L., Waisman, E.M., Levesques, D., Weis, J.J.: A comparison of perturbative schemes and integral equation theories with computer simulations at high pressures. J. Chem. Phys. **85**, 2187 (1986)
70. Twu, C.H., Gubbins, K.E.: Thermodynamics of polyatomic fluid-mixtures -II. Chem. Eng. Sci. **33**, 879 (1978)
71. van Thiel, M., Ree, F.H.: Properties of carbon clusters in TNT detonation products: Graphite-diamond transition. J. Appl. Phys. **62**, 1761–1767 (1987)
72. van Thiel, M., Ree, F.H.: Accurate high-pressure and high-temperature effective pair potentials for the system N_2-N and O_2-O. J. Chem. Phys. **104**, 5019–5025 (1996)
73. Viecelli, J.A., Ree, F.H.: Carbon clustering kinetics in detonation wave propagation. J. Appl. Phys. **86**, 237–248 (1999)
74. Viecelli, J.A., Bastea, S., Glosli, J.N., Ree, F.H.: Phase transformations of nanometer size carbon particles in shocked hydrocarbons and explosives. J. Chem. Phys. **115**, 2730 (2001)
75. Vitello, P., Fried, L.E., Glaesemann, K.R., Souers, C.: Kinetic modeling of slow energy release in non-ideal carbon rich explosives. In: Thirteenth Symposium (International) on Detonation, pp. 465–475, Norfolk (2006)
76. Wagner, W., Pruss, A.: The IAPWS formulation 1995 for the thermodynamic properties of ordinary water substance for general and scientific use. J. Phys. Chem. Ref. Data **31**, 387 (2002)

77. White, W.B., Johnson, S.M., Dantzig, G.B.: Chemical equilibrium in complex mixtures. J. Chem. Phys. **28**, 751 (1958)
78. Williams, G.O., Lebowitz, J.L., Percus, J.K.: Equivalent potentials for equations of state of fluids of nonspherical molecules. J. Chem. Phys. **81**, 2070 (1984)
79. Wu, C.J., Fried, L.E., Yang, L.N., Goldman, N., Bastea, S.: Catalytic behavior of dense hot water. Nat. Chem. **1**, 57 (2009)
80. Zaug, J.M., Fried, L.E., Abramson, E.H., Hansen, D.W., Crowhurst, J.C., Howard, W.M.: Measured sound velocities of H_2O and CH_3OH. High-Pressure Res. **23**, 229–233 (2003)
81. Zaug, J.M., Bastea, S., Crowhurst, J.C., Armstrong, M.R.: Photoacoustically measured speeds of sound of liquid HBO_2: Semi-empirical modeling of boron-containing explosives. J. Phys. Chem. Lett. **1**, 2982–2988 (2010)
82. Zeldovich, I.B.: On the question of energy use of detonation combustion. J. Tech. Phys. **10**, 1453 (1940)
83. Zeldovich, I.B., Kompaneets, A.S.: Theory of Detonation. Academic, New York (1960)
84. Zel'dovich, Y.B., Raizer, Y.P.: Physics of Shock Waves and High Temperature Hydrodynamics Phenomena. Academic, New York (1966)
85. Zhang, F.: Detonation of gas-particle flow. In: Zhang, F. (ed.) Heterogeneous Detonation, Chap. 2, p. 91. Springer, Heidelberg (2009)

2

Steady One-Dimensional Detonations

Andrew Higgins

2.1 Introduction

While treatments of detonation wave propagation using control volume analysis, such as the Chapman–Jouguet (CJ) detonation solution presented in the prior chapter, are very successful in predicting the steady-state, equilibrium properties of detonations, they provide no information about the limits of detonation propagation or the dynamics of detonation waves. Addressing these issues necessitates investigating the structure of the detonation front. To illustrate this point, consider an extremely dilute concentration of fuel in air (e.g., 0.1% of methane in air by volume). If this mixture is entered into a thermochemical equilibrium code, a unique equilibrium CJ detonation solution will be generated. In practice, however, such a dilute mixture is highly unlikely to be able to support detonation wave propagation, since the low post-shock temperatures from the weak leading shock front would result in very slow reaction rates or no perceptible reaction at all. Even if exothermic reaction occurred, the long reaction zone would make the detonation susceptible to momentum and heat transfer losses at the boundaries of the wave (e.g., friction and cooling between the flow within the detonation front and the tube walls in the case of a confined detonation), which would ultimately quench the detonation. Thus, the issue of defining detonability limits naturally leads to an investigation of the reaction zone length.

To clarify the relation between the equilibrium solutions from the previous chapter and the detailed structure of the wave investigated in this chapter: The equilibrium solutions using the CJ criterion will accurately predict the detonation velocity and downstream thermodynamic state *if* a detonation occurs well inside its detonability limits and without losses. Such a control-volume analysis, however, cannot predict whether or not detonation will occur in a particular scenario. In other words, equilibrium calculations cannot be used to define detonability limits. Nor can equilibrium solutions be used to examine the initiation, transient dynamics, or stability of a detonation wave.

F. Zhang (ed.), *Shock Wave Science and Technology Reference Library, Vol. 6*: 33
Detonation Dynamics, DOI 10.1007/978-3-642-22967-1_2,

A one-dimensional model for the structure of a steady detonation wave was developed independently by Zeldovich [106], von Neumann [58,59], and Döring [23] in the 1940s, although the concept of a detonation wave as an initial shock wave that initiates a combustion reaction was suggested as early as the end of the nineteenth century by Mikhel'son [57] and Vieille [98]. For the complete history of the development of detonation wave physics, see [5, 40, 48, 55]. The key assumption of the Zeldovich–von Neumann–Döring (ZND) model is that the leading shock front does not result in an immediate change in the chemical composition of the mixture; the chemistry is "frozen" across the leading shock front. The post-shock flow then reacts or "relaxes" to the final equilibrium solution determined by the CJ condition. The assumption that no reactions occur within the leading shock wave is well supported by the fact that shock waves are extremely thin, typically only a few mean free paths thick, while chemical reactions only occur occasionally upon molecular collisions. Thus, in gases, the chemical reactions occur on timescales many orders of magnitude greater than the shock, such that the chemistry across the shock is effectively frozen. For condensed-phase explosives, decomposition of the explosive molecule may begin on the scale of the shock front itself, but the energy release still occurs on a scale much longer than that of the shock, so the ZND representation remains applicable. The development and applications of the ZND framework are the subject of this chapter.

The ZND models of wave structure are represented schematically in Fig. 2.1. In a gaseous detonation, translational and rotational equilibrium is established extremely quickly following the arrival of the shock front. Vibrational excitation takes longer, but is still rapid compared to the overall reaction zone length and occurs with negligible chemical reaction. The majority of the reaction zone is dominated by a nearly thermally neutral process of chain initiating and chain branching chemical reactions, during which time a "soup" of radical species is built up. Significant exothermic reaction only occurs with recombination to product species (typically CO_2, H_2O, N_2, etc.), which happens relatively quickly in most reactive systems due to the feedback between heat evolution and the exponential dependence of reaction rates on temperature, finally leading to equilibrium at the sonic plane. The unsteady expansion of products beyond the sonic plane matches the product flow to the downstream boundary condition; this flow is treated in Appendix A.1.

For condensed phase (solid and liquid) explosives, the process is more complex and cannot as easily be classified into "nonreacting shock" and "reaction." Translational equilibrium is again established quickly on a length scale comparable to the explosive molecule itself, but vibrational excitation is a complex process owing to the polyatomic nature of most molecular explosives. The transfer of energy from low frequency to high frequency modes of molecular vibration is necessary to break molecular bonds (an endothermic process). The full details of this process are only now becoming accessible theoretically and experimentally. Recombination via chemical kinetics followed by vibrational de-excitation to stable, equilibrium products (again, CO_2, H_2O,

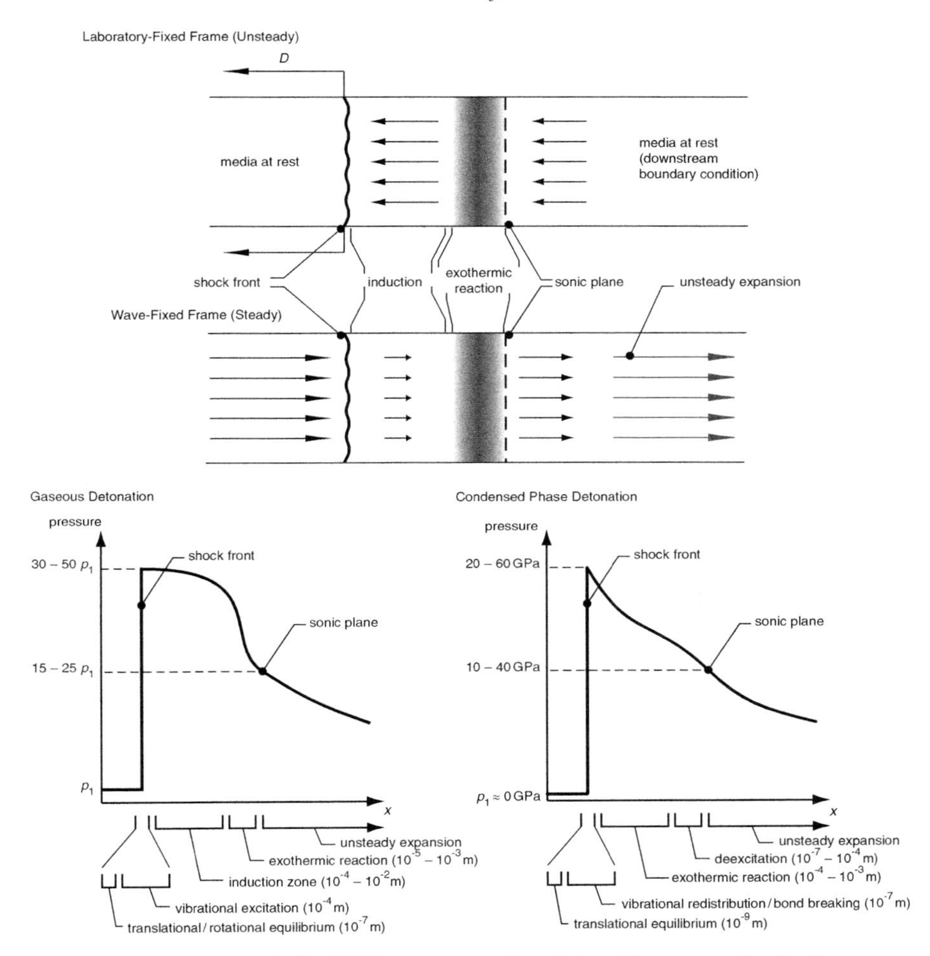

Fig. 2.1. Schematic of the ZND structure of a detonation wave, for both gaseous detonation at atmospheric conditions and condensed phase detonation

N_2, etc.) occurs as well as slower condensation and diffusion-controlled processes such as carbon particulate formation. A current picture of the Non-Equilibrium ZND (NEZND) structure of condensed phase detonation waves has been presented by Tarver [90, 92] and in Chap. 6 of this volume.

This picture is further complicated by the fact that all known gaseous detonations are hydrodynamically unstable (Chap. 3 of this volume) and this instability is manifested as a transient, three-dimensional cellular structure (see Chaps. 3–5 in this volume). Condensed-phase explosives are similarly complex. Most liquid explosives display a structure of transverse waves that are believed to be similar to those in gaseous detonation, only on a much finer scale. Solid explosives as used in most applications are polycrystalline, resulting in the reaction zone structure being controlled by spatially localized reaction centers resulting from the heterogeneous nature of the media. All of these features mean that the steady, one-dimensional model considered in this chapter is highly idealized and typically cannot be used to make direct, quantitative predictions. Nevertheless, the general view of a prompt shock followed by a relatively long, thermally neutral induction process, and concluded with a rapid exothermic reaction, remains a valid picture for many detonation waves. Moreover, the steady, one-dimensional model of a detonation is of enormous value as an analytically tractable treatment of detonation waves, readily amenable to investigation and solution. More sophisticated models (i.e., transient and multidimensional) remain computationally challenging. Properly resolved one-dimensional numerical simulations of a gaseous detonation with realistic chemistry, for example, are at the boundary of what is feasible with state-of-the-art computers [67], and multidimensional simulations with fully resolved, detailed chemistry and molecular transport processes will likely remain beyond computational ability for some time.

2.2 ZND Model for Perfect Gas

For the ZND model of the reaction zone, it is necessary to integrate the differential form of the conservation laws coupled to a reaction rate equation, starting at the immediate post-shock or *von Neumann state*. The steady form of the conservation of mass, momentum, and energy written in conservative form

$$\mathrm{d}\left(\rho u A\right) = 0 \tag{2.1}$$

$$\mathrm{d}\left(p + \rho u^2\right) = -\rho u^2 \frac{\mathrm{d}A}{A} \tag{2.2}$$

$$\mathrm{d}\left(h + \frac{1}{2}u^2\right) = 0 \tag{2.3}$$

can be rewritten in differential form as

$$\frac{\mathrm{d}\rho}{\rho} + \frac{\mathrm{d}u}{u} + \frac{\mathrm{d}A}{A} = 0 \tag{2.4}$$

$$\mathrm{d}p + \rho u \mathrm{d}u = 0 \tag{2.5}$$

$$c_p \mathrm{d}T - \Delta q \mathrm{d}\lambda + u \mathrm{d}u = 0. \tag{2.6}$$

Here, λ is a *reaction progress variable*, where $\lambda = 0$ corresponds to unreacted mixture and $\lambda = 1$ is the fully reacted mixture. This system can be thought of as gas "A" that reacts exothermally to form "B" (i.e., global reaction $\mathrm{A} \rightarrow \mathrm{B}$) with both gases being calorically perfect with the same heat capacity and molecular weight. The heat of reaction is $\Delta q = h^\circ_{f\mathrm{A}} - h^\circ_{f\mathrm{B}}$ and λ is the mass fraction of species "B." This simplified treatment of a reacting flow is essentially Rayleigh flow or "Simple T_o-change flow" for a perfect gas, as developed in texts on compressible flow [74]. The enthalpy incorporates this heat release term similar to the enthalpy of formation in a real reacting system:

$$h = c_p T - \lambda \Delta q. \tag{2.7}$$

Finally, a differential form of ideal gas law ($p = \rho RT$) is also required

$$\frac{dp}{p} = \frac{d\rho}{\rho} + \frac{dT}{T}. \tag{2.8}$$

By eliminating variables and introducing the sound speed $c^2 = \left(\frac{\partial p}{\partial \rho}\right)_s = \gamma RT$, the differential change in velocity is related to differential changes in the reaction progress variable and area

$$\frac{du}{u} = \frac{\frac{\Delta q}{c_p T} d\lambda - \frac{dA}{A}}{1 - \frac{u^2}{c^2}}. \tag{2.9}$$

Introducing the flow Mach number ($M = \frac{u}{c}$), this result can be written as

$$\frac{du}{u} = \frac{\frac{\Delta q}{c_p T} d\lambda - \frac{dA}{A}}{1 - M^2}. \tag{2.10}$$

If the flow is adiabatic or nonreacting ($\Delta q = 0$ or $d\lambda = 0$), then a familiar result for isentropic compressible flow that relates the change in flow velocity to a differential change in area is obtained.

$$\frac{du}{u} = -\frac{\frac{dA}{A}}{1 - M^2}. \tag{2.11}$$

For the remainder of this section, only constant area flow ($dA = 0$) with exothermic heat release ($\Delta q > 0$) will be considered

$$\frac{du}{u} = \frac{\frac{\Delta q}{c_p T} d\lambda}{1 - M^2}. \tag{2.12}$$

Comparing (2.11) and (2.12), it can be seen that heat addition, much like converging area in steady isentropic flow, drives the flow toward sonic by accelerating a subsonic flow or decelerating a supersonic flow. The presence of the $1 - M^2$ term in the denominator raises the question as to what happens when the flow reaches Mach 1. In order to prevent a singularity in (2.12), the differential change in the reaction progress variable $d\lambda$ must also be zero as the flow becomes sonic. Usually, this condition corresponds to equilibrium when the flow composition has finished evolving. Again, there is an analogy to isentropic compressible flow, where sonic flow ($M = 1$) can only occur at a locally constant cross-sectional area ($dA = 0$).

The rate at which chemical reactions release heat into the flow can be modeled by an Arrhenius expression

$$\dot{\lambda} = k(1-\lambda)\exp\left(-\frac{E_a}{RT}\right), \tag{2.13}$$

where the $1-\lambda$ term ensures that the reaction rate decreases to zero as the initial reactant is entirely consumed. The activation energy E_a in the Arrhenius expression will be seen to have a determining role in the structure of the reaction zone.

This rate equation refers to the reaction of a fluid element or particle in the flow. It can be related to the spatial derivative of λ using the particle derivative:

$$\dot{\lambda} = \frac{D\lambda}{Dt} = \frac{\partial\lambda}{\partial t} + u\frac{\partial\lambda}{\partial x}. \tag{2.14}$$

Since the wave-fixed frame is assumed to be steady state in the ZND model, $\frac{\partial\lambda}{\partial t} = 0$, so

$$\frac{\mathrm{d}\lambda}{\mathrm{d}x} = \frac{1}{u}\dot{\lambda}. \tag{2.15}$$

The governing differential equation for the flow velocity through the reaction zone becomes

$$\frac{\mathrm{d}u}{\mathrm{d}x} = \frac{\frac{\Delta q}{c_p T}\dot{\lambda}}{1-M^2}. \tag{2.16}$$

Equations (2.13), (2.15), and (2.16) form a coupled set of ordinary differential equations. In addition, the temperature must be computed at any point in the flow given a value of $u(x)$ and $\lambda(x)$. This can be done via the energy equation

$$T(x) = \frac{1}{c_p}\left(c_p T_1 + \lambda(x)\Delta q + \frac{1}{2}\left(u_1^2 - u(x)^2\right)\right), \tag{2.17}$$

where T_1 and u_1 are the initial temperature and velocity of the flow approaching the detonation before encountering the shock wave ($u_1 = D$). The sound speed (as required to compute Mach number) is given by $\sqrt{\gamma RT}$. This set of (2.13), (2.15)–(2.17) can be numerically integrated, starting from the post-shock von Neumann state ($x = 0$), which is determined by the normal shock relations. The Mach number of the leading shock front can be computed using the relations for a CJ detonation, as given by

$$M_{\mathrm{CJ}} = \sqrt{(\gamma+1)\frac{\Delta q}{c_p T_1} + 1 + \sqrt{\left[(\gamma+1)\frac{\Delta q}{c_p T_1} + 1\right]^2 - 1}}. \tag{2.18}$$

Alternatively, the value of the initial shock Mach number can be guessed and iterated upon until a numerical solution for the reaction zone is found that satisfies the requirement that the reaction reaches equilibrium ($\lambda = 1$) as the flow becomes sonic. Either of these approaches will yield the same solution, unless a "pathological" case is encountered, as discussed in the following sections.

Numerical integration of the reaction zone structure was carried out for three cases, $\frac{E_a}{RT_1} = 10$, 25, and 50, and the results plotted in Fig. 2.2. For these calculations, $Q = \frac{\Delta q}{c_p T_1} = 10$ and $\gamma = 1.2$. The value of k in the rate equation (2.13) was set to unity (alternatively, since k has units of 1/s, it is possible to define a nondimensional time $\hat{t} = kt$). Note that, given the form of the rate equation (2.13), the equilibrium state of $\lambda = 1$ will, in principle, never be reached and can only be approached asymptotically. Thus, it is difficult to define a reaction zone length where the heat release is complete. Rather, some other definition of reaction zone length must be used. Common choices are the half-reaction zone length (variously defined as where $\lambda = 0.5$, where half of the heat addition occurs, or where temperature reaches the midpoint

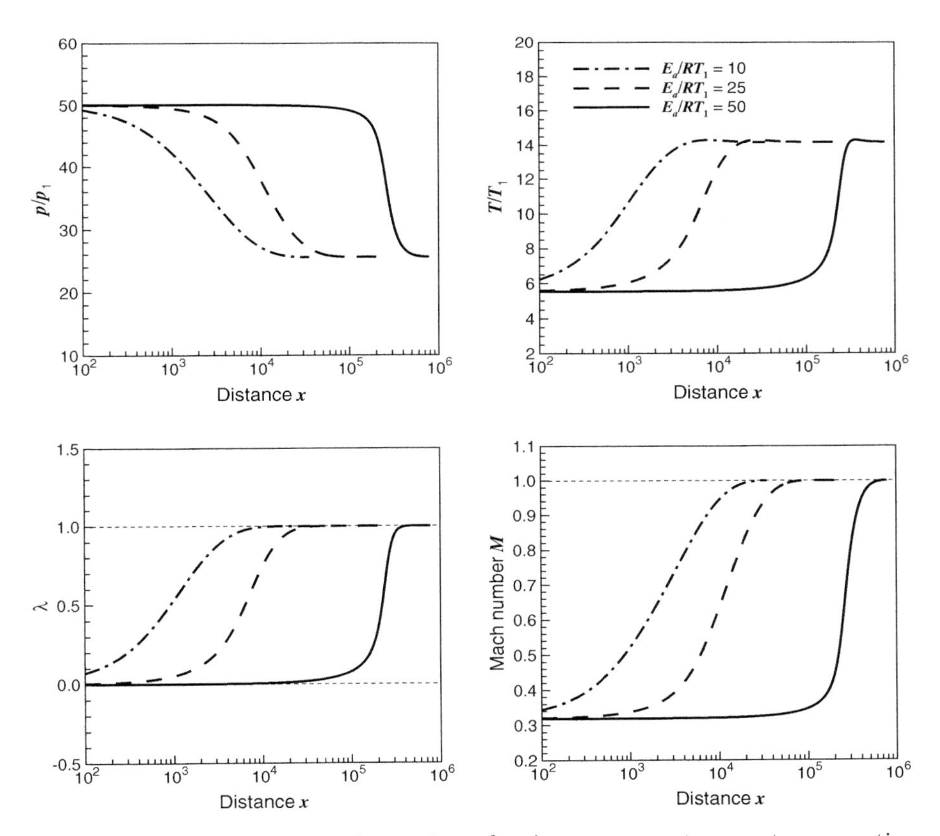

Fig. 2.2. Structure of ZND detonation, showing pressure, temperature, reaction progress variable, and flow Mach number for various values of activation energy $\frac{E_a}{RT_1}$, $Q = 10$, $\gamma = 1.2$. Note the use of log scale for x-axis

between T_{vN} and T_{CJ}), the location of the maximum reaction rate, or the inflection point of the temperature profile.

As activation energy is increased from $\frac{E_a}{RT_1} = 10$ to $\frac{E_a}{RT_1} = 25$, the reaction zone increases by an order of magnitude in length. Increasing from $\frac{E_a}{RT_1} = 25$ to $\frac{E_a}{RT_1} = 50$ results in an additional two orders of magnitude increase in reaction zone length. The length scale of these reaction zones is somewhat arbitrary, due to taking the value of k as unity. What is more significant is the increasing *sharpness* of the reaction profile as activation energy is increased. A lower activation energy results in a relatively gradual increase in temperature beginning immediately after the shock, while the high activation energy results in a long plateau of nearly constant conditions followed by a comparatively rapid energy release. This feature of high activation energy reflects the underlying temperature sensitivity of chemical reactions that must be activated and is one of the most significant and reoccurring themes in the development of detonation theory.

The temperature profiles in Fig. 2.2 feature maxima that occur before the end of the reaction zone; this is most clearly visible in the temperature profile for the $\frac{E_a}{RT_1} = 50$ case due to the steepness of the profile but it is present in all cases plotted. Beyond this maximum, exothermic heat release in the flow has the effect of lowering the temperature. This counterintuitive result is a consequence of the heat release resulting in greater flow acceleration and pressure/density reduction than its contribution to raising the static temperature of the flow. This temperature decrease with heat addition occurs when the Mach number is in the range $\frac{1}{\sqrt{\gamma}} < M < 1$; this result is well established in compressible one-dimensional flow with heat addition (Rayleigh flow).

It is interesting to explore what happens if a calculation of the reaction zone structure is initialized with a non-CJ detonation velocity. This is done in Fig. 2.3 for $\frac{E_a}{RT_1} = 25$, where the solution was initialized with the post-shock state for shock Mach numbers 20% greater and 20% less than the CJ Mach number. For the stronger shock case, the reaction zone structure can be solved for, and as the reaction progress variable reaches $\lambda = 1$, the flow is seen to remain subsonic. This solution corresponds to a strong detonation, i.e., the branch of the product Hugoniot curve above the CJ detonation point on the (p, v) plane. This case is also referred to as an overdriven detonation, since the detonation is being forced to propagate at a speed greater than the CJ speed, and the heat release is insufficient to choke the flow. If the solution is initialized with a sub-CJ detonation velocity, the numerical integration encounters a singularity as the flow becomes sonic while the reaction rate is still finite ($\lambda < 1$) in (2.16). In this case, the heat release is sufficient to choke the flow before adding the full heat release, and further heat addition to a sonic flow is not permitted in a steady solution. This singularity will appear for any initial condition that is even infinitesimally less than the CJ speed. A similar result is obtained if the wave speed is fixed and the heat release Q increased incrementally. The appearance of a mathematical singularity may

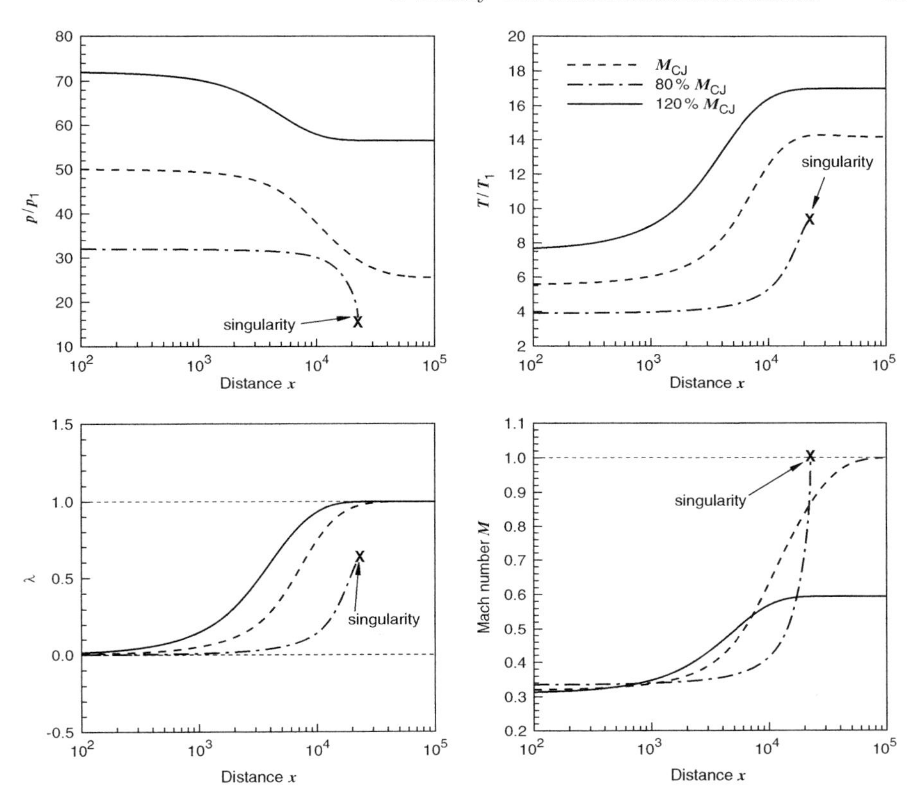

Fig. 2.3. Structure of ZND detonation for reaction zone structure initialized with a CJ detonation, a detonation overdriven at 120% M_{CJ}, and a shock at 80% M_{CJ} showing pressure, temperature, reaction progress variable, and flow Mach number for $\frac{E_a}{RT_1} = 25$, $Q = 10$, $\gamma = 1.2$. The sub-CJ shock does not correspond to a solution of conservation laws, resulting in a singularity appearing in the numerical integration

appear alarming, but recall that the CJ detonation is the minimum velocity wave consistent with the governing conservation laws. The fact that a singularity is encountered in numerically integrating through the reaction zone initialized with a speed less than the minimum speed simply reflects that an attempt is being made to solve a flow using initial conditions for which no steady solution exists.

2.3 Pathological Detonations

Instead of the one step reaction A $\rightarrow$ B, consider a two-step reaction with A reacting to form B, which in turn reacts to form C

$$\text{Reaction 1:} \quad \text{A} \rightarrow \text{B} \tag{2.19}$$

$$\text{Reaction 2:} \quad \text{B} \rightarrow \text{C} \tag{2.20}$$

with the first reaction being exothermic ($\Delta q_1 > 0$), the second reaction being endothermic ($\Delta q_2 < 0$), and the overall reaction exothermic ($|\Delta q_2| < \Delta q_1$). The solution of the reaction zone structure in this case becomes a more complicated problem. Specifically, initializing the differential equations for the reaction zone using the CJ detonation velocity based on the final, equilibrium heat release of the overall reaction ($\Delta q_{\mathrm{tot}} = \Delta q_1 + \Delta q_2$) results in a singularity being encountered as the integration proceeds. In this case, it is necessary to guess and iterate on the value of detonation velocity in order to find a wave velocity that permits the governing differential equations to yield a well-behaved solution. This solution, called the *eigenvalue detonation solution*, may occur whenever there are competing effects influencing the flow in the detonation reaction zone. The particular scenario encountered here, in which a detonation velocity greater than the CJ velocity based on the total heat release is obtained due to an endothermic reaction, is called *pathological detonation*.

To explore the structure of a detonation with competing exothermic and endothermic reactions, the system A → B → C will be considered with the two reaction steps governed by Arrhenius rate laws

$$\dot{\lambda_1} = k(1 - \lambda_1)\exp\left(-\frac{E_{a1}}{RT}\right) \qquad \dot{\lambda_2} = k(\lambda_1 - \lambda_2)\exp\left(-\frac{E_{a2}}{RT}\right), \tag{2.21}$$

where λ_1 and λ_2 are the reaction progress variables for the reaction A → B and B → C, respectively. The mass fractions of A, B, and C are given by

$$Y_{\mathrm{A}} = 1 - \lambda_1 \qquad Y_{\mathrm{B}} = \lambda_1 - \lambda_2 \qquad Y_{\mathrm{C}} = \lambda_2. \tag{2.22}$$

This set is a highly simplified analog to a real, multi-step chemical kinetic mechanism, where the reaction must pass through an intermediate species B in order to produce the final product C.

The differential equation governing the reaction zone is now

$$\frac{\mathrm{d}u}{\mathrm{d}x} = \frac{\frac{1}{c_p T}(\Delta q_1 \dot{\lambda_1} + \Delta q_2 \dot{\lambda_2})}{1 - M^2}, \tag{2.23}$$

which is the two-step analog to (2.16). Equation (2.17) must be similarly modified to account for the two-step reaction.

This system was proposed by Fickett and Davis [19] and its dynamics further explored by Sharpe and Falle [75, 77]. For this system, the activation energy and heats of reaction are assigned as follows:

$$\frac{E_{a1}}{RT_1} = 25 \qquad \frac{E_{a2}}{RT_1} = 25 \tag{2.24}$$

$$Q_1 = \frac{\Delta q_1}{c_p T_1} = 20 \qquad Q_2 = \frac{\Delta q_2}{c_p T_1} = -10. \tag{2.25}$$

Note that the overall reaction has a net heat release of $Q_{\mathrm{tot}} = Q_1 + Q_2 = 10$, the same as the system considered previously. Thus, the CJ detonation Mach number based on the equilibrium heat release is $M_{\text{CJ}_{\text{equil}}} = 6.781$. Initializing the integration of (2.23) with the von Neumann state based on this shock Mach number results in a singularity being encountered, as shown in Fig. 2.4. It is necessary to increase the shock Mach number to $M_{\text{CJ}_{\text{eigen}}} = 7.207$ (or approximately 6% greater than the CJ speed based on equilibrium) in order to find a solution that does not encounter a singularity within the reaction zone structure. This value can only be found by trial and error integration of the governing ordinary differential equations. The term Chapman–Jouguet is still used to refer to this solution, since it features a sonic surface consistent with Jouguet's original criterion. To differentiate these two cases, they will be referred to as the CJ equilibrium solution and the CJ eigenvalue solution. Finally, to eliminate any confusion as to why the presence of an *endothermic* reaction results in a *greater* than equilibrium detonation velocity, note that the detonation velocity associated with the first stage exothermic reaction alone would be $M_{\text{CJ}}(Q_1) = 9.486$. The endothermic reaction reduces the effective heat release and detonation velocity in comparison to this value, but the pathological behavior results in a detonation velocity that is greater than the CJ equilibrium solution based on the total heat release.

Numerical integration of ordinary differential equations with a potential singularity embedded within the solution is a notoriously difficult problem. It may be necessary to intervene in the numerical algorithm used in order to select a solution, since upon encountering the sonic point, the differential equations become indeterminate. Alternatively, it is possible to start at the sonic point and integrate forward, toward the shock. Even in this case, iteration is still required to find a solution consistent with the initial conditions upstream of the wave, since the thermodynamic state of the sonic point is not known a priori.

For the solution of the governing ordinary differential equations initialized with $M_{\text{CJ}_{\text{eigen}}} = 7.207$, the solution for the reaction zone structure is shown in Fig. 2.4. Coincident with the sonic point, the local heat release rate is seen to go to zero due to an exact balance between the rate of exothermic heat addition and endothermic heat removal. This condition is referred to as the *generalized Chapman–Jouguet condition*, a term first used by Eyring et al. [27]. The condition of the local heat release rate being zero behaves like a quasi-equilibrium condition in (2.23), enabling the solution to pass smoothly through the sonic point and proceed downstream as a supersonic flow (recall that heat removal accelerates a supersonic flow away from sonic). This case is shown as the heavy, solid line in Fig. 2.4.

The sonic point of the generalized CJ condition is a *saddle point*, so in principle it is also possible for the solution at the sonic point to return to the subsonic branch under the influence of subsequent endothermic reaction, as shown by the thin, dashed lines in Fig. 2.4. This solution, however, features a nonphysical "kink" in the flow properties. Thus, it could be suggested that the solution that passes smoothly from the subsonic to the supersonic branch is the correct one, but this cannot be proven rigorously. As mentioned previously,

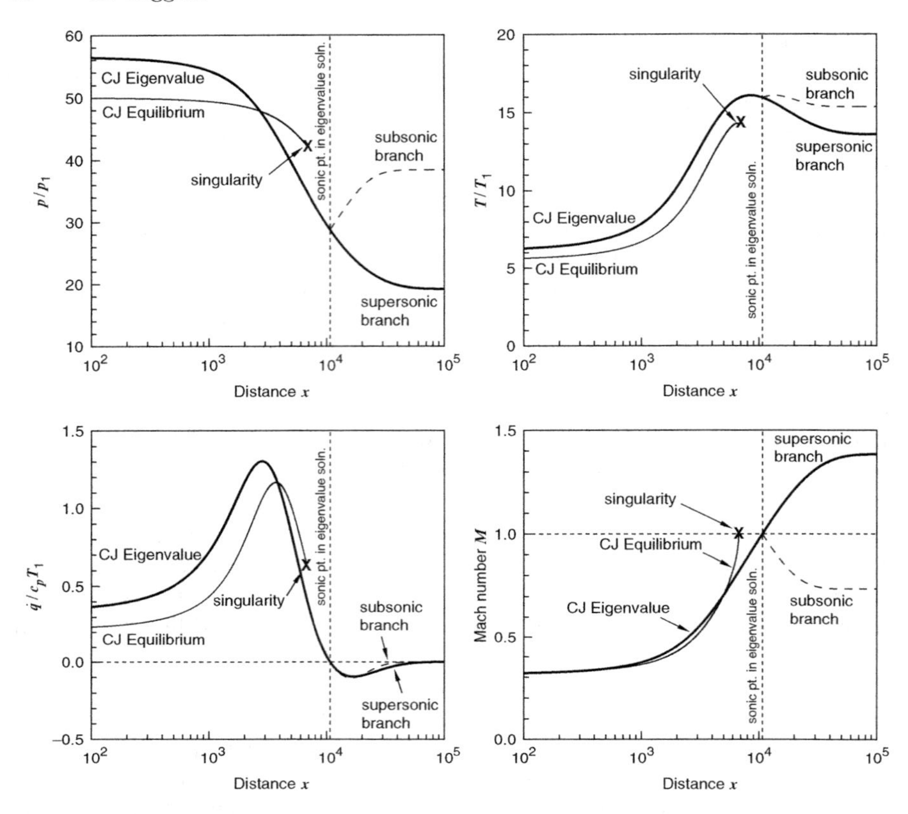

Fig. 2.4. Reaction zone structure of ZND detonation with a two-step, exothermic/endothermic system, showing pressure, temperature, heat release rate, and flow Mach number for $\frac{E_{a1}}{RT_1} = \frac{E_{a2}}{RT_1} = 25$, $Q_1 = 20$, $Q_2 = -10$, $\gamma = 1.2$. The reaction zone initialized with a shock based on the CJ equilibrium solution encounters a singularity in the numerical integration. The eigenvalue solution can pass smoothly through the sonic point, resulting in supersonic flow at the end of the reaction zone (weak detonation)

similar indeterminacy occurs in compressible isentropic flow when an initially subsonic flow becomes sonic at an area minimum (i.e., at a throat). If the requirement that the flow properties at the throat vary smoothly is imposed, the flow should transition to a supersonic solution locally at the throat, and from there, the flow can either continue supersonically or return to the subsonic branch via a shock wave, depending on the downstream boundary condition (e.g., back pressure).

For both the detonation with competing reactions and isentropic flow with area change, the issue of resolving which branch of the solution following the sonic point is the correct solution is ultimately determined by considering the downstream boundary condition. Establishing how the flow exiting the sonic

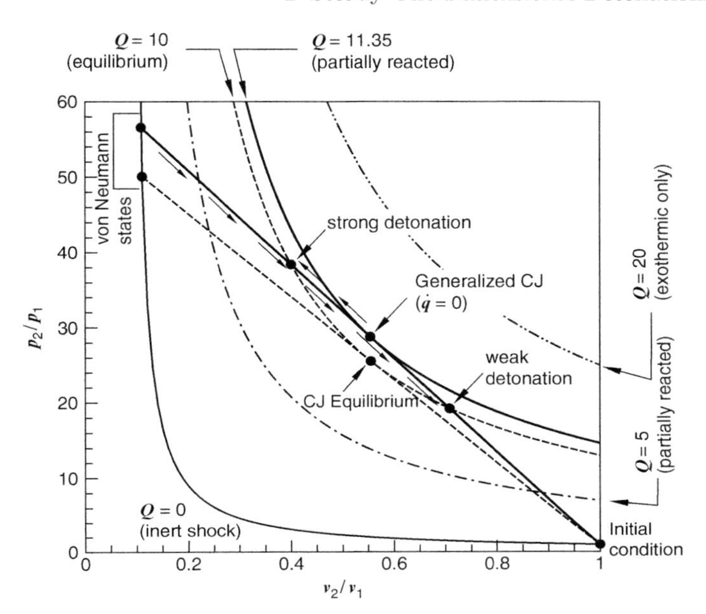

Fig. 2.5. Equilibrium and eigenvalue solutions for a two-step, exothermic/endothermic system visualized in the (p, v) plane ($\frac{E_{a1}}{RT_1} = \frac{E_{a2}}{RT_1} = 25$, $Q_1 = 20$, $Q_2 = -10$, $\gamma = 1.2$)

surface of such a detonation matches with a downstream condition imposed by a piston becomes quite involved as a number of possible scenarios need to be considered. The reader is referred to the book by Fickett and Davis [19] for a complete exposition on this problem. Fickett and Davis also consider the case of two exothermic reactions, which turns out to be qualitatively similar to a single exothermic reaction and does not exhibit pathological behavior. For the most frequently encountered case of a downstream piston at rest (corresponding to a detonation in a close-ended tube, for example), the flow passes smoothly to the supersonic solution beyond the sonic point. In this case, the leading edge of the rarefaction wave lags behind the sonic surface and the end of the reaction zone, resulting in a region of uniform supersonic flow between the end of the reaction zone and the leading rarefaction that will increase in size as the detonation propagates. These scenarios are discussed further in Appendix A.1.

The eigenvalue solution can also be interpreted in the (p, v) plane with the use of partially reacted Hugoniots, as first proposed by von Neumann [58].[1] In Fig. 2.5, the Hugoniot and Rayleigh line for the equilibrium solution are shown as dashed lines. The Hugoniots for the value of heat release at an intermediate

[1] Formally, a Hugoniot is a locus of possible equilibrium end states for a steady wave, so a "partially reacted Hugoniot" is a misnomer.

value ($Q = 5$) and at the sonic point of the eigenvalue solution ($Q = 11.35$) are plotted as dashed-dotted and solid lines, respectively. The Rayleigh line for the eigenvalue solution is seen to be tangent to the solid Hugoniot curve at the point where the net rate of heat release is zero (the generalized CJ condition). In actuality, there is a continuous series of partially reacted Hugoniots, and the correct one to use for the detonation solution cannot be determined until the kinetic rate equation has been integrated, as was done above. Starting from the von Neumann point, as the reaction progresses, the solution proceeds down the Rayleigh line as indicated by the directional arrows. At the sonic point of the eigenvalue solution, the solution is indeterminate. As endothermic reactions dominate over the exothermic beyond the sonic point, the solution may continue down the Rayleigh line (supersonic branch), or may proceed back up the Rayleigh line (subsonic branch). Either branch of the solution eventually reaches the equilibrium Hugoniot. The upper intersection point can be recognized as a *strong detonation* and the lower intersection as a *weak detonation*. As discussed above, the weak solution is more likely to be realized, and thus pathological detonations are an example of a weak detonation solution. If the reaction zone calculation is initialized with the post-shock state corresponding to the CJ equilibrium solution (dashed straight line), then as the tangency condition (sonic flow) is encountered for the first time, the net heat release rate is still positive. As net exothermic reaction continues, there is no longer an intersection between the Rayleigh line and the partially reacted Hugoniot, resulting in the singularity encountered in Fig. 2.4. Thus, it is not possible to construct a trajectory of partially reacted states to reach the CJ equilibrium solution. Detonations for which the equilibrium CJ detonation solution may not be the correct solution (or permissible solution) due to competing effects in the reaction zone may occur in systems with real chemistry as well (Sect. 2.5).

2.4 Detonations with Source Terms

In order to discuss limits to detonation propagation, losses must be introduced into the governing conservation equations. In a system without losses and governed by an Arrhenius-type reaction rate, detonation propagation is always possible due to the following mechanism. An Arrhenius-governed system will inevitably react to equilibrium, and the temperature increase from even a weak shock wave will accelerate the reaction, resulting in a wave of exothermic energy release traveling with the shock some distance behind it. In the case of a very slow reaction, this front of exothermicity may be spatially separated from the shock by vast distances, but in the absence of losses such as heat transfer or friction, the energy released will eventually feed into supporting the shock, ultimately resulting in a detonation. (Recall that a CJ detonation is the minimum velocity allowed for a steady, compressive combustion wave;

lower velocity combustion waves are not permitted by the conservation laws.) This concept is sometimes referred to in the Russian detonation literature as *Khariton's principle*, namely that any media capable of exothermic reaction is capable of supporting detonation wave propagation in the absence of losses [72]. Thus, any discussion of the limits to detonation propagation must include the effects of losses.

The steady, one-dimensional mass, momentum, and energy equations including friction and heat transfer source terms in the wave-fixed reference frame are

$$\mathrm{d}\left(\rho u\right) = 0 \tag{2.26}$$

$$\mathrm{d}\left(p + \rho u^2\right) = f \mathrm{d}x \tag{2.27}$$

$$\mathrm{d}\left[\rho u \left(h + \frac{1}{2}u^2\right)\right] = q \mathrm{d}x + f u_1 \mathrm{d}x, \tag{2.28}$$

where f is a source term of momentum and q is a source term of energy. The velocity u_1 is the velocity of the wall in the wave-fixed frame, which is equal in magnitude to the velocity of the wave in the laboratory-fixed frame, D. As written here, f and q are the volumetric source terms, with units of $[\mathrm{N\,m^{-3}}]$ and $[\mathrm{W\,m^{-3}}]$, respectively. A mass source term could also be included to account for, for example, mass loss into a porous wall; however, this effect is usually treated by introducing area divergence into the governing equations, as discussed in Sect. 2.6. These momentum and energy sources will be related to wall friction and heat transfer coefficients below. Note the appearance of the friction term f in the energy equation, representing the work done by friction. The significance of this term will be elaborated upon in Sect. 2.4.2 below. Following the development of Sect. 2.2, a differential equation for the flow velocity in the reaction zone can be found

$$\frac{\mathrm{d}u}{\mathrm{d}x} = \frac{\frac{\Delta q \dot{\lambda}}{c_p T} + \frac{q}{\rho c_p T} + \frac{f(u_1(\gamma-1)-\gamma u)}{\rho c^2}}{1 - M^2}. \tag{2.29}$$

Note that if $f = 0$ and $q = 0$, (2.29) reverts back to (2.16). Heat transfer to the wall is a heat loss ($q < 0$), so inspecting (2.29) reveals that the effect of heat loss is similar to that of endothermic reactions studied in the previous section. Heat losses will result in the detonation propagating at speeds less than the ideal CJ detonation velocity (i.e., a detonation without losses), and determining the solution will necessitate iterating on the propagation velocity until a regular solution of the reaction zone structure can be found. The effect of friction is not as intuitive, due to the terms of mixed sign involving f in (2.29); however, it will be shown that friction also results in a velocity deficit in comparison to the ideal CJ velocity.

2.4.1 Source Terms

For a detonation in a tube with wall friction, the source term f represents the frictional force at the wall spread over the entire cross-sectional area of the tube. Thus, f is given by

$$f = \frac{\tau_w P}{A}, \tag{2.30}$$

where τ_w is the shear stress at the wall and P the wetted perimeter of the tube. Introducing the hydraulic radius $r_h = \frac{2A}{P}$ and the skin friction coefficient

$$c_f = \frac{\tau_w}{\frac{1}{2}\rho u_{\text{rel}}^2}, \tag{2.31}$$

where u_{rel} is the velocity relative to the wall, and this term becomes

$$f = \rho(u - u_1)|u - u_1|\frac{c_f}{r_h}. \tag{2.32}$$

The absolute value operator ensures that friction always opposes the motion of the flow relative to the wall. Of course, there exists an enormous literature on correlations for the friction coefficient c_f, relating it to Reynolds number, surface roughness, etc. For highly turbulent flows ($Re > 10^6$) typically encountered in detonation waves, the friction coefficient is usually in the range of 0.01–0.001.

The heat source term is given by

$$q = -h_c(T - T_w)\frac{P}{A}, \tag{2.33}$$

where h_c is the heat transfer coefficient and T_w is the wall temperature. The minus sign reflects that heat transfer out of the system is a loss. The heat transfer coefficient may be related to the friction coefficient c_f via the Reynolds analogy, which states that for a turbulent flow the transport of heat and momentum are via the same underlying mechanism, as follows:

$$h_c = c_p\rho|u_1 - u|\frac{c_f}{2}. \tag{2.34}$$

The relative velocity between the flow and the wall is used since the main mechanism of heat transfer is forced convection. The Reynolds analogy is expected to be valid in the case of a media with a ratio of viscous to thermal transport, as expressed by the Prandtl number $\text{Pr} = \frac{\mu c_p}{k}$ (μ is viscosity, k thermal conductivity), that is of order unity. For most gases ($\text{Pr} \approx 0.8$), thus, this analogy is expected to hold. This assumption allows the heat transfer and friction source terms to be expressed in terms of a single nondimensional parameter, c_f. Using these expressions in (2.29), the differential equation for velocity becomes

$$\frac{\mathrm{d}u}{\mathrm{d}x} = \frac{\frac{\Delta q}{c_p T}\dot{\lambda} - \frac{c_f}{r_h}|u_1 - u|\left(\frac{T - T_w}{T}\right) - \frac{c_f}{r_h}\frac{(u_1 - u)|u_1 - u|(\gamma u - (\gamma - 1)u_1)}{c^2}}{1 - M^2}. \tag{2.35}$$

This equation can now be integrated to obtain the structure of the reaction zone with friction and heat losses. Note that the friction coefficient c_f always appears in combination with the hydraulic radius r_h. Also note that (2.17), which relates temperature to the flow velocity, can no longer be used due to the source term in the energy equation. A separate differential equation for temperature must be coupled to the solution of (2.35); such an equation can be found by using the differential equation of state (2.8).

2.4.2 Work Done by Friction

The presence of a work term due to friction in the energy equation and the fact that this term makes a positive contribution to energy (similar in sign to exothermic heat release) may appear counterintuitive and has been the source of some apprehension in implementing the ZND model for detonations with friction. This confusion is compounded by the fact that in one-dimensional compressible adiabatic flow with friction (i.e., Fanno flow), this term is absent [74]. This matter was explored at some length by Tanguay and Higgins [88] and is briefly recounted here in hopes of clarifying this issue.

Traditional Fanno flow in a tube is formulated in the laboratory-fixed reference frame, where the tube walls are stationary. Within the one-dimensional framework, the fluid in the tube moves at a single velocity at a given cross-sectional area along the tube, and thus the shear force τ_w at the wall would do work on the fluid (just as friction acting on a block sliding across a stationary table top does work on the block). However, this work term is excluded from the energy equation in Fanno flow in recognition of the no-slip boundary condition known to apply in viscous fluids. Thus, excluding the work term in the Fanno relations is a means to incorporate the multidimensional no-slip condition into a one-dimensional model.

In the ZND model of a detonation, the model is formulated in the wave-fixed reference frame, where the walls will be moving at speed D (the same speed at which the wave propagates into quiescent gas in the laboratory-fixed reference frame). In this case, the shear force τ_w acting on the fluid at the wall is undergoing a displacement at a rate D in the same direction as the shear force acts, so positive work is being done. As a result of this energy input by friction, the velocity deficits observed in detonation wave propagation with friction are not as great as would be computed by just including friction losses in the momentum equation alone (although a velocity deficit compared to a detonation with no friction still exists).

The question naturally arises, where does this energy come from? This question is addressed via a simple thought experiment proposed by Tanguay and Higgins and illustrated schematically in Fig. 2.6. Beginning with the propagating detonation wave in the laboratory-fixed reference frame (Fig. 2.6a), the wave can be rendered steady by transforming into a coordinate system attached to the wave (Fig. 2.6b). Now, suppose that it is desired wish to study

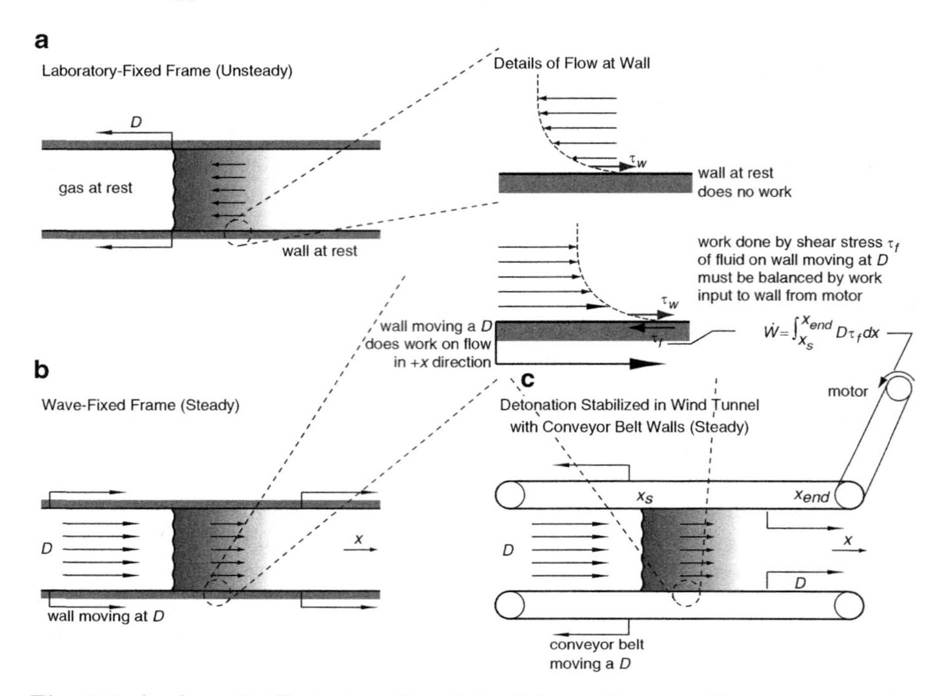

Fig. 2.6. A schematic illustrating the origin of the work term in the energy equation for detonations with friction. (**a**) Laboratory-fixed (unsteady) reference frame, in which no work is done by friction. (**b**) In the wave-fixed (steady) reference frame, in which the wall shear stress does work on the fluid due to displacement of the force. (**c**) Detonation stabilized in a wind tunnel with moving, conveyor-belt walls that equal the speed of the upstream flow. Work must be input to the conveyor belt via a motor to offset the drag due to friction acting on the conveyor belt

a detonation wave propagating in a tube under steady conditions in the laboratory. This could be done by stabilizing the detonation in a wind tunnel, where a flow of explosive gas is fed into the test section at a velocity equal to the detonation velocity, stabilizing the wave in the test section. In order to perfectly simulate a detonation propagating in a stationary tube, it would be necessary to make the walls of the wind tunnel move at the detonation velocity as well; this could be done by making the walls out of a conveyor belt, as shown in Fig. 2.6c. From the perspective of the detonation wave, the three arrangements in Fig. 2.6 are identical. For the detonation stabilized in the wind tunnel, in the region of the detonation reaction zone, the fluid will act via shear stress to slow down the conveyor belt, and thus to maintain a steady picture, a source of work must be put into the conveyor belt (e.g., by using an electric motor) which is then transferred to the fluid. Thus, while there is no work input from friction to the unsteady detonation wave in Fig. 2.6a, there *is* a work input from friction to the steady detonation wave in Fig. 2.6c, and therefore in Fig. 2.6b as well. This situation, where work is present in one reference frame and not in another, is commonly encountered in mechanics.

As a final proof, it is possible to start from a laboratory-fixed reference frame and, using the full unsteady Euler equations with friction, simulate a propagating detonation wave that moves through the computational domain. In this reference frame, no work is done since the no-slip boundary condition is assumed to apply and the walls are at rest, although there is a loss term in the momentum equation due to friction. Such a calculation was done by Dionne et al. [22], and the resulting wave propagation was shown by Tanguay and Higgins [88] to agree with the steady-state analysis where the work term was included in the energy equation and to significantly disagree with the steady-state analysis where the work term was excluded. Thus, in order to properly model a steady detonation in the one-dimensional ZND framework, the work input term due to friction must be included in (2.28).

2.4.3 Effect of Friction

While the Reynolds analogy, as discussed in the previous section, suggests that heat transfer will always be accompanied by friction, and vice versa, it is advantageous to examine each of these effects individually. Examining friction by itself is an approximation to a detonation occurring in an obstacle-laden channel or in an explosive gas mixture filling the interstitial spaces of a solid porous media, where momentum losses may dominate over heat transfer losses.

The effect of friction alone was studied by guessing the detonation wave velocity and then iterating on the governing ODE (2.35) with the heat transfer term removed until an eigenvalue solution could be obtained. The wave velocity found by this method is plotted in Fig. 2.7 as a function of the friction coefficient to hydraulic radius ratio $\frac{c_f}{r_h}$, along with profiles of the reaction wave structure for a few select cases. An activation energy of $E_a = 32$ was used for this figure, and the half reaction zone length of a detonation without losses, denoted $L_{1/2}$, was used to nondimensionalize the hydraulic radius and the distance from the shock front. For the case of no friction ($c_f = 0$), the same solution as studied in Sect. 2.2 is obtained, which propagates at a Mach number of 6.781. The addition of a momentum loss due to friction results in a wave velocity that is less than the CJ velocity without losses. As the value of the friction coefficient is increased (or the tube radius decreased), a lower post-shock temperature results in the reaction zone length becoming longer, making the wave even more susceptible to momentum losses. The fact that the sonic point is encountered before the complete energy is released also contributes to the increasing velocity deficit; however, this effect can be shown to be negligible in comparison to the influence of the momentum losses on the detonation velocity. As the value of $\frac{c_f L_{1/2}}{r_h}$ approaches ≈ 0.01, the decrease in wave velocity becomes increasingly steep until a *turning point* is encountered,

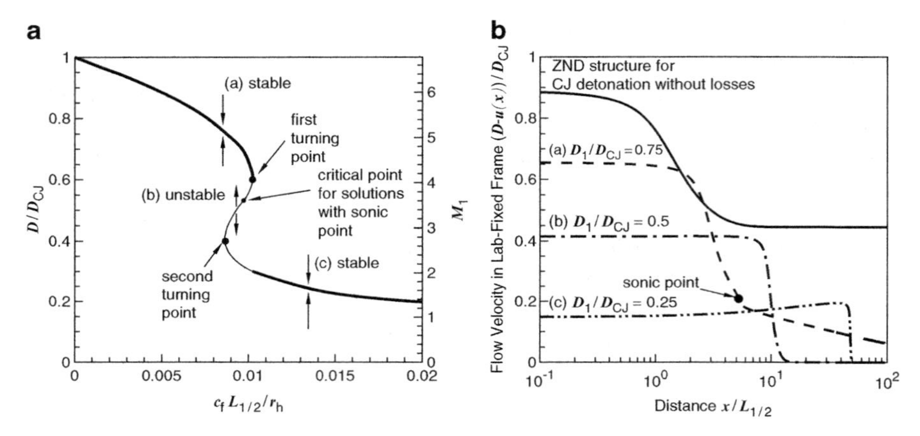

Fig. 2.7. ZND detonation with friction (no heat loss) for $\frac{E_a}{RT_1} = 32$, $Q = 10$, and $\gamma = 1.2$. (**a**) Wave velocity and Mach number as a function of the friction coefficient c_f. (**b**) Structure of the wave, showing as the flow velocity relative to the fixed wall. Hydraulic radius r_h and distance x are nondimensionalized by the half reaction zone length of detonation without losses $L_{1/2}$

at which point the wave is propagating at about 60% of the CJ velocity without losses for the particular activation energy studied here. Further decreasing the wave velocity below the turning point gives rise to the second branch of the solution curve in Fig. 2.7 that is denoted with a thin line, and for a given value of $\frac{c_f L_{1/2}}{r_h}$, the solution curve becomes multivalued in this region. As discussed below, the branch of the solution with a positive slope (i.e., increasing wave velocity with increasing friction) is believed to be unstable and cannot be physically realized.

As the detonation wave velocity is decreased further, at a certain point the flow velocity exiting the wave equals the velocity of flow approaching the wave. At velocities below this critical condition, a sonic point can no longer be located. The detonation wave velocity at this critical point can be solved analytically (see Appendix A.2) as

$$M_{\text{crs}} = \frac{D_{\text{crs}}}{c_1} = \sqrt{\gamma Q + 1}, \tag{2.36}$$

where "crs" denotes the critical sonic point. As viewed from the laboratory-fixed frame, this solution corresponds to a front of constant volume explosion conditions moving into unreacted mixtures at the sound speed of the combustion products. This result is independent of the details of the reaction zone structure such as the reaction rate.

To continue the solution curve below the critical sonic point, a new criterion is adopted, namely that the flow exiting the wave must be at rest in the laboratory-fixed reference frame. This corresponds to the condition of a detonation propagating in a close-ended tube that imposes zero flow velocity at the end of the reaction zone ($D - u = 0$), as shown in cases (b) and (c) in Fig. 2.7b. That the downstream boundary condition now influences the wave propagation reflects the fact that the flow exiting the wave is entirely subsonic with respect to the wave. As the wave velocity decreases further, a second turning point is encountered. Further increasing the friction coefficient $\frac{c_f}{r_h}$ results in a decreasing detonation wave velocity. The length of detonation wave reaction zone in this regime becomes very long, about two orders of magnitude longer than the ideal detonation wave without losses, for the case shown in Fig. 2.7c owing to the low post-shock temperatures ($M_1 = 1.5$ for case (c)). A wave with such a long reaction zone length is unlikely to be observed experimentally; however, it may qualitatively correspond to the low velocity "choking regime" as discussed in Sect. 2.4.5 below.

The stability of the various branches can now be considered. Since discussion of the stability of a detonation wave implies its transient response to a perturbation, it is outside the scope of the steady analysis presented in this chapter. However, following Zeldovich and Kompaneets [107], it is possible to present a heuristic argument for why the region of the solution curve with a positive slope (i.e., increasing velocity with increasing friction coefficient) should not be physically realizable, as follows. It can be shown that steady reaction zone structures initialized with conditions to the left of the solution curve correspond to a situation where the effect of heat release is greater than the effect of friction in comparison to the solution on the curve. In regions to the right of the curve, friction dominates over heat release, resulting in the flow remaining subsonic through the reaction zone. Thus, if a wave starting on the solution curve at point (a) in Fig. 2.7 was artificially perturbed downward to a lower velocity, attempts to solve for the reaction zone structure would result in a singularity as the heat release prematurely brings the flow to sonic. It can be conjectured that in a transient simulation, this perturbation that increases the influence of heat release in comparison to friction would have the effect of accelerating the wave, returning it to the solution curve. Likewise, artificially increasing the wave velocity at point (a) would result in friction having an increased influence, such that the flow velocity remains subsonic through the entire reaction zone (no sonic point), making the wave susceptible to rarefactions from behind that would decelerate the wave back to the solution curve. Thus, the upper branch of the solution curve in Fig. 2.7 can be hypothesized as a stable attractor.

Once past the turning point, where the slope of the detonation velocity with respect to friction becomes positive, the scenario is reversed, and a perturbation in wave velocity downward brings the wave into a region where the influence of heat release is diminished as compared to friction, reducing the wave speed further. A perturbative increase in wave speed would increase the influence of heat release over friction, resulting in the wave accelerating away from this branch of the solution curve. Thus, it is concluded that this branch of the solution is unstable. Once below the critical sonic point, a sonic condition can no longer be found, but similar considerations apply. To the left of the solution curve below the critical sonic point, the flow velocity leaving the wave is negative (meaning, into the end wall), which would have the effect of accelerating the wave due to the reflected compression generated. To the right of the solution curve, the flow velocity is away from the end wall, and the resulting rarefaction generated would decelerate the wave. Thus, the region of the curve with positive slope is also unstable below the critical point as well.

The behavior conjectured here was verified by numerical simulations of the transient problem performed by Dionne et al. [20,22]. Numerical solutions of the unsteady one-dimensional Euler equations with momentum and energy source terms initialized with the steady, ZND structures found here in the region of positive slope (i.e., between the two turning points) quickly developed an instability after propagating only a few half reaction zone lengths and promptly jumped to the upper branch of the solution and then propagated indefinitely. Thus, the upper branch of the solution appears to be stable, although for sufficiently high activation energy, the numerical simulations of Dionne et al. exhibited pulsations such that the wave speed oscillated around the steady-state value found here. The simulations of Dionne et al. also found that solutions initialized with the steady ZND structure for the multivalued region of the solution curve below the second turning point were also unstable and promptly jumped to the upper branch of the solution (while our ad hoc analysis above would suggest that waves in this region could be stable). Only for values of the friction coefficient large enough to bring the solution curve into a region where it is again single valued did Dionne et al. find stable solutions in their transient simulations. Thus, only the sections of the solution curve shown with a thick line in Fig. 2.7a are believed to correspond to steady (or quasi-steady), realizable waves.

The backward "S" shape of the wave velocity as a function of the friction coefficient will be seen to be characteristic of all detonations with losses and, in a larger picture, flame models incorporating losses as well [85], provided the energy release rate has the strong temperature dependence of activated reactions (i.e., Arrhenius kinetics). The upper turning point is usually identified as the point of wave failure or abrupt transition to a regime of lower propagation velocity, with the other branches of the solution in the multivalued region being unstable. This behavior reflects the observed failure of detonation waves due to losses, wherein a deficit in velocity grows with increasing losses until an abrupt drop to a low velocity detonation (LVD) is observed, corresponding to encountering the first turning point and dropping onto the lowest branch of the solution. For lower values of activation energy E_a, it is theoretically possible for the wave velocity to remain a single valued function of friction and to decrease smoothly and monotonically; however, these values of E_a are not representative of real detonable gases. Reaction rate models with low thermodynamic state sensitivity appear in the modeling of condensed-phase explosives, and in these cases the detonation velocity can decrease monotonically without a critical point as well.

2.4.4 Effect of Heat Loss

The effect of heat transfer alone is examined in Fig. 2.8a, where the eigenvalue velocity of detonation is shown for different values of activation energy. The detonation velocity is plotted as a function of the heat transfer parameter, which is represented by the product of the friction coefficient c_f (which determines the heat transfer coefficient via the Reynolds analogy, even though friction has been switched off for these particular calculations) and the ratio of half reaction zone length (for a CJ detonation) to the tube hydraulic radius. Qualitatively, the results are very similar to those in which friction is examined in isolation (i.e., without heat transfer) as was shown in Fig. 2.7. Specifically, in the region to the left of the solution curve, heat generation is greater than heat loss, such that a singularity is encountered, and the region to the right of the solution curve is where heat loss dominates, such that the flow remains subsonic. Again, this feature of the solution can be used to suggest that only the upper branch of the solution is realizable. Thus, the turning point (denoted as a dot in Fig. 2.8a) is the critical value of heat transfer and detonation velocity at which detonation failure occurs. The different solution curves for different values of activation energy show that greater values of E_a, indicating a greater temperature sensitivity of the reaction zone, cannot tolerate large velocity deficits from the CJ velocity before they fail, while lesser

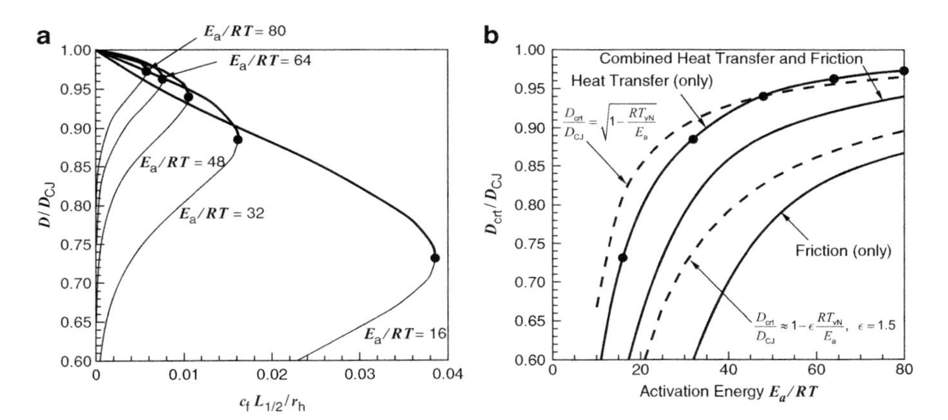

Fig. 2.8. (**a**) Detonation velocity as a function of the heat transfer parameter, with the stable branch shown as a *heavy line*, the nonphysical branch as a *thin line*, and the critical turning points denoted as *dots*, for different values of activation energy. (**b**) Detonation velocity at the critical turning point as a function of activation energy for detonations with different loss mechanisms. *Solid curves* are derived from iteration of numerical integrations of the reaction zone structure to find the eigenvalue detonation solution at the critical turning point, with the *dots* denoting critical points found in (**a**). *Dashed curves* derive from an activation energy asymptotic analysis [107, 109, 110]

values of E_a can exhibit larger velocity deficits. The dependence of the critical detonation velocity upon activation energy is shown in Fig. 2.8b.

For the case of heat transfer without friction losses, Zeldovich and Kompaneets [107] were able to solve analytically for the dependence of the velocity deficit from the CJ value by exploiting the fact that flow properties initially vary slowly behind the shock wave and can be taken as constant until the temperature becomes close to the value at the end of the reaction zone and the heat release occurs extremely rapidly. This approach has since been formalized into a methodology known as *activation energy asymptotics*. Zeldovich and Kompaneets found the following relation for the velocity at the critical turning point conditions

$$\frac{D_{\mathrm{crt}}}{D_{\mathrm{CJ}}} = \sqrt{1 - \frac{1}{\left(\frac{E_a}{RT_{\mathrm{vN}}}\right)}}, \tag{2.37}$$

where "crt" denotes critical turning point. The post-shock temperature of the CJ detonation without losses can be used for the von Neumann temperature T_{vN}. This relation can be approximated as

$$\frac{D_{\mathrm{crt}}}{D_{\mathrm{CJ}}} \approx 1 - \frac{1}{2}\frac{1}{\left(\frac{E_a}{RT_{\mathrm{vN}}}\right)}. \tag{2.38}$$

This curve is shown in Fig. 2.8b as a dashed line and exhibits remarkably good agreement with the value of critical velocity obtained via iteration upon numerical integration of the reaction zone structure in order to find the critical conditions, particularly for larger values of activation energy E_a.

Also shown in Fig. 2.8b are the values of critical velocity at the turning point for the case with friction alone (as was considered in Sect. 2.4.4) and the combined effect of friction and heat transfer. Detonations with frictional losses alone are able to tolerate larger velocity deficits for the same activation energy; this feature is attributed to the work input to the energy equation from friction, as discussed in Sect. 2.4.2, which has the effect of assisting to sustain the reaction in comparison to heat transfer losses. Detonation models including both friction and heat transfer exhibit intermediate velocity deficits at critical conditions, between friction and heat transfer individually. Zeldovich et al. [109, 110] suggested a general form of the velocity deficit relation

$$\frac{D_{\mathrm{crt}}}{D_{\mathrm{CJ}}} \approx 1 - \epsilon\frac{1}{\left(\frac{E_a}{RT_{\mathrm{vN}}}\right)}, \tag{2.39}$$

where ϵ has suggested values in the range of 0.9–1.8 for detonations with combined heat transfer and friction for physically relevant values of E_a ($\epsilon = 1.5$ is shown in the figure), with $\epsilon \rightarrow 0.5$ as $E_a \rightarrow \infty$ and thus converging to (2.38). As seen from Fig. 2.8b, the values of activation energy must be made very large (i.e., beyond values that correspond to real reactive systems) in order to see this convergence. In Sect. 2.6, the same relation relating velocity deficit to activation energy will be found to apply to detonations with front curvature due to lateral flow divergence from yielding confinement.

Detonations with large frictional and heat transfer losses, as would be encountered in porous media for example, have been extensively studied theoretically by Sivashinsky and colleagues in recent years [6–9, 11, 35, 43]. Their work has further elucidated the structure of the reaction front and classified the possible modes of propagation, particularly low velocity (including subsonic) regimes. Their modeling, similar to that presented here, used a single-step Arrhenius kinetics reaction mechanism. As discussed in the next subsection, this reaction model is likely not relevant to the actual mechanism of burning in reactive waves with large losses. Models that attempt to incorporate the contribution of turbulent combustion to the reaction mechanism by using greatly exaggerated values of transport properties (e.g., effective turbulent diffusivity, etc.) may provide a future direction to link these models with experimental results [11].

2.4.5 Experiments with Losses

Gaseous detonations in channels that exhibit large velocity deficits due to momentum and heat transfer losses to wall roughness or obstacles have been extensively studied and are referred to as *quasi-detonations*. Quasi-steady propagation at velocities as low as 50% of the ideal CJ velocity has been observed. Reviews of results of these studies can be found in [32] and Chap. 7 of [51]. It is unlikely that planar shock-initiated homogeneous reactions contribute significantly to the energy release in such detonations. Calculations performed with detailed chemistry show that the reaction rates for the low shock velocities involved are much too slow to result in exothermic reactions on the timescales of the observed quasi-detonations. A seminal study by Shchelkin [78] suggested that shock reflection off obstacles in the tube may generate local hot spots that initiate reaction and thus enable the detonation to continue propagating at speeds for which a planar shock would not be of sufficient strength to sustain propagation. Much of the framework for explaining detonation phenomena in gaseous and condensed phase explosives has since been built around this "hot spot" idea. In addition, it is likely that interactions of the post-shock flow with the obstacles, the boundary layer that forms on the tube wall, and the turbulent nature of the quasi-detonation front itself all contribute to burning the mixture and sustaining the front. The mechanism of propagation of very high speed ($>300\,\mathrm{m\,s^{-1}}$) turbulent flames has not been convincingly elucidated and currently comprises a "no man's land" between premixed turbulent combustion and detonation, which includes the transition between the two (deflagration to detonation transition) [14, 52]. As such, models to address the reaction mechanism are ad hoc and semi-empirical in nature

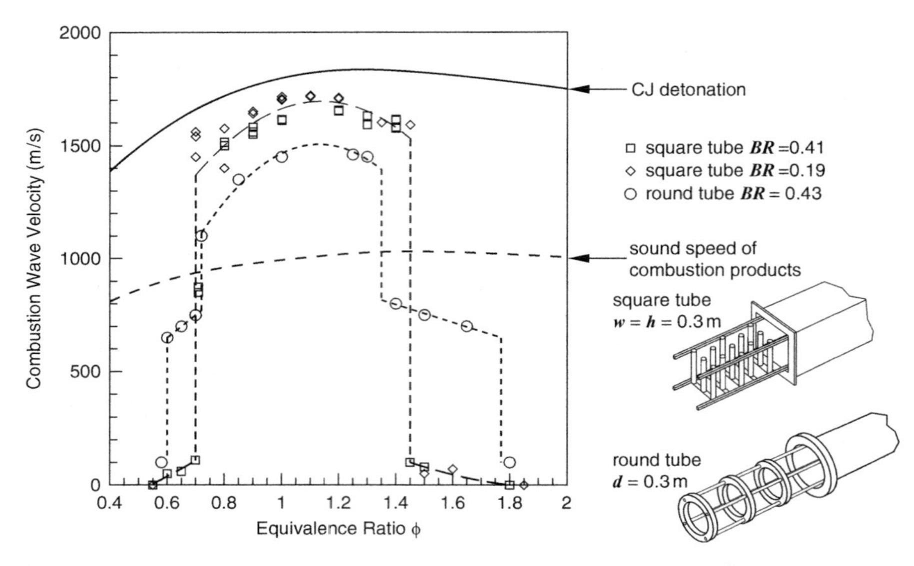

Fig. 2.9. Experiments examining detonations with propane/air mixtures in obstacle-laden channels (1 atm initial pressure). As the mixture equivalence ratio becomes sufficiently rich or lean, the longer reaction zone thickness increases momentum losses until detonation wave undergoes an abrupt transition to a lower velocity choking regime (*circular tube*) or low velocity flame (*square tube*) [12, 65]

and will not be discussed here; see [32,109,110] for a discussion of models that attempt to reproduce the reaction zone of detonations in rough tubes.

The main contribution to understanding quasi-detonation propagation by the type of model presented here is qualitative in nature, particularly the appearance of a critical velocity that can lead to the abrupt transition to a regime of lower propagating velocity. As an example, consider the experimental results of Peraldi et al. [65] and Chao and Lee [12] (see Fig. 2.9), which show the propagation velocity of detonation in propane/air mixtures in obstacle-laden channels . The specific geometry was either a square channel with staggered cylindrical pillars or a circular tube with annular rings, with the obstacle spacing approximately one channel height or tube diameter and with a blockage ratio denoted BR, defined as the cross-sectional area of the obstacle divided by the cross-sectional area of the channel or tube. As the mixture equivalence ratio ϕ was made rich or lean in comparison to stoichiometry, the reaction zone length increased, in turn increasing the relative momentum and heat transfer losses to the detonation wave. For the circular tube, as the equivalence ratio reached a value of $\phi \approx 0.8$ or $\phi \approx 1.4$, there was an abrupt transition to a lower velocity regime of propagation comparable to the sound speed in the combustion products (sometimes called the "choking regime"). This behavior is qualitatively similar to encountering the turning point of the "backwards S" curve in Fig. 2.7. These findings, however, are geometry dependent, as seen in the results with the square tube with staggered pillars, which while exhibiting a critical velocity did not exhibit a transition to the low velocity choking regime. Similar studies of detonations in gaseous mixtures filling the interstitial spaces in packed beds of inert spherical beads by Makris et al. [53, 54] and Pinaev and Lyamin [66] failed to identify a critical velocity for a large number of different explosive mixtures, with a continuous spectrum of propagation velocity from the ideal CJ velocity down to 30% of the CJ velocity. It is likely that the role of turbulent mixing-driven combustion waves accounts for the geometry-depending velocity deficit and critical behavior that cannot be predicted by the simple homogeneous reaction mechanisms with exponential temperature dependence used in this chapter.

2.5 Systems with Real Chemistry

In order to have a more accurate model for real systems, it is necessary to take into account the details of the chemical reactions, rather than simply treating the energy release of reaction as external heat addition as was done in prior sections. In this section, a model of a steady, one-dimensional detonation in a reacting system of ideal gases will be developed, following the developments found in [67, 73]. This analysis begins with the same differential form of the conservation equations (2.1)–(2.3) and additional conservation equations for each individual chemical species

$$\mathrm{d}\left(\rho_i u A\right) = W_i \dot{\omega} A \mathrm{d}x \tag{2.40}$$

since species will be produced and consumed while chemical reactions proceed. The i subscript in (2.40) denotes a particular chemical element or compound. The $\dot{\omega}$ term is the rate of species production via chemical reaction (units: kmol/m^3-s), and W_i is the molecular weight of species i. It is convenient to introduce the mass fraction of a species $Y_i = \frac{\rho_i}{\rho}$, where ρ is the density of the mixture. Under the assumption of the Dalton model of partial pressure, the ideal gas law applies to each individual species $\left(p_i = \rho_i \frac{R_u}{W_i} T\right)$. The average molecular weight of the mixture (which will, in general, not remain constant as the mixture reacts) is given by

$$\frac{1}{W} \equiv \sum_i \frac{Y_i}{W_i}. \tag{2.41}$$

The enthalpy term h contains both the sensible enthalpy and the latent enthalpy of formation. Thus, it is not necessary to introduce a heat release term into the energy equation to account for exothermic chemical reaction. Specifically, h is given by

$$h = \sum_i \left(h^\circ_{f_i} + \int_{298.15}^{T} c_{p_i}(\theta)\mathrm{d}\theta \right) Y_i, \tag{2.42}$$

where $h^\circ_{f_i}$ is the enthalpy of formation of species i at the reference state (298.15 K and 1 atm). The enthalpy of formation is the energy absorbed (or released) when a given species is synthesized from the most stable form of its constituent elements. By definition, h°_f for the stable form of the elements at standard conditions ($T_{\text{ref}} = 298.15$ K, $p_{\text{ref}} = 1$ atm) is zero. A positive enthalpy of formation indicates that a species absorbs energy in its formation; a negative enthalpy of formation indicates that energy is released in the formation of the species. This assumption provides a consistent reference state that takes into account the energy absorbed (or liberated) as chemical bonds are broken (or formed) in a reacting system. The integral term in (2.42) gives the sensible enthalpy change relative to the reference state. Here, θ is introduced as a dummy variable of integration. The numerical values for enthalpy of formation and specific heat as a function of temperature may be obtained from the JANAF thermochemical tables [13], which have been conveniently curve fit with polynomials by [34, 56].

In order to formulate an ODE that can be integrated through the ZND reaction zone, the differential form of the governing conservation equations and the equation of state are used

$$\frac{\mathrm{d}\rho}{\rho} + \frac{\mathrm{d}u}{u} + \frac{\mathrm{d}A}{A} = 0 \tag{2.43}$$

$$\mathrm{d}p + \rho u \mathrm{d}u = 0 \tag{2.44}$$

$$\mathrm{d}h + u\mathrm{d}u = 0 \tag{2.45}$$

$$\frac{\mathrm{d}p}{p} = \frac{\mathrm{d}\rho}{\rho} - \frac{\mathrm{d}W}{W} + \frac{\mathrm{d}T}{T}. \tag{2.46}$$

The $\mathrm{d}h$ term can be expanded as follows:

$$\mathrm{d}h = \sum_i \left(h^\circ_{f_i} + \int_{298.15}^{T} c_{p_i}(\theta)\mathrm{d}\theta \right) \mathrm{d}Y_i + \sum_i Y_i \mathrm{d}\left(\int_{298.15}^{T} c_{p_i}(\theta)\mathrm{d}\theta \right). \tag{2.47}$$

The first term can be denoted $-\mathrm{d}q$

$$\mathrm{d}q \equiv -\sum_i \left(h^\circ_{f_i} + \int_{298.15}^{T} c_{p_i}(\theta)\mathrm{d}\theta \right) \mathrm{d}Y_i. \tag{2.48}$$

This is the usual definition of *enthalpy of combustion* (also called *heat of reaction*), which is the difference between the enthalpy of the products and the reactants at a fixed temperature and pressure.

The differential in the second term of (2.47) can be written as

$$\mathrm{d}\left(\int_{298.15}^{T} c_{p_i}(\theta)\mathrm{d}\theta \right) = c_{p_i}\mathrm{d}T. \tag{2.49}$$

Note that in bringing the differential under the integral, c_{p_i} is a monovariant function of temperature (i.e., Leibniz's Rule for differentiation under the integral sign is not required).

The mass-fraction weighted sum of these terms gives the mixture average heat capacity at constant pressure with frozen chemical composition c_{p_f}

$$c_{p_f} \equiv \left(\frac{\partial h}{\partial T}\right)_{p,Y_i} \equiv \sum_i Y_i c_{p_i}. \tag{2.50}$$

The subscript f is used for properties that derive from partial derivatives where the chemical composition is held fixed.

Thus, the differential from of the conservation of energy can be written very compactly as

$$c_{p_f} \mathrm{d}T - \mathrm{d}q + u\mathrm{d}u = 0. \tag{2.51}$$

Eliminating $\mathrm{d}\rho$, $\mathrm{d}P$, and $\mathrm{d}T$ from the differential form of the conservation relations and equation of state

$$\frac{\mathrm{d}u}{u} = \frac{\frac{\mathrm{d}q}{c_{p_f}T} - \frac{\mathrm{d}W}{W} - \frac{\mathrm{d}A}{A}}{1 - \left(\frac{1}{\frac{R_u}{W}} + \frac{1}{c_{p_f}}\right)\frac{u^2}{T}}. \tag{2.52}$$

Further introducing the frozen heat capacity at constant volume

$$c_{v_f} \equiv \left(\frac{\partial e}{\partial T}\right)_{v,Y_i} \equiv \sum_i Y_i c_{v_i} \tag{2.53}$$

and that the frozen heat capacities are related by

$$c_{p_f} - c_{v_f} = \frac{R_u}{W}. \tag{2.54}$$

Introducing a frozen composition isentropic exponent γ_f

$$\gamma_f = \frac{c_{p_f}}{c_{v_f}}. \tag{2.55}$$

Note that, in general, $c_p - c_v \neq R$ for a reacting gas [60]. The relations here (2.54) are only valid for the special case where the composition is assumed frozen. Introducing the frozen speed of sound (discussed in Sect. 2.5.4 below)

$$c_f = \sqrt{\gamma_f \frac{R_u}{W} T} \tag{2.56}$$

then the result found above (2.52) can be written as

$$\frac{\mathrm{d}u}{u} = \frac{\frac{\mathrm{d}q}{c_{p_f}T} - \frac{\mathrm{d}W}{W} - \frac{\mathrm{d}A}{A}}{1 - \left(\frac{u}{c_f}\right)^2}. \tag{2.57}$$

Introducing the frozen Mach number as the ratio of the flow velocity to the local frozen speed of sound ($M_f = \frac{u}{c_f}$)

$$\frac{\mathrm{d}u}{u} = \frac{\frac{\mathrm{d}q}{c_{p_f}T} - \frac{\mathrm{d}W}{W} - \frac{\mathrm{d}A}{A}}{1 - M_f^2}. \tag{2.58}$$

The similarity of this relation to (2.12) for the perfect gas model is clear, where the additional $\frac{\mathrm{d}W}{W}$ term reflects the changing chemical composition. In order for the solution to pass through a sonic point in the flow and avoid a singularity, the numerator must simultaneously go to zero, i.e., the following condition must be satisfied

$$\frac{\mathrm{d}q}{c_{p_f}T} - \frac{\mathrm{d}W}{W} - \frac{\mathrm{d}A}{A} = 0 \quad \text{when} \quad M_f = 1. \tag{2.59}$$

This condition is the *generalized CJ criterion* for a detonation with detailed chemical reactions. For a constant area flow, the numerator in (2.58) is sometimes referred to as the *thermicity* parameter, denoted σ [19].

$$\sigma = \frac{\mathrm{d}q}{c_{p_f}T} - \frac{\mathrm{d}W}{W}. \tag{2.60}$$

Thermicity can be defined as the influence of chemical reaction on flow velocity (or, via the momentum equation, pressure) due to both chemical energy release and an increase or decrease in the number of moles present. The appearance of the variable molecular weight term provides another possibility for a pathological (or eigenvalue) detonation: the exothermic heat release may exactly balance the decrease in the number of moles as the mixture reacts, resulting in the thermicity going to zero before equilibrium has been established.

The significance of the appearance of the frozen (as opposed to the) sound speed will be discussed in Sect. 2.5.4. It should be mentioned that, although it is the frozen sound speed that appears in the final equation (2.58), at no point was it assumed that the frozen sound speed is the "correct" sound speed to use. Rather, c_f appeared naturally in the governing equations, specifically in (2.52). It was not necessary to explicitly introduce c_f. In fact, the original ODE's (2.43)–(2.47) could be numerically integrated directly, without any manipulation to produce the form of (2.58). However, integration of this set of ODE's for an arbitrary initial von Neumann state would result in encountering a singularity in the solution in some cases. Further examination of the solution would reveal that this singularity occurs as the local flow velocity equals the frozen sound speed with finite thermicity. Only in the case where the flow velocity equals the frozen sound speed as the thermicity goes to zero (i.e., a saddle point) can the solution pass through the sonic point. The form of (2.58) makes this requirement clear.

2.5.1 Chemical Reaction Rates

In order to complete the model of a steady, one-dimensional detonation with real chemistry, the species source term $\dot{\omega}$ must be specified. For a generic reaction A + B → C + D, the rate at which reactions occur in a unit volume is given by:

$$\dot{\omega} = C_{\mathrm{A}} C_{\mathrm{B}} k_{\mathrm{f}}, \tag{2.61}$$

where C_{A} and C_{B} are the molar concentrations of species A and B, respectively. Their product is proportional to the likelihood of collisions, which are a necessary prerequisite to reactions. The k_{f} term is the forward reaction rate constant given by the Arrhenius form

$$k_{\mathrm{f}} = AT^{n} \exp\left(\frac{E_a}{R_u T}\right). \tag{2.62}$$

Here A is a pre-exponential factor that includes the collision cross section of the atoms or molecules and a steric factor that takes the geometry of the collision into account. The "$\exp\left(\frac{E_a}{R_u T}\right)$" term derives from the Maxwell–Boltzmann distribution of molecular speeds and is proportional to the fraction of atoms or molecules that have a kinetic energy greater than the *activation energy* E_a that is required for a collision to initiate reaction. The values of A, E_a, and the temperature exponent n are usually derived from experimental measurements of reaction rate.

Most elementary reactions are bimolecular, as in this generic example. It is common in combustion applications to approximate the detailed chemical reactions with a global reaction rate (e.g., $CH_4 + 2O_2 \rightarrow CO_2 + 2H_2O$). Even in this case, the Arrhenius form is still used, although the constants A, E_a, and n are now fitting coefficients that may have a limited range of validity.

Usually, it is also necessary to consider the backward reaction as well (i.e., reversible reactions): C + D → A + B. However, the reaction rate constants in this case can be expressed in terms of the equilibrium constant as follows. At equilibrium (A + B ⇌ C + D), the rate of forward and backward reactions must be equal, such that the overall composition remains constant. Thus,

$$C_{\mathrm{A}} C_{\mathrm{B}} k_{\mathrm{f}} = C_{\mathrm{C}} C_{\mathrm{D}} k_{\mathrm{b}} \tag{2.63}$$

$$\frac{k_{\mathrm{f}}}{k_{\mathrm{b}}} = \frac{C_{\mathrm{C}} C_{\mathrm{D}}}{C_{\mathrm{A}} C_{\mathrm{B}}} = K_c, \tag{2.64}$$

where K_c is the equilibrium constant based on concentration. If the stoichiometric coefficients are other than unity (e.g., $\nu_{\mathrm{A}}\mathrm{A} + \nu_{\mathrm{B}}\mathrm{B} \rightarrow \nu_{\mathrm{C}}\mathrm{C} + \nu_{\mathrm{D}}\mathrm{D}$), then

$$\frac{k_{\mathrm{f}}}{k_{\mathrm{b}}} = \frac{C_{\mathrm{C}}^{\nu_{\mathrm{C}}} C_{\mathrm{D}}^{\nu_{\mathrm{D}}}}{C_{\mathrm{A}}^{\nu_{\mathrm{A}}} C_{\mathrm{B}}^{\nu_{\mathrm{B}}}} = K_c. \tag{2.65}$$

The equilibrium constant based on partial pressure K_p is related to the equilibrium constant based on concentrations K_c as follows:

$$K_p = \frac{\left(\frac{p_{\mathrm{C}}}{p_{\mathrm{ref}}}\right)^{\nu_{\mathrm{C}}} \left(\frac{p_{\mathrm{D}}}{p_{\mathrm{ref}}}\right)^{\nu_{\mathrm{D}}}}{\left(\frac{p_{\mathrm{A}}}{p_{\mathrm{ref}}}\right)^{\nu_{\mathrm{A}}} \left(\frac{p_{\mathrm{B}}}{p_{\mathrm{ref}}}\right)^{\nu_{\mathrm{B}}}} = K_c \left(\frac{R_u T}{p_{\mathrm{ref}}}\right)^{(\nu_{\mathrm{C}} + \nu_{\mathrm{D}} - \nu_{\mathrm{A}} - \nu_{\mathrm{B}})}, \tag{2.66}$$

where p_{ref} is the reference pressure used in defining the equilibrium constant based on partial pressures.

Equilibrium constants are derived from thermodynamic data as follows:

$$K_p = \exp\left(-\frac{\nu_{\text{C}}\bar{g}^{\circ}_{\text{fC}} + \nu_{\text{D}}\bar{g}^{\circ}_{\text{fD}} - \nu_{\text{A}}\bar{g}^{\circ}_{\text{fA}} - \nu_{\text{B}}\bar{g}^{\circ}_{\text{fB}}}{R_u T}\right), \tag{2.67}$$

where $\bar{g}^{\circ}_{\text{f}}$ is the Gibbs function on a per mole basis of species i evaluated at the temperature T and at the reference pressure $p_{\text{ref}} = 1\,\text{atm}$. Thus, the equilibrium constants, in being derived from fundamental thermodynamic data, are known with much greater confidence than kinetic rate constants. This approach then ensures that the overall kinetic mechanism is consistent with the higher-confidence equilibrium constants. A consequence of this assumption is that it is possible to start with a completely arbitrary or incorrect reaction mechanism and still reach the correct equilibrium composition of a reacting gas mixture. Thus, correctly reproducing the CJ detonation velocity via a ZND calculation with detailed chemistry should not be taken as validation of the kinetic mechanism.

The overall conversion of fuel and oxidizer into products is described by a kinetic mechanism, consisting of elementary reactions. For most combustible mixtures, it is necessary to consider numerous possible reactions. Even a "simple" system like hydrogen/oxygen is typically modeled with 20 or so elementary reactions, while mechanisms for hydrocarbon fuels (e.g., methane) can consider reactions numbering in the hundreds or thousands.

For a mechanism consisting of $j = 1, 2, \ldots, M$ elementary reactions that describes the reaction of $i = 1, 2, \ldots, N$ species, the overall production rate of species i is given by

$$\dot{\omega}_i = \sum_{j=1}^{M} \nu_{ij} r_j \tag{2.68}$$

$$r_j = k_{\text{f}_j} \prod_{i=1}^{N} C_i^{\nu'_{ij}} - k_{\text{b}_j} \prod_{i=1}^{N} C_i^{\nu''_{ij}}, \tag{2.69}$$

where ν'_{ij} is the stoichiometric coefficient of species i in the forward reaction j and ν''_{ij} is the corresponding stoichiometric coefficient for the backward reaction. If a species is absent from a reaction, its stoichiometric coefficient is zero.

The time rate of change of concentration is often expressed in the combustion literature as

$$\frac{\mathrm{d}C_i}{\mathrm{d}t} = \dot{\omega}_i. \tag{2.70}$$

This expression is strictly only valid for a constant volume (density) reaction. In general, reacting flows are variable density, so species concentrations can change due to changes in density as well as reaction, and thus (2.70) is not valid. This point is rarely expressed in most combustion textbooks, with [87] being a notable exception. For this reason, when working with reactive compressible flows, it is preferable to express species concentration in terms of mass fraction using $C_i = \frac{\rho Y_i}{W_i}$. The rate of change of mass fraction is then given by

$$\mathrm{d}Y_i = \frac{\dot{\omega}_i}{\rho u}. \tag{2.71}$$

The one caveat that must be issued in regards to numerical integration of the reaction zone structure is that chemical kinetic mechanisms tend to be numerically *stiff*. Stiffness in this context refers to the fact that the timescales of different processes occurring in a simulation (in this case, different chemical reactions) can vary greatly, often by many orders of magnitude. For example, it is possible for a chemical species that is not initially present to appear suddenly during the evolution of a reaction due to a chain-branching mechanism, only to be quickly consumed again. If too large of a numerical time step is taken, this feature may be missed or introduce numerical instability into the solution. The enormous computational power of today's computers may make it tempting to simply use a very small time step for the entire numerical integration; however, this approach can lead to the accumulation of numerical error, affecting the solution as well. The best approach is to use a numerical method that senses regions of the solution where a property is varying rapidly and decreases the numerical step size accordingly (e.g., predictor/corrector methods). Fortunately, most current high-level numerical packages have built-in ODE solvers that are designed to handle stiff systems of equations; [62] can be consulted for more details regarding implementing numerical methods for stiff reactive systems.

2.5.2 Carbon Monoxide/Oxygen System

As a sample calculation, the one-dimensional steady structure of a carbon monoxide/oxygen detonation will be computed. This system has the advantage that it is described by a very simple mechanism consisting of just three reactions considering four species:

$$\mathrm{CO + O + M \rightleftharpoons CO_2 + M} \tag{2.72}$$

$$\mathrm{CO + O_2 \rightleftharpoons CO_2 + O} \tag{2.73}$$

$$\mathrm{O + O + M \rightleftharpoons O_2 + M.} \tag{2.74}$$

Here M is a generic third body (i.e., M can be CO, O_2, CO_2, or O). The values of the reaction rate constants are given in Table 2.1. Experimentally, this system is not very easy to realize (beyond the toxicity of carbon monoxide) due to the high activation energy of the $\mathrm{CO + O_2 \rightleftharpoons CO_2 + O}$ reaction. In the presence of even trace amounts of hydrogen, OH molecules can form and then promote the formation of CO_2 more readily via the reaction $\mathrm{CO + OH \rightleftharpoons CO_2 + H}$. The H is then "recycled" back into OH, resulting in a catalytic-like

Table 2.1. Carbon monoxide/oxygen reaction mechanism

Reaction	A	n	E_a (kJ/kmol)
$CO + O + M \rightleftharpoons CO_2 + M$	1.8×10^{4a}	0.00	9,970
$CO + O_2 \rightleftharpoons CO_2 + O$	2.5×10^{9b}	0.00	200,000
$O + O + M \rightleftharpoons O_2 + M$	1.2×10^{11a}	−1.00	0

[a] $m^6 kmol^{-2} s^{-1}$
[b] $m^3 kmol^{-1} s^{-1}$

role of hydrogen in converting the CO into CO_2. Thus, in any real system with even trace contamination of hydrogen, water, or hydrocarbons, the mechanism (2.72)–(2.74) is not the dominant reaction path. For the purposes of a model system to explore numerically, however, the dry oxidation of carbon monoxide is convenient to use.

In order to initialize the calculation of the reaction zone structure at the von Neumann point, it is necessary to specify the propagation velocity of the detonation wave. This was done in two different ways: an *equilibrium solution* based on a control volume enclosing the wave and a *ZND solution* in which the structure of the reaction zone is solved. The equilibrium solution based on a control volume was determined through iteration upon the integral conservation laws (2.1)–(2.3) assuming chemical equilibrium at the exit plane until the minimum wave velocity solution was found. The reaction zone structure was not considered in this case. The speed of sound was not explicitly introduced in these calculations, but the solution found corresponds to sonic outflow (i.e., flow velocity equals the equilibrium speed of sound). This solution methodology can be shown to agree with that used in well-known equilibrium programs such as the NASA CEA program [34,56], and generates the same results within numerical precision.

The ZND-based solution was found by numerically integrating the reaction zone structure using the (2.58)–(2.71). The initial velocity of the wave was iterated upon until a solution that had sonic outflow and that did not encounter a singularity was found. Note that the sonic outflow condition found is that defined by using the equilibrium speed of sound. The outflow with respect to the frozen speed of sound was still subsonic (Mach 0.963). Therefore, this solution does not satisfy the generalized CJ condition (2.59).

The results of the two solution methodologies are compared in Table 2.2. The propagation velocity of the detonation wave found by these two methods agrees to six significant digits, and the flow properties at the exit state agree to within five or six significant digits. Thus, it can be concluded that these two methods find the same solution for the detonation wave.

The structure of the wave is shown in Fig. 2.10. The reaction zone length, as defined by identifying the location of maximum thermicity or the inflection point in temperature, is on the order of 10–15 cm. However, the sonic condition (using the equilibrium sound speed) is only approached asymptotically. Thus,

Table 2.2. Carbon monoxide/oxygen detonation

Initial composition	(moles)	$CO + \frac{1}{2}O_2$	
	p_1 (kPa)	101.325	
Initial state	T_1 (K)	300.000	
	c_1 ($\mathrm{m\,s^{-1}}$)	344.705	
Detonation solution		Equilibrium	ZND
	u_1 ($\mathrm{m\,s^{-1}}$)	1,798.32	1,798.32
	M_{1_f}	5.21698	5.21698
von Neumann state	p_{vN} (kPa)	3,288.67	3,288.67
	p_{vN}/p_1	32.4566	32.4566
	T_{vN} (K)	1,684.98	1,684.98
	u_{vN} ($\mathrm{m\,s^{-1}}$)	311.198	311.198
	c_{vN} ($\mathrm{m\,s^{-1}}$)	787.579	787.579
	M_{vN}	0.395132	0.395132
Chapman–Jouguet state	p_2 (kPa)	1,861.26	1,861.28
	p_2/p_1	18.3692	18.3694
	T_2 (K)	3,523.31	3,523.31
	u_2 ($\mathrm{m\,s^{-1}}$)	977.182	977.177
	c_{f_2} ($\mathrm{m\,s^{-1}}$)	1,014.43	1,014.43
	c_{e_2} ($\mathrm{m\,s^{-1}}$)	977.183	977.183
	M_{f_2}	0.96329	0.96328
	M_{e_2}	1.00000	0.99999

it is not possible to define a hydrodynamic thickness of the wave based of the location of the sonic surface (or, any definition of the total thickness of the wave using the flow Mach number—such as where the flow reaches Mach 0.99 for example—would be arbitrary).

It is of interest to explore if an eigenvalue solution for this problem can be found that does satisfy the generalized CJ solution, in contrast to the equilibrium solution found above. Recall that the equilibrium solution has subsonic outflow (Mach 0.963) with respect to the frozen speed of sound. In order to have the exit flow reach sonic with respect to the frozen speed of sound, it is necessary to decrease the wave speed, thus increasing the post-shock flow velocity and thereby making it possible for the heat release of reaction to bring the flow to frozen sonic velocity c_f. However, if the initial velocity of the wave in the equilibrium solution found above is decreased by just $0.1\,\mathrm{m\,s^{-1}}$, the solution for reaction zone structure encounters a singularity, as shown in Fig. 2.11. The flow velocity reaches the frozen sound speed at $x \approx 1\,\mathrm{m}$, but the thermicity has a positive (nonzero) value that prevents the solution from passing through a saddle point. Decreasing the wave velocity further results in this singularity occurring earlier in the flow. Increasing the wave velocity by $0.1\,\mathrm{m\,s^{-1}}$ above the CJ solution results in the flow remaining subsonic with respect to both the frozen and equilibrium sound speeds (i.e., strong detonation solution). Thus, no eigenvalue solution to the wave structure

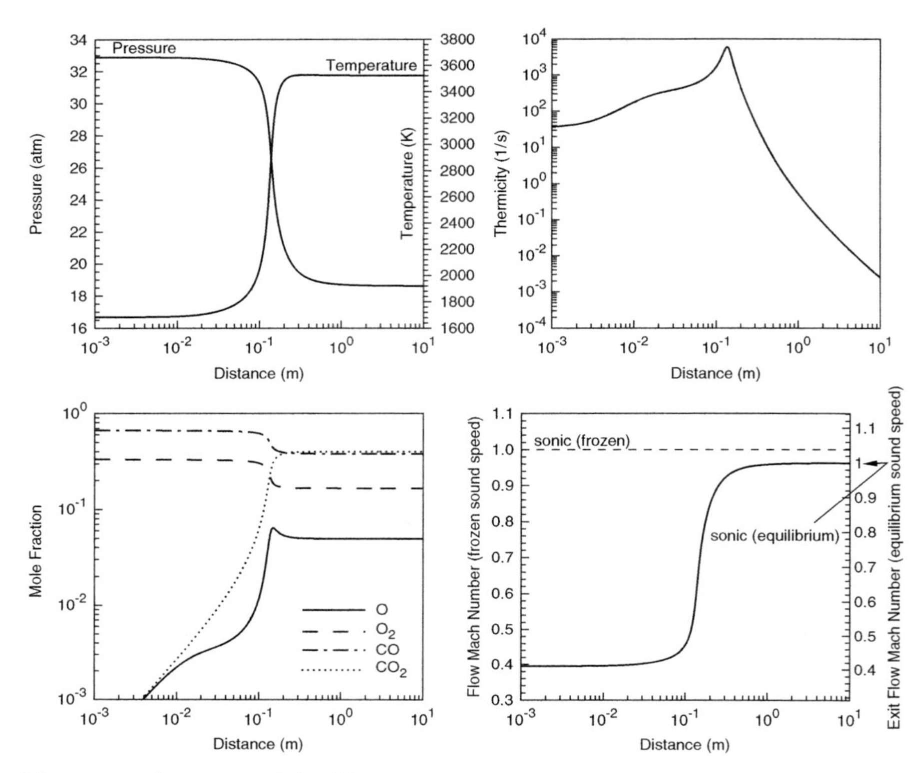

Fig. 2.10. Structure of CO/O_2 detonation, showing pressure, temperature, mole fractions, thermicity, and flow Mach number. Note that Mach number is plotted using the local, frozen speed of sound on the left y-axis, the right y-axis gives the exit flow Mach number using the equilibrium speed of sound

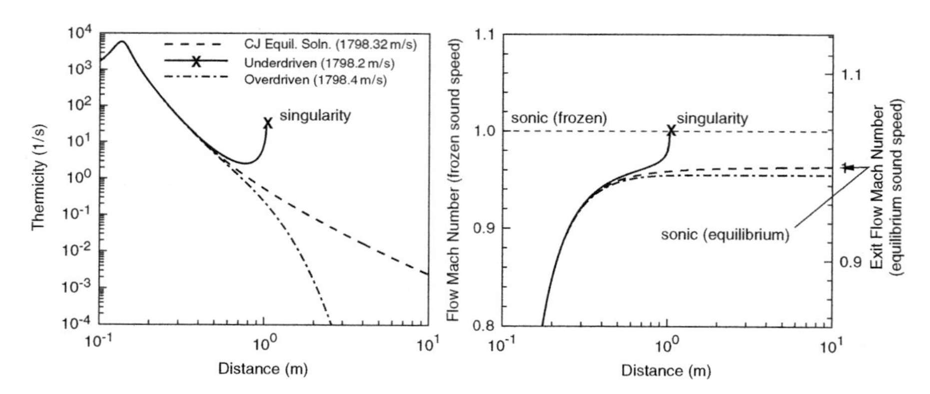

Fig. 2.11. Structure of CO/O_2 detonation, initialized with a velocity slightly above and slightly below the equilibrium CJ solution

exists in this case that permits the solution to satisfy the generalized CJ criterion. This result is due to the fact that the thermicity parameter is always positive and only approaches zero asymptotically. In order for the solution to pass through sonic via a saddle point, the thermicity must pass through zero. Thus, for the carbon monoxide/oxygen system discussed here, there is no ambiguity since the equilibrium and the ZND-based methods generate the same unique solution.

2.5.3 Hydrogen/Chlorine System

Identification of a real combustible mixture that exhibits the type of pathological behavior discussed in Sect. 2.3 is an interesting problem that has been the subject of periodic investigation since the introduction of the ZND model in the 1940s. Zeldovich and Ratner [108] pointed out that the reaction of H_2 and Cl_2 to form HCl can occur via the Nernst chain reaction much more readily than the dissociation of Cl_2, which has a relatively high activation energy. Thus, the highly exothermic reaction forming HCl may be followed by an endothermic dissociation reaction in Cl_2, resulting in the type of exothermic/endothermic reaction necessary for pathological behavior. Subsequent studies of experimental systems exhibiting pathological behavior have tended to focus on this system.

The first detailed chemical kinetic calculations of a ZND detonation in H_2/Cl_2 were performed by Guénoche et al. [37] using the mechanism given in Table 2.3. These calculations are reproduced here. A fuel-lean (chlorine-rich) mixture with fuel equivalence ratio $\phi = 0.66$ at relatively low pressure ($p_1 = 3.33\,\mathrm{kPa}$) was selected here for a sample calculation in order to accentuate the difference between the rapid, exothermic formation of HCl and the slower dissociation of the excess Cl_2. The governing differential equations (2.58)–(2.71) were integrated coupled with the kinetic mechanism in Table 2.3. The initial shock velocity was iterated upon until a solution that

Table 2.3. Hydrogen/chlorine reaction mechanism

Reaction	A ($\mathrm{m^3kmol^{-1}s^{-1}}$)	n	E_a (kJ/kmol)	M
$H + H + M \rightleftharpoons H_2 + M$	1.0×10^{12} [a]	−1.00	0	HCl, Cl_2, Cl
	2.0×10^{13} [a]	−1.00	0	H
	2.0×10^{10} [a]	−0.60	0	H_2
$Cl_2 + M \rightleftharpoons 2Cl + M$	6.15×10^{18}	−2.07	238,815	H_2, Cl_2, HCl, H
	6.15×10^{19}	−2.07	238,857	Cl
$HCl + M \rightleftharpoons H + Cl + M$	6.76×10^{18}	−2.00	427,765	H_2, Cl_2, HCl, H, Cl
$Cl + H_2 \rightleftharpoons HCl + H$	4.80×10^{10}	0.00	22,023	
$H + Cl_2 \rightleftharpoons HCl + Cl$	6.61×10^{8}	0.68	4,564	

[a] $\mathrm{m^6kmol^{-2}\,s^{-1}}$

Table 2.4. Hydrogen/chlorine detonation

Initial composition	(Moles)	$H_2 + \frac{3}{2}Cl_2$	
	p_1 (kPa)	3.33	
Initial state	T_1 (K)	300.00	
	c_1 (m s^{-1})	278.93	
Detonation solution		Equilibrium	Eigenvalue (ZND)
	u_1 (m s^{-1})	1,320.7	1,527.3
	M_{1_f}	4.735	5.476
von Neumann state	p_{vN} (kPa)	86.469	116.06
	p_{vN}/p_1	18.066	34.852
	T_{vN} (K)	1,374.1	1,724.4
	u_{vN} (m s^{-1})	232.96	251.88
	c_{vN} (m s^{-1})	586.81	655.23
	M_{vN}	0.3970	0.3844
Chapman–Jouguet state	p_2 (kPa)	47.073	60.300
	p_2/p_1	14.136	18.108
	T_2 (K)	2,075.94	2,989.09
	u_2 (m s^{-1})	748.4	883.3
	c_{f_2} (m s^{-1})	789.0	883.3
	c_{e_2} (m s^{-1})	748.4	N/A
	M_{f_2}	0.9485	1.0000
	M_{e_2}	1.0000	N/A

passes through the frozen sonic point was identified. This solution is presented in Table 2.4 and Fig. 2.12, showing the thermodynamic properties, species concentrations, thermicity parameter, and the flow Mach number (using the frozen speed of sound). As hypothesized by Zeldovich and Ratner, the H_2/Cl_2 system does exhibit an overshoot in the heat release followed by an endothermic phase. This is most clearly seen by examining the temperature and the thermicity parameter, which passes through zero and becomes negative at the same point where the flow becomes frozen sonic, as required by the generalized CJ condition. (Note the fact that the thermicity becomes negative prevents the use of a log-scale on the y-axis, in comparison to Fig. 2.10.) The history of species concentrations through the reaction zone verifies that the rapid formation of HCl followed by a slower dissociation of Cl_2 into Cl is the source of the exothermic/endothermic nature of the reaction. This solution, resulting from iteration upon the reaction zone structure until a trajectory that satisfies the generalized CJ condition was found, is the eigenvalue solution.

The eigenvalue solution found above is compared to the equilibrium solution, such as would be found via the NASA CEA [34, 56] or other chemical equilibrium software, in Table 2.4. A significant difference between the two solutions is found, with a 15% discrepancy in the detonation propagation velocities with the equilibrium velocity being lower. This result is unlike that

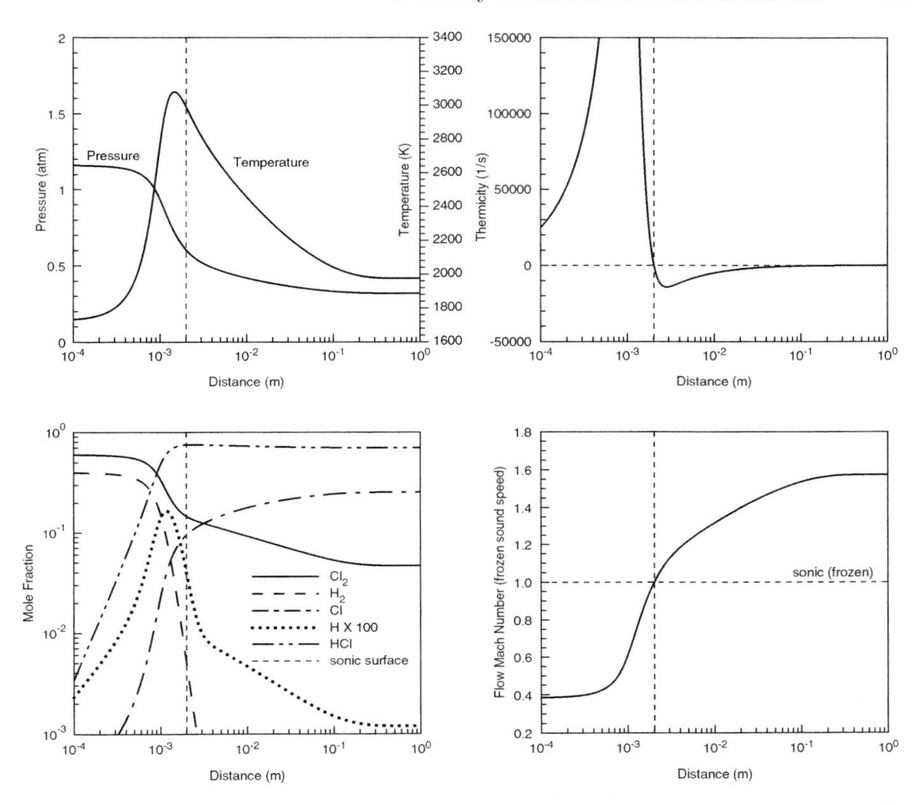

Fig. 2.12. Structure of H_2/Cl_2 detonation, showing pressure, temperature, mole fractions, thermicity, and flow Mach number. The location of the sonic point is denoted as a *vertical dashed line*

obtained with CO/O_2 detonations considered in the prior section, for which the equilibrium and ZND approaches were seen to generate the same solution (within numerical precision). If a ZND calculation of the H_2/Cl_2 reaction zone is initialized with the post-shock von Neumann conditions for the CJ equilibrium solution, as is done in Fig. 2.13, the calculation encounters a singularity. The reaction zone is now longer due to the lower post-shock temperature, but when the exothermic reaction begins to release significant heat into the flow, it results in sonic flow being encountered before the thermicity going to zero, hence causing the singularity. Qualitatively, this is identical to the behavior seen in Sect. 2.3 with an artificially constructed two-step endothermic/exothermic system. Indeed, any attempt to initialize a solution with a shock velocity even incrementally below the eigenvalue solution for the H_2/Cl_2 system results in a singularity. Solutions initialized with a faster shock remain subsonic with respect to the frozen speed of sound (i.e., the strong detonation solution). Finally, note that it is not possible to define an equilibrium sound speed for the (nonequilibrium) sonic point of the eigenvalue solution.

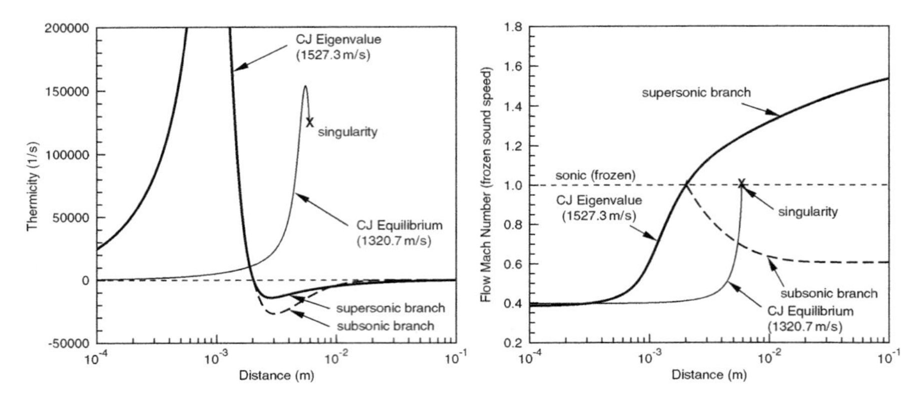

Fig. 2.13. Structure of H_2/Cl_2 ($\phi = 0.666, p_1 = 3.33\,\mathrm{kPa}$) detonation, initialized with CJ equilibrium (resulting in a singularity) and eigenvalue solutions. Both the supersonic and subsonic downstream branches of the eigenvalue solution are shown

The experimental realization of this pathological behavior in the H_2/Cl_2 system has been attempted several times, starting with Zeldovich and Ratner [108], followed by [1, 46]. Comparing these different experiments indicates some inconsistent results, likely due to initiation transients or unstable waves caused by finite tube size effects. While measurement of detonation velocity is conceptually very easy (consisting of just recording the time of arrival at two spatially separated locations), in practice great care must be exercised that the measured propagation velocity is not influenced by the initiation mechanism or near-limit effects due to using too small a detonation tube. The most recent and careful study is that of Dionne et al. [21], who used a long (12-m-long) tube with 5-cm inside diameter. The final 8 m of tube was instrumented with four pressure sensors to verify that a stable detonation wave had been initiated by the powerful spark used. The results of measurements of the detonation propagation velocity are shown in Fig. 2.14, giving the wave velocity as a function of initial pressure for a fuel lean mixture ($\phi = 0.66$). Also shown are the detonation velocity as predicted by the CJ equilibrium solution and the eigenvalue solution as calculated via iteration upon the ZND structure. Note that the velocity of the eigenvalue solution exhibits a very nearly flat dependence on pressure, due to the fact that it only depends upon the exothermic phase of HCl formation and independent of the subsequent dissociation, which is isolated from the front via the sonic point. The equilibrium solution shows a gradual decrease in velocity due to the increased dissociation at lower pressure resulting in less complete combustion and thus less total heat release. This equilibrium behavior is typical of most mixtures.

The experimental data appear to match the equilibrium solution at higher pressure, but as the pressure is decreased, the observed detonation velocities remains constant or slightly increases, moving away from the equilibrium solution and toward the eigenvalue solution. This pressure independence of

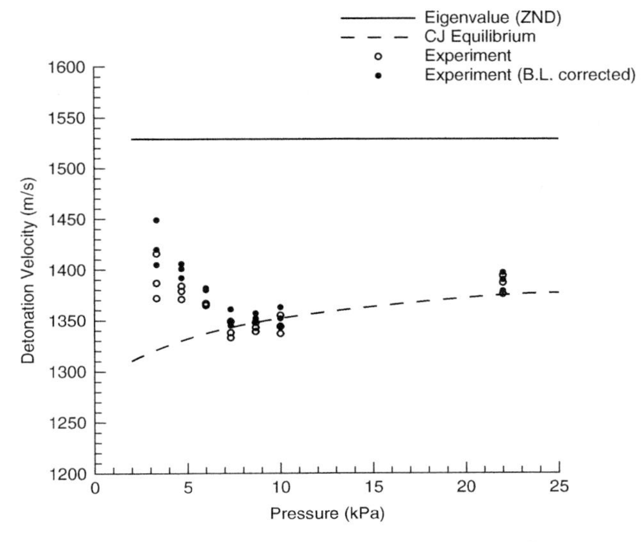

Fig. 2.14. Comparison of experimental measurements of detonation velocities in H_2/Cl_2 ($\phi = 0.66$), CJ equilibrium, and eigenvalue (ZND-based) solutions. Experimental data are shown as *open symbols*, velocities corrected for boundary layer effects are shown as *solid symbols*

H_2/Cl_2 detonation velocity was first experimentally noted in 1934 by Sokolik and Shchelkin [82]. In actuality, the experimental detonation velocity would be expected to decrease as pressure is decreased, even if the experimental detonations correspond to the eigenvalue solution, since the longer reaction zone at lower pressure should experience increased losses associated with the tube walls. For this reason, Dionne et al. corrected their experimental data, extrapolating the measured velocities to what would be expected in an effectively infinite diameter tube (i.e., the ideal CJ velocity) by using the Fay–Dabora model [18,28]. These corrected data are also shown in Fig. 2.14 and moves the data toward the predictions of the eigenvalue solution, particularly at lower pressures. Note that the largest deviation from the equilibrium solution occurs at $p_1 = 3.33\,\text{kPa}$ (the experimental result being about 10% greater than the equilibrium solution), which is why this condition was selected for the sample calculations shown in Figs. 2.12 and 2.13. Dionne et al. also studied mixtures at stoichiometric ($\phi = 1$) and rich ($\phi = 1.5$) equivalence ratios as well, which also exhibited pathological behavior, although to a lesser degree [21].

While the results with H_2/Cl_2 are not as convincing a demonstration of a pathological detonation as might be hoped, some discussion of the comparison between experiment and theory in Fig. 2.14 is in order. For perspective, experimentally observed detonation velocities are usually 1–2% below the equilibrium solution, provided care has been exercised to isolate the initiation transients and near-limit effects discussed above, so the discrepancy of an observed velocity 10% greater than the equilibrium velocity is experimentally significant. In addition, the ZND model neglects the fact that the detonation zone structure is influenced by the unsteady longitudinal pulsations and transverse wave interactions that comprise the cellular structure of the shock front, as is the case in all known detonable mixtures including H_2/Cl_2. Finally, the chemical kinetics model used (Table 2.3) is relatively primitive, dating from the 1970s, and likely should be updated to include vibrationally excited states, as discussed in [91].

Future investigations of pathological behavior deriving from systems with exothermic/endothermic behavior might also consider gaseous hydrogen azide (HN_3), which has been observed to support detonation propagation velocities approximately 7.5% greater than the CJ equilibrium solution [63]. A further recommendation to future studies would be to focus on measurement of properties such as pressure or temperature, which, while being a greater diagnostic challenge, are much more sensitive to the differences in detonation solution (equilibrium versus eigenvalue), as can be seen from Table 2.4. Detonation velocity, which has only a square root dependence on energy release ($D_{CJ} \sim \sqrt{Q}$), is a comparatively insensitive parameter. Another intriguing possibility to realize a pathological detonation is a system that exhibits a large molar decrement. Returning to (2.58), note that the thermicity (numerator) can be brought to zero via an increasing molecular weight (rather than, or in combination with, endothermic reactions). In other words, a mixture that reacts to produce a product composition with fewer moles than the reactants may also exhibit pathological behavior. In pure gaseous systems, this condition is difficult to realize, since the elevated temperatures of reaction result in dissociation and low molecular weight products in comparison to the reactants. However, two-phase reactive systems (e.g., a solid fuel in a gaseous oxidizer) exist that might bring about this condition. Many metals (aluminum, magnesium, zirconium) have refractory oxidation products that should remain largely in the condensed phase even at adiabatic flame temperatures, resulting in a system exhibiting a large molar decrement. The use of a two-phase system introduces a number of additional complexities that might further challenge the interpretation of the results, such as a lack of mechanical or thermal equilibrium between the phases; two-phase detonations are discussed further in [111,112]. Finally, it is likely that many condensed-phase detonations have late endothermic phases to their reaction mechanisms, making them also candidates for pathological behavior. A recognizable example is an emulsion-based blasting compound with large "prills" of ammonium nitrate or large, reactive metal particulates that are only intended to react to completion during the expansion of the post-detonation products [30]; such a system almost certainly exhibits pathological behavior in comparison to solutions that assume complete equilibrium at the sonic point. The large uncertainties in the product equations of state (necessary for accurate equilibrium calculations) and a lack of equilibration between phases, however, make unambiguous identification of deviations from the equilibrium solution difficult for condensed-phase explosives. For a further, contrasting assessment of the likelihood of pathological detonations, see [91].

2.5.4 Frozen and Equilibrium Sound Speed

An issue that arises in the application of the CJ criterion and its generalization to nonideal detonations is which sound speed should be used in defining the sonic condition. In a chemically reacting flow, there are two limits to sound wave propagation, namely frozen and equilibrium. The frozen sound speed corresponds to the case of a very high-frequency sound wave (or, alternatively, very slow reactions) where the mixture does not have time to react over the duration of the acoustic wave and the mixture composition remains fixed. The equilibrium sound speed applies for very low frequency sound (or, very fast reactions) in which case the mixture can react quickly enough to remain in chemical equilibrium throughout the isentropic compression or rarefaction pulse that comprises the acoustic wave. It can be shown that the frozen speed of sound is always greater than the equilibrium speed of sound [19]. In real reacting gases, sound waves propagate in between these two limiting speeds, which are typically about 5% different in combustion products. The issue of which sound speed should be used in applying the CJ criterion was discussed intensively in the 1950s and was never satisfactorily resolved [10,24,36,44,102–104]. In addition to detonations, the issue of which sound speed to use in defining the flow regime arises in other nonequilibrium reactive flows as well, such as in hypersonics [4, 99].

To explore this issue, consider a planar detonation in the absence of losses and having a positive thermicity that proceeds asymptotically toward equilibrium. When solving for the ZND structure, the integration proceeds down the Rayleigh line from the von Neumann point and reaches the same equilibrium sonic point that is found by applying the classical CJ condition to the equilibrium Hugoniot. In this case, as was found with the carbon monoxide/oxygen system considered in Sect. 2.5.2, there is no conflict between the classical CJ condition and the ZND-based solution. Viewed on the (p, v) plane, detonations at sub-CJ velocities do not reach the equilibrium Hugoniot and thus no steady solution can be obtained (this is the scenario wherein a singularity appears), while detonations traveling faster than the CJ equilibrium solution will have a reaction zone trajectory that proceeds down the Rayleigh line and stops when it intersects the equilibrium Hugoniot at the upper intersection point (strong detonation with subsonic flow). Thus, the CJ equilibrium solution is the only steady solution to feature a sonic point, and that sonic point is defined by the equilibrium sound speed.

Since the CJ solution in this case has equilibrium sonic outflow, the flow through the entire reaction zone remains subsonic in comparison to the frozen speed of sound (see Table 2.2), and it could be argued that such a solution would be susceptible to high frequency rarefactions catching up to the wave from behind and disrupting the reaction zone. However, as a detonation propagates, the gradient of expansion behind the wave becomes more gradual, and thus the frequency of rarefaction waves might be expected to approach the low (equilibrium) limit. Further, while the initial leading edge of a centered rarefaction would propagate at the frozen speed of sound, this acoustic wave is quickly attenuated (decays exponentially) and the majority of the rarefaction is governed by the equilibrium speed of sound. This aspect of rarefactions propagating in reactive flow was explored in detail by Wood and Parker [103]. If this is the case, then the equilibrium solution would agree with the conceptual picture of the CJ detonation being isolated from downstream disturbances by a sonic plane.

For a system with pathological heat release, such as the hydrogen/chlorine system considered in Sect. 2.5.3, it is not possible to obtain a solution for the reaction zone structure that is initialized with the CJ equilibrium value of detonation speed (see Fig. 2.13). This case can be visualized on the (p, v) plane by the inability of a sequence of intermediate Hugoniots to provide a path along the Rayleigh line to a state of tangency to the equilibrium Hugoniot (see Fig. 2.5). The only permitted steady solution with a sonic point is the eigenvalue solution with a frozen sonic point embedded in the solution and an exit flow that, with respect to the equilibrium Hugoniot, is a weak detonation. Similarly, in any system with losses such as heat transfer or friction, as discussed in Sect. 2.4 or with laterally divergent flow, as will be discussed in Sect. 2.6, it is the frozen sound speed that defines the saddle point of the eigenvalue solution.

This situation gives rise to an apparent paradox, wherein the limiting solution as the loss mechanism decreases to zero may not converge to the ideal, planar solution, since the former is defined by the frozen speed of sound and the latter uses the equilibrium speed of sound. For example, for the case of a detonation in a tube with losses at the walls, as the radius of the tube increases to infinity, the solution for the detonation velocity will not agree with the planar solution (i.e., $\lim_{r\to\infty} D \neq D_{r=\infty}$). In order to resolve this apparent inconsistency, a possible strategy would be to solve the full, unsteady Euler equations without imposing a particular criterion, and see which solution evolves (this approach was done by Sharpe and Falle [75, 77] for pathological detonations and by Dionne et al. [22] for detonations with friction and heat transfer, as discussed in Sect. 2.4.3). Such a calculation was performed by Sharpe [76] for a model system with a single-step reversible chemistry originally proposed by Fickett and Davis [19] (Fickett and Davis considered a system with zero activation energy, while Sharpe considered a system with low activation energy that gave a stable ZND solution). Sharpe [76] showed that, for a planar detonation initiated by an overdriven blast wave in the above system, after a long propagation time, the wave velocity and reaction zone structure approached that of an equilibrium sonic CJ detonation. However, if even an infinitesimal degree of loss was introduced (curvature, in the case of Sharpe's calculations), the long-term evolution of the solution converged toward the eigenvalue solution with a frozen sonic point where thermicity went to zero (note that Sharpe included the effect of area divergence into his definition of thermicity). Thus, it appears that the counterintuitive scenario discussed above, where the solution in the limit of losses going to zero does not converge to the ideal CJ equilibrium solution, is indeed the case.

Since all real detonations are multidimensional and transient, making the definition of a sonic point or plane in the flow problematic, the equilibrium versus frozen sound speed issue is now regarded as being somewhat academic. The issue does occasionally arise in connection with the hydrogen/oxygen system (or hydrogen/air), which was speculated by Fickett and Davis [19] to exhibit pathological behavior due to molar decrement. Differences between the CJ equilibrium solution and a solution obtained by iterating upon a ZND calculation for hydrogen/air were found by Klein et al. [45] and attributed to the comparatively slow, endothermic dissociation of water resulting in slightly negative values of thermicity at the end of the reaction zone. In this case, the solution is of the eigenvalue type with a frozen sonic point. Discrepancies between ZND and CJ equilibrium solutions for hydrogen/oxygen were also noted by [67] but not explored further. Hydrogen/oxygen was not selected as a sample reactive system in this chapter for precisely this reason, namely, that it may exhibit weakly pathological behavior depending upon the details of the particular kinetics mechanism used.

2.6 Detonations with Divergent Flow

Detonation waves (both condensed-phase and gaseous) propagating in finite-sized charges will be subject to losses resulting from divergent flow at the periphery of the charge. The pressures within the reaction zone of any condensed-phase detonation (typically, 10–30 GPa) greatly exceed the yield strength of any confining material, resulting in an outward expansion of the detonation products as the confinement gives way. The loss in axially directed momentum in the flow of detonation products causes the detonation to propagate at a slower velocity than would be obtained in a rigidly confined charge or, equivalently, in a planar detonation in an infinite-sized charge (i.e., an infinite-sized charge is "self-confined" and is expected to detonate at the ideal CJ velocity). The lower propagation velocity additionally lowers the post-shock temperature/pressure and consequently the chemical reaction rates, while the diverging flow can cause the sonic surface to occur earlier in the reaction zone, such that less energy release is available to drive the detonation wave. The divergent flow also necessitates that the leading shock front is no longer planar and becomes curved. These effects become progressively more significant as the dimension of the charge is decreased until eventually the detonation wave fails altogether. With condensed-phase explosives, this failure process can be spectacularly abrupt; it is not uncommon to recover nearly pristine, unreacted explosive only one charge diameter beyond the point at which the critical dimension is first encountered. In gaseous explosives contained in rigid tubes, near-limit behavior is more complex, since the wave can interact with the tube wall in order to sustain propagation in transient modes known as spinning and galloping detonations (see discussion in Chap. 7 of [51]).

The velocity deficit in finite-sized charges and the existence of a critical dimension (usually, the diameter of a cylindrical charge) are heavily utilized in research on detonation waves in condensed-phase explosives as a means to probe the reaction zone structure and thermodynamic properties of the detonation products. Due to the extremely short reaction timescales in condensed-phase detonations (typically, sub-microsecond to nanosecond), the opaque nature of detonating media, and the extreme pressures generated, in situ measurements of any detonation properties are extraordinarily difficult. As a result, almost all models for condensed-phased detonations utilize data derived from the velocity deficit and/or the curvature of the shock front as the charge diameter is decreased, or measurement of the critical diameter at which detonation failure occurs. The radial expansion of detonation products in a cylindrical charge confined with a ductile material (e.g., copper) is also the basis of most semi-empirical models for the equation of state of the detonation products.

For detonations in gas-phase media on the order of atmospheric pressure, perfectly rigid confinement is possible, for example, using a steel-walled tube. However, the periphery of the gaseous charge still influences the detonation wave via friction and heat transfer to the wall. In Sect. 2.4, these effects were treated as volumetric losses that were spread uniformly across the cross section of the tube. In reality, the influence of the wall is conveyed to the detonation flow via a boundary layer region near the wall. The growth of the boundary layer results in a diverging flow in the reaction zone, qualitatively similar to the diverging flow that occurs in a condensed-phase detonation due to yielding confinement. It may appear counterintuitive that growth of a boundary layer along the wall results in a *diverging* flow in the detonation reaction zone; note that in the wave-fixed reference frame, the walls have the effect of accelerating the flow while also cooling and increasing the density of gas in the boundary layer. By continuity, this effect requires that the core of the flow diverges outward to account for the mass effectively removed by the boundary layer. Boundary layers of this type are sometimes called *negative boundary layers.* Within the context of a quasi-one-dimensional approximation, detonations with divergent flow can be treated with the reaction zone equations derived previously in Sect. 2.2. In particular, (2.10), reproduced here slightly modified

$$\frac{\mathrm{d}u}{\mathrm{d}x} = \frac{\frac{\Delta q}{c_p T}\dot{\lambda} - u\frac{1}{A}\frac{\mathrm{d}A}{\mathrm{d}x}}{1 - M^2}, \qquad (2.75)$$

models the effect of diverging flow via the $\frac{1}{A}\frac{\mathrm{d}A}{\mathrm{d}x}$ term. In this section, the steady reaction zone structure of detonation waves with diverging flow is further explored with a particular emphasis on models that permit the area divergence term to be quantified.

2.6.1 Stream Tube Divergence

If the stream tube defining the reactive flow in the detonation is a two-dimensional slab with yielding confinement on the top and bottom surfaces only or an axisymmetric cylindrical tube with yielding confinement on the circumference, the area of the stream tube is given by

$$\text{2-D Slab:} \quad A(x) = t(x)\, w \tag{2.76}$$

$$\text{Axisymmetric Tube:} \quad A(x) = \frac{1}{4}\pi d(x)^2 \tag{2.77}$$

for the 2D and axisymmetric geometries, respectively, where $t(x)$ is the slab thickness, w is the slab width, and $d(x)$ is the tube diameter (Fig. 2.15). The area divergence term is then

$$\text{2-D Slab:} \quad \frac{1}{A}\frac{\mathrm{d}A}{\mathrm{d}x} = \frac{t'(x)}{t(x)} \tag{2.78}$$

$$\text{Axisymmetric Tube:} \quad \frac{1}{A}\frac{\mathrm{d}A}{\mathrm{d}x} = 2\frac{d'(x)}{d(x)}, \tag{2.79}$$

where the prime ($'$) denotes differentiation with respect to the position x. As will be discussed below, the slope of the stream tube boundary in simple models is often given by a local matching of pressure and flow direction between the CJ (or von Neumann) state and the confinement and is assumed to remain constant through the reaction zone. If this is the case, then the slope will be the same in both 2D and axisymmetric geometries for a given shock or detonation velocity. If the magnitude of area divergence is small compared to the initial cross-sectional area of the stream tube (i.e., $t(0) >> L\, t'(x)$,

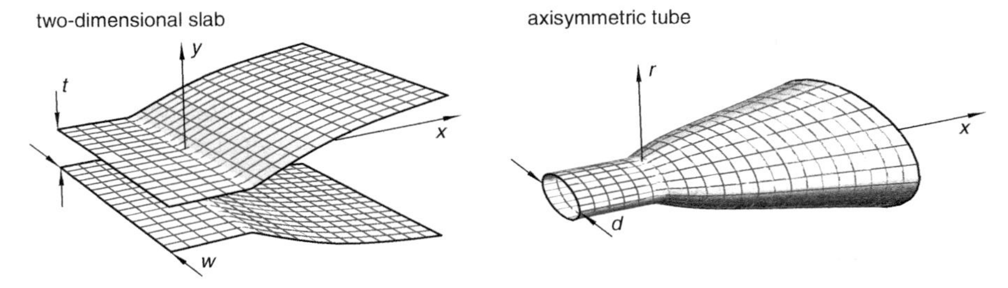

Fig. 2.15. Stream tubes in two-dimensional slab and axisymmetric, cylindrical geometries enclosing the reaction zone of a detonation wave. Note that the magnitude of area divergence and relative length of the reaction zone have been greatly exaggerated for this schematic

$d(0) >> L\, d'(x)$) where L is the length of the reaction zone, then the area divergence terms differ by a numerical factor of *two* for the two geometries

$$\text{2-D Slab:} \quad \frac{1}{A}\frac{\mathrm{d}A}{\mathrm{d}x} \approx \frac{t'(x)}{t(0)} \tag{2.80}$$

$$\text{Axisymmetric Tube:} \quad \frac{1}{A}\frac{\mathrm{d}A}{\mathrm{d}x} \approx 2\frac{d'(x)}{d(0)}. \tag{2.81}$$

This means that the solution for the reaction zone structure in the two geometries should be the same, provided the diameter of the axisymmetric cylinder is scaled by *twice* the thickness of the 2D slab.

Early models for detonation in finite-diameter charges by Jones [41] and Eyring et al. [27] used phenomenological descriptions for the stream tube area divergence (i.e., nozzle flow models). While not rigorous, these models are of interest in their providing a physical picture of the flow fields that result from the boundary of the charge expanding outward as the explosive reacts. The models developed by Jones and elaborated upon by Erying et al. assumed a Prandtl–Meyer (P–M) expansion fan originating where a normal detonation encountered the edge of the charge. The initial state was taken as the CJ state for the ideal, constant area detonation (note that using the subsonic von Neumann state is not an option since the P–M function is only defined for supersonic flow). For an unconfined charge, the P–M fan at the edge was matched to a stream tube containing the core of the detonation flow. For a heavily confined charge, the P–M fan was matched to an oblique shock transmitted into the confining material to determine the divergence angle, which was taken as constant. Eyring et al. went on to develop heuristic models incorporating the fact that, if the flow at the boundary of the charge is diverging outward following the leading shock, the shock front itself cannot be flat and must be oblique at the boundary. In other words, the leading shock front is curved (normal on the central axis of the charge and increasingly oblique toward the edges), similar to the meniscus of a liquid surface in a capillary tube. Eyring et al. were careful to point out that front curvature is a necessary consequence of diverging flow; curvature and flow divergence are different manifestations of the same phenomenon, not separate effects that should be superimposed in models. Theoretical investigations into the 1960s continued to presuppose functional forms for the stream tube area profile, often out of analytic convenience rather than from physical considerations [26, 100].

2.6.2 Radial Flow Derivative

Beginning with the work of Wood and Kirkwood [101], a more rigorous approach to solve for the reaction zone of a detonation wave with divergent flow was developed that examined the two-dimensional flow (either rectangular or axisymmetric) along the central axial streamline of the reaction zone.

Utilizing the fact that the flow is symmetric about this line, the continuity equation for two-dimensional, steady compressible flow

$$\nabla \cdot \left(\rho \vec{V}\right) = 0 \tag{2.82}$$

can be simplified for rectangular coordinates

$$\frac{\partial (\rho u)}{\partial x} + \frac{\partial (\rho v)}{\partial y} = 0 \tag{2.83}$$

since the transverse flow velocity v is zero along this streamline, as follows:

$$\frac{\partial (\rho u)}{\partial x} + \rho \frac{\partial v}{\partial y} = 0. \tag{2.84}$$

Note that, while the transverse velocity is zero along the central streamline ($v = 0$), the derivative of the velocity in the radial direction has a nonzero value $\left(\frac{\partial v}{\partial y} \neq 0\right)$. Likewise the continuity equation for axisymmetric flow in cylindrical coordinates

$$\frac{\partial (\rho u)}{\partial x} + \frac{1}{r} \frac{\partial (r \rho v)}{\partial r} = 0 \tag{2.85}$$

specialized to the central streamline via l'Hôpital's rule yields

$$\frac{\partial (\rho u)}{\partial x} + 2\rho \frac{\partial v}{\partial r} = 0. \tag{2.86}$$

The x-momentum equation in rectangular

$$\frac{\partial \left(\rho u^2\right)}{\partial x} + \frac{\partial (\rho u v)}{\partial y} = -\frac{\partial p}{\partial x} \tag{2.87}$$

and cylindrical coordinates

$$\frac{\partial \left(r \rho u^2\right)}{\partial x} + \frac{\partial (r \rho u v)}{\partial r} = -r \frac{\partial p}{\partial x} \tag{2.88}$$

both revert (via symmetry) to the familiar form of the momentum equation when applied along the axial streamline

$$\frac{\partial p}{\partial x} + \rho u \frac{\partial u}{\partial x} = 0. \tag{2.89}$$

Returning to continuity, (2.84) and (2.86) can be written as

$$\frac{\partial (\rho u)}{\partial x} + \alpha \rho \frac{\partial v}{\partial z} = 0, \tag{2.90}$$

where z denotes the transverse y-direction in the rectangular geometry and radial r-direction in the cylindrical geometry, and α has the value of 1 and

2 for the rectangular and cylindrical geometries, respectively. By comparing (2.90) with (2.4), the derivative of the transverse/radial velocity term ($\frac{\partial v}{\partial z}$) can be related to the area divergence of a stream tube

$$\frac{1}{A}\frac{\mathrm{d}A}{\mathrm{d}x} = \frac{\alpha}{u}\frac{\partial v}{\partial z}. \tag{2.91}$$

This equivalence can also be demonstrated by considering an arbitrarily small stream tube that encloses the flow along the charge axis, and using the derivative of the radial flow velocity to approximate the divergence of the stream tube boundary. As the stream tube shrinks to the axis, the correspondence (2.91) becomes exact.

2.6.3 Shock Front Curvature

The derivative of the radial flow velocity can also be related to the radius of curvature R of the leading shock front by using the geometric construction in the shock-attached reference frame shown in Fig. 2.16. The flow velocity approaching the shock is the detonation propagation velocity ($u_1 = D$). Note that this construction applies to the 2D slab and axisymmetric geometries equally. Using the fact that the component of flow velocity parallel $u_{\parallel} = D\sin\theta$ to a shock front does not change as the flow crosses the shock, as required by conservation of momentum, it is possible to express the radial component of velocity as

$$v = D\sin\theta\cos\theta - u_{\perp}\sin\theta. \tag{2.92}$$

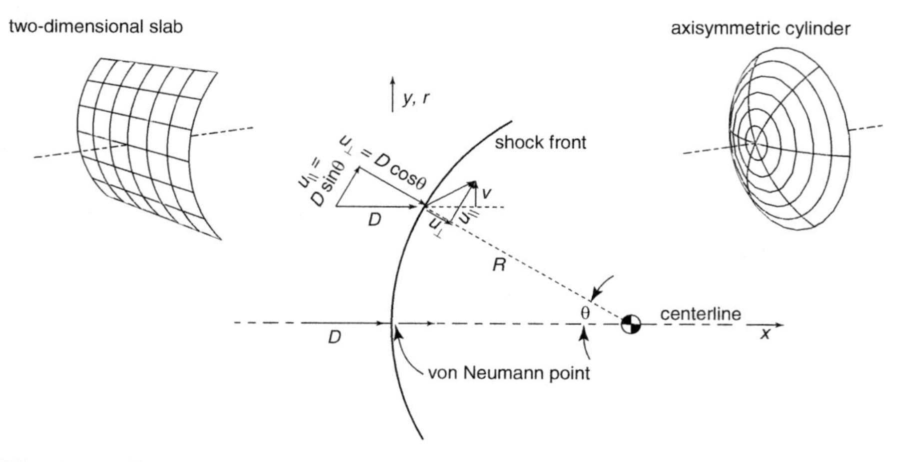

Fig. 2.16. Geometric construction relating shock front radius of curvature R to the derivative of the radial flow velocity $\frac{\partial v}{\partial r}$ or $\frac{\partial v}{\partial y}$ at the centerline immediately behind the shock

Performing a partial differentiation of the velocity v with respect to the angle θ

$$\left(\frac{\partial v}{\partial \theta}\right)_{\text{shock}} = D\left(\cos 2\theta\right) - u_{\perp}\cos\theta - \sin\theta\left(\frac{\partial u_{\perp}}{\partial \theta}\right)_{\text{shock}}, \tag{2.93}$$

where the "shock" subscript denotes that the differentiation was performed along the shock front. The differentiation can be converted to a differential with respect to y or r (denoted as z) as follows:

$$\left(\frac{\partial v}{\partial z}\right)_{\text{shock}} = \left(\frac{\partial v}{\partial \theta}\right)_{\text{shock}}\left(\frac{\partial \theta}{\partial z}\right)_{\text{shock}} = \frac{\left(\frac{\partial v}{\partial \theta}\right)_{\text{shock}}}{R\cos\theta}. \tag{2.94}$$

Taking the limit approaching the central axis

$$\lim_{\theta\to 0}\left(\frac{\partial v}{\partial z}\right) = \frac{D - u(0)}{R}, \tag{2.95}$$

where $u(0)$ is the axial velocity at the von Neumann point along the central axis. Thus, the radius of the curvature of the shock front is directly related to the derivative of the radial flow immediately behind the shock front. This relation applies strictly only at the von Neumann point (i.e., immediately after the leading shock); however, lacking further information, this derivative is often taken as constant through the reaction zone.

The shock radius R can be related to the curvature κ of the shock front as follows:

$$\kappa = \frac{\alpha}{R}, \tag{2.96}$$

where again α is 1 for the two-dimensional slab geometry (cylindrical curvature of the shock front) and 2 for the axisymmetric cylinder geometry (spherical curvature of the shock front). The fact that, for a locally steady detonation front, the shock curvature κ uniquely determines the eigenvalue velocity of propagation provides the means to construct the detonation front shape and trajectory as the detonation propagates, for example, through a complex charge geometry. This technique, called *Detonation Shock Dynamics*, is the subject of Chap. 7 of this volume.

2.6.4 Confinement Interaction via Newtonian Theory

In order to associate these divergent flow models with experimental results or make quantitative predictions of velocity deficits or critical diameter, it is necessary to link the flow divergence to the overall dimension (diameter or thickness) of the charge and the properties of the confinement. For condensed-phase detonations, solving for this interaction can be challenging; this topic is elaborated upon in Chap. 7 of the present volume. As an illustrative numerical example relevant to gaseous detonations, a simple Newtonian model for the interaction of the diverging flow in the reaction zone with the confinement will be developed further in this section. Being a nozzle flow-type model, it does not explicitly include the curvature of the shock front. However, this model (originally proposed by Tsuge et al. [31, 95]) is instructive in that it treats the evolving interaction of the reacting flow with the confinement via an analytically tractable method without resorting to empirical input or an assumed formula for the area divergence.

The Newtonian model for hypersonic flow assumes that a flow encountering an inclined surface loses the component of velocity normal to the surface but retains the tangential velocity. In other words, a flow encountering an inclined surface slides along the surface, and the change in momentum flux of the flow determines that the pressure on the surface must vary as

$$p = p_\infty + \rho_\infty u_\infty^2 \sin^2 \beta, \tag{2.97}$$

where β is the surface inclination angle to the flow with freestream density ρ_∞, pressure p_∞, and velocity u_∞. Originally proposed by Newton, this flow model has been shown to be remarkably accurate in predicting surface pressures for slender bodies in hypersonic flow [4]. The model is invoked here to treat the flow of the confinement material as it encounters the expanding flow of reacting gas, thereby linking the pressure of the reacting flow to the divergence angle of the stream tube enclosing that flow. Since the stream tube cannot support a pressure difference, the reacting flow must locally match the pressure of the confinement and thereby the slope of the stream tube boundary. Using (2.97) for the flow of confinement material

$$p = p_c + \rho_c u_c^2 \sin^2 \beta = p_c + \rho_c u_c^2 \frac{\left(\frac{\mathrm{d}z}{\mathrm{d}x}\right)^2}{1 + \left(\frac{\mathrm{d}z}{\mathrm{d}x}\right)^2}, \tag{2.98}$$

where "c" denotes the properties of the confinement before interaction with the detonation wave and "z" is either the radius r or the half-thickness y of the stream tube. Solving for the slope of the stream tube

$$\frac{\mathrm{d}z}{\mathrm{d}x} = \sqrt{\frac{\frac{p(x)-p_c}{\rho_c u_c^2}}{1 - \frac{p(x)}{\rho_c u_c^2}}}. \tag{2.99}$$

This expression is used to find the area divergence term as a function of the local pressure in the stream tube

$$\text{2-D Slab:} \quad \frac{1}{A}\frac{\mathrm{d}A}{\mathrm{d}x} = \frac{t'}{t} = \frac{y'}{y} = \frac{1}{y}\sqrt{\frac{\frac{p(x)-p_c}{\rho_c u_c^2}}{1 - \frac{p(x)-p_c}{\rho_c u_c^2}}} \tag{2.100}$$

$$\text{Axisymmetric Tube:} \quad \frac{1}{A}\frac{\mathrm{d}A}{\mathrm{d}x} = 2\frac{d'}{d} = 2\frac{r'}{r} = \frac{2}{r}\sqrt{\frac{\frac{p(x)-p_c}{\rho_c u_c^2}}{1 - \frac{p(x)-p_c}{\rho_c u_c^2}}}. \tag{2.101}$$

These expressions for $\frac{1}{A}\frac{\mathrm{d}A}{\mathrm{d}x}$ can be used directly in the ODE governing the reaction zone structure (2.75). Since pressure now appears explicitly, this ODE must be integrated coupled to the momentum equation (2.5) reproduced here

$$\frac{\mathrm{d}p}{\mathrm{d}x} = -\rho u \frac{\mathrm{d}u}{\mathrm{d}x} = -\frac{\dot{m}}{A}\frac{\mathrm{d}u}{\mathrm{d}x} \tag{2.102}$$

which can be related to the local streamtube dimension as follows:

$$\text{2-D Slab:} \quad \frac{\mathrm{d}p}{\mathrm{d}x} = -\rho_1 u_1 \frac{y_1}{y}\frac{\mathrm{d}u}{\mathrm{d}x} \tag{2.103}$$

$$\text{Axisymmetric Tube:} \quad \frac{\mathrm{d}p}{\mathrm{d}x} = -\rho_1 u_1 \left(\frac{r_1}{r}\right)^2 \frac{\mathrm{d}u}{\mathrm{d}x}, \tag{2.104}$$

where ρ_1 and u_1 are the density and velocity upstream of the detonation ($u_1 = D$) and y_1 and r_1 are the initial half-thickness and radius of the explosive, respectively.

This set of ODE's was integrated to obtain the structure of a detonation with yielding confinement. The properties of the explosive were the same that have been used previously in this chapter ($\gamma = 1.2$, $Q = 10$, and a single-step Arrhenius reaction with $\frac{E_a}{RT_1} = 25$). The inert confinement was assumed to be at the same initial pressure as the explosive, but with a density 2.5 times greater than the explosive ($\rho_c = 2.5\rho_1$). This density ratio approximately corresponds to the density ratio of air to stoichiometric hydrogen/oxygen, which will be used as an illustrative experiment later in this section. As expected from the form of the governing ODE (2.75), where the effect of area divergence from the yielding confinement is seen to compete with exothermic heat release, it is necessary to iterate upon the detonation velocity to find an eigenvalue solution that can pass smoothly through the sonic point without encountering a singularity. As was seen previously with the effect of heat transfer and friction, there is a critical amount of loss that the detonation can sustain, in this case, resulting from the explosive charge being too thin and losing too much momentum to the divergence of the flow. For charges thinner than this critical thickness, no steady solution with a sonic point can be found.

The structure of the reaction zone for the critical slab thickness at which failure occurs (i.e., at the critical turning point) is shown in Fig. 2.17 as thick lines in comparison to the ideal CJ solution for an infinite thickness charge (thin lines). A small schematic done to scale is included to show the actual amount of area divergence observed at the critical turning point. All dimensions are normalized by the half reaction thickness of the ideal CJ detonation ($L_{1/2}$). The location of reaction completion or the sonic surface cannot be defined for the ideal CJ detonation. The sonic surface for the finite-size charge with area divergence, however, can be determined from the eigenvalue solution. Note that the area divergence from the shock to the sonic plane is moderate (≈30% increase in area) and that, even for the smallest charge thickness that can support detonation, the thickness of the explosive slab is still more than three times the length of the detonation front (i.e., the distance from the shock to the sonic surface). In comparison to the ideal CJ solution, the location of the peak in exothermicity and half reaction length have more than quadrupled due to the lower post-shock temperature, and the exothermicity is still finite as the solution passes through sonic. At the sonic surface

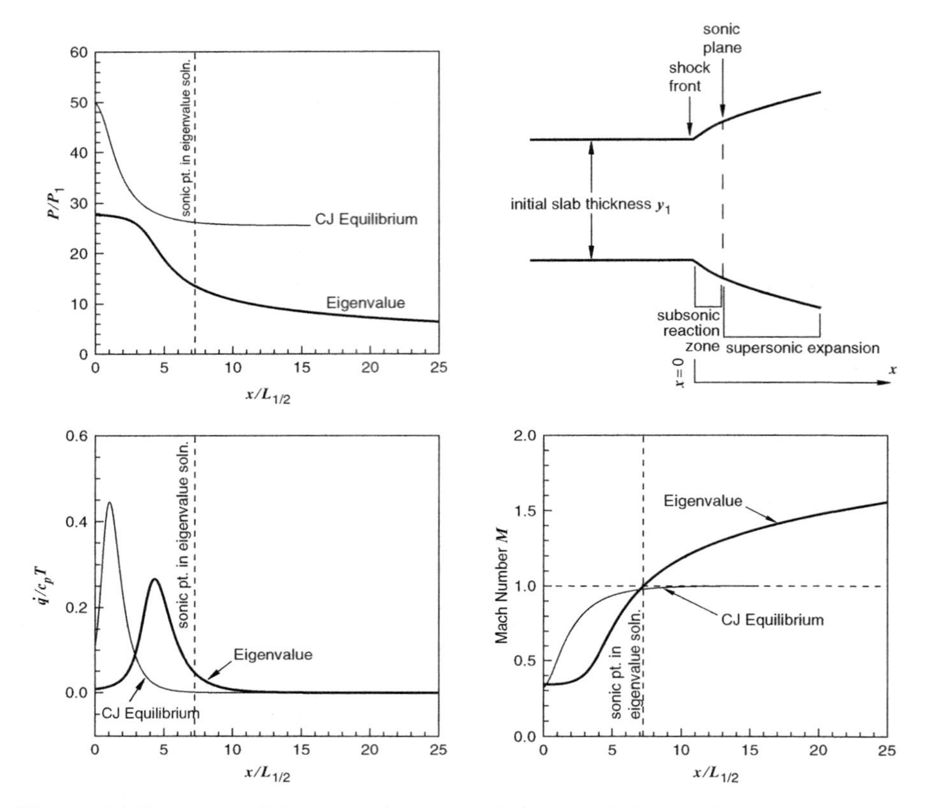

Fig. 2.17. Structure of detonation in a two-dimensional slab at the critical turning point (critical slab thickness), showing pressure, heat release rate, and flow Mach number ($\frac{E_a}{RT_1} = 25$, $Q = 10$, $\gamma = 1.2$, and confinement $\rho_c = 2.5\rho_1$). The ideal CJ solution (without flow divergence) is also shown as a *thin line* for each variable. A silhouette of the quasi-1D stream tube is also shown to scale

approximately 96% of the available chemical energy has been released. This effect in isolation would only result in an approximately 2% velocity deficit. Thus, the observed velocity deficit (about 25% below CJ) is predominately due to a loss in directed axial momentum from the area divergence and is not due to the sonic surface isolating chemical energy release from the shock front.

The eigenvalue velocity is plotted (normalized by the ideal CJ velocity) as a function of the charge diameter or thickness in Fig. 2.18. Following the convention of the condensed-phase explosives literature, the detonation velocity is plotted as a function of the inverse thickness ($\frac{1}{2t}$) or inverse diameter ($\frac{1}{d}$). The inverse of the dimension is used so that extrapolation of the velocity to the y-axis intercept should yield the ideal (infinite diameter) CJ velocity. The factor of two is used in plotting the results as a function of thickness due

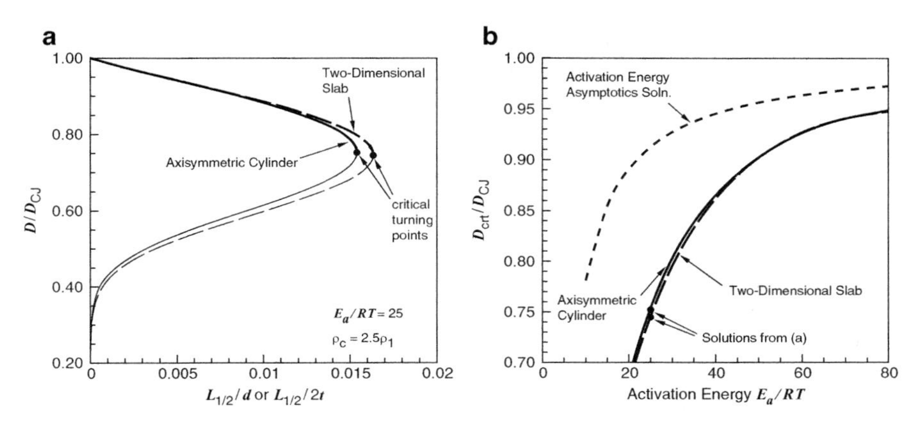

Fig. 2.18. (**a**) Detonation velocity as a function of the charge diameter (axisymmetric cylindrical charge) or thickness (two-dimensional slab charge), with the stable branch shown as a *heavy line*, the nonphysical branch as a *thin line*, and the critical turning points denoted as *dots*. (**b**) Detonation velocity at the critical turning point as a function of activation energy. The activation energy asymptotic analysis of [38, 45, 105] is also shown

to the approximate 2:1 scaling expected between the slab and axisymmetric geometries, as discussed in Sect. 2.6.1. For a sufficiently large initial thickness or diameter, two detonation solutions are found, similar to detonations with heat transfer and friction losses discussed in Sect. 2.4. Again, the lower branch, which exhibits a nonphysical increase in detonation velocity with decreasing thickness, is ruled out as unstable. As the thickness or diameter is decreased, the velocity of the upper (physical) branch decreases until a critical turning point is encountered; any further decrease in the explosive charge dimension is expected to result in detonation failure. The approximate 2:1 scaling expected between results with a 2D slab charge and an axisymmetric cylindrical charge is confirmed.

As activation energy increases, the magnitude of velocity deficit that the detonation can tolerate before failure decreases, as well as the relative amount of area divergence and the effective "thickening" of the wave that can be sustained. The effect of activation energy on the critical thickness and diameter of two-dimensional slabs and axisymmetric tubes was studied systematically using the Newtonian model for diverging flow interaction with confinement in Fig. 2.18b. For values of activation energy representative of real detonable mixtures ($\frac{E_a}{RT_1} = 30 - 80$), the critical velocity occurs at 80–95% of the ideal CJ detonation velocity.

Similar to the solution for detonations with heat loss by Zeldovich and Kompaneets discussed in Sect. 2.4.4, it is possible to perform an asymptotic analysis of the reaction zone structure in the limit of very high activation energy. Such analysis has been done by He and Clavin [38], Yao and

Stewart [105], and Klein et al. [45]. The analysis by [38] and [45] yields a result for the critical velocity identical to that previously found by Zeldovich and Kompaneets for detonations with heat loss (2.38), reproduced here

$$\frac{D_{\mathrm{crt}}}{D_{\mathrm{CJ}}} \approx 1 - \frac{1}{2}\frac{1}{\left(\frac{E_a}{RT_{\mathrm{vN}}}\right)}. \tag{2.105}$$

The analysis of [105] produces a slightly more complex expression dependent upon γ, which nonetheless is equivalent to (2.105). This relation is plotted in Fig. 2.18 as a dashed line. While exhibiting qualitatively similar trends to the results obtained by integrating the governing ODEs to find the eigenvalue detonation velocity, quantitative agreement is only found for very high values of activation energy (where the asymptotic analysis is intended to apply).

2.6.5 Comparisons to Experiment

Experiments with columns of detonable gas (hydrogen/oxygen) confined by inert gas (nitrogen) have been performed by [18, 83, 84, 97]. Some characteristic images of successful propagation and failure of propagation are shown in Fig. 2.19; further experiments are discussed in Chap. 4 of this volume. The expansion of the detonation products and the oblique shock being driven into the confining gas are clearly visible, as is the curvature of the shock front. Quantitative comparison of the measured detonation velocity in these finite-diameter columns of detonable gas with ZND-type calculation using the Newtonian model for interaction with the confinement (as was done in Sect. 2.6.4, but with a more detailed chemistry model) was done by Tsuge et al. [31, 95] and showed general agreement with the experiments. The experimental data, however, had relatively large error bars, and more accurate velocity measurements would necessitate measuring propagation in longer columns of detonable gas, which is a difficult experiment to prepare.

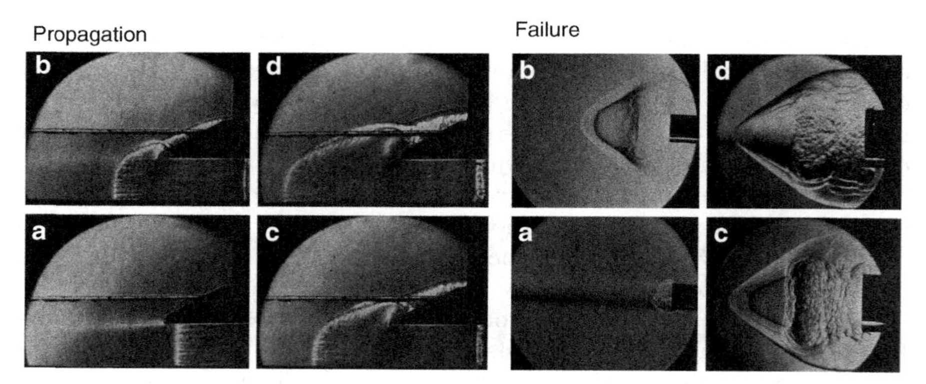

Fig. 2.19. Experimental photographs of detonations in finite-diameter, unconfined columns of hydrogen/oxygen [83]

Perhaps the most impressive demonstration linking the ZND model to an experimental result is the study of gaseous detonations in porous-walled tubes and channels of Radulescu [68, 70]. Due to the flow into the porous walls, the reaction zone of the detonation in such a tube or channel experiences a divergent flow, qualitatively similar to the divergence resulting from yielding confinement or boundary layers discussed earlier in this section. This divergence results in velocity deficits and, for a sufficiently small tube diameter, failure of the detonation. The experimental measurements of the velocity deficit and failure diameter in mixtures of hydrogen/oxygen and acetylene/oxygen with large amounts of argon dilution exhibited good correlation with a ZND calculation with detailed chemistry (similar to that in Sect. 2.5) which includes the effect of flow divergence (as is done in this section). The radial flow derivative (or, equivalently, the area divergence) was estimated by assuming that the flow into the porous wall was choked (sonic) at the pore openings, with the net outflow being scaled by the porosity of the wall. Comparisons of this model demonstrated good agreement with the experimental results, including prediction of the critical diameter at which detonation failure was observed (see Fig. 2.20). What is more, studies in porous-walled tubes and porous-lined

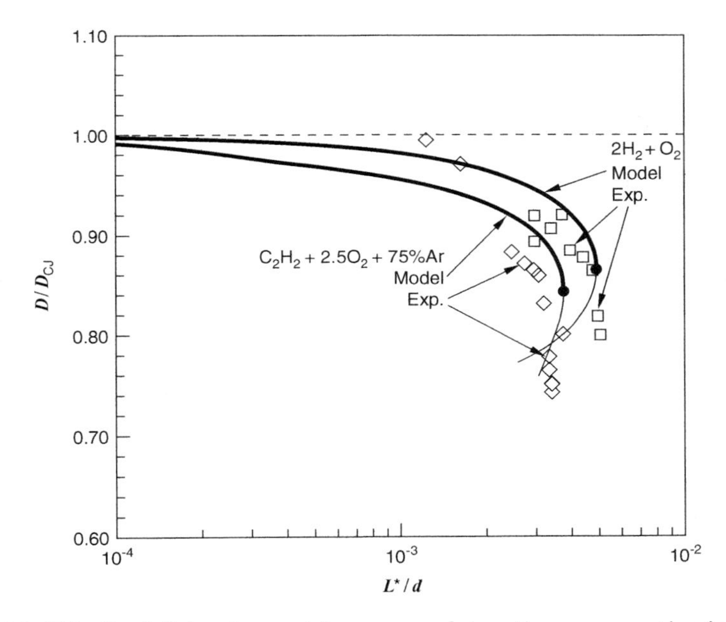

Fig. 2.20. Velocity deficits observed for gaseous detonations propagating in porous-walled tubes, compared to a ZND model with area divergence due to flow into the tube walls. The tube diameter d is normalized by the computed reaction zone length of the ZND detonation with divergence, L^*. In the experiments, initial pressure was changed in order to vary reaction zone length while the tube diameter was constant [68]

channels (with porous surfaces only on the upper and lower wall, with the side walls being impermeable) recovered the 2:1 scaling between the two geometries, as predicted by weak area divergence and front curvature theory. Interestingly, however, this excellent agreement was only obtained in hydrogen/oxygen mixtures and acetylene/oxygen mixtures with large amounts of argon dilution. While these mixtures exhibit the cellular structure characteristic of all gaseous detonations, the cellular structure in these particular mixtures is remarkably regular and is believed to be a piecewise laminar ZND-type structure, with the transverse waves not contributing significantly to the burning mechanism within the front. When these experiments were repeated in acetylene/oxygen mixtures with little or no argon dilution, which exhibit an irregular cellular structure, the agreement with the ZND calculations and the 2:1 scaling between geometries no longer occurred (i.e., the detonation continued to propagate in smaller diameter tubes and with larger velocity deficits for which the ZND calculation would predict failure, and the scaling between diameter and channel height became 1:1 in these cases). These results strongly suggests that detonation propagation in mixtures with irregular cellular structure, which are known to exhibit significant amounts of burning in the transverse waves of the cells, is controlled by the cellular structure of the wave and can no longer be described by a laminar, ZND-type model.

2.7 Concluding Remarks

The appeal of the ZND model lies in its ability to be easily implemented via numerical integration of ordinary differential equations with high confidence in the results (e.g., nearly all of the figures in this chapter were generated with a relatively few lines of code in *Mathematica* utilizing the built-in ODE solver NDSolve). The limitations of the ZND model come from its assumptions of an essentially laminar, one-dimensional structure and steadiness. As the remainder of this book exemplifies, the real dynamics of detonation waves is dominated by multidimensional and transient effects. Ultimately, even multidimensional, transient simulations of detonations must be averaged back onto a one-dimensional representation in order to be digested, such as in the recent study by Radulescu et al. [71].

Besides the presence of multidimensional structure and transient dynamics, other "breakdowns" of the ZND model can occur if the reaction zone is no longer spatially resolved from the shock front. The scenario of reactions occurring within the shock front itself due to "ultrafast" chemistry was explored via Direct Simulation Monte Carlo calculations by Anderson and Long [2, 3]. The detonation structures observed were identified as weak detonations that proceeded from the initial state along the Rayleigh line directly to the product Hugoniot. While detonations of this type are improbable in real gases, since such very reactive mixtures are unlikely to be sufficiently metastable to be explored experimentally, similar phenomena might exist in polycrystalline condensed-phase explosives due to the dispersed nature of the shock and localized reactions at hot spots. Evidence of the lack of a von Neumann spike has been reported in experiments in which laser light, reflected off the interface between high explosives (e.g., RDX, HMX, etc.) and a water window, was used to measure the particle velocity as the detonation emerged. The laser interferometry measurements showed a monotonically increasing velocity (corresponding to a monotonically increasing pressure) through the shock/reaction complex as the detonation crossed the interface [47,96], although these results remain controversial. Outside of these exceptions, the ZND framework remains the foundation upon which transient and multidimensional treatments of detonations are built.

A principal theme of this chapter has been that any energetic material is capable of exhibiting detonation wave propagation and that only in the presence of losses at the boundaries can limits to detonation propagation be meaningfully defined (Khariton's principle). This proposition has motivated the consideration of detonations in media that are so extremely dilute that they would not seem detonable. The interstellar medium is one such example [94]. Some models for waves of star formation treat them as detonation waves, wherein the energy released by supernovae or newly ignited stars drives shock waves into the interstellar media, compressing the media and promoting further nucleation of stars (i.e., self-propagating star formation). The spiral arms of galaxies such as our own Milky Way have been speculated to be just such waves [16, 29, 33, 81]. Other possible astrophysical appearances of detonation waves are reviewed by Oran [61].

Another reoccurring theme in this chapter is that factors competing with the effect of exothermic energy release to accelerate the flow in the reaction zone (e.g., friction, endothermic chemical reaction, area divergence, etc., all of which act to decelerate the flow) result in the flow passing through a sonic point (saddle condition) at a nonequilibrium state, qualitatively similar to the converging/diverging nozzle of compressible flow. The solution for the reaction zone in this case is an eigenvalue of the governing differential equations, in contrast to the classical equilibrium CJ theory. It is interesting to note that essentially the same dynamics occur in other natural phenomena and technological applications. The acceleration of the initially subsonic solar wind leaving the sun's surface to supersonic flow into interstellar space without passing through a physical throat (i.e., the flow is entirely divergent) was hypothesized in a seminal study by Parker [64] to result from a saddle point condition where the divergence of the flow balances a body force term resulting from the sun's gravity. An application more closely related to detonation flow fields is the dual-mode ramjet/scramjet [17]. In this aerospace propulsion concept, a single engine geometry is able to realize both subsonic and supersonic combustion regimes, depending on the flight Mach number of the vehicle. In ramjet mode, the engine would exploit the combined effect of exothermic reaction and area divergence to continuously accelerate the flow from subsonic to supersonic without having to pass through a local minimum in area (i.e., a nozzle throat). Similar "transonic" combustion regimes are believed to occur in ram accelerators [39,73]. A dynamical systems analysis of the critical points in these type of flows was recently performed by De Sterck [86].

Researchers who have labored to develop the framework for understanding detonation waves should derive satisfaction that their efforts may find a wider range of application than the narrow field for which that framework was originally developed.

Acknowledgments

This chapter was developed out of discussions with Jimmy Verreault, Oren Petel, François-Xavier Jetté, Patrick Batchelor, and David Mack. Vincent Tanguay contributed to the analysis of the inclusion of the work done by friction in detonations and the Taylor wave analysis in the appendix. Jean-Philippe Dionne's doctoral dissertation provided a template for much of this chapter. Jenny Chao and Matei Radulescu are thanked for sharing their experimental data. Fan Zhang and Craig Tarver provided helpful and insightful commentary.

A.1 Appendix A: Gasdynamics of Detonation Products

The existence of sonic flow at the exit plane of a detonation, which comprises the classical Chapman–Jouguet condition and its generalization to nonideal detonations, means that the detonation wave is decoupled from the expansion of the detonation products in the wake of the wave. Thus, the details of the expansion do not need to be explicitly considered in solving for the detonation wave properties or structure. The expansion of the gas that has been processed by the detonation (i.e., the burned gas) can be of interest in its own right for some applications. Consideration of how the detonation exit state is matched with the downstream expansion can also provide some guidance in determining the possible non-CJ exit states that might be realizable. The gasdynamics of the unsteady expansion of detonation products was treated by G.I. Taylor in a seminal work [93], and the solution he found is often referred to as the *Taylor wave*. In this appendix, the Taylor wave flow field solution is briefly developed using the method of characteristics for the planar case (Sect. A.1.1), matching of the Taylor wave with different branches of detonation solutions is then examined (Sect. A.1.2), and finally a more formal similarity solution for the cylindrical and spherical case (Sect. A.1.3) is developed.

A.1.1 Planar Detonation

Taylor's treatment of the dynamics of the products emerging from a planar detonation begins by assuming that the detonation wave propagates at a steady speed, with the products leaving the wave at constant conditions. It is further assumed that there are no shock waves in the flow downstream of the detonation. Thus, since all the flow originates at the same state (denoted state "2" to be consistent with the notation in this chapter) and remains isentropic, the entire flow has the same value of entropy (i.e., the flow is homentropic). This means that, at every point in the flow, $u' - c$ characteristic waves originating from the detonation but propagating in the opposite direction as the wave have the same value of the Riemann invariant:

$$C_- = u' - \frac{2}{\gamma - 1}c = u'_2 - \frac{2}{\gamma - 1}c_2, \tag{2.106}$$

where the prime ($'$) on u' is used to denote that it is the velocity in the unsteady, laboratory-fixed reference frame, in order to differentiate it from the steady flow velocity used throughout this chapter. From (2.106), it can be shown that at every point in the products, u' and c are linearly related. This result means that $u' + c$ characteristics (along which $C_+ = u' + \frac{2}{\gamma-1}c =$ constant) that propagate in the detonation products toward the wave must have constant values of u' and c and are therefore straight lines in the (x, t) plane.

One possible solution that can be constructed with straight $u' + c$ characteristics is a centered rarefaction fan originating at the origin. The detonation is also assumed to originate at the origin and has negligible thickness compared to the domain of interest. Thus, the flow pattern becomes self-similar, as seen in Fig. 2.21. At any point inside the centered rarefaction, the value

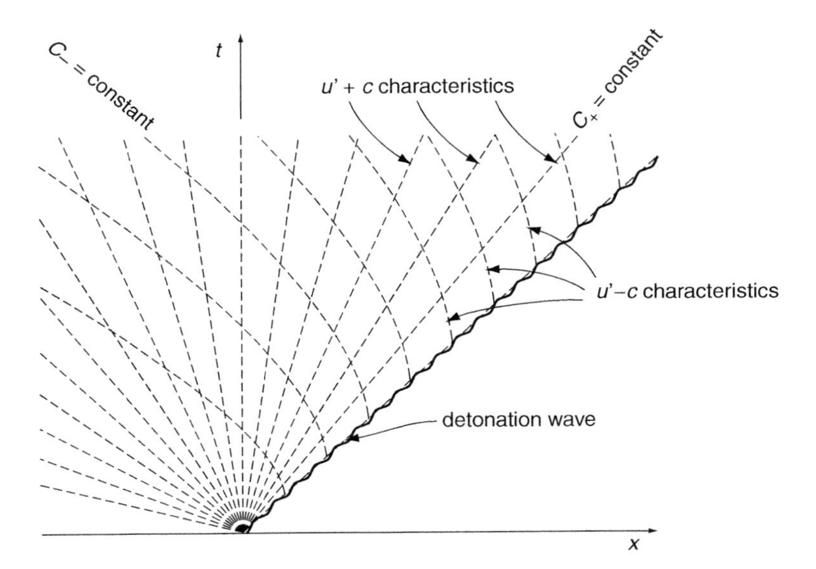

Fig. 2.21. Centered rarefaction solution for the expansion of products behind a detonation wave, showing the $u' + c$ and $u' - c$ characteristics

of $u' + c$ must equal the value of $\frac{x}{t}$ at that point. Using this property, along with the constant C_- characteristics emanating from the detonation, the flow velocity and sound speed can be solved for:

$$u'(x,t) = \frac{2}{\gamma+1}\left(\frac{x}{t} - c_2\right) + \frac{\gamma-1}{\gamma+1}u_2' \tag{2.107}$$

$$c(x,t) = \frac{\gamma-1}{\gamma+1}\left(\frac{x}{t} - u_2'\right) + \frac{2}{\gamma+1}c_2. \tag{2.108}$$

Note that position and time always appear as the combination $\frac{x}{t}$, verifying that this is a similarity solution. The other thermodynamic properties (pressure, density, temperature) can be determined from the sound speed by using the isentropic relations. The velocity profile behind the wave is plotted in Fig. 2.22. For this particular plot, the conditions at the detonation exit plane were taken as the CJ conditions in the limit of large heat release

$$\lim_{Q\to\infty}\frac{u_2'}{D} = \frac{1}{\gamma+1}, \quad \lim_{Q\to\infty}\frac{c_2}{D} = \frac{\gamma}{\gamma+1},$$
$$\lim_{Q\to\infty}\frac{\rho_2}{\rho_1} = \frac{\gamma+1}{\gamma}, \quad \lim_{Q\to\infty}\frac{p_2}{\rho_1 D^2} = \frac{1}{\gamma+1}, \tag{2.109}$$

where a value of $\gamma = 1.2$ was used for Fig. 2.22. Note that the velocity in the centered rarefaction fan goes to zero and reverses direction, reaching an escape speed of

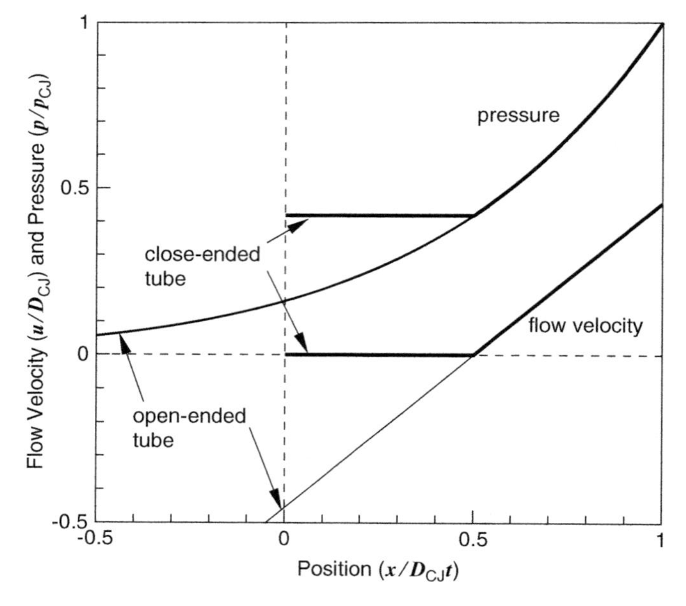

Fig. 2.22. Flow field behind a planar CJ detonation wave (Taylor wave) showing pressure and flow velocity for both close-ended and open-ended tubes ($\gamma = 1.2$)

$$\lim_{c \to 0} \frac{u}{D} = -\frac{1}{\gamma - 1} \tag{2.110}$$

as the flow expands to zero pressure and temperature, with the flow being directed in the $-x$ direction. This solution would correspond to the case where detonation is initiated in a tube with an evacuated semi-infinite region ($x < 0$). For the more familiar case of a detonation initiated at the closed end of a tube, the rarefaction fan can be truncated once the product flow has been brought to rest ($u' = 0$) and then remains uniform from this point to the endwall. The trailing edge of the rarefaction in this case propagates at $c_3 = c_2 - \frac{\gamma-1}{2} u'_2$. The fraction of the tube containing gas at rest can be shown to be

$$\frac{c_3}{D} = 1 - \frac{\gamma + 1}{2} \frac{u'_2}{D}. \tag{2.111}$$

In the limit of a high heat release detonation, this fraction is exactly $\frac{1}{2}$, meaning half the gas behind the detonation is a rest.

When applying the Taylor wave solution to gas-phase detonations propagating in a tube, the flow may be significantly modified by heat transfer and friction to the tube walls, and the solution is no longer self-similar. This problem was treated by [25, 79, 80]; for a more modern treatment, see Radulescu and Hanson [69].

A.1.2 Matching Expansion to Detonation Exit State

How the unsteady expansion downstream of the wave can be matched with the steady detonation front is now examined. In the previous section, it was assumed that the detonation was a CJ detonation, meaning that the products exit the wave at $u'_2 = D_{\mathrm{CJ}} - c_2$. In this case, the leading edge of the centered rarefaction (i.e., the Taylor wave) will have a speed $u'_2 + c_2 = D_{\mathrm{CJ}}$ and thus will remain parallel to the detonation in the (x, t) plane, as shown in Fig. 2.23a. Thus, the sonic surface permits the steady detonation to match the unsteady rarefaction in the products downstream of the detonation.

If a weak detonation with supersonic exit flow is considered ($u_2 > c_2$), then $u'_2 + c_2 < D$ and thus centered rarefactions can never reach the detonation. In this case, an ever increasing, uniform column of gas at the weak detonation exit state will form between the head of the rarefaction and the detonation front. This scenario is illustrated in Fig. 2.23b.

Finally, if a strong detonation is considered, then $u_2 < c_2$ (subsonic exit flow), so that $u'_2 + c_2 > D$, and any $u' + c$ wave will overtake the detonation. In this case, it is not possible to construct a similarity solution with a centered rarefaction. Other solutions can be found to exist for a strong detonation, with a piston-supported detonation being one example, where straight $u + c$ characteristics emanating from the piston face continuously merge with the detonation as shown in Fig. 2.23c. If the piston were suddenly stopped, then the rarefactions generated would begin to interact with the detonation, resulting in its deceleration until a limiting characteristic became parallel with the

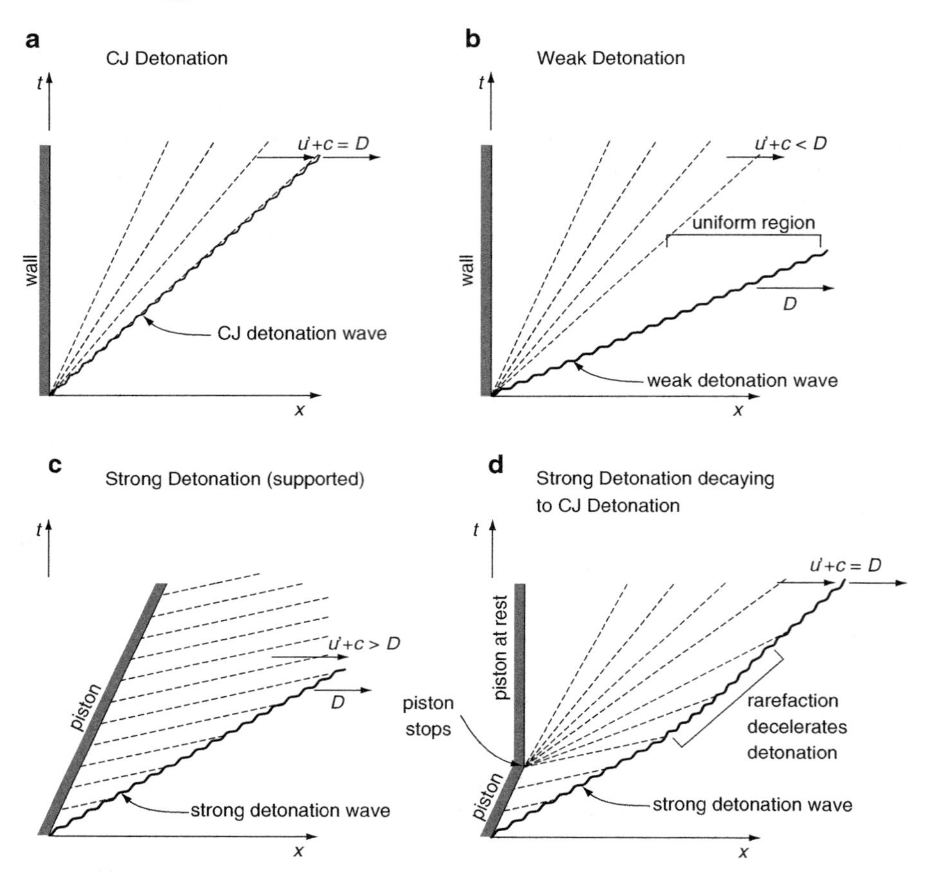

Fig. 2.23. Expansion of combustion products behind a detonation wave for the cases of (**a**) CJ detonation, (**b**) weak detonation, (**c**) strong (piston-supported) detonation, and (**d**) a piston-supported detonation in which the piston stops and the rarefactions generated overtake detonation, decelerating it until the detonation is parallel with $u' + c$ characteristics (CJ detonation)

detonation, establishing a CJ detonation. This simple picture, illustrated in Fig. 2.23d, perhaps provides a more satisfying explanation for the legitimacy of the CJ condition of sonic flow at the exit of an unsupported detonation wave.

A.1.3 Cylindrical and Spherical Detonations

To construct a solution for the dynamics of the products behind a spherical detonation originating from a point (or a cylindrical detonation initiated along a line), the conservation of mass and momentum for spherical and cylindrical symmetry are used as follows:

$$\frac{\partial \rho}{\partial t} + \rho \frac{\partial u}{\partial r} + u \frac{\partial \rho}{\partial r} + \alpha \frac{\rho u}{r} = 0 \tag{2.112}$$

$$\frac{\partial u}{\partial t} + u \frac{\partial u}{\partial r} + \frac{1}{\rho} \frac{\partial p}{\partial r} = 0, \tag{2.113}$$

where $\alpha = 1, 2$ for cylindrical and spherical geometries, respectively (note that for $\alpha = 0$ the equations revert to planar), and u is the velocity in the laboratory-fixed frame. Following the notation of [89], the following nondimensional variables are introduced:

$$\xi = \frac{r}{Dt}, \quad \psi = \frac{\rho}{\rho_2}, \quad \phi = \frac{u}{D}, \quad f = \frac{p}{\rho_2 D^2}. \tag{2.114}$$

Taking ξ as the independent variable of a similarity solution, the governing PDE's become ODE's and the differentials can be solved for

$$\begin{aligned} \phi'(\xi) &= \frac{f'}{\psi(\xi - \phi)} \\ \psi'(\xi) &= \frac{\alpha \psi \phi (\xi - \phi) + \xi f'}{\xi (\xi - \phi)^2}. \end{aligned} \tag{2.115}$$

In order to evaluate f', an equation of state of the form $f(\psi)$ is required. Here, the ideal gas equation for an isentropic process will be used ($\frac{p}{\rho^\gamma} =$ constant); implementation of equations of state more relevant to condensed-phase explosives is considered by [89]. For an ideal gas,

$$f'(\psi) = \left(\frac{\gamma}{\gamma + 1} \right)^2 \psi^{\gamma - 1} \psi'. \tag{2.116}$$

When this expression is used in (2.115) along with (2.109), the differential equations become singular at the exit plane of the CJ detonation (i.e., the denominators of both equations go to zero as $\xi \to 1$). Thus, while flow properties are constant at the exit plane of the detonation, their spatial derivatives would be infinite in the case of a CJ detonation. In order to circumvent this issue, Taylor [93] constructed a local, approximate solution in the vicinity of the detonation front, and then began a numerical integration an arbitrarily small distance from the front (e.g., at $\xi = 0.999$) and integrated inward. Alternatively, it is possible to just arbitrarily assume values slightly perturbed from their CJ values in order to initiate the numerical integration and still obtain reasonable results.

Such a numerical integration of (2.115) was performed for a spherical wave ($\alpha = 2$) and the results shown in Fig. 2.24, where the velocity ratio $\frac{u}{D} = \phi$ and pressure ratio ($\frac{p}{p_{CJ}} = \psi^\gamma$) are plotted for a value of γ representative of a gas-phase detonation ($\gamma = 1.2$). Also shown is the structure of the Taylor wave for a value of $\gamma = 3$. This value of γ is often used as a crude approximation for the products of condensed-phase explosives, and fitting of γ to more sophisticated equations of state usually results in values in this vicinity [89]. From Fig. 2.24,

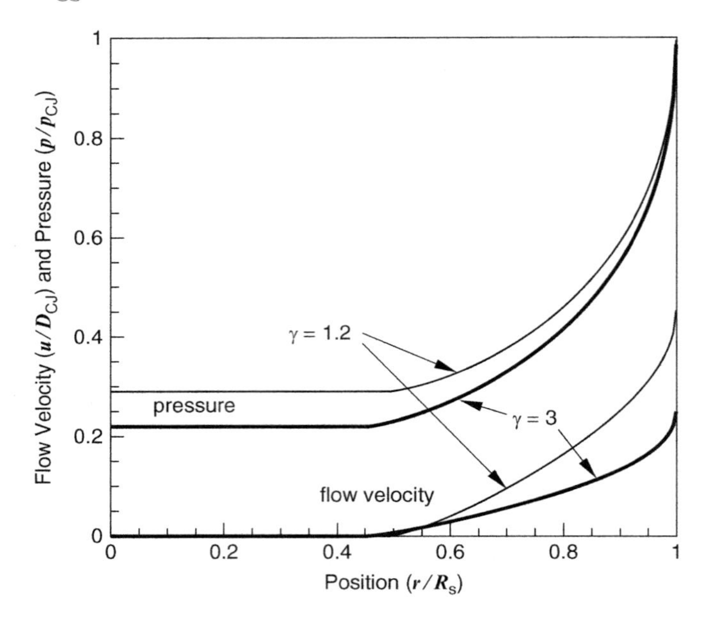

Fig. 2.24. Flow field behind a spherically expanding CJ detonation wave (Taylor wave) showing pressure and flow velocity for the cases of a gas-phase detonation ($\gamma = 1.2$) and condensed-phase detonation ($\gamma = 3$)

it can be seen that the flow reaches zero velocity at approximately halfway between the detonation front and the origin. The solution is truncated at that point, and constant conditions are assumed from there to the center in order to avoid the solution for density and flow velocity asymptoting to nonphysical, negative infinity as the solution approaches the center (note that constant values of density and velocity are also valid solutions of (2.115), so the solution can be truncated at any point and replaced with constant values). The pressure at the center of the spherical wave is approximately one-third to one-fifth of the CJ detonation pressure.

Note that if $\alpha = 0$ is used in (2.115) for the planar geometry, the numerators are zero and either trivial behavior is obtained (i.e., the differentials of flow velocity and density are zero and thus the flow is uniform behind the front) or the denominators must also be zero. Imposing the latter condition can be shown to be equivalent to the planar solution constructed via the Method of Characteristics in Sect. A.1.1.

The fact that the solutions for detonation products in the cylindrical and spherical cases become singular at the detonation front when the CJ conditions are imposed was a source of concern for some early researchers, leading some of them to even suggest that spherically expanding detonations might not exist [15, 42]. Spherical detonations were experimentally obtained as early as 1925 [49], but the issue was still debated into the 1960s. Lee et al. pointed out that initiation of a spherical detonation will always likely require some degree of initial overdrive that will result in the detonation being initially on the strong branch of the solution ($u' + c > D$) and will only asymptotically approach an ideal CJ detonation as the wave expands, so that the issue of the infinite gradients in the solution is avoided [50].

A.2 Appendix B: Critical Sonic Point with Friction

For a detonation with friction (and no heat transfer), the critical condition at which the wave propagation velocity equals the velocity of the combustion products can be solved for analytically as follows [7, 20]. The momentum and energy equations (2.27) and (2.28) can be equated (excluding the heat loss term) via the friction factor

$$\mathrm{d}\left(p + \rho u^2\right) = \mathrm{d}\left[\rho \frac{u}{u_1}\left(h + \frac{1}{2}u^2\right)\right] = f \mathrm{d}x \tag{2.117}$$

and integrated to yield

$$p + \rho\left(u^2 - \frac{u}{u_1}\left(h + \frac{1}{2}u^2\right)\right) = C_1, \tag{2.118}$$

where C_1 is a constant of integration that can be evaluated using conditions upstream of the detonation wave (u_1, p_1, ρ_1, and $h_1 = c_p T_1$). At the critical sonic condition, the flow exiting the wave is sonic and equal to the velocity of the flow approaching the wave (in the laboratory-fixed frame, this is equivalent to the flow that has been processed by the wave being at rest),

$$u_{2_{\mathrm{crs}}} = c_{2_{\mathrm{crs}}} = D_{\mathrm{crs}}, \tag{2.119}$$

where the "2" subscript designates the exit state. The condition that exit velocity is the same as the inflow velocity means that the density is constant across the wave ($\rho_1 = \rho_2$). These relations allow the critical Mach number of propagation to be solved for

$$M_{\mathrm{crs}} = \frac{D_{1_{\mathrm{crs}}}}{c_1} = \sqrt{\gamma Q + 1}. \tag{2.120}$$

Since the density is constant across the wave, this critical sonic condition corresponds to a front of constant volume explosion propagating through the medium with a pressure increase given by

$$\frac{p_{2_{\mathrm{crs}}}}{p_1} = \frac{p_{\mathrm{CV}}}{p_1} = \gamma Q + 1. \tag{2.121}$$

At this critical sonic condition, the flow velocity being zero relative to the wall at the exit of the wave means that friction no longer influences the flow at that point, and thus it is no longer possible to find an eigenvalue solution that passes through a sonic point. At lower wave velocities, the flow will be entirely subsonic from the shock to the rear boundary condition (i.e., no sonic condition can be found). To continue the solution curve in Fig. 2.7 to lower velocities, a different criterion is adopted, namely that the flow velocity must be zero in the laboratory-fixed frame, satisfying a closed end-wall boundary condition. This solution may have some relevance to the so-called "choking regime" of combustion wave propagation that is experimentally observed in combustible mixtures in obstacle laden tubes, as discussed in Sect. 2.4.5.

References

1. Akyurtlu, A.: An investigation of the structure and detonability limits of hydrogen-chlorine detonations, PhD Thesis, University of Wisconsin-Madison (1975)
2. Anderson, J.B., Long, L.N.: Direct Monte Carlo simulation of chemical reaction systems: Prediction of ultrafast detonations. J. Chem. Phys. **118**(7), 3102–3110 (2003)
3. Anderson, J.B., Long, L.N.: Direct simulation of pathological detonations. In: Ketsdever, A.D., Muntz, E.P. (eds.) Rarefied Gas Dynamics. American Institute of Physics Conference Series, vol. 663, pp. 186–193 (2003)
4. Anderson J. Jr.: Hypersonic and High-Temperature Gas Dynamics, AIAA Education Series, 2nd edn., 813pp. AIAA, Restons (2006)
5. Bauer, P.A., Dabora, E.K., Manson, N.: Chronology of early research on detonation wave. In: Kuhl, A.L., Leyer, J.C., Borisov, A.A., Sirignano, W.A. (eds.) Dynamics of Detonations and Explosions: Detonations. Progress in Astronautics and Aeronautics, vol. 133, pp. 3–18. AIAA, VA (1991)
6. Brailovsky, I., Sivashinsky G.: On deflagration-to-detonation transition. Combust. Sci. Technol. **130**(1), 201–231 (1997)
7. Brailovsky, I., Sivashinsky, G.: Hydraulic resistance and multiplicity of detonation regimes. Combust. Flame **122**(1–2), 130–138 (2000)
8. Brailovsky, I., Sivashinsky, G.I.: Hydraulic resistance as a mechanism for deflagration-to-detonation transition. Combust. Flame **122**(4), 492–499 (2000)
9. Brailovsky, I., Sivashinsky, G.: Effects of momentum and heat losses on the multiplicity of detonation regimes. Combust. Flame **128**(1–2), 191–196 (2002)
10. Brinkley S.R. Jr., Richardson J.M.: On the structure of plane detonation waves with finite reaction velocity. Symp. Int. Combust. **4**(1), 450–457 (1953)
11. Bykov, V., Goldfarb, I., Gol'dshtein, V., Kagan, L., Sivashinsky, G.: Effects of hydraulic resistance and heat losses on detonability and flammability limits. Combust. Theory Model. **8**(2), 413–424 (2004)
12. Chao, J., Lee, J.H.S.: The propagation mechanism of high speed turbulent deflagrations. Shock Waves **12**, 277–289 (2003)
13. Chase, M.W., Curnutt, J.L., Downey, J.R., Jr., McDonald, R.A., Syverud, A.N., Valenzuela, E.A.: Janaf thermochemical tables, 1982 supplement. J. Phys. Chem. Ref. Data **11**(3), 695–940 (1982)
14. Ciccarelli, G., Dorofeev, S.: Flame acceleration and transition to detonation in ducts. Prog. Energy Combust. Sci. **34**(4), 499–550 (2008)

15. Courant, R., Friedrichs, K.O.: Supersonic Flows and Shock Waves. Interscience, New York (1948)
16. Cowie, L.L., Rybicki, G.B.: The structure and evolution of galacto-detonation waves: some analytic results in sequential star formation models of spiral galaxies. ApJ **260**(2), 504–511 (1982)
17. Curran, E.T., Heiser, W.H., Pratt, D.T.: Fluid phenomena in scramjet combustion systems. Annu. Rev. Fluid Mech. **28**(1), 323–360 (1996)
18. Dabora E.K., Nicholls J.A., Morrison R.B.: The influence of a compressible boundary on the propagation of gaseous detonations. Symp Int Combust **10**(1), 817–830 (1965)
19. Davis, W.C., Fickett, W.: Detonation. University of California Press, Berkeley (1979)
20. Dionne, J.-P.: Numerical study of the propagation of non-ideal detonations, PhD Thesis, McGill University, Montreal (2000)
21. Dionne, J.P., Duquette, R., Yoshinaka, A., Lee, J.H.S.: Pathological Detonations in H_2-Cl_2. Combust. Sci. Technol. **158**(1), 5–14 (2000)
22. Dionne, J.-P., Ng, H.D., Lee, J.H.S.: Transient development of friction-induced low-velocity detonations. Proc. Combust. Inst. **28**(1), 645–651 (2000)
23. Döring, W.: Über den detonationsvorgang in gasen. Ann. Phys. **435**(6–7), 421–436 (1943)
24. Duff, R.E.: Calculation of reaction profiles behind steady-state shock waves. I. Application to detonation waves. J. Chem. Phys. **28**(6), 1193–1197 (1958)
25. Edwards, D.H., Brown, D.R., Hooper, G., Jones, A.T.: The influence of wall heat transfer on the expansion following a C-J detonation wave. J. Phys. D: Appl. Phys. **3**(3), 365–376 (1970)
26. Erpenbeck, J.J.: Steady quasi-one-dimensional detonations in idealized systems. Phys. Fluids **12**(5), 967–982 (1969)
27. Eyring, H., Powell, R.E., Duffy, G.H., Parlin, R.B.: The stability of detonation. Chem. Rev. **45**(1), 69–181 (1949)
28. Fay, J.A.: Two-dimensional gaseous detonations: Velocity deficit. Phys. Fluids **2**(3), 283–289 (1959)
29. Freedman, W.L., Madore, B.F., Mehta, S.: Galactic detonation waves numerical models illustrating the transition from deterministic to stochastic. ApJ **282**(2), 412–426 (1984)
30. Frost, D., Zhang, F.: Slurry detonation. In: Zhang, F. (ed.) Heterogeneous Detonation, Shock Wave Science and Technology Reference Library, vol. 4, pp. 169–216. Springer, Berlin (2009)
31. Fujiwara, T., Tsuge, S: Quasi-onedimensional analysis of gaseous free detonations. J. Phys. Soc. Jpn. **33**(1), 237–241 (1972)
32. Gelfand, B.E., Frolov, S.M., Nettleton, M.A.: Gaseous detonations – a selective review. Prog. Energy Combust. Sci. **17**(4), 327–371 (1991)
33. Gerola, H., Seiden, P.E.: Stochastic star formation and spiral structure of galaxies. ApJ **223**, 129–135 (1978)
34. Gordon, S., McBride, B.J.: Computer program for computation of complex chemical equilibrium compositions, rocket performance, incident and reflected shocks, and Chapman-Jouguet detonations. Technical Report NASA SP-273, NASA (1971)
35. Gordon, P.V., Sivashinsky, G.I.: Pressure diffusivity and low-velocity detonation. Combust. Flame **136**(4), 440–444 (2004)

36. Gross, R., Oppenheim, A.K.: Recent advances in gaseous detonation. ARS J. **29**, 173–179 (1959)
37. Guenoche, H., Le Diuzet, P., Sedes, C.: Influence of the heat-release function on the detonation states. In: Bowen, J.R., Manson, N.N., Oppenheim, A.K. (eds.) Gasdynamics of Detonations and Explosions. Progress in Astronautics and Aeronautics, vol. 75, pp. 387–407. AIAA, VA (1981)
38. He, L., Clavin, P.: On the direct initiation of gaseous detonations by an energy source. J. Fluid Mech. **277**, 227–248 (1994)
39. Hertzberg, A., Bruckner, A.P., Knowlen, C.: Experimental investigation of ram accelerator propulsion modes. Shock Waves **1**, 17–25 (1991)
40. Heuze, O.: 1899–1909: The Key Years of the Understanding of Shock Wave and Detonation Physics. AIP Conf. Proc. **1195**(1), 311–314 (2009)
41. Jones, H.: A theory of the dependence of the rate of detonation of solid explosives on the diameter of the charge. Proc. R. Soc. Lond. Ser A. Math. Phys. Sci. **189**(1018), 415–426 (1947)
42. Jouguet, E.: Mécanique des explosifs, étude de dynamique chimique. O. Doin & Fils, Paris (1917)
43. Kagan, L., Sivashinsky, G.: The transition from deflagration to detonation in thin channels. Combust. Flame **134**(4), 389–397 (2003)
44. Kirkwood, J.G., Wood, W.W.: Structure of a steady-state plane detonation wave with finite reaction rate. J. Chem. Phys. **22**(11), 1915–1919 (1954)
45. Klein, R., Krok, J.C., Shepherd, J.E.: Curved quasi-steady detonations: Asymptotic analysis and detailed chemical kinetics. Technical Report FM95-04, Graduate Aeronautical Laboratories, California Institute of Technology (1995)
46. Knystautas, R., Lee, J.H.: Detonation parameters for the hydrogen-chlorine system. In: Kuhl, A.L., Bowen, J.R., Leyer, J.C., Borisov, A.A. (eds.) Dynamics of Explosions. Progress in Astronautics and Aeronautics, vol. 114, pp. 32–44. AIAA, VA (1988)
47. Kolesnikov, S., Utkin, A.: Nonclassical steady-state detonation regimes in pressed TNETB. Combust. Explos. Shock Waves **43**, 710–716 (2007)
48. Krehl, P.O.K.: History of Shock Waves, Explosions and Impact: A Chronological and Biographical Reference. Springer, Berlin (2009)
49. Laffitte, P.: Recherches expérimentales sur l'onde explosive et l'onde de choc. Ann. Phys. **10**(4) 623–634 (1925)
50. Lee, J.H., Lee, B.H.K., Shanfield, I.: Two-dimensional unconfined gaseous detonation waves. 10th Symp. (Int.) Combust. **10**(1), 805–815 (1965)
51. Lee J.H.S.: The Detonation Phenomenon. Cambridge University Press, Cambridge (2008)
52. Lee, J.H.S., Moen, I.O.: The mechanism of transition from deflagration to detonation in vapor cloud explosions. Prog. Energy Combust. Sci. **6**(4), 359–389 (1980)
53. Makris, A.: The Propagation of Gaseous Detonations in Porous Media. PhD Thesis, McGill University, Montreal (2003)
54. Makris, A., Shafique, H., Lee, J., Knystautas, R.: Influence of mixture sensitivity and pore size on detonation velocities in porous media. Shock Waves **5**(1), 89–95 (1995)

55. Manson, N., Dabora, E.: Chronology of research on detonation waves: 1920–1950. In: Kuhl, A.L., Leyer, J.C., Borisov, A.A., Sirignano, W.A. (eds.) Dynamic Aspects of Detonations. Progress in Astronautics and Aeronautics, vol. 153, pp. 3–39. AIAA, VA (1993)
56. McBride, B.J., Gordon, S.: Computer program for calculation of complex chemical equilibrium compositions and applications II. Users manual and program description. Technical Report NASA RP-1311, NASA (1996)
57. Mikhel'son, V.A.: On the normal ignition velocity of explosive gaseous mixtures. Scientific Papers of the Moscow Imperial University on Mathematics and Physics **10**, 1–93 (1893)
58. von Neumann, J.: Theory of detonation waves. Technical Report OSRD-549, National Defense Research Committee (1942)
59. von Neumann, J.: Theory of Detonation Waves, John von Neumann: Collected Works, 1903–1957, vol. 6, pp. 178–218. Pergamon Press, Oxford (1963)
60. Nikolaev, Yu.A., Fomin, P.A.: Analysis of equilibrium flows of chemically reacting gases. Combust. Explos. Shock Waves **18**(1), 53–58 (1982)
61. Oran, E.S.: Astrophysical combustion. Proc. Combust. Inst. **30**(2), 1823–1840 (2005)
62. Oran, E.S., Boris, J.P.: Numerical Simulation of Reactive Flow. Cambridge University Press, Cambridge (2001)
63. Paillard, C., Dupre, G., Lisbet, R., Combourieu, J., Fokeev, V.P., Gvozdeva, L.G.: A study of hydrogen azide detonation with heat transfer at the wall. Acta Astronaut. **6**(3–4), 227–242 (1979)
64. Parker, E.N.: Dynamics of the interplanetary gas and magnetic fields. ApJ **128**, 664–676 (1958)
65. Peraldi, O., Knystautas, R., Lee, J.H.: Criteria for transition to detonation in tubes. 21st Symp. (Int.) Combust. **21**(1), 1629–1637 (1988)
66. Pinaev, A.V., Lyamin, G.A.: Fundamental laws governing subsonic and detonating gas combustion in inert porous media. Combust. Explos. Shock Waves **25**(4), 448–458 (1989)
67. Powers, J., Paolucci, S.: Accurate spatial resolution estimates for reactive supersonic flow with detailed chemistry. AIAA J. **43**, 1088–1099 (2005)
68. Radulescu, M.I.: The propagation and failure mechanism of gaseous detonations: Experiments in porous-walled tubes. PhD Thesis, McGill University, Montreal (2003)
69. Radulescu, M.I., Hanson, R.K.: Effect of heat loss on pulse-detonation-engine flow fields and performance. J. Propul. Power **21**(2), 274–285 (2005)
70. Radulescu, M.I., Lee, J.H.S.: The failure mechanism of gaseous detonations: Experiments in porous wall tubes. Combust. Flame **131**(1–2), 29–46 (2002)
71. Radulescu, M.I., Sharpe, G.J., Law, C.K., Lee, J.H.S.: The hydrodynamic structure of unstable cellular detonations. J. Fluid Mech. **580**, 31–81 (2007)
72. Rozing, V.O., Khariton, Yu.B.: The detonation cutoff of explosive substances when the charge diameters are small. Dokl. Akad. Nauk. SSSR **26**(4), 360–361 (1940)
73. Sasoh, A., Knowlen, C.: Ram accelerator operation analysis in thermally choked and transdetonative propulsive modes. Trans. Jpn. Soc. Aeronaut. Space Sci. **40**(128), 130–148 (1997)
74. Shapiro, A.H.: The Dynamics and Thermodynamics of Compressible Fluid Flow, vol. 1. Ronald Press, New York (1953)

75. Sharpe, G.J.: Linear stability of pathological detonations. J. Fluid Mech. **401**, 311–338 (1999)
76. Sharpe, G.J.: The structure of planar and curved detonation waves with reversible reactions. Phys. Fluids **12**(11), 3007–3020 (2000)
77. Sharpe, G.J., Falle, S.A.E.G.: One-dimensional nonlinear stability of pathological detonations. J. Fluid Mech. **414**, 339–366 (1999)
78. Shchelkin, K.I.: Influence of the tube walls roughness on the onset and propagation of detonation in gases. Zh. Eksp. Teor. Fiz. **10**, 823–827 (1940)
79. Sichel, M., David, T.S.: Transfer behind detonations in H_2-O_2 mixtures. AIAA J. **4**(6), 1089–1090 (1966)
80. Skinner, J.H., Jr.: Friction and heat-transfer effects on the nonsteady flow behind a detonation. AIAA J. **5**(11), 2069–2071 (1967)
81. Sleath, J.P., Alexander, P.: A new model of the structure of spiral galaxies based on propagating star formation – I. The galactic star formation rate and Schmidt Law. Mon. Not. R. Astron. Soc. **275**, 507–514 (1995)
82. Sokolik, A.S., Shchelkin, K.I.: Detonation in gas mixtures: The influence of pressure on the velocity of a detonation wave. Acta Physicokhimika SSSR **1**, 311–317 (1934)
83. Sommers, W.P.: Gaseous detonation wave interactions with nonrigid boundaries. ARS J. **31**, 1780–1782 (1961)
84. Sommers, W.P., Morrison, R.B.: Simulation of condensed-explosive detonation phenomena with gases. Phys. Fluids **5**(2), 241–248 (1962)
85. Spalding, D.B.: A theory of inflammability limits and flame-quenching. Proc. R. Soc. Lond. Ser A. Math. Phys. Sci. **240**(1220), 83–100 (1957)
86. De Sterck, H.: Critical point analysis of transonic flow profiles with heat conduction. SIAM J. Appl. Dyn. Syst. **6**, 645–662 (2007)
87. Strehlow, R.A.: Fundamentals of Combustion. International Textbook Co., Scranton, PA (1968)
88. Tanguay, V., Higgins, A.J.: On the inclusion of frictional work in nonideal detonations. In: Shepherd, J.E. (ed.) 20th International Colloquium on the Dynamics of Explosions and Reactive Systems: Extended Abstracts and Technical Program (2005). ISBN 1600490018, 9781600490019 (CD-ROM)
89. Tanguay, V., Higgins, A.J., Zhang, F.: A simple analytical model for reactive particle ignition in explosives. Propellants Explos. Pyrotech. **32**(5), 371–384 (2007)
90. Tarver, C.M.: Multiple roles of highly vibrationally excited molecules in the reaction zones of detonation waves. J. Phys. Chem. A **101**(27), 4845–4851 (1997)
91. Tarver C.M.: On the existence of pathological detonation waves. AIP Conf. Proc. **706**, 902 (2004). doi: 10.1063/1.1780383
92. Tarver, C., Forbes, J., Urtiew, P.: Nonequilibrium Zeldovich-von Neumann-Doring theory and reactive flow modeling of detonation. Russ. J. Phys. Chem. B Focus Phys. **1**(1), 39–45 (2007)
93. Taylor, G.I.: The dynamics of the combustion products behind plane and spherical detonation fronts in explosives. Proc. R. Soc. Lond. Ser A. Math. Phys. Sci. **200**(1061), 235–247 (1950)
94. Trubachev, A.: Detonation waves in interstellar gas. Combust. Explos. Shock Waves **33**, 72–76 (1997)

95. Tsuge, S., Furukawa, H., Matsukawa, M., Nakagawa, T.: On the dual properties and the limit of hydrogen-oxygen free detonations waves. Astronaut. Acta **15**, 377–386 (1970)
96. Utkin, A.V., Kolesnikov, S.A., Pershin, S.V.: Effect of the initial density on the structure of detonation waves in heterogeneous explosives. Combust. Explos. Shock Waves **38**(5), 590–597 (2002)
97. Vasil'ev, A.A., Zak, D.V.: Detonation of gas jets. Combust. Explos. Shock Waves **22**(4), 463–468 (1986)
98. Vieille, P.: Rôle des discontinuités dans la propagation des phénomènes explosifs. C. R. Acad. Sci., **130**, 413–416 (1900)
99. Vincenti, W.G., Kruger, C.H.: Introduction to Physical Gas Dynamics. Krieger, Malabar, FL (1965)
100. Wecken, F.: Non-ideal detonation with constant lateral expansion. In: Fourth Symposium (International) on Detonation, pp. 107–116 (1965)
101. Wood, W.W., Kirkwood, J.G.: Diameter effect in condensed explosives. The relation between velocity and radius of curvature of the detonation wave. J. Chem. Phys. **22**(11), 1920–1924 (1954)
102. Wood, W.W., Kirkwood, J.G.: On the existence of steady-state detonations supported by a single chemical reaction. J. Chem. Phys. **25**(6), 1276–1277 (1956)
103. Wood, W.W., Parker, F.R.: Structure of a centered rarefaction wave in a relaxing gas. Phys. Fluids **1**(3), 230–241 (1958)
104. Wood, W.W., Salsburg, Z.W.: Analysis of steady-state supported one-dimensional detonations and shocks. Phys. Fluids **3**(4), 549–566 (1960)
105. Yao, J., Stewart, D.S.: On the normal detonation shock velocity-curvature relationship for materials with large activation energy. Combust. Flame **100**(4), 519–528 (1995)
106. Zel'dovich, Ya.B.: On the theory of the propagation of detonations on gaseous system. Zh. Eksp. Teor. Fiz. **10**(5), 542–568 (1940)
107. Zel'dovich, Ya.B., Kompaneets, A.S.: Theory of Detonation. Academic, New York (1960)
108. Zel'dovich, Ya.B., Ratner, S.B.: Zh. Eksp. Teor. Fiz. **11**, 170 (1941)
109. Zel'dovich, Ya.B., Gel'fand, B.E., Kazhdan, Ya.M., Frolov, S.M.: Detonation propagation in a rough tube taking account of deceleration and heat transfer. Combust. Explos. Shock Waves **23**, 342–349 (1987)
110. Zel'dovich, B., Borisov, A.A., Gelfand, B.E., Frolov Ya, S.M., Mailkov, A.E.: Nonideal detonation waves in rough tubes. In: Kuhl, A.L., Bowen, J.R., Leyer, J.C., Borisov, A.A. (eds.) Dynamics of Explosions. Progress in Astronautics and Aeronautics, vol. 114, pp. 211–231. AIAA, VA (1988)
111. Zhang, F.: Detonation in reactive solid particle-gas flow. J. Propuls. Power **22**, 1289–1309 (2006)
112. Zhang, F.: Detonation of gas-particle flow. In: Zhang, F. (ed.) Heterogeneous Detonation, Shock Wave Science and Technology Reference Library, vol. 4, pp. 87–168. Springer, Berlin (2009)

3

Detonation Instability

Hoi Dick Ng and Fan Zhang

3.1 Introduction

Detonations are a complex spatial-temporal unstable phenomenon. In theory, the complete dynamics of this phenomenon requires a time-dependent multi-dimensional solution that describes the inherent instability of the detonation wave. The effect of instability is a fundamental factor on the dynamic behavior of detonation waves and thus governs practical properties of an explosive mixture such as detonation limits, critical tube diameter, or initiation energy. Since the pioneering work by Erpenbeck [42–44] and Fickett and Wood [51] in the 1960s, significant progress has been made over the past few decades in detonation stability theory through either analytical or numerical approaches. Excellent reviews of many important contributions can be found in the review papers by Short [152], Stewart and Kasimov [164], and Powers [125]. While a substantial body of literature exists, detonation stability analysis follows generally three basic approaches: linear stability analysis, asymptotic modeling, and direct numerical simulation. Each approach has its own advantages and disadvantages. Although linear stability analysis can provide information on the unstable modes and neutral stability boundaries in different parameter spaces, the linear theory deals only with weak disturbances and therefore cannot be used to fully describe the inherent nonlinear nature of real detonation waves. By invoking different limits, asymptotic analysis can provide a better understanding of the mechanism leading to nonlinear instability. However, the multiscale nature of the detonation restricts such analysis to only simple models with simplified chemical kinetic rate laws. With a significant advance in scientific computing, the solution of the reactive Euler equations or Navier–Stokes equations can be approximated with high fidelity by numerical simulations. Although today's numerical simulations can provide the full nonlinear instability of one- or two-dimensional detonation structures, interpretation or analysis of the numerical results requires better approaches. The numerical resolution issues restrict the simulation to be performed only in small domains and a high-resolution three-dimensional simulation with detailed chemistry

F. Zhang (ed.), *Shock Wave Science and Technology Reference Library, Vol. 6*: Detonation Dynamics, DOI 10.1007/978-3-642-22967-1_3,

still remains a challenge to perform. In detonation stability theory, there no doubt remain many unanswered questions, in particular to clarify further the origin of these instabilities and their role in order to understand the dynamic behavior of unsteady multidimensional cellular detonations. The objective of this chapter is to focus on some basic mathematical formulation in detonation stability theory and to provide basic knowledge of the physical phenomena through modeling and numerical simulations, which might help one to understand the fundamental concepts and to highlight some future challenges.

In Chap. 2, the steady detonation structure was introduced. Although the ZND theory provides the basic steady structure of a detonation wave, in general the laminar steady structure is seldom observed experimentally. Almost all detonation fronts in explosive mixtures are unstable and possess cellular, three-dimensional transient structures, particularly in low density molecular explosives such as gases, with intense velocity and pressure fluctuations in the reaction zone as illustrated in the experimental picture shown in Fig. 3.1. Since typical chemical reaction rates for common combustible mixtures have an extreme sensitivity to the temperature (i.e., rate law in an Arrhenius form), small variations in the shock strength produce large variations in reaction rates in the flow directly behind the shock. The changes in reaction rates due to the initial disturbances then result in variations in the shock strength. This creates an unstable feedback loop that results in the spontaneous and nonlinear instability of propagating detonations. Various instabilities in the flow field associated with the chemical heat release generate disturbances that act on the detonation front and cause the propagation to be unsteady and multidimensional. Their mutual interactions form the classical triple shock Mach interaction configuration [10, 27]. The trajectories of these triple points can be recorded on a smooth surface coated with a carbon soot deposit and have a characteristic fish scale or cell pattern as the detonation propagates past it. Hence, the one-dimensional ZND detonation is nearly always hydrodynamically unstable, and the multidimensional cellular structure is a manifestation

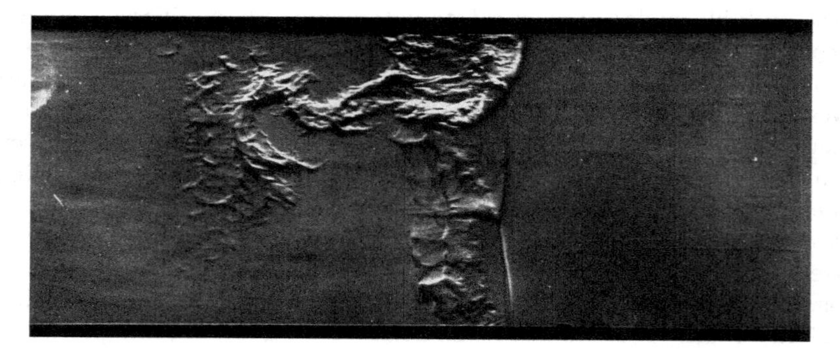

Fig. 3.1. Flow structure in a multidimensional cellular gaseous detonation, with transverse waves, turbulence, and unburned gas pockets [133]

of the instability of the detonation caused by the time-dependent coupling between the chemical reactions and the fluid dynamic flow field.

3.1.1 Linear Stability Analysis

The classical linear hydrodynamic stability analysis of the steady solution is the first step in investigating the stability of detonation waves. The basic approach is to impose small (in general, multidimensional) perturbations to the solution and study if the amplitude of such perturbations grows or decays with time. The assumption of small perturbations permits the equations to be linearized and then integrated. The results of the linear stability analysis define the region of instability in the parameter spaces and give a good estimate of the growth rate of perturbations on the steady solution and an estimate of the temporal and spatial period of the solution being sought. An exact linear stability analysis was pioneered by Erpenbeck [42, 43]. In his Laplace transform formulation, stability or instability was determined in a point-wise fashion for a given set of parameter values and neutral stability boundaries were interpolated. Lee and Stewart [81] adopted a formal normal mode analysis for the one-dimensional linear detonation stability problem, using a shooting method to numerically determine unstable mode growth rates, frequencies, and perturbation eigenfunction structures. Short and Stewart [157] subsequently extended this normal-mode approach to include two-dimensional transverse linear disturbances for cellular detonation instability, while Sharpe's modification emphasized Chapman–Jouguet detonations by determining the asymptotic solution near the CJ point [139]. Beside the classical works listed above, linear stability analysis has also been carried out with curvature [186], in circular tubes [76], pathological detonations [140], and with the effect of equation of state (EOS) and chemistry [17,56,88,89,154,160,161]. Although these theories provide exact stability spectra to the linear detonation stability problem, as with most hydrodynamic stability analyses, the dispersion relationship is rather involved and difficult to solve analytically. Solutions to the system of stability equations cannot be found without resorting to numerical computations, and thus they do not reveal the underlying physics determining the observed stability spectrum, i.e., the mechanism of linear instability. In addition, the linear theory deals only with weak amplitude disturbances, in contrast to what is seen in experiments. The role of linear stability at best would be to provide information from stability theory in order to validate codes before their use to compute the full, complex two- or three-dimensional nonlinear structure.

3.1.2 Asymptotic Theory and Modeling

The goal of asymptotic analysis is to elucidate the essential physical mechanism of detonation instability or to provide some basic instability understanding through the derivation of reduced evolution equations. The work by Zaidel [193] on the square-wave model approximation to the detonation structure and later the work by Abouseif and Toong [2] can be considered as the early asymptotic treatment of the detonation stability problem. Using a model with a large-activation-energy induction zone coupled with a finite rate heat release zone, the ad hoc treatment considered by Abouseif and Toong [2] allows an approximate linear stability analysis, and the model also describes a basic wave-interaction mechanism behind a pulsating instability [50].

Rational asymptotic investigations of the linear detonation stability have been carried out by several researchers. In general, the basic idea behind formal asymptotic modeling is to either invoke a limiting decomposition of the underlying steady wave or to make assumptions about the characteristic frequency and wavelength of the instability [152]. The asymptotic strategy was generally formulated using the following limits or a combination thereof: the limit of large activation energy in the steady detonation wave structure [18, 19], Newtonian limit where the ratio of specific heats is close to unity [150], limit of large overdrive, and/or the limit of weak effective heat release [22, 24, 62, 153, 158]. Asymptotic approximation using these limits allows an analytical dispersion relation to be obtained which governs the linear stability spectrum of a detonation to small-amplitude perturbations.

Nonlinear asymptotic theories have been developed to formulate reduced models and nonlinear evolution equations for propagating detonations. Madja [94] derived a simplified model for detonation waves, which is analogous to the Burger equation with a reactive source term. Clavin et al. [23, 24] carried out an asymptotic analysis using the combination of a high degree of overdrive, weak heat release, Newtonian limit, and the assumption of high-temperature sensitivity at the shock. With these combined limits, the flow behind the shock can be taken to be quasi-isobaric (nearly constant pressure) to the leading order of the solution where the effect of acoustics is minimized and a nonlinear integral differential equation that governs the overdriven detonation shock front speed $D(t)$ was derived. The associated detonation instability mechanism essentially originates from the time delay between the shock velocity and the profile variations in the heat-release zone. In the same spirit, Calvin and Williams [25] have also considered the limit of weak heat release using an integral equation formulation for detonations with small overdrive extending to the near Chapman–Jouguet regime.

Using weak curvature with large activation energy, Yao and Stewart [191] obtained a nonlinear detonation shock evolution equation that can reproduce the dynamics of both pulsating and cellular detonation instabilities (also in [165]). A one-dimensional evolution equation for two-step chain-branching-like reaction kinetics was obtained by Short [151] and Short and Sharpe [156]. In addition to these formal asymptotic studies, weakly nonlinear stability of both one and two-dimensional overdriven detonations was studied by several researchers, e.g., [14, 44, 45, 95]. There also exist other analytical theories for interpreting experimental observations and estimating the cell size, such as the early studies by Strehlow and Fernandes [166] and Barthel [8] and later by Majda [94] and Bourlioux [12].

Although these theories provide good qualitative understanding of detonation instability, they are not without limitations as they are mostly valid only in linear regimes or their derivations are based on asymptotic approximation in simpler models, and hence the parametric space where the equations apply is again restricted. The inclusion of more complex and realistic chemistry models into asymptotic theories for detonation instability requires further effort.

3.1.3 Numerical Simulation

With current computational techniques, the unsteady reactive Euler equations can be solved numerically with given initial and boundary conditions to investigate the full nonlinear dynamics of detonations. The numerical description of unstable detonations provides detailed information on the flow field that cannot be obtained by experimental measurements. It also has the advantage that stability in one, two, and three dimensions can be separately investigated, and thus facilitates the analysis of the numerical results.

After the pioneering numerical work of the characteristics method by Fickett and Wood [51], the time-dependent 1D pulsating detonations were more thoroughly studied by Abouseif and Toong [2] and Moen et al. [106] using a finite difference technique. They respectively reproduced the salient stability predictions of Erpenbeck [43]. Since then, one-dimensional pulsating detonation simulations have been improved significantly in quality and accuracy. Using a finite-volume numerical approach with shock front-tracking and mesh refinement, Bourlioux et al. [12,15] obtained accurate numerical results which were in close agreement with the prediction of the linear stability analysis of Lee and Stewart [81], (i.e., the neutral stability boundaries and the lowest-frequency single-period limit cycle). They also showed that nonlinear effects quickly become important as the detonation parameters became far from the stability boundary, and irregular oscillations were observed. Recent advances in computational algorithms (adaptive mesh refinement and higher order schemes) and numerical strategies (computing in a shock-attached frame, shock fitting, or shock tracking) permit high-resolution numerical simulation of 1D detonations [66,77,109,141,143,155]. Such simulations make it possible to examine the long-time evolution of nonlinear longitudinal instability and explore the propagation mechanism far from the stability limit. Investigations of the nonlinear oscillatory behavior of 1D detonations provide a first step in elucidating the nature of the self-sustained cellular detonation structure and interpreting the instability mechanism. Although recent works also take into consideration the effect of chemistry efficiently using reduced or detailed kinetics [29,88,110,132,136,155,192], particular care has to be taken with the numerical resolution of the simulations to ensure different instability features are well resolved, especially with detailed chemistry where various multiscale instability modes can exist.

To understand the mechanism of nonlinear longitudinal pulsation, recent studies focused on the flow field results obtained from 1D detonation simulations [28, 29, 86, 159]. In particular, the work of Leung et al. [86] analyzed in detail the flow dynamics and reconstructed the simulation results in terms of characteristics, allowing the trajectory of pressure waves and particle paths to be followed. Equivalent to the work of Alpert and Toong [5] and McVey and Toong [99] on the exothermic hypersonic flow around a blunt projectile, the characteristic analysis allowed to work out a complete picture of the mechanism that leads to different pulsating regimes.

Multidimensional simulations have been performed since the early 1980s, notably by Taki and Fujiwara [170] and Oran et al. [72, 116–118]. Other examples of two-dimensional simulations were reported by [13, 31, 53–55, 145, etc.]; and examples of three-dimensional detonations are illustrated in [32,37,52,176, 177, 189]. The majority of these simulations invoke simple chemistry models using a single-step Arrhenius reaction rate law. In recent years, however, multidimensional simulations using more complex reaction models have also been carried out. The two-step chemical kinetic description is a widely used model that has been developed to retain the essential features of chain-branching chemistry. A main advantage of the use of a two-step chain-branching kinetic model is that the temperature-dependent induction and the exothermic reaction zone are decoupled into two separate stages. Different kinetic parameters in the two-step model can be determined by fitting data from constant volume explosion calculations computed using detailed chemical kinetic mechanisms, and generally very good agreement of the reaction profile can be obtained, especially for H_2–O_2-diluent mixtures [26, 60, 156, 162]. In fact, both reduced and detailed kinetics of the hydrogen–oxygen reaction were implemented in recent simulations of two- and three-dimensional unstable detonations [46, 67, 88, 119]. Attempts were also made to include the viscosity effect using Navier–Stokes equations [118]. Real detonations, however, may not be represented when the Reynolds number is low; sufficiently high Reynolds number and numerical resolution (associated with an embedded numerical viscosity) are necessary.

In general, multidimensional numerical simulations reproduced, qualitatively, the cell dynamics (i.e., the structure of the detonation front consists of the transient incident shock, Mach stem, and transverse reflected wave) as revealed by experiments. By tracking the maximum pressure contours or heat released, these studies reproduce qualitatively the triple point trajectories imprinted on smoked foils from experiments. It has been demonstrated that even when using a one-step Arrhenius kinetics model that increasing the activation energy of the combustible mixture results in more irregular structures characterized by stronger triple points, larger variations of the local shock velocity inside the detonation cell, and a higher frequency of appearance and disappearance of triple points [53, 54]. This result indicates how the reaction sensitivity influences the cellular dynamics of detonation. A high degree of instability results in the formation of unburned pockets, as first observed computationally by Oran et al. [118]. The unburned pockets behind the shock front have an important implication on the detonation propagation mechanism. With high numerical resolution, Radulescu et al. [133] showed that an unstable detonation of a methane–oxygen mixture is unable to propagate due to the lack of molecular or turbulent transport in the Euler simulation to burn these pockets via diffusion. Hence, the combustion of these pockets is not due to the shock compression alone and must rely on a turbulent diffusional transport mechanism. The existence of unburned pockets that get swept downstream results in an increase in the effective length of the reaction zone of the detonation wave compared to the corresponding one-dimensional ZND structure, which leads to the concept of a hydrodynamic thickness in the chemical reaction processes [41, 84, 163, 180].

Although multidimensional unstable detonations can now be described numerically, the analysis of such a large amount of complex data is difficult, thus limiting the value of their study at present. Development of statistical approaches to analyze the cell pattern obtained from simulations is necessary [147]. As examples, Radulescu et al. [134] applied Favre-averaging to quantities in time and in the transverse spatial direction to the cellular detonation to elucidate the concept of hydrodynamic thickness. Austin and Shepherd [7] analyzed the dependence of the reaction rates within the reaction zone structure on the distribution of shock strengths and on the time-variability, or time scales of these cellular pulsations. Statistical approaches may be combined with definitions of macroscopic parameters used to analyze different components observed within the detonation structure. The use of nonlinear theory to describe the reaction front may lead to new insights on the instability of the detonation wave [113, 123]. Research also focuses on the triple-point shear layer, which plays a prominent role in the propagation of gaseous detonation waves and the formation of localized ignition promoted by the Kelvin–Helmholtz instability [97]. Recently Mach and Radulescu [93] revealed a novel mechanism of cell formation in unstable gaseous detonation. Their work suggests that new cells can be formed by the hydrodynamic mechanism of shock bifurcation during the Mach reflection. The results of these analyses have brought about new challenges and have stimulated proposals for the study of the cellular structure of detonation waves in a different direction.

Because of computational limitations, many simulations are still confined to a narrow computational domain, and the cell characteristics can be influenced by the channel width due to mode-locking. The time-scale of the cellular evolution can be very long as compared to the reaction zone time-scale, thus requiring a very long computation time. Much of these simulations also suffer from a dependence on the resolution of the numerical computation, particularly when detailed chemistry is implemented. Both the ability to achieve reliable results from multidimensional simulations and the analysis of these results in order to improve the understanding of the complex detonation instability phenomena remain a challenge.

3.2 Basic Formulation for Linear Stability Analysis

Linear stability analysis of detonation deals with the evolution of small perturbations to the 1D steady detonation wave structure. To be reviewed here are the basic steady planar detonation wave solutions and the formal normal-mode analysis for the problem of linear detonation stability of a simple depletion one-step Arrhenius reaction model and an ideal gas EOS as presented in the works by Lee and Stewart [81], Short and Stewart [157], and Bourlioux [12].

Generally, linear hydrodynamic stability analysis has the following steps: First, a steady-state solution must be established onto which small time-dependent perturbations are imposed. The perturbed solution is then substituted into the governing equations which can then be linearized by retaining the first order terms. To look at instability, the normal mode solution is sought with $\mathrm{Re}(\alpha)$ as the growth rate and $\mathrm{Im}(\alpha)$ as the frequency, where α refers to the eigenvalues. From $\mathrm{Re}(\alpha)$, the sign will determine the stability, i.e., if $\mathrm{Re}(\alpha) > 0$ results in exponential growth, the solution is unstable; in contrast $\mathrm{Re}(\alpha) < 0$ gives an exponential decay of the perturbation, implying a stable solution; and $\mathrm{Re}(\alpha) = 0$ means no growth/decay implying a neutral stability boundary. The normal mode analysis involves solving a set of ordinary differential equations for the solution with an appropriate boundary condition. The result is an eigenvalue problem for α. Thus, the goal is to solve a dispersion relation of the form:

$$\mathbf{F}(\alpha, k, l, \mathbf{R}) = 0, \tag{3.1}$$

where k and l are wavenumbers. $\mathbf{R}$ is a set of bifurcation parameters for the detonation stability problem.

3.2.1 Governing Equations

The standard model for detonation stability analysis uses the reactive inviscid Euler equations which can be written as:

$$\begin{aligned} \frac{D\tilde{v}}{D\tilde{t}} - \tilde{v}\nabla \cdot \tilde{\mathbf{u}} &= 0 \\ \frac{D\tilde{\mathbf{u}}}{D\tilde{t}} + \tilde{v}\nabla\tilde{p} &= 0 \\ \frac{D\tilde{e}}{D\tilde{t}} + \tilde{p}\frac{D\tilde{v}}{D\tilde{t}} &= 0 \\ \frac{D\lambda}{D\tilde{t}} &= \tilde{r}, \end{aligned} \tag{3.2}$$

where the variables $\tilde{v}, \tilde{\mathbf{u}}, \tilde{p}, \tilde{e}, \tilde{r}$ and λ are the specific volume $\tilde{v} = 1/\tilde{\rho}$, velocity, pressure and specific internal energy, reaction rate and reaction progress variable, respectively. The label "$\sim$" refers to dimensional quantities. For a simple model of an ideal gaseous mixture, the caloric and ideal thermal equations of state (EOS) are, respectively:

$$\tilde{e} = \frac{\tilde{p}\tilde{v}}{(\gamma - 1)} - \lambda\tilde{Q} \quad \text{and} \quad \tilde{p} = \tilde{\rho}R\tilde{T} \tag{3.3}$$

and the chemistry is modeled by a one-step Arrhenius simple depletion form given by:

$$\tilde{r} = \tilde{k}\,(1 - \lambda)\exp\left(-\tilde{E}_a/\tilde{R}\tilde{T}\right). \tag{3.4}$$

It is often convenient to rewrite the energy equation in the following alternative form:

$$\tilde{v}\frac{D\tilde{p}}{D\tilde{t}} + \gamma p\tilde{v}\nabla \cdot \tilde{\mathbf{u}} = (\gamma - 1)\,\tilde{Q}\tilde{r}. \tag{3.5}$$

3.2.2 Shock Relations

The governing equations of motion are completed by the conservation laws across the leading shock front. Given a shock with velocity $\mathbf{D}$, the shock relations can be written as:

$$\begin{aligned}
\frac{1}{\tilde{v}}\left(\tilde{\mathbf{u}} - \tilde{\mathbf{D}}\right) \cdot \mathbf{n}\big|_s &= \frac{1}{\tilde{v}}\left(\tilde{\mathbf{u}} - \tilde{\mathbf{D}}\right) \cdot \mathbf{n}\big|_0 = \tilde{m} \\
\tilde{p}_s - \tilde{p}_0 &= -\tilde{m}^2\left(\tilde{v}_s - \tilde{v}_0\right) \\
\tilde{e}_s - \tilde{e}_0 &= \frac{1}{2}\left(\tilde{p}_s + \tilde{p}_0\right)\left(\tilde{v}_0 - \tilde{v}_s\right) \\
\lambda_s &= \lambda_0 \\
\tilde{\mathbf{u}} \cdot \boldsymbol{l}_i\big|_s &= \tilde{\mathbf{u}} \cdot \boldsymbol{l}_i\big|_0,
\end{aligned} \tag{3.6}$$

where subscript "0" refers to the upstream properties and s denotes the state behind the shock. $\tilde{m}$ is the constant mass flux across the shock. $\mathbf{n}$ is the unit vector normal to the shock and $\boldsymbol{l}_i$ are the unit vectors that lie tangent to the shock. For a steady planar shock wave, the normal shock relations with the ideal gas EOS are:

$$\begin{aligned}
\frac{\tilde{v}_s^*}{\tilde{v}_0^*} &= \frac{2 + (\gamma - 1)\,M_0^{*2}}{(\gamma + 1)\,M_0^{*2}} \\
\frac{\tilde{p}_s^*}{\tilde{p}_0^*} &= \frac{\gamma + 1 + 2\gamma\left(M_0^{*2} - 1\right)}{(\gamma + 1)} \\
\frac{\tilde{u}_s^* - \tilde{D}^*}{\tilde{D}^*} &= \frac{\tilde{v}_s^*}{\tilde{v}_0} = \frac{(\gamma + 1)\,M_0^{*2}}{2 + (\gamma - 1)\,M_0^{*2}} \\
M_s^{*2} &= \frac{2 + (\gamma - 1)\,M_0^{*2}}{2\gamma M_0^{*2} - (\gamma - 1)},
\end{aligned} \tag{3.7}$$

where superscript "$*$" refers to the steady solution.

3.2.3 Steady Planar ZND Detonation Solution

Assuming a steady one-dimensional planar detonation wave to travel in the positive x-direction into an ambient state, i.e., $\tilde{p}_0, \tilde{v}_0, \tilde{u}_0, \lambda_0 = 0$, with detonation velocity $\tilde{D}^*$ and defining a steady shock-attached coordinate $\tilde{X} = \tilde{x} - \tilde{D}^* t$, such that $\tilde{U}^* = \tilde{u}^* - \tilde{D}^*$ where $\tilde{U}^*$ is the particle velocity relative to the wave, the conservation conditions can be written as:

$$\begin{aligned} \frac{\tilde{U}^*}{\tilde{v}^*} &= \frac{\tilde{U}_s^*}{\tilde{v}_s^*} = \tilde{m}^* \\ \tilde{p}_s^* - \tilde{p}^* &= -\tilde{m}^{*2} \left(\tilde{v}_s^* - \tilde{v}^*\right) \\ \tilde{e}_s^* - \tilde{e}^* &= \frac{1}{2} \left(\tilde{p}_s^* + \tilde{p}^*\right) \left(\tilde{v}^* - \tilde{v}_s^*\right). \end{aligned} \tag{3.8}$$

By defining $P = \tilde{p}^*/\tilde{p}_s^*$ and $V = \tilde{v}^*/\tilde{v}_s^*$ the last two equations yield:

$$\begin{aligned} P - 1 &= -\gamma M_s^{*2} (V - 1) \\ P (V - \Gamma) &= 1 - \Gamma V + 2\Gamma\beta\lambda, \end{aligned} \tag{3.9}$$

where $\Gamma = \frac{\gamma-1}{\gamma+1}$, $M_s^{*2} = \frac{\tilde{U}_s^{*2}}{\tilde{c}_s^{*2}} = \frac{\tilde{U}_s^{*2}}{\gamma \tilde{p}_s^* \tilde{v}_s^*}$ and the energy $\beta = \frac{\tilde{Q}}{\tilde{p}_s^* \tilde{v}_s^*}$. Eliminating P from the above two equations leads to the following quadratic equation:

$$\gamma M_s^{*2} V^2 - (1 + \Gamma) \left(1 + \gamma M_s^{*2}\right) V + \Gamma + 1 + \gamma M_s^{*2} \Gamma + 2\Gamma\beta\lambda = 0. \tag{3.10}$$

Solving the above quadratic equation yields:

$$V = \frac{1}{(\gamma + 1) M_s^{*2}} \left[1 + \gamma M_s^{*2} \pm \left(1 - M_s^{*2}\right) \left\{1 - \frac{2\left(\gamma^2 - 1\right) M_s^{*2} \beta\lambda}{\gamma \left(1 - M_s^{*2}\right)^2}\right\}^{1/2}\right]. \tag{3.11}$$

The lower sign is for the strong overdriven detonation. For $\lambda = 0$, the above equation recovers the post-shock condition $V = 1$. It is convenient to define the following parameters:

$$a = \frac{1 + \gamma M_s^{*2}}{\gamma + 1} \qquad b = \frac{2\left(\gamma^2 - 1\right) M_s^{*2}}{\gamma \left(1 - M_s^{*2}\right)^2} \tag{3.12}$$

and the ZND detonation structure can then be described by (in the steady shock-attached frame):

$$\begin{aligned} P &= \frac{\tilde{p}^*}{\tilde{p}_s^*} = a + (1 - a)\sqrt{1 - b\beta\lambda} \\ V &= \frac{\tilde{v}^*}{\tilde{v}_s^*} = \frac{1}{\gamma M_s^{*2}} \left[\gamma a - (1 - a)\sqrt{1 - b\beta\lambda}\right] \\ \frac{\tilde{U}^*}{\tilde{c}_s^*} &= \frac{1 - P}{\gamma M_s^*} + M_s^*. \end{aligned} \tag{3.13}$$

The scaled Chapman–Jouguet detonation velocity (Mach number) $M_{\text{CJ}}^* = \frac{\tilde{D}_{\text{CJ}}}{\tilde{c}_0}$, determined by the velocity of the fully reacted flow ($\lambda = 1$) being sonic is given by:

$$M_{\text{CJ}}^* = \sqrt{1 + \frac{(\gamma^2 - 1)\, Q}{2\gamma}} + \sqrt{\frac{(\gamma^2 - 1)\, Q}{2\gamma}}, \tag{3.14}$$

where

$$Q = \frac{\gamma \tilde{Q}}{\tilde{c}_0^2}.$$

For a piston-supported or overdriven detonation wave, the detonation overdrive, f, is defined by:

$$f = \left(\frac{M^*}{M_{\text{CJ}}^*} \right)^2. \tag{3.15}$$

3.2.4 Non-dimensionalization

It is convenient to scale the variables and introduce dimensionless variables for the governing equations. Variables were made dimensionless with respect to the post-shock state [81, 157], i.e.:

$$p = \frac{\tilde{p}}{\tilde{p}_s} \quad T = \frac{\tilde{T}}{\tilde{T}_s} \quad v = \frac{\tilde{v}}{\tilde{v}_s} \quad u = \frac{\tilde{u}}{\tilde{c}_s}. \tag{3.16}$$

The scaling for length is the steady half-reaction length $\tilde{l}_{1/2}$, the distance behind the shock where the reaction progress variable λ reaches half, and for time $\tilde{l}_{1/2}/\tilde{c}_s$. The scaled activation energy and heat release quantities θ and β are defined as:

$$\theta = \frac{\gamma \tilde{E}_a}{\tilde{c}_s^2} \qquad \beta = \frac{\gamma \tilde{Q}}{\tilde{c}_s^2}. \tag{3.17}$$

As used in many other early studies [42, 43], it is also convenient to define an alternative dimensionless activation energy E and heat release Q that are scaled with respect to the ambient pre-shock sound speed $\tilde{c}_0^2$,

$$E_a = \frac{\gamma \tilde{E}_a}{\tilde{c}_0^2} \qquad Q = \frac{\gamma \tilde{Q}}{\tilde{c}_0^2}. \tag{3.18}$$

With the above non-dimensionalization, the condition immediately behind the shock gives:

$$p^* = v^* = T^* = 1,\ U^* = M_s^* \text{ and } \lambda^* = 0. \tag{3.19}$$

The spatial structure of the detonation (distance behind the shock) is then determined by numerical quadrature of the first order rate equation:

$$X = \int_0^{\lambda^*} \frac{U^*}{r^*} \mathrm{d}\lambda^* = \frac{1}{k} \int_0^{\lambda^*} \frac{U^*}{1 - \lambda^*} \exp\left(\theta / p^* v^*\right) \mathrm{d}\lambda^*. \tag{3.20}$$

By choosing the rate constant $k = \tilde{k}\tilde{l}^*_{1/2} \,/\, \tilde{c}_s^2$, the length scale of the reaction zone becomes unity ($X = 1$) when $\lambda^* = 1/2$, from the integral:

$$k = \int_0^{1/2} \frac{U^*}{1-\lambda^*} \exp\left(\theta / p^* v^*\right) \mathrm{d}\lambda^* . \tag{3.21}$$

3.2.5 Linear Stability Analysis Formulation

The general linear stability analysis problem is formulated by first defining an unsteady shock-attached frame of reference. For a two-dimensional disturbance:

$$X = x - D^* t - \psi\left(y, t\right), \tag{3.22}$$

where ψ denotes a small perturbation to the shock position superimposed to the steady trajectory. The velocity U, V relative to the steady frame becomes:

$$\begin{aligned} U &= u_1 - D^* \\ V &= u_2 . \end{aligned} \tag{3.23}$$

The various operators written in the shock-attached frame are:

$$\frac{\partial}{\partial x} \to \frac{\partial}{\partial X} \quad \frac{\partial}{\partial y} \to \frac{\partial}{\partial y} - \frac{\partial \psi}{\partial y}\frac{\partial}{\partial X} \quad \frac{\partial}{\partial t} \to \frac{\partial}{\partial t} - \left(D + \frac{\partial \psi}{\partial t}\right)\frac{\partial}{\partial X} \tag{3.24}$$

and in matrix form, the governing equation becomes:

$$\begin{aligned} &\frac{\partial z}{\partial t} + A\frac{\partial z}{\partial X} + B\frac{\partial z}{\partial y} - \dot{\psi}\frac{\partial z}{\partial X} - B\frac{\partial \psi}{\partial y}\frac{\partial z}{\partial X} = c \\ &z = \left(v\; U\; V\; p\; \lambda\right)^T \\ &A = \begin{pmatrix} U & -v & 0 & 0 & 0 \\ 0 & U & 0 & v/\gamma & 0 \\ 0 & 0 & U & 0 & 0 \\ 0 & \gamma p & 0 & U & 0 \\ 0 & 0 & 0 & 0 & U \end{pmatrix} B = \begin{pmatrix} V & 0 & -v & 0 & 0 \\ 0 & V & 0 & 0 & 0 \\ 0 & 0 & V & v/\gamma & 0 \\ 0 & 0 & \gamma p & V & 0 \\ 0 & 0 & 0 & 0 & V \end{pmatrix} c = \begin{pmatrix} 0 \\ 0 \\ 0 \\ (\gamma - 1)\,\beta r / v \\ r \end{pmatrix} . \end{aligned} \tag{3.25}$$

The perturbations to the steady-state structure are written in the normal mode form:

$$z = z^* + z' \mathrm{e}^{\alpha t} \mathrm{e}^{\mathrm{i}ky} \text{ and } \psi = \psi' \mathrm{e}^{\alpha t} \mathrm{e}^{\mathrm{i}ky}, \tag{3.26}$$

where z^* is the unperturbed reference steady solution and z' is the perturbation. z' and α are complex numbers. $\mathrm{Re}(\alpha)$ is the disturbance growth rate, $\mathrm{Im}(\alpha)$ is the disturbance frequency, and k is the disturbance wavelength in the transverse direction.

Substituting these perturbations into the full governing equation (3.25) and performing the expansion, a system of first-order linearized differential equations for the complex perturbation eigenfunction z' are obtained:

$$\alpha z' + A^* z'_X + ikB^* \cdot z' + C^* \cdot z' - \psi' (\alpha + ikB^*) z^*_X = 0$$
$$A^* z'_X + (C^* + \alpha I + ikB^*) \cdot z' - \psi' (\alpha + ikB^*) z^*_X = 0$$

$$C^* = \begin{pmatrix} -U_X & v_X & 0 & 0 & 0 \\ p_X/\gamma & U_X & 0 & 0 & 0 \\ 0 & V_X & 0 & 0 & 0 \\ \frac{(\gamma-1)\beta}{v}\left(\frac{r}{v} - r_v\right) & p_X & 0 & \gamma U_X - \frac{(\gamma-1)\beta r_p}{v} & -\frac{(\gamma-1)\beta r_\lambda}{v} \\ -\lambda_v & \lambda_X & 0 & -r_p & -r_\lambda \end{pmatrix}^* . \tag{3.27}$$

The shock conditions for the linearized perturbations are determined from the linearized Rankine–Hugoniot shock relationship. The Mach number of the unsteady wave relative to the ambient condition is defined as:

$$M_0 = \frac{\left(D^* - \dot{\psi}\right) \tilde{c}^*_s}{\tilde{c}^*_0} = M^*_0 \left(1 - \frac{\dot{\psi}}{D^*}\right) = \left(1 - \frac{\dot{\psi}\tilde{c}^*_s}{\tilde{D}^*}\right). \tag{3.28}$$

Using the mass conservation across the shock $\frac{\tilde{U}^*_s}{\tilde{v}^*_s} = \frac{\tilde{D}^*}{\tilde{v}^*_0}$ and the Mach number at the shocked state $M^*_s = \frac{\tilde{U}^*_s}{\tilde{c}^*_s}$, $M_0 = M^*_0(1 - \delta)$ where $\delta = \frac{1}{M^*_s}\frac{\tilde{v}^*_s}{\tilde{v}^*_0}\dot{\psi}$. From the Rankine–Hugoniot (RH) relationship derived previously (3.7):

$$\begin{aligned} \frac{\tilde{v}_s}{\tilde{v}_0} &= \frac{2 + (\gamma - 1) M_0^2}{(\gamma + 1) M_0^2} \\ v_s &= \frac{\tilde{v}_0}{\tilde{v}^*_s}\frac{\tilde{v}_s}{\tilde{v}_0} \\ &= \frac{\tilde{v}_0}{\tilde{v}^*_s}\frac{2 + (\gamma - 1) M_0^2}{(\gamma + 1) M_0^2} \approx \frac{\tilde{v}_0}{\tilde{v}^*_s}\frac{2 + (\gamma - 1) M_0^{*2}(1 - 2\delta)}{(\gamma + 1) M_0^{*2}(1 - 2\delta)} \\ &\approx \frac{\tilde{v}_0}{\tilde{v}^*_s}\left[\frac{2 + (\gamma - 1) M_0^{*2}}{(\gamma + 1) M_0^{*2}} + \frac{4\delta}{(\gamma + 1) M_0^{*2}}\right]. \end{aligned}$$

The above expansion yields:

$$v_s \approx 1 + \frac{\tilde{v}_0}{\tilde{v}^*_s}\frac{4\delta}{(\gamma + 1) M_0^{*2}} = 1 + \frac{4\dot{\psi}}{(\gamma + 1) M_0^{*2} M^*_s}. \tag{3.29}$$

Equivalently, following the same procedure, the linearized RH equations for the pressure and particle velocity are:

$$p_s \approx 1 - \frac{\tilde{p}_0}{\tilde{p}^*_s}\frac{4\gamma M_0^{*2}\delta}{(\gamma + 1)} = 1 - \frac{4\gamma}{(\gamma + 1)} M^*_s \dot{\psi} \tag{3.30}$$

$$U_s \approx M^*_s + \left(\frac{2\left(1 + M_0^{*2}\right)}{(\gamma + 1) M_0^{*2}}\right)\dot{\psi} \tag{3.31}$$

The initial values of z' at the perturbed shock position are therefore determined from the above linearized Rankine–Hugoniot shock relationships:

$$\begin{aligned}
v' &= \frac{4}{(\gamma+1)\, M_0^{*2} M_s^*} \alpha\psi' \\
p' &= -\frac{4\gamma M_s^* \alpha\psi'}{(\gamma+1)} \\
U' &= \frac{2\left(1+M_0^{*2}\right)}{(\gamma+1)\, M_0^{*2}} \alpha\psi' \\
V' &= (D^* - M_s^*)\, ik\psi' \\
\lambda' &= 0.
\end{aligned} \tag{3.32}$$

Transforming the linearized equations to the reaction coordinate $s = \lambda^*(X)$:

$$\frac{\partial}{\partial X} = \frac{r^*}{U^*}\frac{\partial}{\partial s}. \tag{3.33}$$

Normalizing the perturbation z' with respect to the shock value by setting:

$$\xi = S^{-1}\frac{z'}{\psi'}, \tag{3.34}$$

where the matrix S is given by:

$$S = \begin{pmatrix}
\dfrac{4}{(\gamma+1)\, M_0^{*2} M_s^*} & 0 & 0 & 0 & 0 \\
0 & \dfrac{2\left(1+M_0^{*2}\right)}{(\gamma+1)\, M_0^{*2}} & 0 & 0 & 0 \\
0 & 0 & (D^* - M_s^*) & 0 & 0 \\
0 & 0 & 0 & -\dfrac{4\gamma}{(\gamma+1)} M_s^* & 0 \\
0 & 0 & 0 & 0 & 1
\end{pmatrix}. \tag{3.35}$$

The linearized equations (3.27) become:

$$\frac{\partial \xi}{\partial s} = -\alpha F^* \xi - G^* \xi - ikH^* \xi + \alpha h^* + ikj^* \tag{3.36}$$

with real matrices:

$$\begin{aligned}
F^* &= \frac{U^*}{r^*} S^{-1} A^{*-1} S \\
G^* &= \frac{U^*}{r^*} S^{-1} A^{*-1} C^* S \\
H^* &= \frac{U^*}{r^*} S^{-1} A^{*-1} B^* S \\
h^* &= S^{-1} A^{*-1} \frac{\partial z^*}{\partial s} \\
j^* &= S^{-1} A^{*-1} B^* \frac{\partial z^*}{\partial s}
\end{aligned} \tag{3.37}$$

and the shock conditions are given by:

$$\xi(0) = \begin{pmatrix} \alpha \\ \alpha \\ ik \\ \alpha \\ 0 \end{pmatrix}. \tag{3.38}$$

3.2.6 Radiation Boundary Condition

In order to solve the dispersion relation, it is required to provide one additional boundary condition, applied to the end of the reaction zone, as a closure condition to the system, i.e.,

$$\alpha U' - ikU^{*\infty}V' - \frac{v^{*\infty}}{\gamma c^{*\infty}} p' \sqrt{\alpha^2 + k^2 c^{*\infty 2}\left(1 - M^{*\infty 2}\right)} = 0, \tag{3.39}$$

where the ∞ superscript refers to the fully burned gas. The detailed derivation of this closure condition is given in Stewart and Kasimov [164] and Lee and Stewart [81]. Essentially, this expression (3.39) can be derived from the acoustic analysis of the flow at the end of the reaction zone. The mathematical approach is to decompose the linear acoustics around the equilibrium state into a family of waves that travel at the real characteristic speeds. This condition is then obtained by insisting that the solution does not depend on the wave family that travels toward the shock. This implies that no perturbation at the end of the reaction zone influences the evolution of the reaction zone [164]. It is worth noting that Lee and Stewart [81] also presented a more generalized closure condition in which perturbations to the reaction rate are included in the boundary condition. However, provided that the radiation condition is applied at sufficiently large distance into the burnt zone and hence the reaction rate is exponentially small at the end of the reaction zone, the boundary condition given here is typically sufficient.

The boundary (or closure) condition is generally referred to as the acoustic radiation or causality condition of Buckmaster and Ludford [18]. For the ideal steady detonation structure using a simple reaction model as considered in the present analysis, the steady reaction zone length is infinite, i.e., the equilibrium point of the detonation is at $x = \infty$, where $\lambda^* \to 1$ in an exponential manner. The boundary condition can be interpreted as a causality condition, that any perturbation that travels upwind to interfere with the detonation shock from the rear of the flow can have no causal influence on the stability since such disturbances would take an infinite time to disturb the entire reaction zone inside the detonation structure.

It is worth noting that the linear stability analysis of detonation flows corresponding to a more general model, e.g., reaction rates that produce finite length reaction zones and influenced by the magnitude of the detonation overdrive and reaction order, must be studied separately. It is necessary to conduct an asymptotic analysis of the eigenmode structure and examine in detail appropriate spatial boundedness closure conditions at the end of the steady reaction zone of the eigenvalue system, i.e., the requirement that only spatially bounded solutions to the stability perturbations equations are allowed when considering unstable modes. Such formal analysis is carefully examined by Sharpe [139] and Short et al. [161].

3.2.7 Numerical Examples

The result is a boundary value problem that is bounded by the linearized shock relation and the closure condition. For a given set of initial thermodynamic and chemical parameters characterizing the steady detonation structure, the complex eigenvalues α and eigenfunction $\boldsymbol{z}'(x)$ are determined by a numerical shooting technique based on that given in [81], a two-point boundary-value solution technique utilizing a standard two-variable Newton–Raphson technique to iterate upon α in order to satisfy the acoustic radiation condition. This involves splitting the stability equation into real and imaginary parts, taking an initial guess for $\alpha = \alpha_\mathrm{r} + \mathrm{i} \cdot \alpha_\mathrm{i}$, integrating the dispersion equation from the perturbed shock condition into the burnt zone and iterating by a Newton root finding procedure on the initial guess for α until the boundary condition (acoustic radiation condition) is met in the burnt zone. Different sets of stability diagrams can be generated by changing the different bifurcation parameters or wavenumber k, then calculating α. These stability diagrams from the linear stability analysis explain the general trends in the complex two-dimensional linear stability response of a steady planar detonation wave for the three important bifurcation parameters: the detonation overdrive, the chemical heat release, and the activation energy of the Arrhenius chemical reaction.

For example, Fig. 3.2a shows the k-spectrum of the single low-frequency unstable mode for a low activation energy case with $E_a = 12.5$, $Q = 50$, $\gamma = 1.2$ and $f = 1.0$ found in [157]. For this particular set of parameters of the steady detonation, the linear stability analysis shows that the mode is one-dimensional stable to longitudinal disturbances, having a low-wavenumber neutral stability point $\mathrm{Re}(\alpha) = 0$ at $k = 0.15$. It attains a maximum growth rate, $\mathrm{Re}(\alpha) = 0.159$, at $k = 1.10$ and is short wavelength stable for $k > 2.51$. It is interesting to remark that a random initial disturbance in a sufficiently wide channel will potentially excite a large range of transverse disturbances, with locally dominant growth around the wavelength associated with the maximum growth rate of the unstable mode. Hence, attempts have been made to relate this wavelength of the most unstable mode (maximum growth rate) to the onset of the nonlinear cellular detonation patterns observed numerically and experimentally [157]. By increasing the activation energy E_a, a second higher frequency, two-dimensional unstable mode appears. Figure 3.2b shows the unstable k-spectrum for $E_a = 27.5$, $Q = 50$, $\gamma = 1.2$ and $f = 1.0$ which contains two unstable modes, with the maximum growth rate associated with the lowest frequency mode having $\mathrm{Re}(\alpha) = 0.310$ at $k = 0.86$ (or wavelength

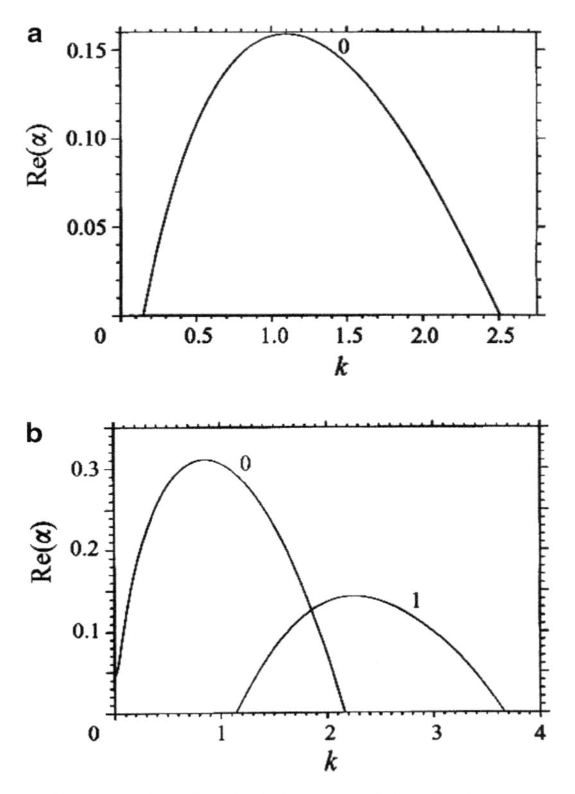

Fig. 3.2. Stability diagram showing $\mathrm{Re}(\alpha)$ versus k for the two-dimensional unstable modes with (**a**) $E_a = 12.5$, $Q = 50$, $\gamma = 1.2$ and $f = 1.0$ and (**b**) $E_a = 27.5$, $Q = 50$, $\gamma = 1.2$ and $f = 1.0$ [157]

$W = 2\pi/k = 7.35$). The lowest frequency mode is one-dimensionally unstable, while the higher frequency mode has the finite band of unstable wavenumbers, $1.71 < W < 5.56$. The detonation is stable to disturbances with wavelengths $W < 1.71$. By increasing the activation energy further, it is thus expected that several high-frequency unstable modes will be present. The fact that these higher frequency modes could be excited with locally dominant growth around the wavelength associated with their corresponding maximum growth rates points to the emergence of an irregular cellular structure, and the complexity of the linear stability spectrum would appear to be consistent with experimental results [13].

Apart from the disturbance growth rate and frequencies associated with transverse wavenumbers, in hydrodynamic stability analysis neutral stability boundaries are often constructed using a combination of the stability parameters: specific heat ratio γ, normalized chemical energy release Q, normalized activation energy E_a/RT_s, and overdrive f, as well as the transverse wave

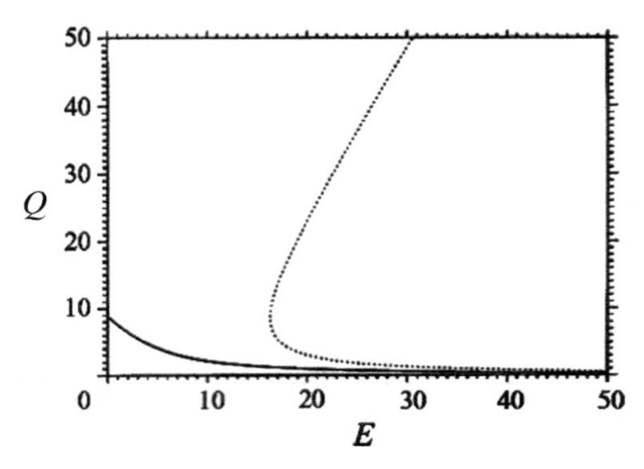

Fig. 3.3. Two-dimensional (*solid line*) and one-dimensional (*dotted line*) neutral stability boundaries for $\gamma = 1.2$, $f = 1.0$ in Q and E_a parameter space. Regions to the right of the boundaries are unstable, and those to the left are stable [157]

number. From these plots such as Fig. 3.3, the region of stability can be identified and the effect of different bifurcation parameters on the stability of the ZND detonation can be studied.

An interesting analysis was presented by Eckett [38, 39] to better represent the stability results and the neutral linear stability diagrams for one-dimensional planar CJ detonations obtained using the normal mode stability analysis method. Typically, the neutral stability curves vary in different parameters planes ($Q-E_a$, $Q-\gamma$ planes, etc.). However, Eckett showed that if the CJ Mach number M_{CJ} is used as a parameter (which is a function of overdrive and heat release Q), the neutral stability boundaries in $\theta_{\mathrm{CJ}} - M_{\mathrm{CJ}}$ plane, where θ_{CJ} is the activation energy normalized by the shock temperature T_s:

$$\theta = \frac{\tilde{E}_a}{R\tilde{T}_s} \tag{3.40}$$

essentially collapse to a single curve, independent of γ (see Fig. 3.4). For strong detonations with large values of M_{CJ}, the neutral stability curve asymptotes to a constant value of $\theta_{\mathrm{CJ}} \approx 4.74$. In this regime, the stability of the wave is then a function of θ_{CJ} only, an example of the dominant effect of θ for the Arrhenius reaction rate model. This curve provides useful information and has been used qualitatively to determine if a mixture is stable, slightly unstable, or highly unstable [7].

3.3 Nonlinear Instability Simulation and Analysis

Unlike hydrodynamic linear stability analysis, numerical simulations of the chemically reactive gasdynamics equations allow the full nonlinear dynamics of detonations to be investigated and provide more details on the entire flow field inside and beyond the detonation front. Numerical simulation also allows one

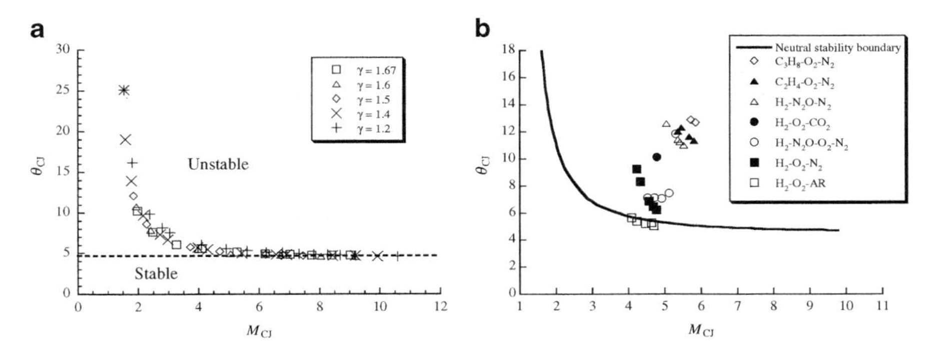

Fig. 3.4. (**a**) Neutral stability curve for planar CJ detonations with one-step Arrhenius kinetic model [38,39]. (**b**) Categorization of detonation front structure using the neutral stability curve for planar CJ detonations with a one-step Arrhenius kinetic model [7]

to systematically study the dynamics of detonation structure from simple one-dimensional instability—longitudinal pulsation—to multidimensional cellular instability. Using numerical simulations, the development of detonation can also be studied by using arbitrary initial conditions to see if the established detonation wave, after its transient development, is stable or unstable in the long-time solution.

3.3.1 Mathematical Model

For an ideal detonation model, the full unsteady nonlinear dynamics of detonations are modeled by solutions of the inviscid compressible, time-dependent Euler equations, which are expressed as the conservation equations for mass, momentum, and energy:

$$\begin{aligned}
&\frac{\partial \rho}{\partial t} + \nabla \cdot (\rho \mathbf{V}) = 0 \\
&\frac{\partial (\rho \mathbf{V}_i)}{\partial t} + \nabla \cdot (\rho \mathbf{V}_i \mathbf{V}) + \frac{\partial p}{\partial \mathbf{x}_i} = 0 \\
&\frac{\partial (\rho E)}{\partial t} + \nabla \cdot ((\rho E + p)\, \mathbf{V}) = 0 \\
&\frac{\partial (\rho Y_i)}{\partial t} + \nabla \cdot (\rho Y_i \mathbf{V}) = \rho \Omega_i (\rho, p, Y_i)\,,
\end{aligned} \tag{3.41}$$

where ρ is the mass density, $\mathbf{V} = (u,\ v,\ w)$ is the velocity, p is the pressure, Y_i is the mass fraction of a reactant, and E is the total energy per unit mass, which is the sum of internal energy and kinetic energy:

$$E = e(\rho, p, Y_i) + \frac{1}{2}\mathbf{V}^2. \tag{3.42}$$

To complete the governing equations, an EOS of the form $e(\rho,\ p,\ Y_i)$ and a chemical reaction mechanism are required to describe the chemical production/consumption rates $\Omega_i(\rho,\ p,\ Y_i)$ for species i. The effect of chemical kinetics is thus modeled in the mass conservation of mixture species by introducing different rate laws for the reactions. The liberation of chemical energy, which is reflected in the change of the internal energy of the mixture $e(\rho,\ p,\ Y_i)$, affects the gasdynamics of the flow, defining the structure of a detonation wave.

In the simplest case, the chemical kinetics are represented by a one-step Arrhenius rate law *Reactant* $\rightarrow$ *Product*. The reaction progress is given by an Arrhenius rate law:

$$\frac{\mathrm{d}\beta}{\mathrm{d}t} = -k\beta \exp\left(\frac{-E_a}{T}\right). \tag{3.43}$$

Such a model has been primarily considered in many detonation simulations due to its simplicity, yet retains some important features such as the temperature sensitivity given by the global activation energy. Considering only two spatial dimensions, the governing equations can be written in nondimensional conservative form as:

$$\frac{\partial \mathbf{U}}{\partial t} + \frac{\partial \mathbf{F}(\mathbf{U})}{\partial x} + \frac{\partial \mathbf{G}(\mathbf{U})}{\partial y} = \mathbf{S}(\mathbf{U}), \tag{3.44}$$

where the conserved variable $\mathbf{U}$, the convective fluxes $\mathbf{F}$, $\mathbf{G}$, and reactive source term $\mathbf{S}$ are, respectively,

$$\mathbf{U}=\begin{pmatrix}\rho\\ \rho u\\ \rho v\\ \rho E\\ \rho\beta\end{pmatrix} \mathbf{F}(\mathbf{U})=\begin{pmatrix}\rho u\\ \rho u^2+p\\ \rho uv\\ (\rho E+p)u\\ \rho u\beta\end{pmatrix} \mathbf{G}(\mathbf{U})=\begin{pmatrix}\rho v\\ \rho uv\\ \rho v^2+p\\ (\rho E+p)v\\ \rho v\beta\end{pmatrix} \mathbf{S}(\mathbf{U})=\begin{pmatrix}0\\ 0\\ 0\\ 0\\ \rho\Omega\end{pmatrix} \tag{3.45}$$

with

$$E = \frac{p}{(\gamma-1)\rho} + \frac{(u^2+v^2)}{2} + \beta Q, \quad T = \frac{p}{\rho}$$
$$\Omega = -k\beta \exp\left(\frac{-E_a}{T}\right).$$

In the equations, ρ, u, v, p, T, and E are density, particle velocities, pressure, temperature, and total energy per unit mass, respectively. The variable β is used hereafter as the reaction progress variable, which varies between 1 (for unburned reactant) and 0 (for product). The mixture is assumed to be an ideal gas that is also calorically perfect (with constant specific heat ratio γ). The parameters Q and E_a represent the nondimensional heat release and activation energy, respectively. These variables have been made dimensionless with reference to the uniform unburned state ahead of the detonation front.

$$\rho = \frac{\tilde{\rho}}{\tilde{\rho}_0}, \; p = \frac{\tilde{p}}{\tilde{p}_0}, \; T = \frac{\widetilde{T}}{\widetilde{T}_0}, \quad u, \tilde{v} = \frac{\tilde{u}, \tilde{v}}{\sqrt{\widetilde{R}\widetilde{T}_0}}, \; Q = \frac{\widetilde{Q}}{\widetilde{R}\widetilde{T}_0}, \; E_a = \frac{\widetilde{E}_a}{\widetilde{R}\widetilde{T}_0}. \quad (3.46)$$

The pre-exponential factor k is an arbitrary parameter that merely defines the spatial and temporal scales. It is chosen such that the half-reaction length $L_{1/2}$, i.e., the distance required for half the reactant to be consumed in the steady ZND wave, is scaled to unit length. Hence,

$$x = \frac{\widetilde{x}}{\widetilde{L}_{1/2}}, \; y = \frac{\tilde{y}}{\widetilde{L}_{1/2}}, \; t = \tilde{t}\frac{\sqrt{\widetilde{R}\widetilde{T}_0}}{\widetilde{L}_{1/2}}, \; k = \tilde{k} \cdot \frac{\widetilde{L}_{1/2}}{\sqrt{\widetilde{R}\widetilde{T}_0}}. \quad (3.47)$$

3.3.2 Numerical Methods and Validations

Given that the governing equations written in conservative forms are hyperbolic, there exist an extensive number of conservative schemes that can approximate the solutions (see, e.g., [175]). In the dissertations of Bourlioux [12] and Quirk [129], details of numerical methods particularly suited for detonation simulations are discussed. Adaptive mesh refinement is now commonly used to improve spatial resolution. The grid refinement strategy is usually based on the density gradient or reaction rate. Some state-of-the art numerical algorithms also couple a high order scheme (WENO) with shock front tracking or shock-fitting to improve the order accuracy around the shock front [12, 66]. Different initial conditions can be used to start the numerical computation, for example, by using the ZND solution, a blast wave profile [98, 107], a piston-driven initiation [141], or a high-pressure balloon. Using arbitrary initial conditions allows the verification of detonation formation and ascertains if the long-time asymptotic solution is stable or unstable. The effect of numerical resolution, numerical domain size, and back boundary conditions are carefully examined by Sharpe and Falle [143] and Hwang et al. [68]. It is worth noting that the development of instability and the natural stability mode may take an extremely long time to achieve (tens of thousands of the characteristic ZND reaction times), and the numerical computation must be carried out for a sufficiently long time. A common approach to reduce the numerical integration expense is to consider a grid moving with the wave at the CJ speed [20, 53, 54]. However, care must be taken in implementing the rear boundary condition, and a sufficiently large domain must be used in order to accommodate the full hydrodynamic reaction zone structure of the detonation so that the disturbances from the rear boundary cannot influence the front and thus the development of the instability. Another improved simulation setup is to integrate the governing equations in a shock-attached frame with a "nonreflecting" boundary condition as recently proposed by Kasimov and Stewart [77].

Before a numerical simulation can truly be trusted, some validation of the numerical method and its formulation, especially the grid resolution, must be

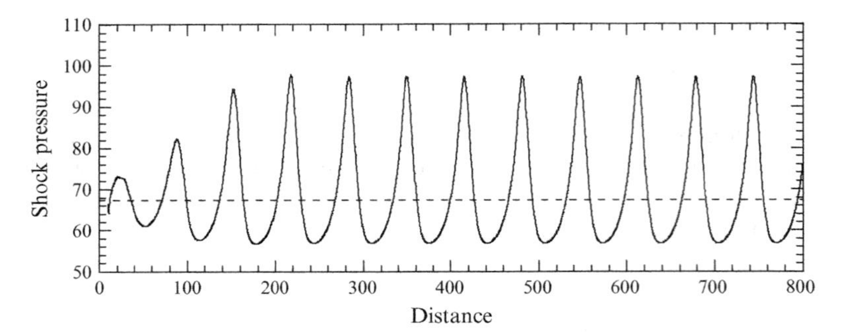

Fig. 3.5. Pressure behind the shock front versus position for the overdriven detonation with $Q = 50$, $\gamma = 1.2$, $E_a = 50$ and $f = 1.6$, using 20 grid points per $L_{1/2}$. The *dashed line* indicates the steady-state ZND value of the von Neumann pressure p_{vN}

performed. A canonical problem for detonation code testing used by many researchers [15, 65, 68, 129] considers the dimensionless parameters with the values $Q = 50$, $\gamma = 1.2$, $E_a = 50$ and overdriven factor $f = 1.6$ (i.e., $f = (D/D_{\mathrm{CJ}})^2$ where D is the detonation velocity of the equivalent steady ZND detonation and D_{CJ} is the Chapman–Jouguet detonation velocity). In the numerical study, the simulations can be initialized by imposing the corresponding steady ZND solution onto the computational grid to see if the instability caused by numerical perturbations grows or decays. The approach is similar to linear stability analysis where small disturbances were imposed on the steady solution. According to exact linear stability analyses [15, 81], the corresponding ZND profile has a single instability mode and the nonlinear manifestation of this instability is a regular periodic pulsating detonation [15, 64]. Figure 3.5 shows a numerical example of the leading shock pressure versus position plot. Note that there is no perturbation applied to the ZND initial condition, but the instability grows quickly from the numerical startup error (e.g., adjustment in the shock capturing). After the transient development, it correctly predicts the single instability mode of the detonation front.

As suggested by Hwang et al. [68], a convergence study of the magnitude of the peak pressure reached behind the overdriven detonation during the limit-cycle pulsations can be used to assess the numerical code. The results of various numerical schemes [15, 68, 121, 129, 175] are presented in Fig. 3.6, showing the peak shock pressure versus the relative mesh spacing (i.e., a relative mesh spacing of 0.5 corresponds to a resolution of 20 grid points per $L_{1/2}$). From this graph, it can be seen that a good numerical scheme should converge to approximately the peak pressure value of ~98.6 as first predicted by Fickett and Wood [51]. Besides the convergence in peak pressure, the period of oscillation for the pulsating detonations can also be verified [68]. Figure 3.7 compares the period of pressure oscillation with the relative mesh spacing obtained by a number of different schemes. The period of oscillation can be determined by taking the average of several cycles between successive pressure

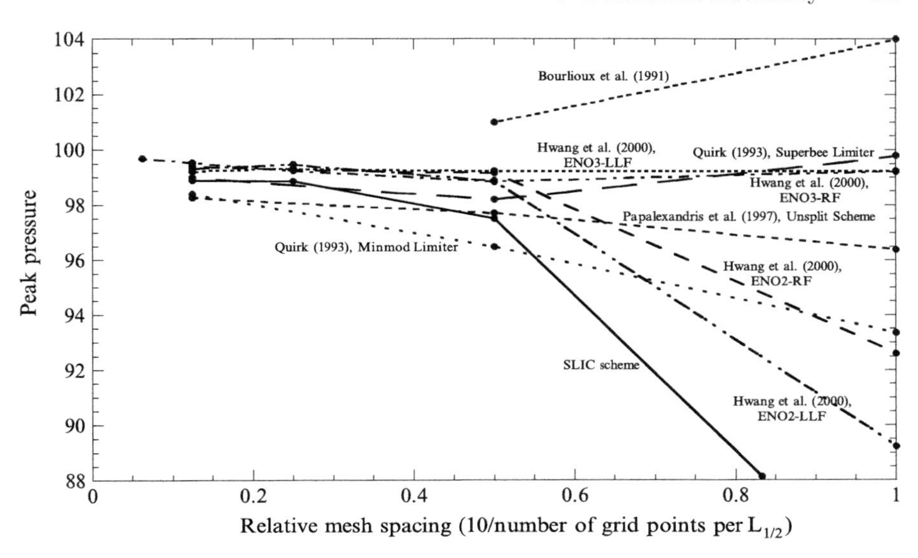

Fig. 3.6. Comparison of peak pressure behind the shock front as a function of relative mesh spacing among different numerical schemes for an overdriven pulsating detonation ($Q = 50$, $\gamma = 1.2$, $E_a = 50$ and $f = 1.6$) [68]

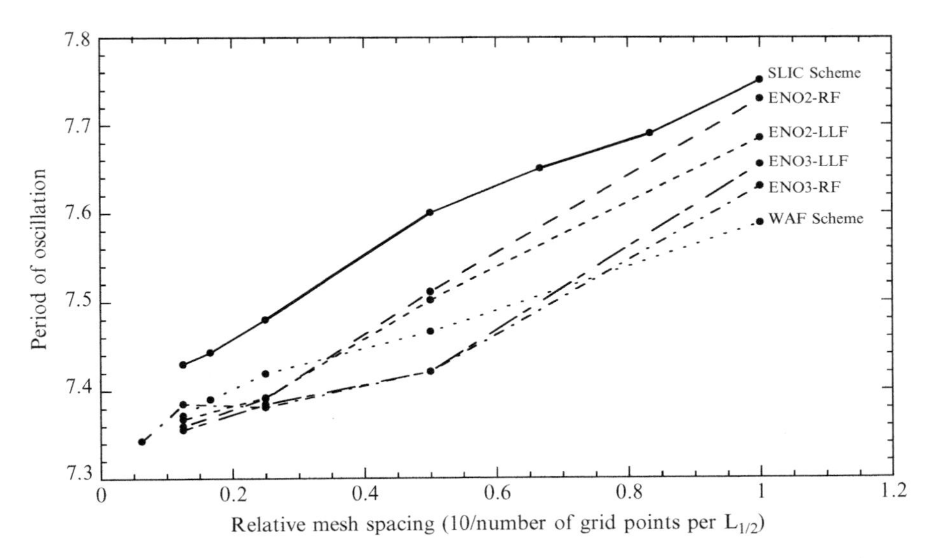

Fig. 3.7. Comparison of the period of pressure oscillation as a function of relative mesh spacing obtained by different numerical schemes for an overdriven pulsating detonation ($Q = 50$; $\gamma = 1.2$; $E_a = 50$ and $f = 1.6$) [68]

peaks. For most standard numerical schemes, a resolution of about 20 grids per $L_{1/2}$ should generally be used to accurately capture the peak pressure behind the leading pulsating detonation front.

Another common one-dimensional canonical test is to determine the stability boundaries where the stable solution bifurcates to an unstable mode. Any reliable numerical formulation with good temporal and spatial accuracy should be able to capture the location of the neutral stability boundary which can be obtained exactly from linear stability analysis. For example, some sample results of the early transient development of a 1D detonation wave obtained from numerical simulations are shown in Fig. 3.8. The leading shock pressure behind the evolving 1D CJ detonation wave normalized by the von Neumann pressure p_{vN}, i.e., the post-shock pressure in the corresponding steady ZND solution, is plotted for different activation energies E_a. For very low activation energy ($E_a < 25.26$), the oscillation due to the initial perturbation damps out with time, and the propagating detonation is stable after a long-term evolution (Fig. 3.8a,b). The stability boundary, above which there is a gradual increase in the amplitude of leading shock pressure oscillations with time, is around $E_a = 25.26$–25.27 for the simulation parameter set considered (heat release $Q = 50$, $\gamma = 1.2$, $f = 1$). A detonation wave with activation energy above this boundary is thus unstable, as illustrated in Fig. 3.8c. The stability limit value obtained from the numerical simulation is generally in good agreement with the prediction of the linear analysis, i.e., $E_a = 25.28$ [139,143]. It is worth noting that due to the slow convergence of the final steady asymptotic solution, especially near the stability limits, simulations must be carried out for a sufficient time and with enough numerical resolution to determine properly the bifurcation from a stable one-dimensional detonation to a pulsating unstable detonation.

For multidimensional compressible reactive flow, there exist exact solutions for validating numerical algorithms (e.g., see [126]). However, it is rather difficult to propose a benchmark problem for two-dimensional detonation simulations; and the validation and comparison used in previous studies are usually qualitative, although there exist some attempts to systematically compare numerical solutions with two-dimensional linear stability results [13,145,172]. Two-dimensional detonation waves are characterized by an ensemble of interacting transverse waves that sweep laterally across the leading shock front of the detonation wave and contain different small-scale features. To resolve all the various length scales involved in the problem, a very high resolution is required, as pointed out in [142].

For two-dimensional detonation code validation, a standard test often considered in prior studies [13,65,129] is a 2D detonation with parameters set to $Q = 50$, $\gamma = 1.2$, $E_a = 10$, and $f = 1.2$. This corresponds to the case of high energy release and low activation energy, producing a regular cellular pattern with a complex structure of transverse waves. The same degree of numerical resolution as utilized by Bourlioux and Majda [13], i.e., 24 cells per $L_{1/2}$, is used in the present study so that direct comparisons can be made. The solution for the planar steady ZND wave is imposed as the initial condition and

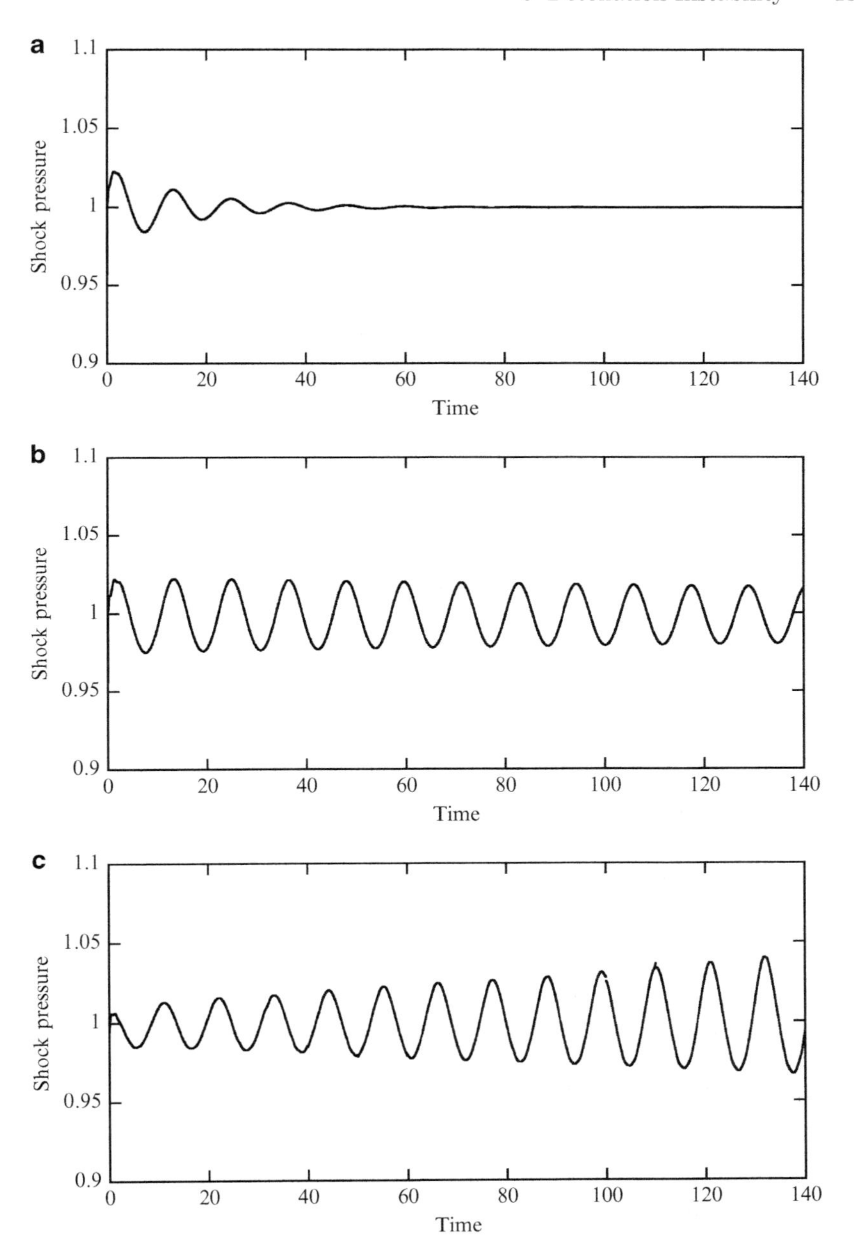

Fig. 3.8. Leading shock pressure history for activation energies close to the stability limit (**a**) $E_a = 24.00$; (**b**) $E_a = 25.24$; and (**c**) $E_a = 25.28$ with $Q = 50$, $\gamma = 1.2$ and $f = 1.0$) [109]

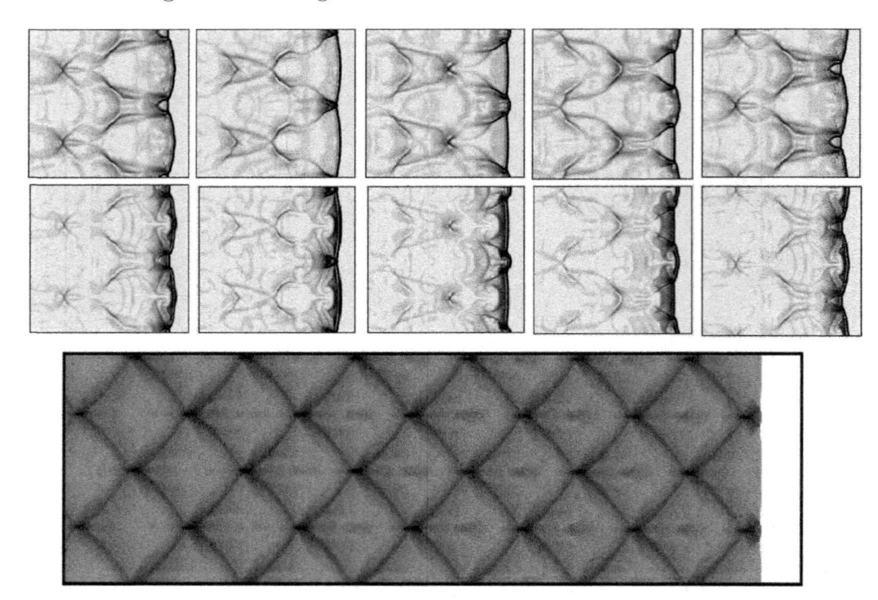

Fig. 3.9. Two-dimensional simulation of an overdriven detonation with $Q = 50$, $\gamma = 1.2$, $E_a = 10$ and $f = 1.2$, using 24 grid points per $L_{1/2}$ and channel width = 10 half reaction lengths (shown twice). (**a**) Sequence of five Schlieren-type plots showing the pressure (*top*) and density (*bottom*) flow field behind the shock; and (**b**) numerical smoked foil record

perturbed by introducing a small curvature into the front to accelerate the growth of transverse instabilities. The width of the computational domain is ten half-reaction zone lengths. Periodic boundary conditions are used along the top and bottom boundaries.

Any numerical technique should be able to capture the salient characteristics of the flow field behind the detonation as shown in the Schlieren-type or gradient plots of pressure and density given in Fig. 3.9a. The interactions of incident shocks, Mach stems, and transverse waves form a characteristic cell pattern, producing detonation cells. The cell pattern and the cell width can be compared as shown in Fig. 3.9b. Figure 3.9b is a typical numerical smoked foil showing the time-integrated maximum pressure contour from the simulation, which corresponds to the trajectories of the triple shock interactions (i.e., triple points). The obtained transverse characteristic length scale, i.e., the cell size, has a value of 10 $L_{1/2}$ for the given channel width. For other systematic verification and numerical analyses, an attempt was also carried out by Choi et al. [21] to systematically examine and discuss the effect of several numerical issues, including the effect of grid size, time step, computational domain, and boundary conditions, on the simulation of detonation cellular structure with variable stability regimes ranging from weakly to highly unstable detonations. Such a study provides some guidelines for numerical requirements needed to achieve valid solutions for detonation structures.

3.3.3 Nonlinear Dynamics of One-Dimensional Unstable Detonations

To elucidate the unstable structure of detonations, we begin by addressing the simplest configuration of one-dimensional, time-dependent dynamics of planar detonations with simplified one-step Arrhenius chemistry. Although advances in scientific computing now permit direct numerical simulations of multidimensional cellular detonation structures, the one-dimensional unsteady detonation should retain many of the essential physical features of multidimensional detonations, and yet is much simpler to investigate. In the one-dimensional treatment, the dynamic structure manifests itself through longitudinal pulsation.

The dynamics of pulsating CJ detonations have been widely studied in the literature using the dimensionless parameters with the values $Q = 50$, $\gamma = 1.2$ while varying activation energy E_a as a control parameter. The global activation energy, as shown in the linear stability analysis section, is a significant bifurcation parameter to characterize the stability of the detonation wave, and, indeed, a wide spectrum of 1D longitudinal instability dynamics can result from varying the global activation energy. ZND profiles computed using two different activation energies are shown in Fig. 3.10. In both cases, temperature begins to rise immediately behind the leading shock front. From these curves, it can be seen that increasing E_a has the effect of steepening the temperature profile of the steady ZND structure. Hence, the wave should be more sensitive to temperature perturbation at higher E_a, which will lead to different instability phenomena. One interesting remark which can be made from these ZND profiles is that by having an extremely high value of activation energy E_a, the thermally neutral induction zone can be roughly approximated. However, a large E_a will also result in a very thin heat release zone. Therefore, although convenient for analysis, a drawback of one-step Arrhenius kinetics is that

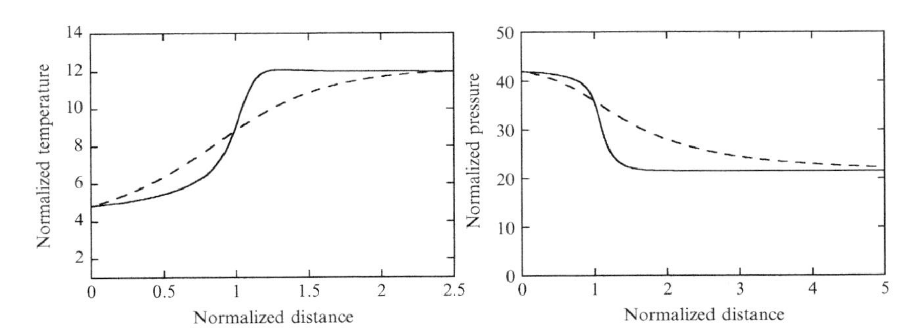

Fig. 3.10. Steady ZND detonation profiles for a mixture with $Q = 50$, $\gamma = 1.2$, $f = 1.0$ and two values of activation energies $E_a = 25.0$ (*dashed lines*); and $E_a = 50.0$ (*solid lines*)

independent variation of the two characteristic zone lengths is not possible.

Discussed here are very long-term high-resolution numerical simulations of an idealized pulsating detonation using single-step Arrhenius kinetics. As pointed out by Sharpe [141], it is important to allow the detonation to run for thousands of half-reaction times to ensure that the final nonlinear behavior of detonation propagation has been reached. After the initiation transient has passed, the "saturated" nonlinear dynamics of the pulsating front can be analyzed. As shown in Fig. 3.8, above the value of E_a for the neutral stability boundary, the detonation loses its stability and begins to oscillate; different modes of oscillation (signified by different values of peak amplitudes of the oscillation) begin to appear as the value of activation energy is continuously increased.

In fact, the computational results are essentially a time-series showing the evolution of the one-dimensional pulsating detonation front. In order to gain further insight into the nonlinear dynamics of these numerical results, Ait Abderrahmane et al. [3] analyzed such numerical data using classical nonlinear dynamics time-series analysis by reconstructing the attractors (or a multidimensional phase portrait) and studying their geometrical characterization. As described in Packard et al. [120], considering a variable $x(t)$, say shock pressure, is measured or numerically calculated at equal sampling time T_s, a time series can be written in the following form:

$$\{x(t_0), x(t_0+T_s), x(t_0+2T_s), \ldots, x(t_0+NT_s)\}. \tag{3.48}$$

This measured quantity $x(t)$ is presumed to represent one of n state variables which completely describe a dynamical system whose real trajectories lie on a d-dimensional ($d \leq n$) attractor in phase space. Using methods such as the Taken's time-delay embedding technique [96, 169], it is possible to *embed* the above time series and reconstruct the attractor in an m-dimensional embedding space by mean of the following (embedding) vectors:

$$\begin{aligned}
y_1 &= (x(t_0), x(t_0+\tau), \ldots, x(t_0+(m-1)\tau))^T \\
y_2 &= (x(t_0+l), x(t_0+l+\tau)), \ldots, x(t_0+l+(m-1)\tau)^T \\
&\ldots \\
y_s &= (x(t_0+(s-1)l), x(t_0+(s-1)l+\tau), \ldots, \\
&\quad x(t_0+(s-1)l+(m-1)\tau))^T,
\end{aligned} \tag{3.49}$$

where τ represents the *delay-time* or the *lag* between successive components of an embedding vector, and l is the number of sampling intervals between the first components of successive vectors [4, 16, 120, 138]. To provide further characterization of the attractors reconstructed from the time series, techniques such as the well-known *Grassberger–Procaccia algorithm* [58] and the interactive method by Albano et al. [4] that take into consideration the inherent noise within the data [1, 173] can be used to determine the dimension of the attractors, i.e., a measure of complexity of the attractor. In general, the correlation dimension can be used to segregate between deterministic and stochastic chaos. When the correlation dimension converges toward a constant value with the increase of the embedding dimension, m, the system is considered deterministic. Alternatively, the correlation dimension is considered as a lower bound of the number of state variables needed to describe the related dynamics. If the correlation dimension does not converge to a constant value but its value increases proportionally to the increase of the embedding dimension, m, then the system is considered to be a stochastic process where there exists an arbitrarily large number of degrees of freedom [4, 58, 138].

Both the numerical time-series results and their corresponding attractors are shown in Fig. 3.11. For an activation energy of 27.0, the instability manifests as a periodic oscillation and the attractor that depicts such a bifurcation in phase space is the limit cycle shown in Fig. 3.11a. Increasing the value of the activation energy to 27.40, the system of equations that models this detonation has undergone a second Hopf bifurcation. The time series of the pressure exhibits a two-period oscillation and a double limit cycle can be seen in the attractor shown in Fig. 3.11b. These two figures show that for any given initial condition, the system undergoes two oscillations before it comes back to its initial state. Increasing further the value of the control parameter to 27.8 and 27.82, the period-doubling bifurcation process repeats itself and yields attractors with four and eight cycles, respectively; see Fig. 3.11c, d. It is worth noting that higher modes of oscillations were also captured by Henrick et al. [66] using a shock-fitting numerical algorithm with a high order WENO scheme.

The appearance of high-mode oscillations via an infinite sequence of bifurcations eventually leads to an interesting phenomenon called chaos, a concept that has been well established in nonlinear dynamics. For $E_a = 28.0$, the results show an attractor which looks weakly chaotic; see Fig. 3.11e. Such an attractor is referred to in the nonlinear dynamics literature as a "noisy limit cycle"; it arises after the saturation of a doubling period cascade [11]. This attractor is formed with 2^n disjointed periodic cycles. Within each of these cycles the dynamics is chaotic, hence the cycles are called noisy. Increasing the activation energy slightly further to 28.2, the attractor become a three-period cycle, see Fig. 3.11f. It is a well-known result in the nonlinear mathematics literature that period-three implies chaos [87]. According to nonlinear theory, this three-period cycle will undergo, similar to the previous periodic cycles, a cascade of period doublings that leads to chaos again. Hence, the three-period cycle attractor is a periodic window inserted within a chaotic region. As expected, by further increasing the value of the activation energy to 28.3, the attractor becomes again chaotic. This time the corresponding attractor, shown in Fig. 3.11g, exhibits more chaotic dynamics. Shown in Fig. 3.11h is the time series and attractor that correspond to the value of the activation energy equal to 30.00.

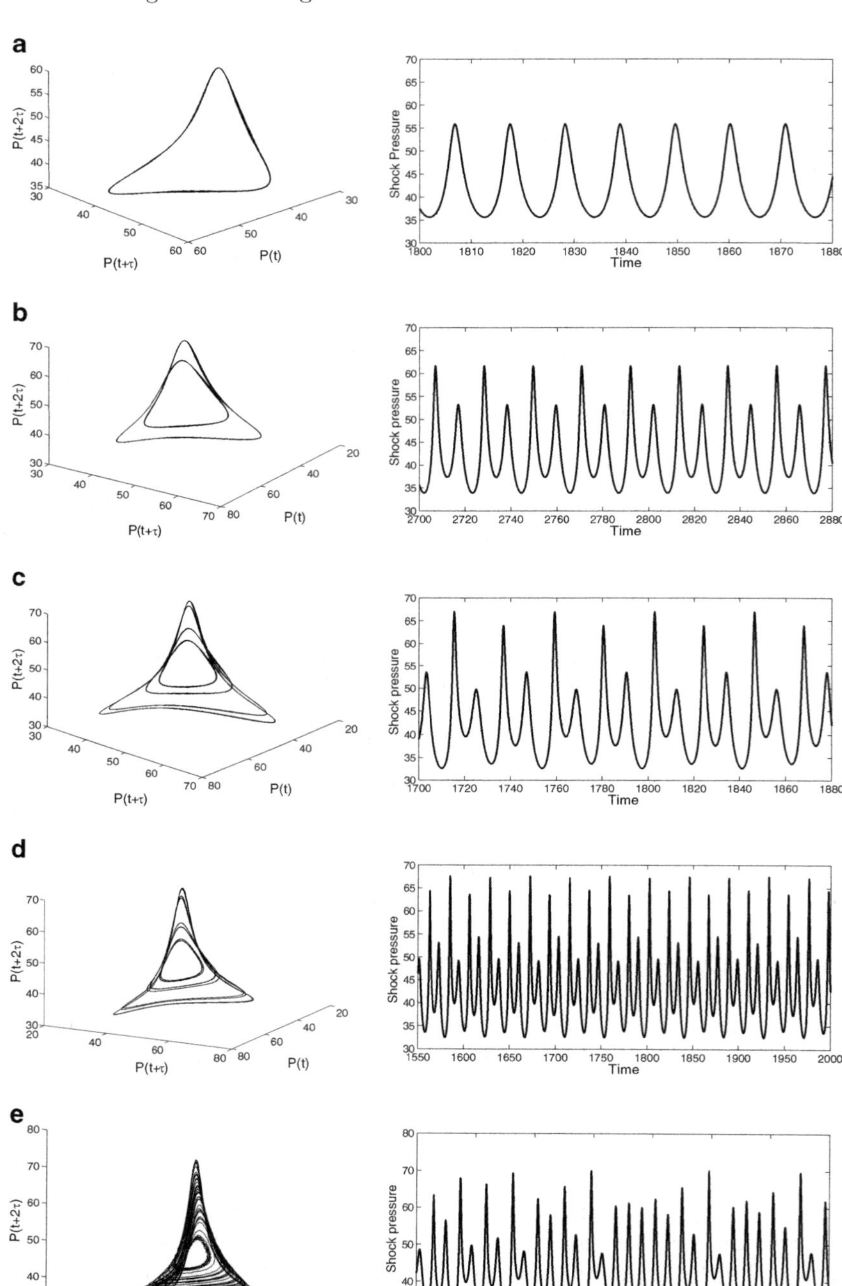
a
b
c
d
e
P(t+2τ)
P(t+τ)
P(t)
Shock Pressure
Shock pressure
Time

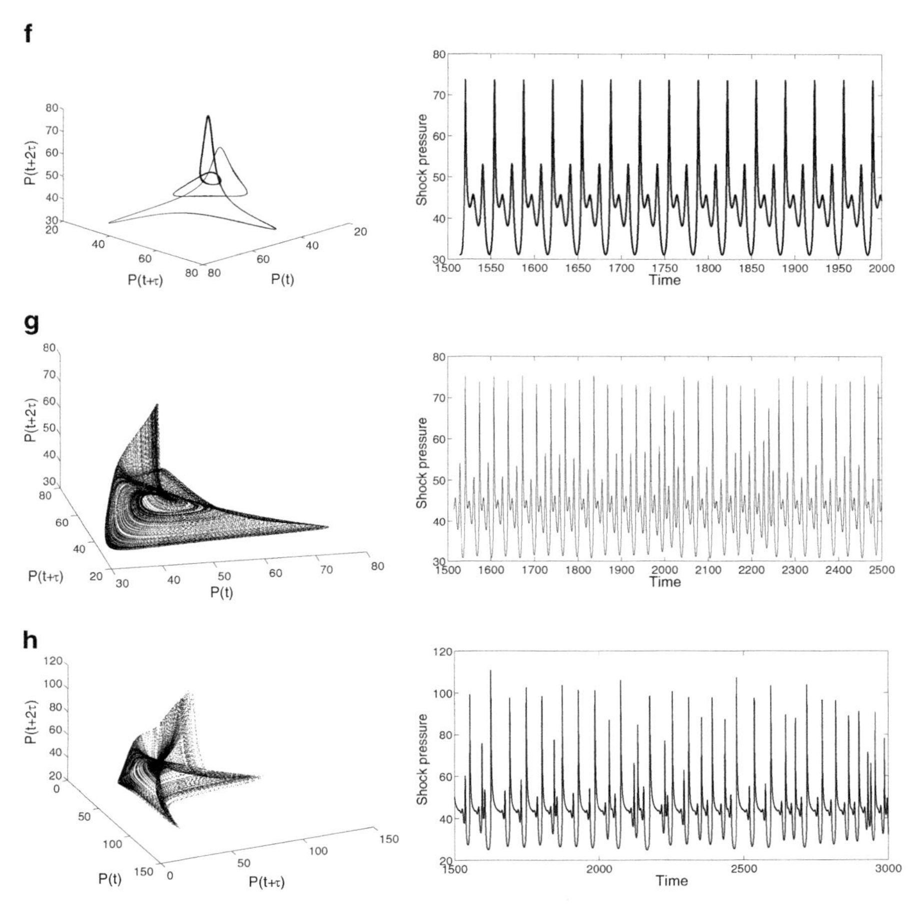

Fig. 3.11. Time-series solution showing the one-dimensional detonation front evolution and its attractor for different activation energies $E_a = 27.00$; 27.40; 27.80; 27.82; 28.00; 28.20; 28.30 and 30.00 [3]

Another powerful technique commonly used to distinguish the regularity of time-series data is to compute the power spectral density (PSD). The PSD describes how the power (or variance) of a time series is distributed with frequency. Mathematically, it is defined as the Fourier Transform of the autocorrelation sequence of the time series. The PSD of the leading shock pressure history for four different activation energies and modes ($E_a = 27.0$; 27.40, 27.80. and 30.0) is shown in Fig. 3.12. For the first three activation energies, the plots show very large peaks at dominant frequency modes and small amplitude peaks at their second harmonics, which are exactly two times the fundamental frequency. In other words, the energy is concentrated at this frequency leading to a regular oscillation. On the other hand, with a large

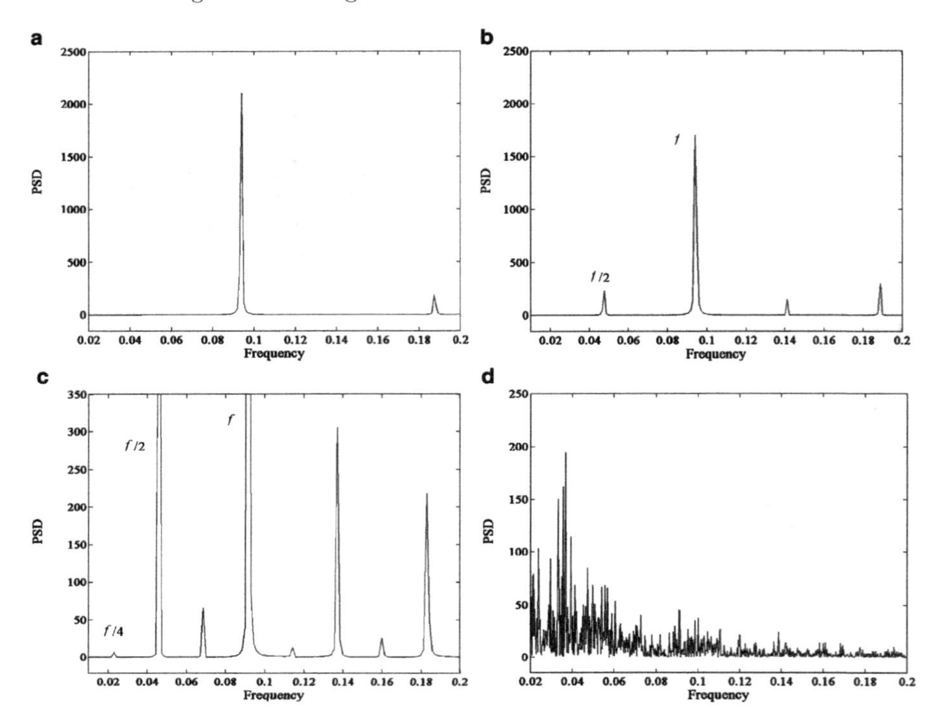

Fig. 3.12. Power spectral density (PSD) calculated using time-series solution of the one-dimensional detonation wave with (**a**) $E_a = 27.00$; (**b**) 27.40; (**c**) 27.80; and (**d**) 30.00 [109]

activation energy ($E_a = 30.0$), there are no isolated power spikes, meaning the energy is distributed over a wide range in the frequency spectrum. This again indicates the existence of chaos in the system. It is worth noting that a broadband power spectrum is typical not only for external noise (experimental or numerical), but also for complex nonlinear dissipative systems that exhibit low dimensional deterministic chaos.

Once a chaotic phenomenon is identified, it is useful to obtain some quantitative geometric characterization. It is also important to segregate this phenomenon between deterministic and stochastic chaos. The geometrical characterization, i.e., the correlation dimension of the attractors of 1D pulsating detonations, is summarized in Table 3.1. The dimension of the attractors is around one for the limit cycles in Fig. 3.11a–d. The dimension of the attractors increases after the saturation of the double-period cascade to 2 and 2.93, respectively, for the activation energies 28.0 and 28.1. These two high dimensions for the "noisy limit cycles" confirm the presence of chaotic dynamics within the limit cycles. After that, the correlation dimension decreases toward one (period-three limit cycle) when the activation energy was increased to 28.2; this confirms the existence of a periodic window in the bifurcation param-

Table 3.1. Correlation dimensions for the attractors calculated using Grassberger and Procaccia method [58]

Attractor for E_a	D_2
27.00	1.0087
27.40	0.9910
27.80	1.0153
27.83	1.0343
28.00	2.0004
28.10	2.9317
28.20	1.0466
28.30	1.5953
30.00	2.8404

eter range. The increase in correlation dimension that follows the increase of the control parameter E_a to 28.3 and 30 indicates that the three-period limit cycle has undergone another series of period doubling cascades that yield chaos (the dimensions of such attractors are non-integer). The convergence of the correlation dimension toward a non-integer constant value of 2.84 at a high value of the control parameter $E_a = 30$ indicates that the dynamics of the detonation is deterministic chaos and not stochastic.

Another quantitative way to determine the existence of chaos is provided by the Lyapunov exponent [40]. This quantity is a measure of sensitive dependence on initial conditions. It quantifies the exponential rate of divergence of initially neighboring phase-space trajectories and estimates the degree of chaos in a system. For any time series in a dynamical system, the presence of a positive, largest characteristic exponent indicates that the trajectories comprising an attractor diverge over a long time average, which is characteristic for chaotic behavior. This divergence describes the rapid loss of the system's memory of its previous history as time evolves. There are several algorithms for estimating the positive Lyapunov exponent directly from a time series [40, 73, 135, 190], etc.

Following Rosenstein et al.'s approach, Fig. 3.13 shows some typical plots of the natural log of divergence versus evolution time of initially close state-space trajectories for the case of $E_a = 28.17$ and for the large activation energy $E_a = 30.00$. Different curves for each plot correspond to different embedded dimensions assumed for the system [135]. The linear region means that the divergence of nearest neighbors is exponential, where the slope can be interpreted as a measure of the largest Lyapunov exponent. In both cases, a positive Lyapunov exponent is found, which indicates the presence of chaos in the system and that the system is sensitive to the initial conditions. Note that the magnitude also provides a measure of the degree of chaos, i.e., the rate of divergence. In a more general sense, the Lyapunov exponent also provides another way to characterize whether systems are likely to be dynamically irregular (if positive) or not (otherwise).

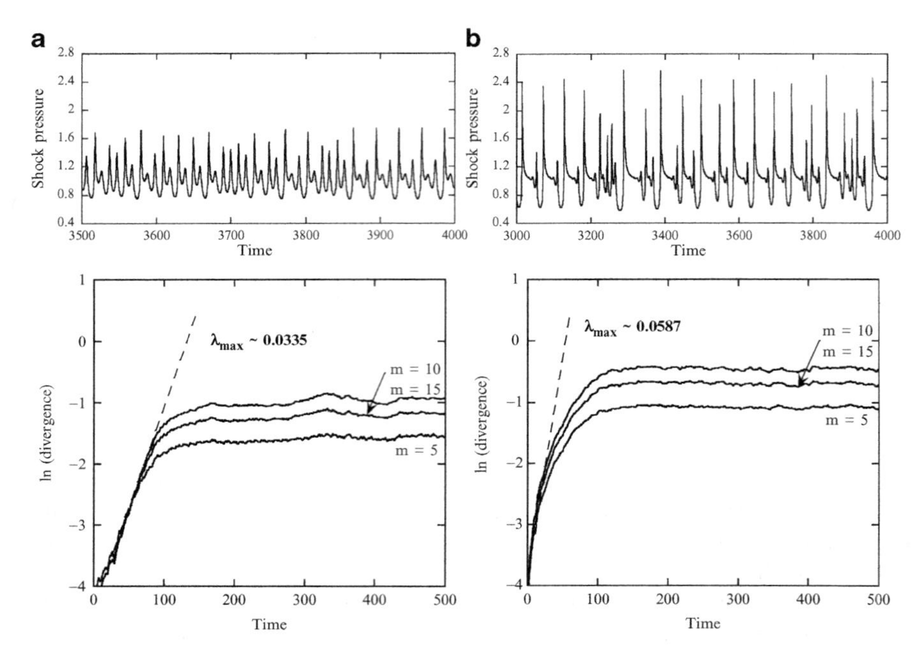

Fig. 3.13. Calculation of the largest Lyapunov exponent with different embedded dimensions for the case of activation energies (**a**) $E_a = 28.17$; and (**b**) $E_a = 30.00$ [109]

In nonlinear dynamics theory, finding and characterizing the routes leading to chaos have been the main focus of research. Three universal routes to chaos were typically identified, i.e., Feigenbaum, intermittency, and quasi-periodic routes [11]. To depict the transition of different modes in a pulsating detonation, a bifurcation diagram plotting the peak amplitudes that define the oscillation modes or periods as a function of a varying parameter can be used. A very high resolution version of such bifurcation diagram was obtained by Henrick et al. [66] given in Fig. 3.14. At low activation energy (below the neutral stability limit of 25.265), the detonation front is stable and hence no oscillation amplitude is recorded. Above this limit, the propagating front starts to oscillate and a single peak pressure amplitude can be found, represented by a single point in the bifurcation diagram. The amplitude of the oscillation continues to grow as E_a is increased and eventually a period-doubling bifurcation occurs. Further bifurcation processes follow as the activation energy continues to increase. At each bifurcation i, the appearance of a sub-harmonics with $f/2^i$ where f is the fundamental frequency can also be seen in the PSD plots. (For example, see Fig. 3.12 for the oscillations with period-two and period-four.) This cascade of period-doubling bifurcations can eventually lead to a very irregular oscillation with different peak amplitudes.

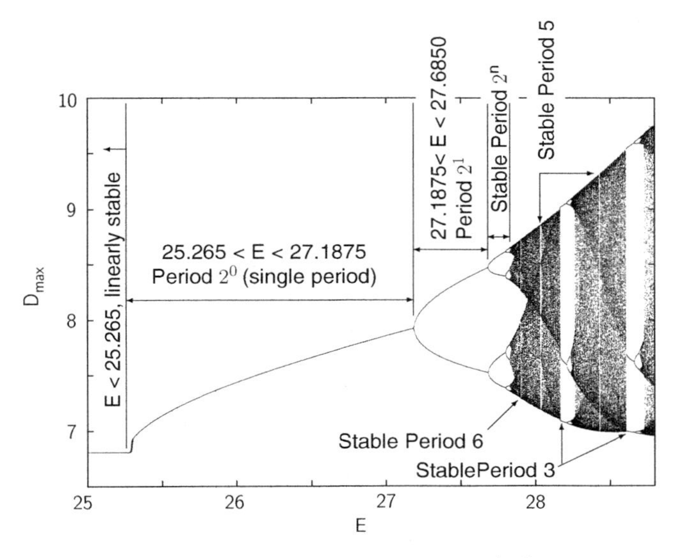

Fig. 3.14. Bifurcation diagram [66]

Table 3.2. Numerically determined bifurcation points and an approximation to Feigenbaum's number (Henrick et al. [66])

n	E_n	$E_{n+1} - E_n$	δ_n
0	25.265 ± 0.005	–	–
1	27.1875 ± 0.0025	1.9225 ± 0.0075	3.86 ± 0.05
2	27.6850 ± 0.001	0.4975 ± 0.0325	4.26 ± 0.08
3	27.8017 ± 0.0002	0.1167 ± 0.0012	4.66 ± 0.09
4	27.82675 ± 0.00005	0.02505 ± 0.00025	–

As discussed previously and now shown clearly in the bifurcation diagram, the early stage of the instability transition follows the generic pattern of the period-doubling, i.e., by systematically increasing the activation energy, instability of the detonation front evolves away from a single-mode oscillation to a period-two oscillation, and multi-mode oscillations begin to appear as the value of activation energy is further increased. The route thus follows closely the Feigenbaum scenario, which involves successive subharmonic bifurcations [47–49]. It is interesting to note that the bifurcations come faster and faster until the system becomes aperiodic.

Table 3.2 shows the values of activation energy where transitions occur. To further characterize the bifurcation diagram and provide evidence for the validity of the Feigenbaum scenario describing the 1D pulsating detonation phenomenon, these values are used to determine the Feigenbaum number [47], which is defined as the ratio between the spacing of successive bifurcations, i.e.:

$$\delta = \frac{\mu_i - \mu_{i-1}}{\mu_{i+1} - \mu_i}, \tag{3.50}$$

where μ_i is the location of the bifurcation point. For equivalent bifurcation processes occurring in the logistic model as well as other simple nonlinear dynamical systems, Feigenbaum discovered a universal feature that the bifurcations should be occurring at a ratio that converges to a natural constant that is approximately 4.669 in the asymptotic limit where the number of bifurcations goes to infinity, independent of the particular system. It may be worthwhile to point out that this convergent Feigenbaum cascade of bifurcations has also been observed in a number of experimental studies of different types: Rayleigh-Bénard flow in a flat convective layer of liquid heated from below, circular Couette flow, acoustic cavitation noise, a driven nonlinear semiconductor oscillator, shallow water waves in a resonator, optical turbulence in a hybrid optically bistable system, the Belousov–Zhabotinskii reaction in a well-stirred flow reactor, etc. [61]. Some values of the Feigenbaum number from the bifurcation diagram are shown in Table 3.2, which also appear to converge to this universal value. Overall this result appears to support the conjecture that the transition process of the pulsating detonation front closely follows Feigenbaum's picture of a simple nonlinear model with fewer degrees of freedom.

From a detailed nonlinear dynamic analysis with the aid of phase portraits, correlation dimensions, power spectrums, and bifurcation diagrams, it was shown that when the control parameter (i.e., activation energy) was systematically increased, 1D Chapman–Jouguet (CJ) detonations undergo a series of Feigenbaum bifurcations (period-doubling cascade) before chaotic behavior sets in. A closer examination of the results demonstrated the existence of chaos in the system. The analysis also showed that the resulting chaos following the route of period-doubling is deterministic.

Mathematically, chaos means the system contains an infinite number of periodic orbits. A more physical definition is aperiodic dynamics in a deterministic system demonstrating sensitive dependence on initial conditions [74, 90]. For a chaotic system, a small deviation in the initial conditions will result in an exponentially growing departure in its specific temporal behavior. Equivalently, a small change in the pulsating structure at a given starting time would result after a short period of time in a completely different oscillatory history (or specific trajectory through the phase space). There are several regions in the bifurcation diagram where highly aperiodic oscillations are found. These results can be further analyzed for the existence of deterministic chaos in the system.

The chaotic variability thus makes the assessment of numerical grid convergence difficult as differences observed on different grids are due to the nature of chaos rather than to any inherent lack of convergence. Because of the chaotic nature of the phenomenon, to examine grid convergence, very long sample time series must be looked at over many runs for unstable chaotic detonation front propagation, and statistical ensembles must be considered to properly describe the physical dynamics using a nonlinear model. In higher dimensions, the effect of chaos may play a prominent role in determining the natural cell size or the cell pattern for unstable detonations thus leading to a range of equally probable cell size distributions [147].

3.3.4 Nonlinear One-Dimensional Oscillator Model

The remarkable resemblance between a simple nonlinear dynamical system and 1D pulsating detonations has been previously recognized by several authors [12, 143, 155, 199], etc. Results from nonlinear dynamics and chaos analysis, such as the Feigenbaum's period-doubling sequence as the activation energy increases, have indicated a strong similarity between 1D pulsating detonations and simple nonlinear mechanical systems. This suggests that it may be possible that the dynamics of the pulsating detonation front can be simply modeled by a nonlinear oscillator equation. This problem can perhaps provide a useful analogy for the one-dimensional detonation to gain further insight on how the oscillatory structure is influenced by different factors such as chemical kinetics.

Such an analogy has been attempted previously by Zhang et al. [199]. Their formulation considers a control volume as shown in Fig. 3.15. It moves with the detonation complex and is bounded by the leading shock front and a rear boundary. A suitable choice for that rear boundary can be defined by the limiting characteristic and its location can be estimated from a numerical simulation. Despite the unstable nature of the pulsating detonation, the existence of the rear information boundary can be demonstrated from Fig. 3.16. It gives the $x' - t$ diagram showing the $u' + c$ characteristic lines behind the leading shock front at which the coordinate is fixed (i.e., u' is the particle velocity in the shock-attached frame) for $E_a = 27$. From this plot, despite the unstable oscillatory nature of the propagating front, the domain of influence of the shock front is essentially bounded by a limiting characteristic and the

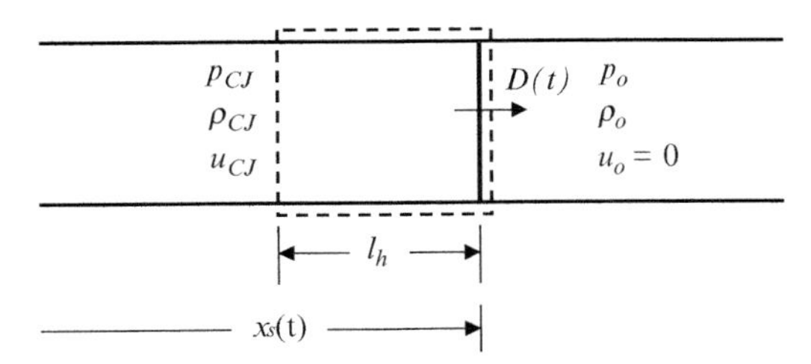

Fig. 3.15. Control volume for the formulation of the nonlinear oscillator model

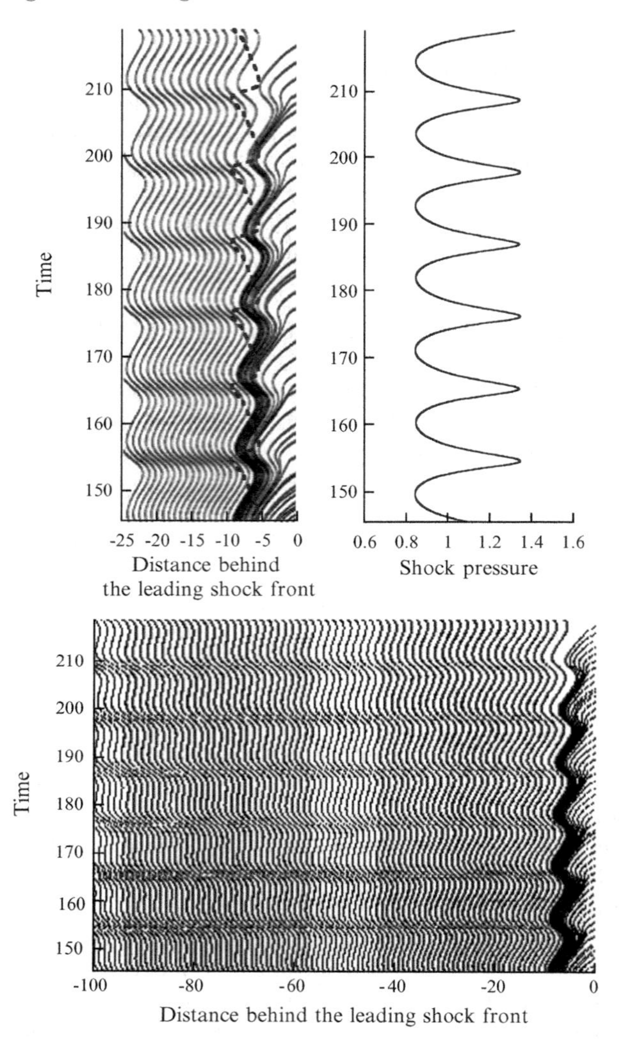

Fig. 3.16. Characteristic plot showing the limiting $u' + c$ characteristics as a rear boundary and the location of the end of the reaction zone

shock. The limiting characteristic acts as a boundary such that the characteristic lines of any trailing nonsteady expansion in the downstream flow field are incapable of penetrating and influencing the detonation structure [76–78]. This information boundary thus ensures the self-contained and autonomous nature of the unstable detonation. The notion of this limiting characteristic can have significant implications on the analytical treatment of unsteady detonation waves as it may provide a closure condition for the boundary-value problem encountered from the derivation of the evolution equation using asymptotic analysis [76–78]. On the same plot, the location where the reaction progress variable has been depleted to $\beta = 10^{-5}$ is also indicated by the dashed line. It is noted that the location where the chemical reaction is essentially complete does not necessarily coincide with the limiting characteristic. The distance of the rear boundary from the leading shock is sometimes found to be longer than the reaction zone length. This implies that, unlike the steady one-dimensional ZND detonation, it is not sufficient to require the termination of chemical reactions to define the self-contained system for an unstable detonation. Thermodynamic and hydrodynamic equilibrium may require a longer span to be achieved. It is thus important to note that a meaningful autonomous structure must include both the thermodynamic and hydrodynamic equilibrium within it.

The distance between the leading shock and the limiting characteristic is referred to as the hydrodynamic thickness $\tilde{l}_h$. Experimentally, the distance from the leading shock to the rear information boundary in an average sense is generally referred to as hydrodynamic thickness of real cellular detonations [85, 163]. Recent studies also suggest that the hydrodynamic thickness may perhaps provide a suitable length scale that characterizes the detonation structure within which the various chemical and gasdynamic mechanisms responsible for sustaining the unstable detonation are contained. It should have significant implications for understanding the propagation and developing models of real multidimensional detonations, which are always nonsteady and highly unstable [84, 134].

Nevertheless, in the derivation of the nonlinear oscillator model, the exact location of this rear boundary is not required, and it is sufficient to just assume such a rear boundary exists. Assume that the time-averaged velocity of the detonation front is in good agreement with the steady-state CJ value $\tilde{D}_{\mathrm{CJ}}$, i.e.,

$$\tilde{D}_{\mathrm{CJ}} = \frac{1}{\tilde{\tau}} \int_0^{\tilde{\tau}} \tilde{D}\left(\tilde{t}\right) \mathrm{d}\tilde{t}. \tag{3.51}$$

Therefore, the instantaneous shock front location $\tilde{x}_s$ can be given by:

$$\tilde{x}_s = \tilde{D}_{\mathrm{CJ}} \cdot \tilde{t} + \tilde{F}\left(\tilde{t}\right), \tag{3.52}$$

where $\tilde{F}(\tilde{t})$ denotes the displacement of the shock front from its mean trajectory and satisfies:

$$\frac{1}{\tilde{\tau}} \int_0^{\tilde{\tau}} F\left(\tilde{t}\right) \mathrm{d}\tilde{t} = 0. \tag{3.53}$$

Differentiating the shock front trajectory gives the shock velocity as:

$$\tilde{D} = \dot{\tilde{x}}_s\left(t\right) = \tilde{D}_{\mathrm{CJ}} + \dot{\tilde{F}}\left(\tilde{t}\right), \tag{3.54}$$

where $\dot{\tilde{F}}(\tilde{t})$ is the fluctuating velocity with respect to its time-averaged part, $\tilde{D}_{\mathrm{CJ}}$. The fluctuation of the shock trajectory $\tilde{F}(\tilde{t})$ is chosen as the dependent variable in the oscillator equation to be formulated.

In a shock-attached coordinate system, the conservation equations of mass, momentum, and energy for the control volume shown in Fig. 3.15 can be written as:

$$\begin{aligned}
&\tfrac{\mathrm{d}}{\mathrm{d}\tilde{t}}\int_0^{\tilde{l}_h} \tilde{\rho}\mathrm{d}\tilde{x} + \tilde{\rho}_{\mathrm{CJ}}\left(\tilde{D}-\tilde{u}_{\mathrm{CJ}}\right) - \tilde{\rho}_0\tilde{D} = 0\\
&\tfrac{\mathrm{d}}{\mathrm{d}\tilde{t}}\int_0^{\tilde{l}_h} \tilde{\rho}\left(\tilde{D}-\tilde{u}\right)\mathrm{d}\tilde{x} + \tilde{\rho}_{\mathrm{CJ}}\left(\tilde{D}-\tilde{u}_{\mathrm{CJ}}\right)^2 - \tilde{\rho}_0\tilde{D}^2 = -\tilde{p}_{\mathrm{CJ}} + \tilde{p}_0 + \int_0^{\tilde{l}_h}\tilde{\rho}\tfrac{\mathrm{d}\tilde{D}}{\mathrm{d}\tilde{t}}\mathrm{d}\tilde{x}\\
&\tfrac{\mathrm{d}}{\mathrm{d}\tilde{t}}\int_0^{\tilde{l}_h} \tilde{\rho}\left[\tilde{e}_i + \tfrac{\left(\tilde{D}-\tilde{u}\right)^2}{2}\right]\mathrm{d}\tilde{x} + \tilde{\rho}_{\mathrm{CJ}}\left(\tilde{D}-\tilde{u}_{\mathrm{CJ}}\right)\left[\tilde{e}_{i,\mathrm{CJ}} + \tfrac{\left(\tilde{D}-\tilde{u}_{\mathrm{CJ}}\right)^2}{2} + \tfrac{\tilde{p}_{\mathrm{CJ}}}{\tilde{\rho}_{\mathrm{CJ}}}\right]\\
&-\tilde{\rho}_0\tilde{D}\left(\tilde{e}_{si,0} + \tfrac{\tilde{D}^2}{2} + \tfrac{\tilde{p}_0}{\tilde{\rho}_0}\right) = \tilde{Q}\int_0^{\tilde{l}_h}\tilde{\rho}\tilde{\Omega}\mathrm{d}\tilde{x} + \int_0^{\tilde{l}_h}\tilde{\rho}\tfrac{\mathrm{d}\tilde{D}}{\mathrm{d}\tilde{t}}\left(\tilde{D}-\tilde{u}\right)\mathrm{d}\tilde{x},
\end{aligned} \tag{3.55}$$

where $\tilde{\rho}, \tilde{u}, \tilde{p}$, and $\tilde{e}_i$ are the density, particle velocity in laboratory frame, pressure, and the specific internal energy, respectively. $\tilde{Q}$ and $\tilde{\Omega}$ represent the amount of heat release and the reaction rate. The subscript "0" refers to the quiescent state and the subscript "CJ" denotes the Chapman–Jouguet equilibrium state. By combining with the continuity and momentum equations, the energy equation can be rearranged as follows:

$$\frac{\mathrm{d}}{\mathrm{d}\tilde{t}}\int_0^{\tilde{l}_h} \tilde{\rho}\tilde{e}\mathrm{d}\tilde{x} + \tilde{\rho}_{\mathrm{CJ}}\tilde{e}_{\mathrm{CJ}}\left(\tilde{D}-\tilde{u}_{\mathrm{CJ}}\right) - \tilde{\rho}_0\tilde{e}_0\tilde{D} - \tilde{p}_{\mathrm{CJ}}\tilde{u}_{\mathrm{CJ}} = \tilde{Q}\int_0^{\tilde{l}_h}\tilde{\rho}\tilde{\Omega}\mathrm{d}\tilde{x}, \tag{3.56}$$

where the variable $\tilde{e}$ is the specific total energy as a sum of both the internal and kinetic energy, $\tilde{e} = \tilde{e}_i + \frac{\tilde{u}^2}{2}$. For a steady ZND detonation where $\tilde{D}(\tilde{t}) = \tilde{D}_{\mathrm{CJ}}$, the unsteady integral term drops out and equation reduces to

$$\tilde{\rho}_{\mathrm{CJ}}\tilde{e}_{\mathrm{CJ}}\left(\tilde{D}-\tilde{u}_{\mathrm{CJ}}\right) - \tilde{\rho}_0\tilde{e}_0\tilde{D} - \tilde{p}_{\mathrm{CJ}}\tilde{u}_{\mathrm{CJ}} = \tilde{Q}\int_0^{\tilde{l}_h}\tilde{\rho}^0\tilde{\Omega}^0\mathrm{d}\tilde{x}, \tag{3.57}$$

where the superscript "0" refers to the steady ZND solution. The integral term in the above equation is given by

$$\int_0^{\tilde{l}_h}\tilde{\rho}^0\tilde{\Omega}^0\mathrm{d}\tilde{x} = \tilde{\rho}_0\tilde{D}_{\mathrm{CJ}}, \tag{3.58}$$

which simply states that the rate of reactant depletion inside the control volume is equal to the unburned mass flux entering the control volume for the steady ZND wave. If the steady-state component is subtracted from the energy equation, then

$$\frac{\mathrm{d}}{\mathrm{d}\tilde{t}}\int_0^{\tilde{l}_h}\tilde{\rho}\tilde{e}\mathrm{d}\tilde{x}+\tilde{\rho}_{\mathrm{CJ}}\tilde{e}_{\mathrm{CJ}}\dot{\tilde{F}}=\tilde{Q}\int_0^{\tilde{l}_h}\tilde{\rho}\tilde{\Omega}\mathrm{d}\tilde{x}-\tilde{Q}\tilde{\rho}_0\tilde{D}_{\mathrm{CJ}}, \tag{3.59}$$

where the term $\tilde{\rho}_0\tilde{e}_0\dot{\tilde{F}}$ has been neglected since $\tilde{\rho}_0\tilde{e}_0/\tilde{\rho}_{\mathrm{CJ}}\tilde{e}_{\mathrm{CJ}}$ is of the order $1/M_{\mathrm{CJ}}^2 << 1$. By normalizing the dimensional flow variables, (3.59) becomes:

$$\frac{\mathrm{d}}{\mathrm{d}t}\int_0^{l_h}\rho e\mathrm{d}x+\rho_{\mathrm{CJ}}e_{\mathrm{CJ}}\dot{F}=Q\int_0^{l_h}\rho\Omega\mathrm{d}x-QD_{\mathrm{CJ}}. \tag{3.60}$$

It may also be useful to express both the reaction rate and the total energy as the following nondimensional volume-averaged quantities:

$$\Psi=\int_0^1\frac{\rho\Omega}{D_{\mathrm{CJ}}l_h^{-1}}\mathrm{d}\zeta$$

$$I=\int_0^1\frac{\rho e}{D^2/2}\mathrm{d}\zeta, \tag{3.61}$$

where $\zeta = x/l_h$. Using these definitions, the following oscillator equation can be obtained:

$$l_hI\ddot{F}+\frac{\rho_{\mathrm{CJ}}e_{\mathrm{CJ}}}{D}\dot{F}+\frac{l_hD}{2}\dot{I}=\frac{D_{\mathrm{CJ}}Q}{D}\Psi_1, \tag{3.62}$$

where $\ddot{F}$ and $\dot{F}$ are the nondimensional shock acceleration and the fluctuation of the shock velocity with respect to the steady ZND wave. $\Psi_1 = \Psi - 1$ is the fluctuation of the dimensionless mean rate of the reactant depletion within the control volume. $\dot{I}$ denotes the rate of the fluctuation of the nondimensional total specific energy inside the control volume since:

$$\dot{I}=\frac{\mathrm{d}}{\mathrm{d}t}\left(I-I^0\right)=\frac{\mathrm{d}}{\mathrm{d}t}\left(\int_0^1\frac{\rho e}{D^2/2}\mathrm{d}\zeta-\int_0^1\frac{\rho^0e^0}{D_{\mathrm{CJ}}^2/2}\mathrm{d}\zeta\right)=\frac{\mathrm{d}}{\mathrm{d}t}\int_0^1\frac{\rho e}{D^2/2}\mathrm{d}\zeta. \tag{3.63}$$

This final equation essentially gives a balance relation for the rate of energy fluctuation in the control volume. The first term on the left-hand side of (3.62) denotes the energy fluctuation rate associated with the shock velocity oscillation. The second term corresponds to the unsteady exit condition at the rear boundary. The third term represents the rate of the energy fluctuation within the control volume itself, and the term on the right-hand side denotes the fluctuation of the chemical energy release rate inside the control volume. Note that this equation has the form of a second order nonlinear differential equation in terms of the variable F similar to that describing a nonlinear oscillator.

The validity of the above equation to model the instability pattern of the pulsating detonation can be confirmed by fitting it with the computational results using a gauging procedure. The front evolution equation is integrated in which the coefficients for each term are extracted from the flowfield results of numerical simulations. Figure 3.17 compares the solutions from the full direct numerical simulation as shown in Fig. 3.11 with those obtained from the nonlinear evolution equation. Very good agreement is achieved and different oscillation modes of the front can indeed be recovered by the solution of the oscillator equation.

It is in fact possible to write the evolution equation in a more analogous form of a classical nonlinear oscillator. Assuming that the shock speed is slowly varying compared to the time scale l_h/D_{CJ}, the integrands for a mass element

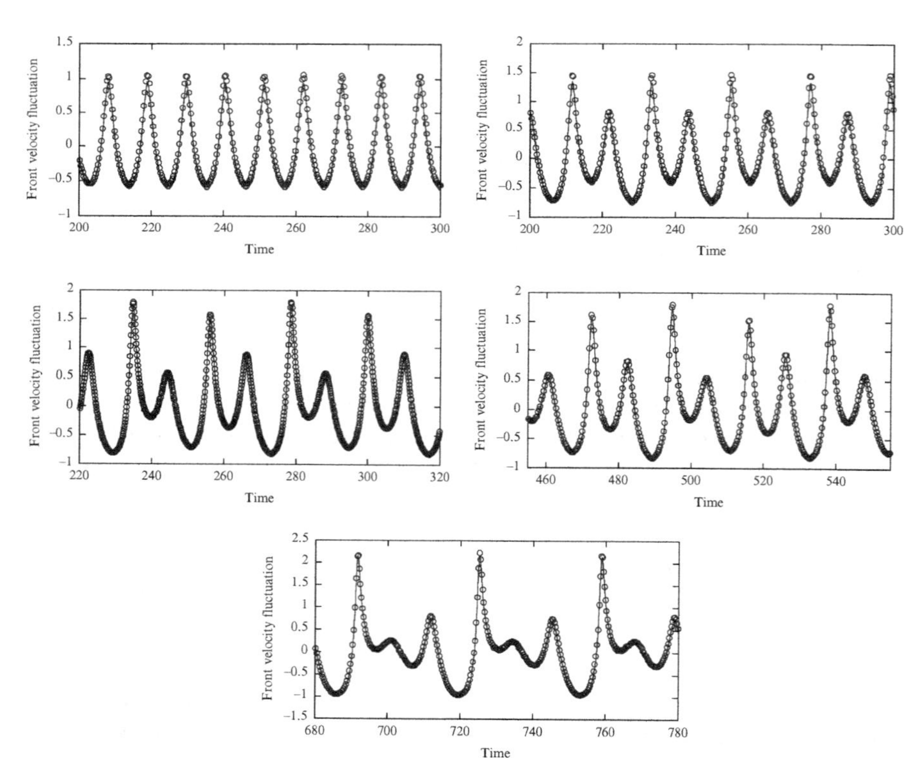

Fig. 3.17. Comparison between the results obtained from the oscillator model (*solid line*) and direct numerical simulations (*data points*) for $E_a = 27.00$; $E_a = 27.40$; $E_a = 27.80$; $E_a = 27.82$; and $E_a = 28.20$

can be expressed, to leading order, in terms of the shock speed at time t',

$$D_{\mathrm{CJ}} + \dot{F}(t') = D_{\mathrm{CJ}} + \dot{F}(t) - (t - t')\,\ddot{F}(t) + O\left[(t - t')\right]. \tag{3.64}$$

In this situation, at any time, the structure of the detonation looks like that of a stationary wave traveling at speed $D_{\mathrm{CJ}} + \dot{F}(t)$, but with corrections due to the wave acceleration $\ddot{F}(t)$. Without carrying out the details here of the necessary integrations, the following asymptotic results should be expected,

$$\frac{\mathrm{d}}{\mathrm{d}t}\int_0^{l_h} \rho e \mathrm{d}x = \Phi(t) = \Phi^0\left(\dot{F}(t)\right) + \Phi^1\left(\dot{F}(t)\right)\ddot{F}(t) + \cdots$$

$$Q\int_0^{l_h} \rho\Omega \mathrm{d}x - QD_{\mathrm{CJ}} = \Psi(t) = \Psi^0\left(\dot{F}(t)\right) + \Psi^1\left(\dot{F}(t)\right)\ddot{F}(t) + \cdots, \tag{3.65}$$

where the higher order corrections of the asymptotic expansions are neglected. The quantities Φ^0 and Ψ^0 are, respectively, the energy and the rate of heat release inside the control volume for a stationary wave traveling at speed $D_{\mathrm{CJ}} + \dot{F}(t)$. Φ^1 and Ψ^1 correspond to the coefficients of the temporal change of the energy and the rate of heat release caused by the wave acceleration, respectively. Substituting the above expansion into the oscillator equation gives the following expression:

$$\Phi^1\left(\dot{F}\right)\dddot{F} + \left(\frac{\mathrm{d}\Phi^0\left(\dot{F}\right)}{\mathrm{d}\dot{F}} + \frac{\mathrm{d}\Phi^1\left(\dot{F}\right)}{\mathrm{d}\dot{F}}\ddot{F} - \Psi^1\left(\dot{F}\right)\right)\ddot{F} + \rho_{\mathrm{CJ}} e_{\mathrm{CJ}}\dot{F} - \Psi^0\left(\dot{F}\right) = 0. \tag{3.66}$$

The above equation is, in essence, the balance relation of the rate of the energy fluctuation recast in the form of a second-order differential equation with respect to $\dot{F}(t)$. Comparing this equation with the classical nonlinear oscillator equation, i.e.,

$$m\ddot{x} + kx = \mu f(x, \dot{x}) \tag{3.67}$$

the analogy is clear. It can be seen that the first term $\Phi^1(\dot{F})\dddot{F}$ can indeed represent the inertia force (i.e., $m\ddot{x}$), the third term $\rho_{\mathrm{CJ}} e_{\mathrm{CJ}}\dot{F} - \Psi^0(\dot{F})$ is similar to the spring or restoring force (i.e., kx) in which $\Psi^0(\dot{F})$ provides a possible "nonlinear restoring force" responsible for different nonlinear instabilities. The term,

$$\left(\frac{\mathrm{d}\Phi^0\left(\dot{F}\right)}{\mathrm{d}\dot{F}} + \frac{\mathrm{d}\Phi^1\left(\dot{F}\right)}{\mathrm{d}\dot{F}}\ddot{F} - \Psi^1\left(\dot{F}\right)\right)\ddot{F} \tag{3.68}$$

represents the self-excited driving force $\mu f(x, \dot{x})$ that provides the damping responsible for the oscillatory behavior. In this form, the key components for the pulsating detonation analogous to the mass, spring, and damper elements in a mechanical oscillator can be identified. Furthermore, from the point of view of the oscillator concept, the effect of chemical kinetics on the self-sustained pulsating detonation and the origin of different unstable temporal patterns can be discussed. The rate of heat release, which is governed by chemical kinetics, essentially affects the key components for the mechanism of the self-sustained oscillation, i.e., the self-exciting driving force represented by the damping term $\left(\frac{\mathrm{d}\Phi^0(\dot{F})}{\mathrm{d}\dot{F}} + \frac{\mathrm{d}\Phi^1(\dot{F})}{\mathrm{d}\dot{F}}\ddot{F} - \Psi^1(\dot{F})\right)\ddot{F}$ and the restoring force $(\rho_{\mathrm{CJ}} e_{\mathrm{CJ}} \dot{F} - \Psi^0(\dot{F}))$. An alternating damping controlled by these two terms is responsible for maintaining the oscillation as observed in the numerical results. Similar to the classical nonlinear dynamical system, the nonlinearity of the restoring force to the driving force embedded in the term $\Psi^0(\dot{F})$ should provide the ingredient for the transition of the instability patterns.

Chemical kinetics also influences the amount of resonant excitation of oscillation as a result of the coupling between the self-excited driving force by the unsteady part of the chemical energy release and the front fluctuation, causing a higher degree of the instability pattern. Resonant oscillation can be achieved when the fluctuations of the heat release rate and pressure are positively correlated (Rayleigh's criterion). In other words, it requires $\Psi(t)$ to be fully in phase with $\dot{F}$, so the product $\Psi(t) \cdot \dot{F} > 0$ holds over the entire oscillatory cycle. The product $\Psi(t) \cdot \dot{F}$ is essentially similar to the power input to the system, and the integral of this product over a cycle describes the self-excited energy required for the oscillating motion of the shock front. From this argument, a generalized resonance-coupling criterion has been proposed:

$$J^* = \frac{1}{\tau}\int_0^{\tau} \frac{\Psi(t)\,\dot{F}}{D_{\mathrm{CJ}}}\mathrm{d}t > 0, \tag{3.69}$$

where J^* has units of power and τJ^* gives the net excess (over the steady ZND value) of the excited energy. From the energy point of view, more chemical energy release must be excited to the shock front in order to maintain a higher instability mode of the detonation front. As shown in Fig. 3.18, the value of J^* remains zero below the neutral stability boundary limit (i.e., $E_a < 25.26$). As the value of activation energy increases away from the stability limit, the value of J^* also increases. The continuous increase of J^* due to different degrees of resonant coupling between the shock fluctuation and heat release rate controlled by chemical kinetics results in the variety of instability patterns ranging from harmonic to irregular chaotic motion of the pulsating front.

This section of the chapter has demonstrated the remarkable resemblance of the pulsating detonation to a classical oscillator. It is desirable to adopt other methodologies developed in nonlinear dynamics that are not considered here to further explore the dynamics of the pulsating detonation. The possible use of a nonlinear oscillator model to interpret the numerical simulations can be emphasized as a new approach to study detonation structure and

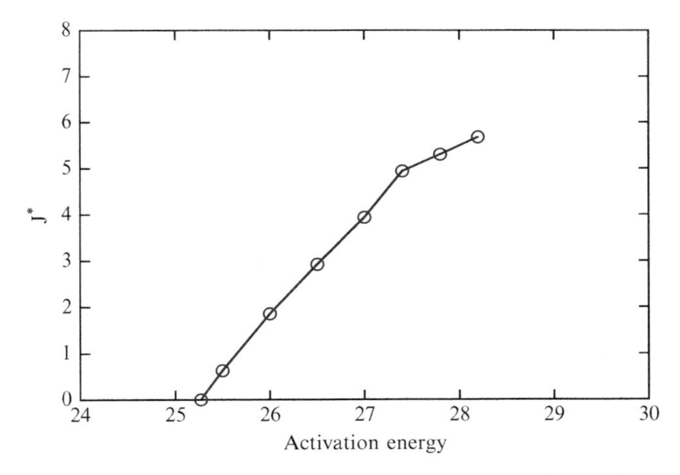

Fig. 3.18. Integral value of the product of the unsteady chemical energy release and shock velocity fluctuation over a cycle period τ for different values of activation energy

different factors that influence its unstable behavior. From the framework of a nonlinear oscillator, the mechanism responsible for developing different irregular oscillatory patterns appears to be related to the resonant excitation of the heat release governed by the chemical kinetics; and the substantial effect of nonlinearity can in turn lead to the transition route of the instability to chaos as observed in classical dynamical systems.

3.3.5 Two-Dimensional Unstable Detonation

Two-dimensional simulations are fairly well established and high-resolution simulations of the cellular detonation structure are now possible. Nevertheless, it is crucial to ensure that the detonation structure is well resolved before making any major conclusion about the phenomenon. Furthermore, the size of the computational domain must be chosen to be sufficiently large in order to minimize the effect of the channel wall leading to mode locking. Recently Sharpe and Radulescu performed a highly resolved simulation with a very large domain to look at the evolution of "natural" cell size [147], and it is of interest to present their results to elucidate some characteristics of the cellular detonation structure.

In Sharpe and Radulescu's simulation, the two-dimensional Euler equations with a single Arrhenius reaction step are solved using a parallelized adaptive mesh refinement technique to improve the resolution at locations of interest (reaction zones, shocks, and slip lines) and a standard upwind numerical scheme with Roe's linearized Riemann solver with the minmod limiter. Consequently, the results are obtained using an effective resolution of 64 points per $L_{1/2}$. The long computational domain (3,000 times of half-reaction zone length) is bounded by free slip solid walls at the top and bottom separated by a width of $96 \cdot L_{1/2}$. The flow state at the left rear boundary is prescribed to be the CJ detonation state and the right transmissive boundary is a free flow extrapolation. The computation is initialized by a ZND solution and perturbed by a small random disturbance to accelerate the development of the cellular instability. The chemical parameters are the same as most one-dimensional pulsating detonation simulations ($Q = 50$, $\gamma = 1.2$), and the activation energy E_a is used as the bifurcation parameter.

To generate "numerical soot foils" that illustrate the cellular detonation pattern (or the trajectories of the triple points of Mach interactions), a common approach is to plot either the time-integrated maximum pressure contour, total heat released at each point, or the shock front position from the numerical simulation. Sharpe and Quirk [146] and Sharpe and Radulescu [147] recently proposed an alternative method to generate such foils by recording the time-integrated maximum vorticity w reached at each grid cell and making use of the sign of the vorticity $w_{\pm}$. As studied by Inaba et al. [69], physically the writing on the soot foil is mainly attributed to the high vorticity at the triple points. Numerically, the advantage of using local maxima in vorticity allows each family of triple point tracks to be extracted separately based on the sign of the vorticity records (one propagating upward and the other propagating downward) so that the cell size (i.e., distance between neighboring triple point tracks of the same family) can be properly defined and analyzed. This significantly facilitates and improves the statistical analysis and determination of cell size as the distance between triple point tracks of same family.

By looking at the vorticity records as shown in Fig. 3.19b, it is perhaps not difficult to see that w_+ and w_- preferentially create the tracks of the upward and downward moving families of triple points, respectively. Although in some

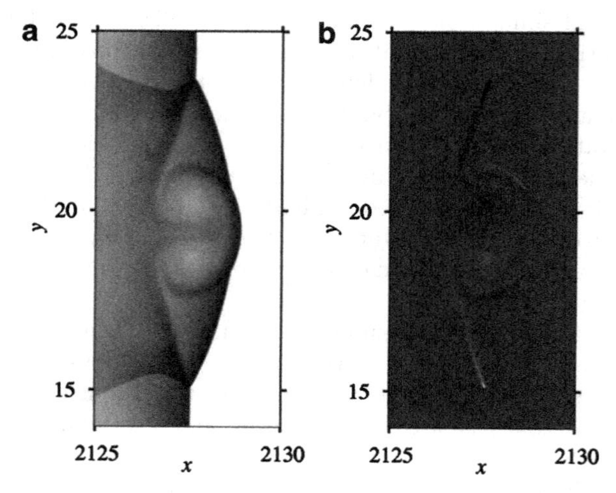

Fig. 3.19. Density and vorticity flow fields within a detonation structure for $E_a = 15$ and 64 pts/$L_{1/2}$ [147]

cases that the opposite family of tracks can still be seen in each $w_{\pm}$ record due to the vorticity generated by the secondary triple points created behind the Mach stem from the forward jets [142], these are much weaker as compared to the main triple point trajectories. Superimposing both vorticity records by plotting $(|w_{+}| + |w_{-}|)/2$ on the same figure, the complete numerical soot foil can be generated as shown in Fig. 3.20a–d.

The "numerical smoked foils" are shown in Fig. 3.20a–d for different activation energies E_a, which essentially control the detonation front evolution by the temperature sensitivity of the reaction, similar to the one-dimensional results. Early in the cell formation process, the cell patterns are rather regular and small in all cases. However, as the detonation wave continues to propagate, instability develops. For low activation energy $E_a = 15$, some transverse waves begin to weaken and eventually disappear. The evolutionary dynamics of the growth nevertheless appears quite slow and the detonation needs to run for a long time before the final size is reached. In the final frame, the cell pattern remains quite regular. For higher activation energies, the growth of instability is faster and the cell pattern becomes more irregular with increasing E_a. The dynamics and the dependence of the regularity of cell pattern on the activation energy are in good accord with experimental observation and stability theory. By increasing the activation energy, the temperature sensitivity of the reaction is greater and large fluctuations of the reaction rate resulting from small gasdynamics disturbances can render the detonation highly unstable.

Sharpe and Radulescu [147] statistically analyzed the cell pattern produced by numerical simulations and systematically determined detonation cell sizes from such simulations. In essential, the first statistical method is based on a probability density function of the main track spacing between neighboring triple points moving in the same direction extracted from the numerical vorticity records. Sharpe and Radulescu also performed a spectral analysis of the numerical vorticity data, using the autocorrelation function. In that method, the cell size is determined based on the peak autocorrelation in the transverse direction by measuring the correlation of the foil when shifted by a fixed amount with the original one.

These combined statistical methods can give useful information about the size and regularity of the cells appearing on given parts of the numerical vorticity records from such simulations. One of the interesting results obtained from their study is that the statistical analysis shows for the first time evidence of "cell size doubling" bifurcations in the ideal one-step chemistry model as the cellular dynamics become more irregular (Fig. 3.21). Such bifurcation appears to be in accord with the dynamics of one-dimensional pulsating instability, but with a shift toward smaller E_a.

Sharpe and Radulescu [147] also carried out a numerical experiment by studying the cellular dynamics subjected to different random initial perturbations. Interestingly, what they found is that there is indeed a "shot-to-shot" variation in the predicted cell size. This result is perhaps equivalent to the chaos observed in highly unstable one-dimensional pulsating detonations. For

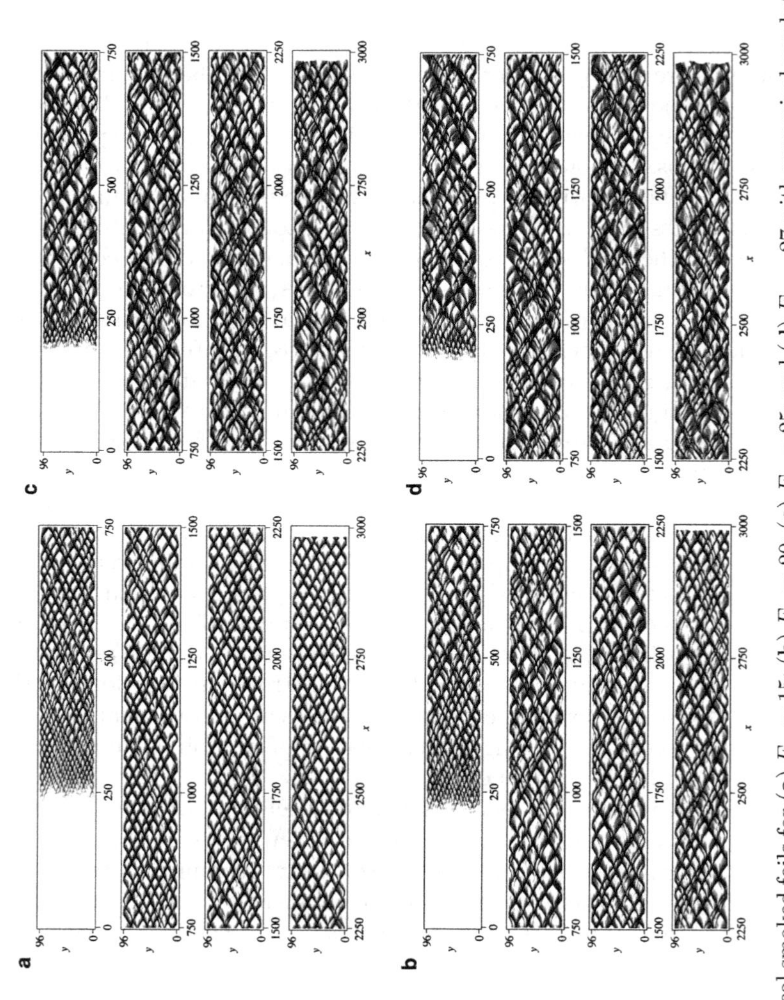

Fig. 3.20. Numerical smoked foils for (**a**) $E_a = 15$; (**b**) $E_a = 20$; (**c**) $E_a = 25$; and (**d**) $E_a = 27$ with numerical resolution of 64 pts/$L_{1/2}$ [147]

moderate activation energy $E_a = 20$, although the long-time cellular dynamics were found to behave qualitatively similar with different initial perturbations in that they consist of constantly emerging and re-merging transverse waves giving cells of varying sizes, statistical analysis indicates that a range of equally probable spacings may exist. Even for a more highly regular case considered by Sharpe and Quirk [146], a range of nonlinear stable final cell sizes could be produced for the same parameter set by using different initial perturbations. Sharpe and Radulescu [147] explained that this variability in the cellular dynamics may be due to the fact that long-lived correlations can exist. For instance, bands of cells maintaining the same spacing can be established over very long distances, and computationally the solution would thus require an extremely long domain or multiple runs to obtain a converged distribution, especially for more irregular cell patterns (as the activation energy increases further), and this problem can be resolved only once such calculations become computationally viable and by applying appropriate statistical analysis.

3.3.6 Detonation Structure Using a Binary Mixture Model

The high resolution numerical computations of detonation structure in the literature such as those performed by Sharpe and coworkers are carried out by assuming a single global EOS. Despite its simplicity, it is often required to select the molecular weight W and ratio of specific heats γ as average values obtained from detailed thermodynamic calculations and then adjust the energy release Q to match the experimentally measured detonation CJ properties of the system [53]. This is, however, a major simplification of the

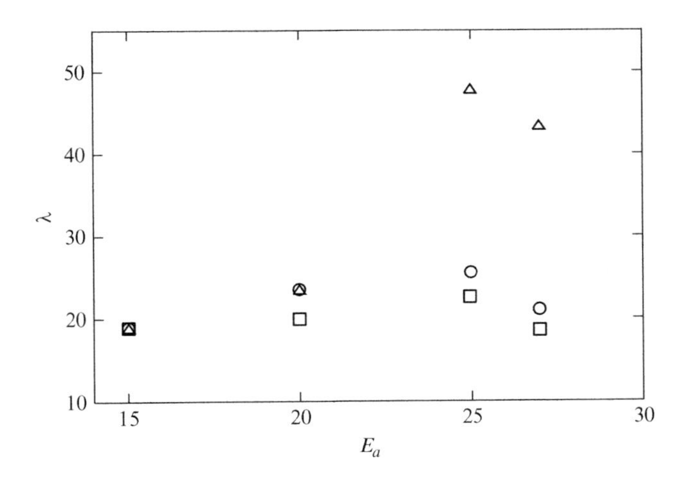

Fig. 3.21. Transverse wave spacing λ normalized by $L_{1/2}$ as a function of activation energy E_a determined from the numerical soot record using two sets of statistical results and different definitions (shown by *different symbols*) [147]

physical problem in most practical cases and can be problematic. For reactive flow, different species are being created or destroyed through chemical reactions, and the multi-gas components could have a significant effect on the thermodynamics of the flow itself. Consider for example a stoichiometric reacting mixture of hydrogen and oxygen; even if assuming a direct conversion from reactants to products with both modeled by the ideal gas EOS, i.e., $2H_2 + O_2 \rightarrow 2H_2O$, then the properties of these two mixtures vary noticeably: for the reactants, $\gamma = 1.4$, mean molecular weight $W = 12$, while for the products $\gamma = 1.2$, $W = 18$. This change in γ implies a factor of two difference in the $(\gamma - 1)$ term appearing in the EOS and hence a marked difference in the behavior of the two gases. In reality, the problem is even more complicated than this since the chemical reaction does not progress as a single step, but involves the conversion of the reactants to various intermediaries before the final state is reached, each of which will also have unique thermodynamic properties. For a simplified condensed-phase combustion model, the variation in the equivalent γ between the liquid/solid reactants and gaseous products can also be very pronounced for detonations in these explosives.

It is recognized from stability analyses that the ratio of specific heats γ is one of the fundamental parameters in detonation dynamics. To better represent the realistic structure and examine the stability properties of detonations, the next level in modeling is to take into account the nature of multicomponent species resulting in the variation of γ. However, implementing a multispecies system within a conservative formulation commonly leads to problems in the numerical solutions, especially those associated with contact discontinuities and shock waves. The difference between the ratios of the specific heats of the two gases (reactants and products) gives rise to numerical artifacts (generation of spurious waves at their interface which propagate through the solution) [9, 185]. The numerical errors typically appear in both the predicted density and pressure, with those in the pressure particularly severe. These errors are often critical when considering reactive flow, since any perturbations in temperature may be amplified by the exponential temperature dependence of the reaction rate term.

Based on the thermodynamically consistent and fully conservative (TCFC) formulation proposed by Wang et al. [185], a reactive binary-mixture model for detonation simulation can be obtained with:

$$\begin{aligned}
\frac{\partial}{\partial t}\left(\frac{\rho}{W}\right) + \nabla\cdot\left(\frac{\rho}{W}\mathrm{V}\right) &= \rho\left[\frac{1}{W_{\mathrm{reac}}} - \frac{1}{W_{\mathrm{prod}}}\right]\frac{\mathrm{d}\beta}{\mathrm{d}t} = \rho\Delta W^{-1}\dot{\omega} \\
\frac{\partial}{\partial t}\left(\frac{\rho\xi}{W}\right) + \nabla\cdot\left(\frac{\rho\xi}{W}\mathrm{V}\right) &= \rho\left[\frac{\xi_{\mathrm{reac}}}{W_{\mathrm{reac}}} - \frac{\xi_{\mathrm{prod}}}{W_{\mathrm{prod}}}\right]\frac{\mathrm{d}\beta}{\mathrm{d}t} = \rho\Delta\left(\xi W^{-1}\right)\dot{\omega},
\end{aligned} \tag{3.70}$$

where $\xi = \gamma/(\gamma - 1)$ and W is the local molecular weight [20]. This formulation is fully conservative throughout the domain and hence shock speeds are predicted correctly. Also, since no special treatment is required at the material interface, this TCFC model can be implemented using any available

Table 3.3. Material and chemical parameters used in the binary mixture model

Cases	Mixture	P_o (atm)	T_o (K)	γ_{react}	W_{react}	γ_{prod}	W_{prod}	Q (MJ kg^{-1})	E_a/R (K)
1	$H_2 + O_2$	1	295	1.404	12.01	1.22	14.474	8.27	11,284
2	$H_2 + O_2 +$ 50% Ar	1	295	1.509	25.977	1.2	29.001	4.264	9,486
3	$H_2 +$ Air	1	295	1.405	20.911	1.17	23.904	5.306	15,497

conservative solver [9]. This simplified model considers the reactant and product gases as two constituents of the system and models the conversion between these by a simple one-step reaction mechanism. This model introduces one extra level of complexity than previous studies using a single fluid approximation, but at the same time allows high resolution simulations to be performed. The benefit of using a binary-mixture formulation is that the individual thermodynamic properties of the combustion reactants and products can be considered without performing any averaging or matching for the computation.

Using this model, the CJ detonation structure and evolution in real mixtures such as hydrogen-based mixtures can be more closely examined. The material and chemical parameters of the model can be obtained using a detailed chemistry calculation for stoichiometric H_2/O_2 mixtures [137] (see Table 3.3). For a binary mixture formulation, the CJ detonation velocity for the multi-gas case can be readily determined as [83]:

$$\eta_{\mathrm{CJ}} = \frac{1}{M_{\mathrm{CJ}}^2} = \frac{\gamma_1}{\gamma_2} + \frac{K}{2} - \sqrt{K\left(\frac{\gamma_1}{\gamma_2} + \frac{K}{4}\right)}, \tag{3.71}$$

where:

$$K = 2\left[\frac{\gamma_1(\gamma_2-\gamma_1)(\gamma_2+1)}{\gamma_2^2(\gamma_1-1)} + \frac{\gamma_1^2(\gamma_2^2-1)}{\gamma_2^2}\left(\frac{Q}{\gamma_1 p_o v_o}\right)\right] \tag{3.72}$$

and the CJ properties can be obtained using the reactive Rankine–Hugoniot relationship given below:

$$\frac{\rho_2}{\rho_1} = \frac{\gamma_1(\gamma_2+1)}{\gamma_2(\gamma_1+\eta_{\mathrm{CJ}})} \quad \frac{u_2}{D} = \frac{\gamma_1 - \gamma_2\eta_{\mathrm{CJ}}}{\gamma_1(\gamma_2+1)} \quad \frac{p_2}{\rho_1 D^2} = \frac{\gamma_1+\eta_{\mathrm{CJ}}}{\gamma_1(\gamma_2+1)}. \tag{3.73}$$

For the single-step Arrhenius model, the pre-exponential constant is chosen to approximate the steady ZND reaction structure obtained using detailed chemistry such that the half reaction length roughly approximates the ZND induction zone length.

A comparison of the temperature profiles calculated using the simple binary reactive gas model with single-step Arrhenius kinetics and the full detailed chemistry model is given in Fig. 3.22. It can be seen that the shocked

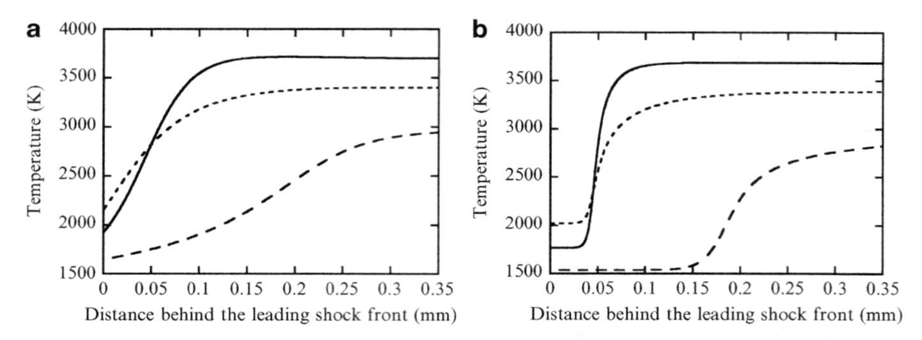

Fig. 3.22. A comparison of ZND temperature profiles calculated using (**a**) the binary mixture model with a single reaction and (**b**) the detailed chemistry. The *solid, dotted* and *dashed lines* correspond to the cases 1, 2, and 3 as shown in Table 3.3, respectively [20]

and CJ states are relatively well approximated with the present multi-gas model. Without using a multicomponent model formulation, an estimated average value of both the heat release Q and the specific heat ratio must be taken, and it is clear that these average values cannot correctly give both the solutions of the vN and CJ states. This multicomponent formulation which takes into account the thermodynamic properties (i.e., the specific heat ratio and molecular weight) of both reactant and products mixtures allows a better model to be made for these two states in the detonation structure, as illustrated in Fig. 3.22. The post-shock conditions and the equilibrium CJ state are important because of their influence on the stability of the detonation structure [81]. It is important to note that only a single step Arrhenius reaction is considered. Again, the induction zone region could not be approximated using single-step Arrhenius kinetics [110]. Significant improvement can be obtained by using a two- or three-step chain-branching kinetic model [26, 88, 107, 155, 156, 162]. The effect of chemical kinetics on the instability will be discussed in the next section.

Figure 3.23 illustrates the cellular detonation structures simulated using the binary reactive model, showing snapshots of Schlieren-type density gradients and reactant mass fraction plots. The results show a weakly unstable structure (displaying regular detonation cells) in the stoichiometric H_2–O_2 mixture with and without Ar dilution, to a strongly unstable or irregular cellular structure in the stoichiometric H_2–Air mixture. The effect of the EOS and the influence of Ar dilution on the specific heat ratio could be the main mechanism explaining the unstable detonation front behavior in these results [80]. For stoichiometric H_2–O_2 without dilution (see Table 3.3), a smaller value of γ_{reac} leads to a smaller shock temperature T_s, and thus a larger reduced activation energy E_a/RT_s. In such cases, instability will occur due to the high temperature sensitivity and an irregular cellular structure will result. With increasing Ar dilution, a higher γ_{reac} leads to a higher shock temperature and

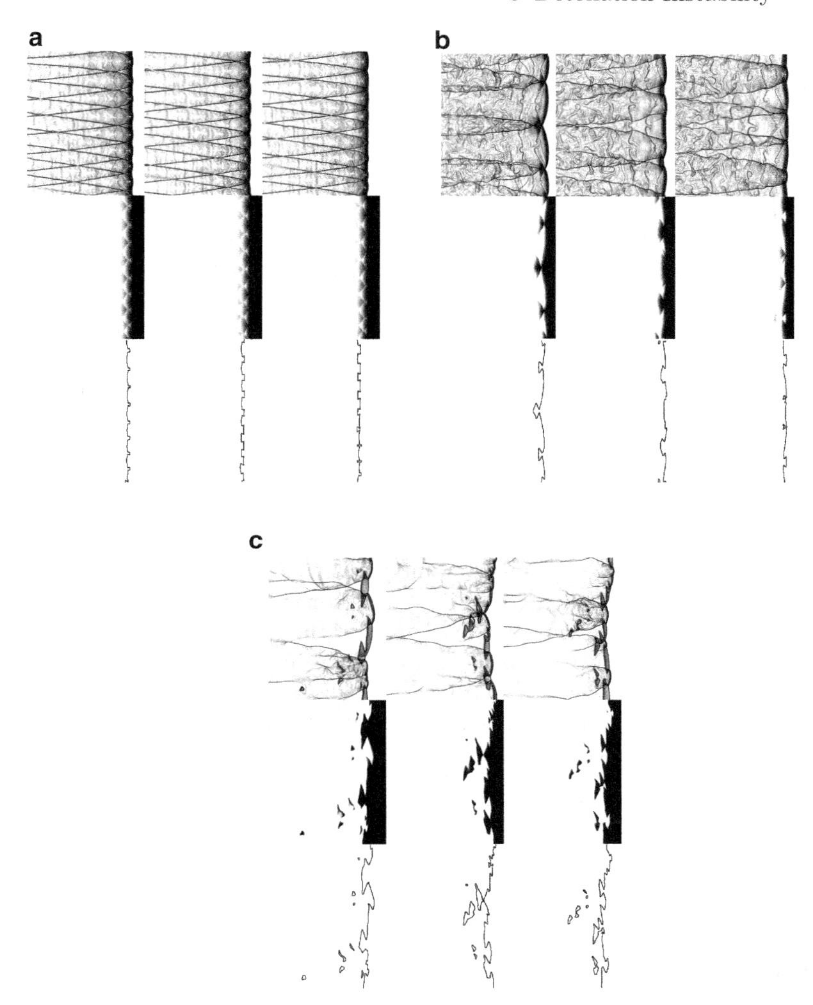

Fig. 3.23. Sequence of three plots of density-Schlieren, reactant mass fraction and its edge-detected interfaces for (**a**) stoichiometric H_2–O_2–50% Ar mixture; (**b**) stoichiometric H_2–O_2 mixture; and (**c**) stoichiometric H_2–Air mixture [20, 113]

thus a weaker sensitivity of the chemical reaction, which leads to a more stable or regular detonation front. For fuel–air mixtures such as stoichiometric H_2–Air, the activation energy E_a is higher and the shock temperature is lower compared to the fuel-oxygen case (see Table 3.3), hence resulting in a more unstable detonation front. These observations are equivalent to those from one-dimensional simulations with varying E_a. For an unstable detonation, as in the case of the stoichiometric H_2–Air mixture, the appearance of unburned pockets of reactants and the compressible, turbulent nature of the flow are quite apparent in Fig. 3.23c. Recent experimental investigation also suggested a characteristic keystone feature in the cellular detonation structure, which arises from the reaction rate sensitivity to shock conditions [124]. Such a keystone feature is also clearly revealed by the multicomponent reactive flow model with single-step Arrhenius chemistry, as shown in the mass fractions plots of Fig. 3.23.

To characterize the degree of instability of the reaction front, the concept of fractal geometry can be applied to quantitatively analyze numerical simulation results [113]. Such analysis is motivated by the work of Pintgen and Shepherd [123] who, using planar laser-induced fluorescence (PLIF) and Schlieren imaging methods, examined the details of the detonation front and the reaction zone structure. They used fractal analysis to determine an average geometrical quantity to define the regularity of the fronts. Their analyses were performed on experimental results, which were subjected to the limitations of the experimental techniques used (i.e., illumination and imaging system qualities) and image interpretation. Nevertheless, they found that for $\theta = E_a/RT_s \leq 6$, the spread of the fractal dimension is small, ranging from 1.05 up to 1.15. For weakly unstable mixtures with intermediate θ, the fractal dimension ranges between 1.05 and 1.4, and for the highest value of $\theta = 12.4$, a maximum fractal dimension of 1.5 is obtained, indicating the increasing degree of corrugation of the reaction front as θ increases.

Note that the concept of a fractal dimension is widely invoked in fluid dynamics, and it has been used for analyzing turbulent flames. In principle, the fractal dimension is a measure of the roughness of the object. For a fractal curve lying in a plane, the fractal dimension is bounded by $1 < D_2 < 2$. The closer the dimension of a fractal is to its upper limit, which is the dimension of space in which it is embedded, the more rough or space-filling it is. For a smooth interface boundary, the fractal dimension D_2 reduces to one. Using imaging processing techniques, the reaction zone interface can be tracked and used for fractal dimension measurement. To estimate the fractal dimension of the reaction front, the box-counting method is used. The box-counting technique is an efficient measurement method for measuring the fractal dimension of digital images and is widely used in turbulence research. The basic procedure is to systematically lay a series of squares or boxes of decreasing side length δ over a set S and count the number of boxes $N_\delta(S)$ that overlaps the set for various small δ. The dimension is the logarithmic rate at which $N_\delta(S)$ increases as $\delta \to 0$, which can be estimated by the slope of the regression line of the plot log $N_\delta(S)$ against $\log(\delta)$:

$$D_2(S) = -\lim_{\delta \to 0} \frac{\log(N_\delta(S))}{\log(\delta)}. \tag{3.74}$$

For stable argon-diluted H_2–O_2 detonation (activation energy $\theta = 4.7$), the flow structure is very regular and it can be expected that the edge detected for the reaction front structure from the reactant mass fraction to have a rather regular interface (see Fig. 3.23). For the undiluted H_2–O_2 case, a more unstable detonation front is observed with $\theta = 6.4$, and the flow structure becomes less

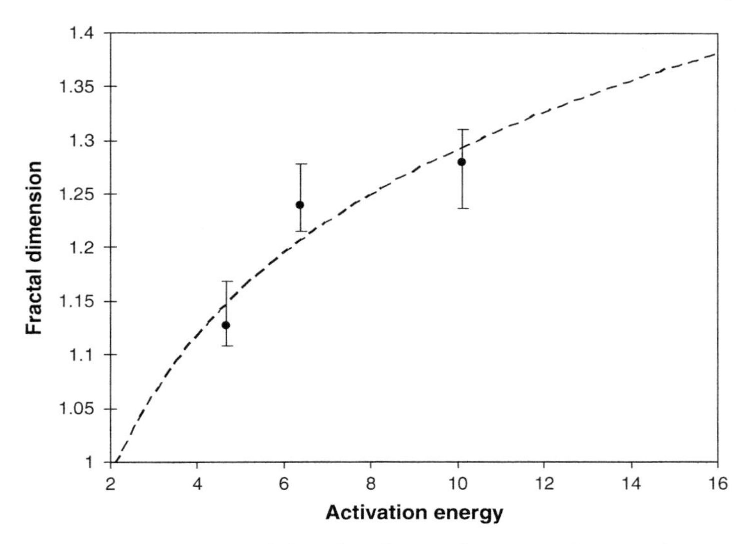

Fig. 3.24. Fractal dimension of the edge-detected reaction front in detonation simulations estimated using the box-counting method as a function of activation energy

regular with more flow disturbances inside the structure. The edge detected for the reaction front structure from the reactant mass fraction also shows a more irregular curve, indicating a more corrugated front. Further increasing the activation energy to $\theta = 10.1$, as in the undiluted H_2–air case, leads to a highly unstable detonation front structure. For this case, the appearance of unburned pockets of reactants and the compressible, turbulent nature of the flow are quite apparent and the edge-detected reaction front is highly irregular.

Figure 3.24 shows the average fractal dimension of the interface boundary $\bar{D}$ of a series of CFD images for each activation energy considered in Fig. 3.23, giving a detonation front from the stable (laminar regime) to the highly unstable (turbulent) case. The error bars indicate the minimum and maximum fractal dimension obtained from a sequence of images at different times for each activation energy value. It is found that the fractal dimension of the interfaces increases from an initial value of about 1.10 ($\theta = 4.7$) and approaches a value of about 1.3 or higher ($\theta = 10.1$). These values agree well with previous studies by Pintgen and Shepherd who analyzed the experimental results from their PLIF measurements of the detonation front. With increasing activation energy in the unstable detonation, it appears that the fractal dimension approaches the experimentally measured value between 1.32 and 1.4 for fully developed turbulence in an axisymmetric jet, turbulent shear flow, and Richtmeyer–Meskhov instability experiments [114]. Such agreement may thus indicate the self-similarity of the turbulent flow structure in the reaction zone behind the detonation front.

3.4 The Effect of Chemistry on Detonation Instability

In the previous section, the nonlinear characteristics of the one-dimensional pulsating and two-dimensional cellular detonation structure were mostly described using a one-step kinetic model. In this simple detonation model, the activation energy is the main parameter which controls the sensitivity of the reaction zone and is the bifurcation parameter which characterizes the stability of the detonation front. However, this may not be a sufficient parameter, as pointed out by many studies. For example, using a more realistic reduced chemical kinetic mechanism, Radulescu et al. [132] carried out numerical simulations of the one-dimensional pulsating detonations in acetylene–oxygen mixtures with different degrees of argon dilution and initial pressures, the results of which are shown in Fig. 3.25. By increasing the argon dilution, the 1D pulsating detonation becomes more "stable" in the sense that the oscillations become more regular and have a smaller amplitude, similar to the case for single-step kinetics with a low value of activation energy. Similar observations were made in experiments with argon dilution which show a more regular cell pattern as the argon dilution is increased.

Zero-dimensional chemical kinetic computations can be used to determine the global activation energy of an equivalent one-step reaction. Its value can be estimated by performing constant-volume explosion calculations. Assuming that the induction time τ_i has an Arrhenius form, i.e.,

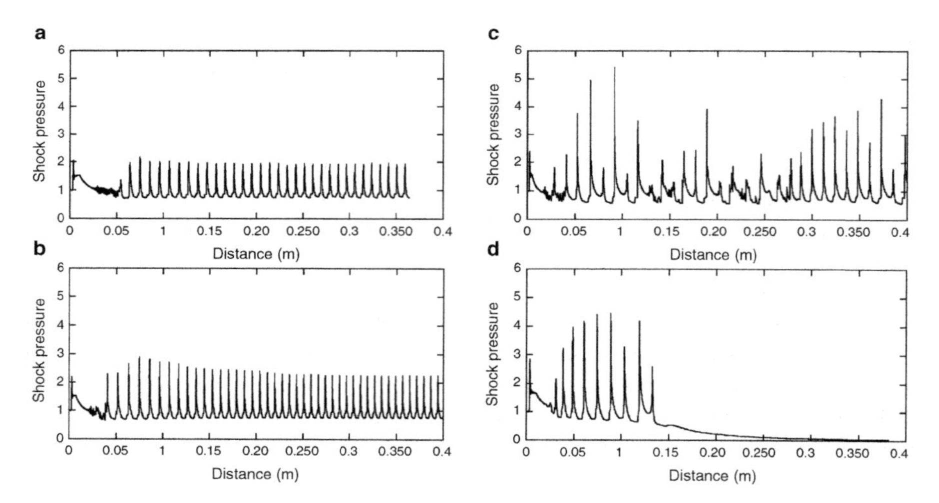

Fig. 3.25. Leading shock pressure evolution for pulsating detonation in stoichiometric acetylene–oxygen mixtures with different degrees of argon dilution. (**a**) 90% Ar at $p_o = 100\,\text{kPa}$; (**b**) 85% Ar at $p_o = 60\,\text{kPa}$; (**c**) 81% at $p_o = 41.7\,\text{kPa}$; (**d**) 70% at $p_o = 16.3\,\text{kPa}$ [132]

Table 3.4. Values of detonation parameters in stoichiometric acetylene–oxygen mixtures with varying argon dilution

Mixture	P_o (kPa)	E_a/RT_s	Δ_I (cm)	Δ_R (cm)	χ
$C_2H_2 + 2.5O_2 + 90\%Ar$	100	5.07	1.92×10^{-2}	5.30×10^{-2}	1.83
$C_2H_2 + 2.5O_2 + 85\%Ar$	60	4.86	1.51×10^{-2}	3.41×10^{-2}	2.15
$C_2H_2 + 2.5O_2 + 81\%Ar$	41.7	4.77	1.52×10^{-2}	3.03×10^{-2}	2.39
$C_2H_2 + 2.5O_2 + 70\%Ar$	16	4.77	2.25×10^{-2}	3.32×10^{-2}	3.24

$$\tau_i = A\rho^n \exp\left(\frac{E_a}{RT}\right). \tag{3.75}$$

The activation temperature E_a/RT_s may be determined by

$$\varepsilon_I = \frac{E_a}{RT_s} = \frac{1}{T_s}\frac{\ln\tau_2 - \ln\tau_1}{\frac{1}{T_2} - \frac{1}{T_1}}, \tag{3.76}$$

where two constant-volume explosion simulations are performed with initial conditions (T_1, τ_1) and (T_2, τ_2). The conditions for states one and two are obtained by considering the effect of a change in the shock velocity of $\pm 1\% D_{CJ}$ [137].

As shown in Table 3.4, the global activation energy increases slightly at a higher argon dilution. In an attempt to explain the effect of argon dilution, the reaction structure of the steady-state ZND model can be analyzed for varying degrees of argon dilution. The effect of argon dilution on detonation is to lower the total energy release of the mixture, which results in the decrease of detonation velocities and thus shock temperatures. However, addition of argon causes an increase in the specific heat ratio of the mixture, leading to the opposite effect of increasing the shock temperature. The results indicate that these two competing effects are approximately balanced, and therefore the shock temperature T_s does not change significantly with argon dilution (i.e., a dilution of 0–81% Ar causes a slight increase in shock temperature of approximately 200 K). Since the shock temperature remains almost constant, the influence of the diluent does not significantly affect the initiation rates. As seen in Fig. 3.26, where different ZND profiles are plotted for large variations of argon dilution, the length of the thermally neutral induction zone does not vary significantly to 81% Ar. Another implication of the small changes in the shock temperature is that the global activation energy (E_a/RT_s), often used in previous studies to describe the stability of detonations, does not vary significantly with argon dilution. Hence, it cannot account for the dramatic changes of stability observed both numerically and experimentally.

It is noted from Fig. 3.26, however, that there is a dramatic change in the reaction zone structure with increasing argon dilution, which leads to an increase of the reaction zone length. The increase, with argon dilution, of the characteristic reaction zone length during which the exothermic recombination processes occur can be explained by considering the changes in the

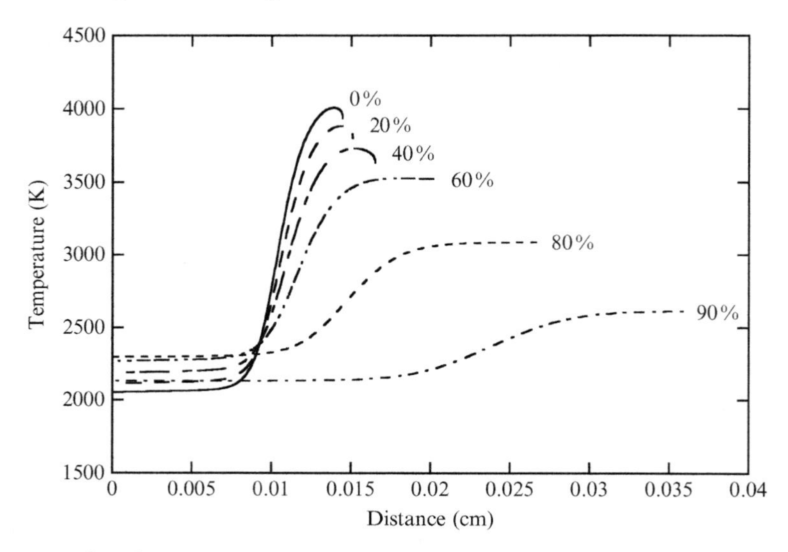

Fig. 3.26. Steady ZND temperature profiles for stoichiometric acetylene–oxygen detonations with different degrees of argon dilution [132]

elementary rates occurring at the end of the chain-branching steps and during the recombination steps of the oxidation scheme. As argon is added, the total heat release (per mole) decreases significantly, resulting in a lower temperature rise in the reaction zone. As a consequence of a lower temperature in the reaction zone, the chemical reaction rates of the exothermic reactions are reduced, leading to an increase in the heat release times. Therefore, as shown by these ZND analyses, the main effect of argon dilution on the ZND structure is to lengthen the reaction zone. Further analysis of the ZND reaction zone structure and discussion on the existence of the second slow heat-release step in these acetylene-oxygen mixtures diluted with different degrees of argon will be presented in Chap. 5. (e.g., see Figs. 5.39 – 5.40.)

Another interesting observation from the numerical results shown in Fig. 3.25 is that for 70% argon dilution, as the detonation wave decays from the initial overdriven state to the CJ average velocity, the frequency of oscillations decreases and the amplitude of the pulsations grows. During the low velocity phase of the oscillation, the leading shock decays to sub-CJ velocities. When the leading shock pressure decays below a critical value, corresponding to $\approx 0.7\,D_{\mathrm{CJ}}$, the detonation wave fails. Analysis of the reaction zone structure during the last cycle of the oscillation reveals that failure occurs due to the complete decoupling of the reaction zone from the leading shock front. This decoupling was found to occur when the temperature behind the leading front drops below $\sim$ 1,300 K. From Fig. 3.27 which shows the ignition delay behind shocks with different strengths in the constant volume approximation, a rapid increase in the induction time below this critical Mach number or shock temperature value can be seen. This dramatic increase of ignition time

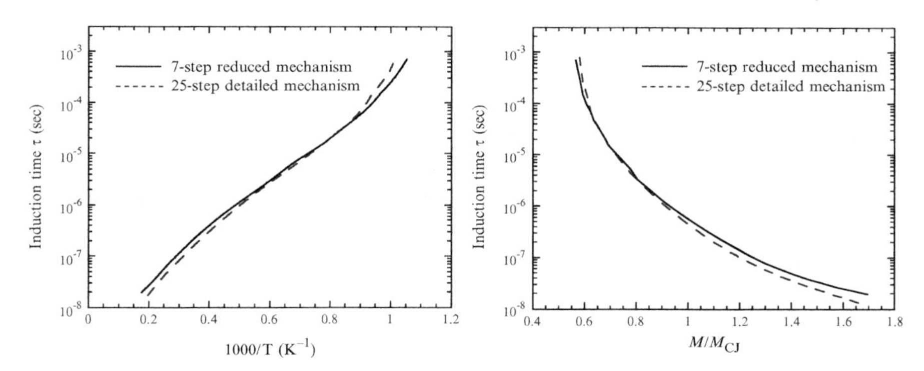

Fig. 3.27. Ignition delay times in stoichiometric acetylene–oxygen–70% Ar mixtures with initial conditions corresponding to different shock strengths [178]

is due to the chain-branching/chain-breaking competition which characterizes the oxidation of acetylene. This critical temperature is close to the chain branching cross-over temperature. Below this critical temperature, the generation of radicals from the chain-branching reaction is found to be too slow to maintain the coupling with the exothermic part of the reaction structure that is required to drive the wave. As a result, the reaction layer decouples from the shock and the detonation quenches. The detonation failure is thus due to the significant decrease in the chain-branching reaction rate.

A single Arrhenius reaction model usually cannot reproduce the above important chemistry features of the reaction structures behind real detonation fronts which are typically governed by chain-branching kinetics. More specifically, one-step Arrhenius kinetics cannot independently vary the two important characteristic zone lengths, i.e., induction length and the main heat release length. The chain-branching reaction structure consists of the induction process and the exothermic stage during which radicals recombine to form products. For example, the reduced kinetic model used by Radulescu et al. [132] and developed by Varatharajan and Williams [178] is given by:

$$
\begin{array}{ll}
C_2H_2 + O_2 \rightarrow CH_2O + CO & (I) \\
CH_2O \rightarrow CO + 2H & (II) \\
C_2H_2 + OH \rightarrow CH_2CO + H & (III) \\
C_2H_2 + 2O_2 \rightarrow 2CO + 2OH & (IV) \\
2H + O_2 \rightarrow H_2O + O & (V) \\
CH_2CO + O + O_2 \rightarrow 2OH + 2CO & (VI) \\
CO + OH \rightarrow CO_2 + H & (VII)
\end{array}
$$

In essence, the seven-step mechanism was systematically derived from a more complex full mechanism by eliminating less significant reactions for the high temperatures prevailing in detonations and applying steady-state and partial equilibrium approximations to further simplify the kinetic description. The resulting mechanism has a chain-branching-thermal explosion character. The seven-step mechanism consists of four important steps during the induction process (I–IV) leading to ignition and three steps describing the exothermic stage following ignition during which radicals recombine to form products (steps V–VII). This reduced seven-step reaction mechanism was found to be very suitable for describing both the induction and exothermic recombination part of the reactions and agrees very well with the full mechanism predictions for zero-order chemical kinetics calculations, as shown in Fig. 3.27, for example [178].

3.4.1 Simplified Chain-Branching Kinetic Model

To clarify the role of chemical kinetics on the unsteady detonation structure, the recent trend has been to use more detailed kinetics models. Although a complex set of chemical kinetic rate equations derived from elementary reactions could, in principle, be solved simultaneously with the reactive Euler equations using currently available computers, interpretation of the large amount of detailed information generated by such numerical simulations becomes a difficult problem. Therefore, it may suffice to use a simplified multistep chemical kinetic model to mimic the chain-branching reactions.

The nonlinear dynamics of one-dimensional pulsating detonations have been recently studied, both asymptotically and numerically, using a two-step reaction model by Short [151] and Short and Sharpe [156]. A slightly extended model [110] that includes a temperature-dependent term in the main exothermic reaction zone can be given by:

$$\Omega_1 = H\left(1-\beta_1\right) \cdot K_{\mathrm{I}} \exp\left[E_{\mathrm{I}}\left(\frac{1}{T_s}-\frac{1}{T}\right)\right], \tag{3.77}$$

where β_1 is the reaction progress variable in the induction period and $H(1-\beta_1)$ is a step function, i.e.,

$$H\left(1-\beta_1\right)\begin{cases} =1 & \text{if } \beta_1<1 \\ =0 & \text{if } \beta_1 \geq 1. \end{cases} \tag{3.78}$$

At the end of induction period, the second step begins to describe the rapid energy release after the chain-branching thermal explosion and the slow heat release in the radical recombination stage. The reaction rate equation for this step is given by:

$$\Omega_2 = -\left(1-H\left(1-\beta_1\right)\right) \cdot \beta_2 K_{\mathrm{R}} \exp\left[-E_{\mathrm{R}}/T\right], \tag{3.79}$$

where β_2 and K_{R} denote the chain-recombination reaction progress variable and rate constant for the heat release process. The local chemical energy that has been released at any instant during the reaction is equal to $q = \beta_2 \cdot Q$, where $Q = Q_2$ is the total chemical energy available in the mixture and released in the second reaction step. For non-dimensionalization, the energy release Q and activation energies E_{I} and E_{R} can be scaled with RT_0. It is appropriate to again introduce here the alternative scaling for the activation energies:

$$\varepsilon_{\mathrm{I}} = \frac{E_{\mathrm{I}}}{\delta} \quad \varepsilon_{\mathrm{R}} = \frac{E_{\mathrm{R}}}{\delta} \text{ with } \delta = \frac{\left[2\gamma M_{\mathrm{CJ}}^2 - (\gamma-1)\right]\left[2 + (\gamma-1) M_{\mathrm{CJ}}^2\right]}{(\gamma+1)^2 M_{\mathrm{CJ}}^2}, \tag{3.80}$$

where δ is the temperature jump across the shock. Therefore, ε is simply the reduced activation energy normalized by the temperature behind the shock of the CJ detonation. For typical hydrocarbon mixtures, the reduced activation energy of the induction stage ε_{I} is large because in the induction zone energy is required to break the strong chemical bonds of the fuel and convert it into radicals. Typical values for ε_{I} usually range from 4 (for H_2–O_2 mixture) to 12 (for heavy hydrocarbon mixtures). In contrast, the second step involves only reactions between energetic free radicals. Therefore, in typical chain-branching reactions, the induction stage generally has a larger activation energy compared to the second stage, i.e.:

$$E_{\mathrm{I}} >> E_{\mathrm{R}} \text{ or } \varepsilon_{\mathrm{I}} >> \varepsilon_{\mathrm{R}}. \tag{3.81}$$

Using the two-step chain-branching model (Fig. 3.28), Short and Sharpe [156] found that the pulsating structure can be influenced by independently changing the main heat release reaction rate or equivalently the length of the main heat release layer. The ZND detonation is stable when its structure is dominated by the heat release zone. As the ratio of the induction length Δ_{I} to the length of main heat release zone Δ_{R} increases, the detonation becomes unstable and different oscillatory behavior can be observed. Short and Sharpe further suggests that a bifurcation boundary between stable and unstable ZND detonations may be found when the ratio of the length of the heat

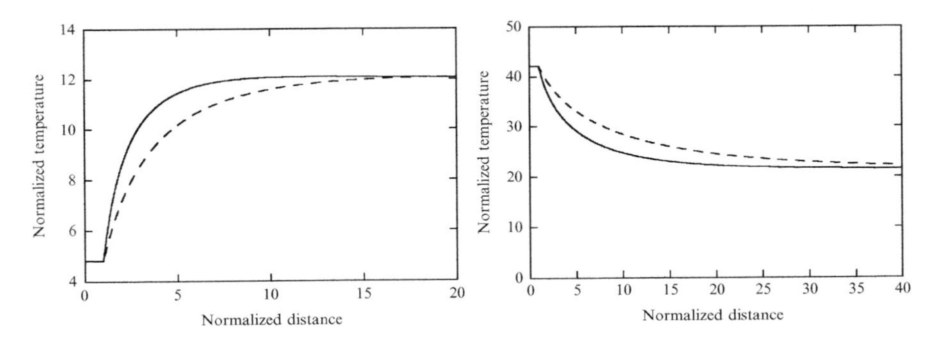

Fig. 3.28. Steady ZND detonation profiles computed using the two-step chain-branching kinetic model for a mixture with $Q = 50$, $\gamma = 1.2$, $\varepsilon_{\mathrm{I}} = 8$, $\varepsilon_{\mathrm{R}} = 1$ and two reaction rate constants $K_{\mathrm{R}} = 0.779$ (*dashed lines*); and $K_{\mathrm{R}} = 1.558$ (*solid lines*) [110]

release layer to that of the induction zone layer is of the order $O(\varepsilon_{\mathrm{I}})$, where ε_{I} is the activation energy in the induction zone. If the ratio $\Delta_{\mathrm{R}}/\Delta_{\mathrm{I}} \leq O(\varepsilon_{\mathrm{I}})$, the ZND detonation is unstable.

To identify a quantitative parameter that can be used to characterize the detonation stability governed by chain-branching kinetics analogous to the activation energy for the case of one-step Arrhenius kinetics, the following stability parameter was proposed [110, 130]:

$$\chi \equiv \varepsilon_{\mathrm{I}} \frac{\Delta_{\mathrm{I}}}{\Delta_{\mathrm{R}}} = \varepsilon_{\mathrm{I}} \Delta_{\mathrm{I}} \frac{\dot{\sigma}_{\max}}{u'_{\mathrm{CJ}}}, \tag{3.82}$$

where again Δ_{I} and Δ_{R} denote the characteristic induction length and reaction length, respectively, ε_{I} is activation energy governing the sensitivity of the induction period. Generally, the induction length Δ_{I} is simply defined as the length of the thermally neutral period. However, there is no standard definition for reaction length. Nevertheless the characteristic reaction length can be defined and approximated using the thermicity parameter. In general, the definition of thermicity is given by:

$$\dot{\sigma} = \sum_{i=1}^{N_s} \left(\frac{W}{W_i} - \frac{h_i}{c_p T} \right) \frac{\mathrm{d}y_i}{\mathrm{d}t}. \tag{3.83}$$

For the present two-step chemical model of ideal gases with constant specific heat ratio, the thermicity can be expressed as [50]:

$$\dot{\sigma} = (\gamma - 1) \frac{Q}{c^2} \frac{\mathrm{d}\beta_2}{\mathrm{d}t}. \tag{3.84}$$

The thermicity $\dot{\sigma}$ basically denotes the normalized chemical energy release rate and has a dimension of 1/time. Hence, the inverse of maximum thermicity $\dot{\sigma}_{\max}$ can be taken as a characteristic time scale for the heat release, which may provide an appropriate choice to define a characteristic reaction time. Using this characteristic reaction time, the reaction length can be estimated by:

$$\Delta_{\mathrm{R}} = \frac{u'_{\mathrm{CJ}}}{\dot{\sigma}_{\max}}, \tag{3.85}$$

where u'_{CJ} is the particle velocity at the CJ plane in shock-fixed coordinates.

In fact, the derivation and physical meaning of this stability parameter χ were equivalently discussed by Soloukhin in a comment to Meyer and Oppenheim's work on the shock-induced ignition problem where these early researchers noted that ignition behind a shock could either occur uniformly (the strong regime) or originate from several exothermic spots (the mild regime) [103, 181, 184]. For the case of detonations, the mild regime can analogously correspond to unstable detonations where instability is caused by large sensitivity of the reaction rate to temperature ([64, 143], etc.), while the strong regime is similar to stable detonations, where chemical reaction is insensitive to perturbations and all the fuel burns uniformly as in the ZND model. Hence, the stability in the reaction zone of gaseous detonations may be comparably linked to the regime of thermal ignition behind shock waves [168].

Following Oppenheim's coherence concept, a requirement of stability in the reaction structure of a detonation can be equivalently formulated [110, 130]. For stability, neighboring particles processed initially by slightly different shock strengths need to release their chemical energy on similar time scales, so that the compression pulses can overlap, or be "coherent" in time and space as to give rise to a single global gas dynamic effect [92, 103, 115]. To meet this requirement, the sensitivity of the chemical induction length (or equivalently time) to changes in shock temperature needs to be small. Otherwise, particles of gas being shocked at different temperatures due to perturbations in the flow will take significantly different times to burn which can lead to the formation of pockets of partly burnt fuel whose burnout at later time eventually causes instability in the reaction zone [143]. For gaseous detonation waves, the induction length can be assumed to have the common Arrhenius form:

$$\Delta_{\mathrm{I}} \propto \exp\left(\frac{E_{\mathrm{I}}}{T_s}\right), \tag{3.86}$$

where T_s is the shocked gas temperature and E_{I} the global activation energy describing the sensitivity of the thermally neutral chemical induction process. Thus, the stability condition requires that:

$$\left|\left(\frac{T_s}{\Delta_{\mathrm{I}}}\right)\left(\frac{\partial \Delta_{\mathrm{I}}}{\partial T_s}\right)\right| = \frac{E_{\mathrm{I}}}{T_s} \equiv \varepsilon_{\mathrm{I}} \tag{3.87}$$

be small, i.e., small activation energy ε_{I}. However, in addition to the effect of sensitivity on the induction process, the characteristic length scale for energy deposition Δ_{R} should play an important role. Note that a relatively long period for energy release (broad power pulses) will still lead to quasi-simultaneous energy deposition and coherence in time and space even if the changes of induction length (or time) are important. This can be illustrated schematically as shown in Fig. 3.29. Therefore, stability can be more properly described by the sensitivity to temperature fluctuations of the characteristic induction length relative to the characteristic exothermic reaction length. Mathematically, a stability parameter, based on this physical concept, of the same form found earlier in this section can be defined, i.e.,

$$\chi = \left|\left(\frac{T_s}{\Delta_{\mathrm{R}}}\right)\left(\frac{\partial \Delta_{\mathrm{I}}}{\partial T_s}\right)\right| = \varepsilon_{\mathrm{I}}\frac{\Delta_{\mathrm{I}}}{\Delta_{\mathrm{R}}} = \varepsilon_{\mathrm{I}}\Delta_{\mathrm{I}}\frac{\dot{\sigma}_{\mathrm{max}}}{u'_{\mathrm{CJ}}}. \tag{3.88}$$

This nondimensional parameter includes the essential terms that influence the characteristics of power pulses or energy release. For small values of χ, it is expected that power pulses originating from neighboring particles will overlap, thus leading to a coherent phenomenon in time and space (see Fig. 3.29a–c). In that case, small disturbances in the flow will not cause significant fluctuations of the energy release in the reaction zone structure, giving a stable or

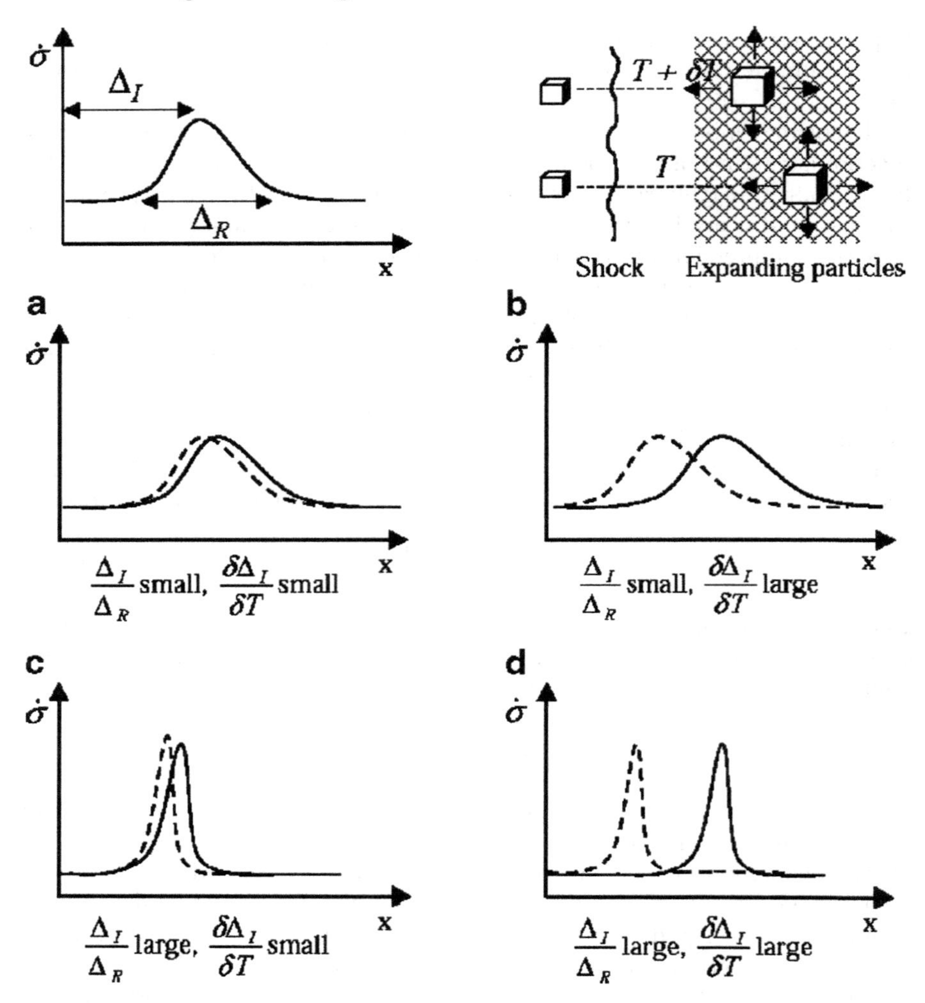

Fig. 3.29. An illustration of the coherence concept between neighboring power pulses, given by the exothermicity profiles for two neighboring gas elements shocked at temperatures differing by δT. (**a**) Small temperature sensitivity, long exothermic reaction length; (**b**) large temperature sensitivity, long exothermic reaction length; (**c**) small temperature sensitivity, short exothermic reaction length; and (**d**) large temperature sensitivity, short exothermic reaction length. Only case (**d**) results in incoherence of power pulses and the development of instability [110, 130]

weakly unstable system. On the other hand, if this parameter is large, as can occur with the conditions given in Fig. 3.29d, the power pulses will not be coherent and this can lead to various gasdynamic fluctuations in the reaction zone structure. This stability parameter therefore describes the scenario in which the incoherence in the energy release of the gas leads to gasdynamic instabilities in the reaction zone.

3.4.2 Neutral Stability Boundaries

A series of numerical computations and a parametric study to examine the long-term nonlinear dynamics of one-dimensional pulsating detonations with the generic two-step reaction model (3.77)–(3.79) were previously performed [110]. The aim of such a parametric study was to identify numerically the neutral stability curve for one-dimensional detonations with chain-branching kinetics based on the parameter χ to characterize the one-dimensional detonation stability. In that study, calculations were performed using a wide combination of parameters $(\gamma, Q, \varepsilon_{\mathrm{I}}, \varepsilon_{\mathrm{R}})$ to mimic the characteristics of any realistic chemical system for various combustible mixtures. For each set of mixture parameters $(\gamma, Q, \varepsilon_{\mathrm{I}}, \varepsilon_{\mathrm{R}})$, computations were performed in which the value of the reaction rate constant K_{R} was systematically varied to determine the critical value at which the detonation becomes unstable. Recall that the change of the reaction rate constant K_{R} in the two-step kinetic model has an effect on the length scale of the heat release zone. The stability parameter χ associated with this critical reaction rate constant K_{R} was then computed which corresponds to the value at the stability boundary.

The computational results can be summarized by constructing a plot in the $\chi - M_{\mathrm{CJ}}$ plane for different mixture parameters as shown in Fig. 3.30. Only the critical value of the stability parameter χ, at which the detonation becomes unstable, is shown for a given set of parameters (γ, Q, ε_{I}, ε_{R}). It is interesting to note that all the data points corresponding to different mixture

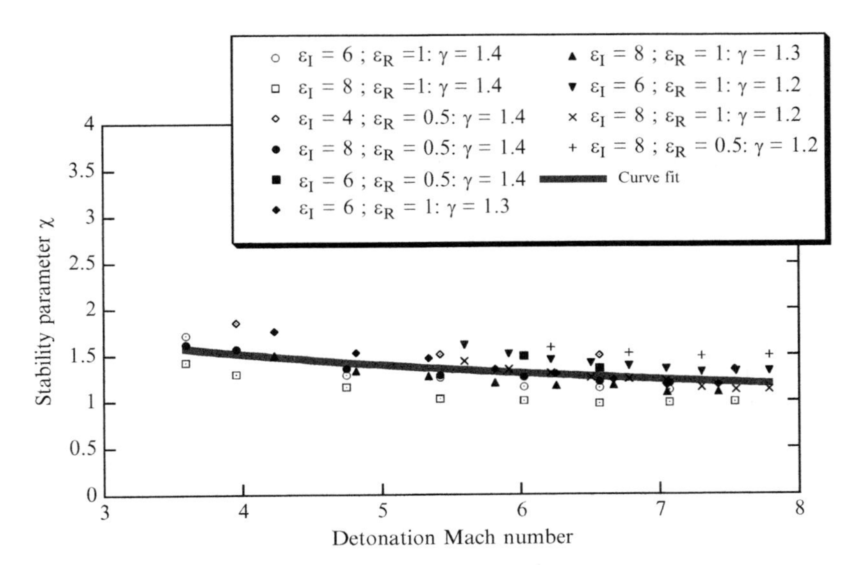

Fig. 3.30. Neutral stability curve obtained using the two-step chain-branching model [110]

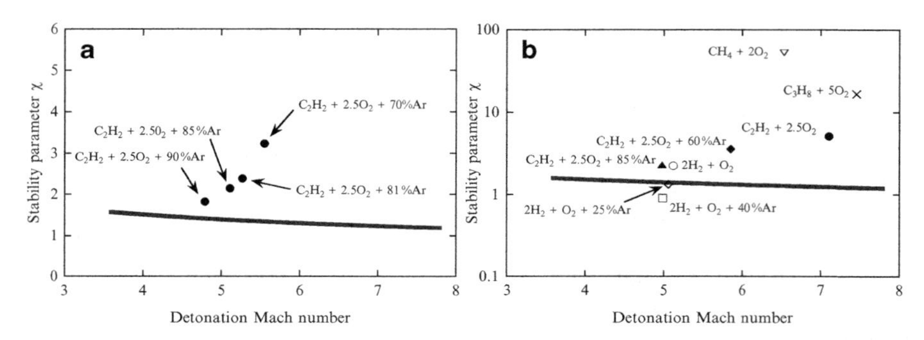

Fig. 3.31. (**a**) Characterization of the effect of argon dilution on the stoichiometric acetylene–oxygen detonation using the stability parameter χ and the neutral stability curve. (**b**) Characterization of the cellular structure of different combustible mixtures using the stability parameter χ and the neutral stability curve [110]

conditions can essentially be collapsed to a single curve (similar to Fig. 3.4 in the linear stability analysis section).

Using the results from the two-step kinetic model, as an example, the loss of stability with decreasing argon dilution in acetylene–oxygen mixtures can be analyzed. The effect of argon dilution can now be associated with the change of the stability parameter χ. Found also in Table 3.4 are the values of χ corresponding to the conditions in the numerical computations. The increase in the stability parameter χ with decreasing argon dilution can thus describe the loss of stability observed in 1D numerical simulations. The values of χ are also plotted relative to the neutral stability curve obtained using the two-step kinetic model in Fig. 3.31a. The parameter χ and the neutral stability curve together provide a quantitative measure of the degree of instability for one-dimensional pulsating detonations, irrespective to the chemical kinetic models.

Values of different detonation parameters including the stability parameter χ for some combustible mixtures are shown in Table 3.5. Also tabulated is the qualitative assessment of the cell regularity as observed experimentally. The value of χ for each mixture is mapped onto the neutral stability curve for comparison in Fig. 3.31b, which shows that for mixtures like H_2–O_2 and C_2H_2–O_2 with argon dilution, the stability parameters lie slightly above the neutral stability boundary. One-dimensional unsteady simulations with detailed chemistry for these mixtures confirm that their propagation can be described by a one-dimensional pulsating mode and the periodic pattern is regular (Figs. 3.32 and 3.33). These mixtures are generally stable, which implies that they should have a highly regular cellular structure comprised of weak transverse waves and are capable of maintaining their propagation by the shock-induced ignition mechanism.

In contrast, C_2H_2–O_2 or H_2–O_2 mixtures without argon dilution or mixtures with hydrocarbon fuels such as methane or propane have large values

Table 3.5. Values of different detonation parameters computed for mixtures at $T_o = 298\,\mathrm{K}$ and $p_o = 0.2\,\mathrm{atm}$

Mixture	ε_I	Δ_I(cm)	Δ_R(cm)	χ	Cell regularity
$CH_4 + 2O_2$	11.84	0.102	2.30×10^{-2}	52.5	Highly irregular
$C_3H_8 + 5O_2$	10.50	1.66×10^{-2}	1.05×10^{-2}	16.6	Highly irregular
$C_2H_2 + 2.5O_2$	4.82	3.82×10^{-3}	3.57×10^{-3}	5.16	Irregular
$C_2H_2 + 2.5O_2 + 60\%$ Ar	4.73	1.19×10^{-2}	1.57×10^{-2}	3.59	Regular
$C_2H_2 + 2.5O_2 + 85\%$ Ar	4.83	5.42×10^{-2}	0.114	2.30	Regular
$2H_2 + O_2$	5.28	2.44×10^{-2}	5.72×10^{-2}	2.25	Regular
$2H_2 + O_2 + 25\%$ Ar	4.91	2.18×10^{-2}	8.08×10^{-2}	1.33	Highly regular
$2H_2 + O_2 + 40\%$ Ar	4.68	2.32×10^{-2}	0.119	0.91	Highly regular

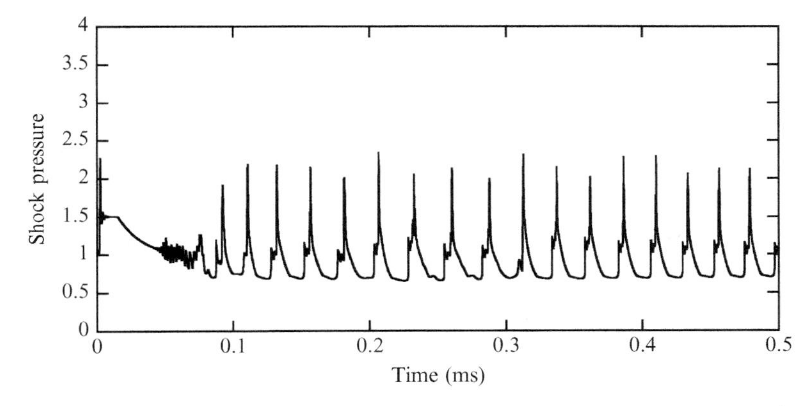

Fig. 3.32. Leading shock pressure history for stoichiometric acetylene–oxygen with 85% Ar dilution at $T_o = 298\,\mathrm{K}$ and $p_o = 0.2\,\mathrm{atm}$ [110]

of χ. The fact that these mixtures are characterized by a highly irregular cellular structure can now be interpreted in terms of the stability parameter χ. The incoherence in the power pulse or energy release originating from neighboring particles, which is expected for large values of χ, leads to gasdynamic instabilities in the reaction zone. One-dimensional shock-ignition is no longer possible to maintain self-propagation. Multidimensional effects such as transverse waves and other turbulent effects are essential for their propagation.

3.4.3 Relation Between One- and Two-Dimensional Detonation Instability

Traditionally, the regularity of the cellular detonation pattern observed on soot foils has been used to classify the structure of detonation waves. Previous studies have suggested that the regularity of cell spacing and ZND detonation stability appear to be intimately interrelated [53]. Mechanically, highly

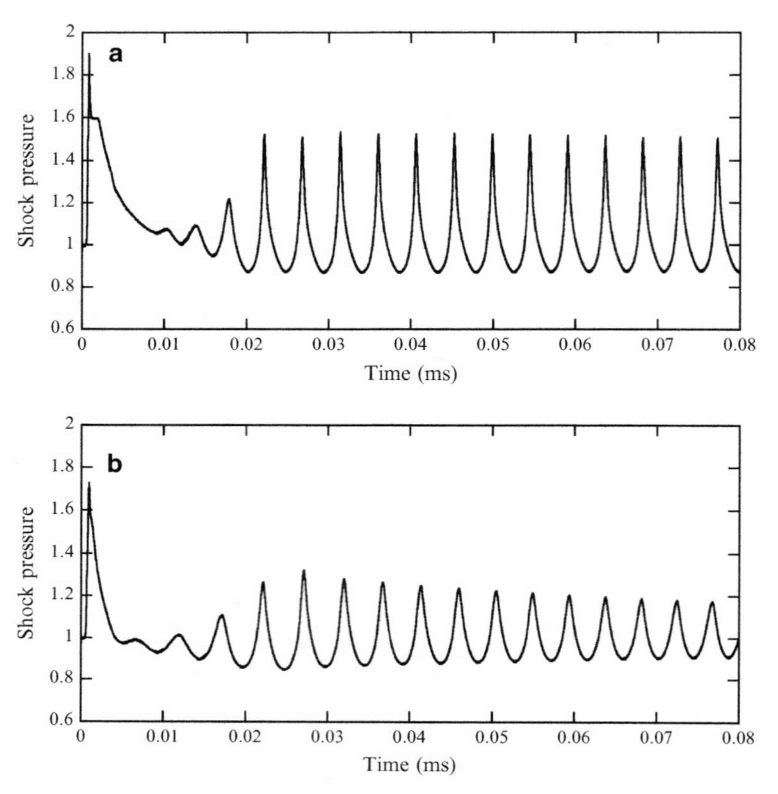

Fig. 3.33. Leading shock pressure history for stoichiometric hydrogen–oxygen with (**a**) 25% Ar and (**b**) 40% Ar dilutions, respectively, at $T_o = 298\,\mathrm{K}$ and $p_o = 0.2\,\mathrm{atm}$ [110]

unstable systems are characterized by relatively strong and irregular transverse waves, while stable systems are characterized by relatively weak and regular transverse waves.

To examine the effect of chemistry, to which the results of one-dimensional pulsating detonations driven by chain-branching chemistry model can be compared with multidimensional cellular detonations, the simulations of two-dimensional cellular detonation waves using the same two-step chain-branching model are shown in Fig 3.34. Similar to the one-dimensional case, the effect of the ratio of the reaction time to the induction time is varied by using the reaction rate constant K_R as a bifurcation parameter. Other values of the initial conditions for the 2D numerical experiments are given by $Q = 23.98$; $\varepsilon_I = 4$; $\varepsilon_R = 1$ and $\gamma = 1.333$, which approximately models the conditions for a H_2–O_2 mixture.

For low values of K_R, a fairly regular cell pattern is observed. By increasing the value of K_R, thus increasing the rate of heat release in the exothermic

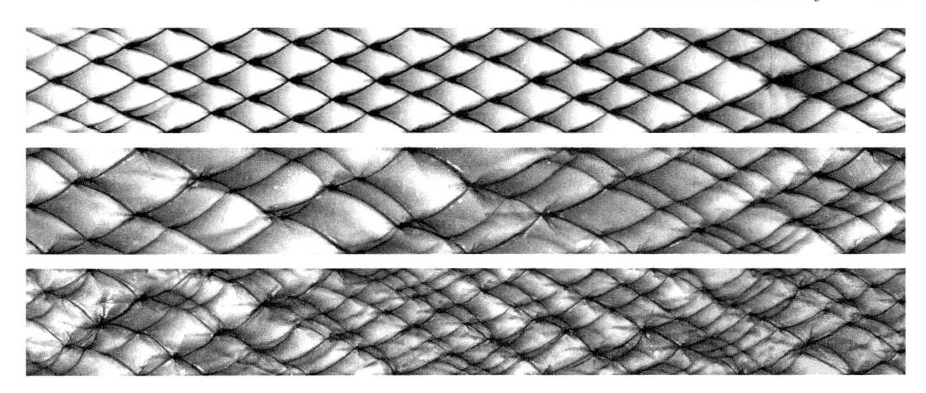

Fig. 3.34. Numerical soot foils obtained from the simulation using the two-step chain-branching kinetic model for $K_{\mathrm{R}} = 0.866$ (*top*); $K_{\mathrm{R}} = 3.46$ (*middle*); and $K_{\mathrm{R}} = 12.9$ (*bottom*)

reaction layer, there is some evidence of bifurcation in the simulated soot traces and the cell pattern becomes less regular. For very large values of K_{R}, the cell behavior becomes highly irregular. Such irregularity can be explained by the presence of more than one dominant frequency of the transverse fluctuation as well as the excitation of higher harmonics, which manifest themselves as finer cell patterns superimposed on the dominant bands. The results of these numerical simulations show a transition structure for unstable cellular detonations, from the regular regime to a highly irregular pattern, as the reaction rate constant K_{R} is systematically increased.

Recall from the one-dimensional pulsating detonation driven by the same two-step chain-branching kinetic model that the effect of increasing K_{R} (i.e., shortening the reaction zone length) was to render the wave unstable. Here, similar observations were made for the two-dimensional case where the degree of cell regularity is similarly influenced by the increase of K_{R}. For the three cases studied above, the stability parameter associated with each reaction rate constant K_{R} can be evaluated. These 2D results can then be mapped onto the $\chi - M_{\mathrm{CJ}}$ plane. By doing so, these different cases can be explained by comparing the location of their associated parameter χ to the neutral stability curve, as shown in Fig. 3.35. From this figure, it is understood that below the neutral stability curve, the cellular structure should be very regular. Above the stability boundary limit, the cell pattern can be described as highly irregular. It is of interest to note that not only can the parameter χ be related to the degree of instability in a 1D pulsating detonation structure, but it can also be considered a fundamental parameter that takes into account the essential chemical kinetic parameters needed to assess the degree of regularity of cellular detonation waves.

As pointed out earlier, there is a strong connection between the stability of ZND detonations with the regularity of multidimensional cellular structure.

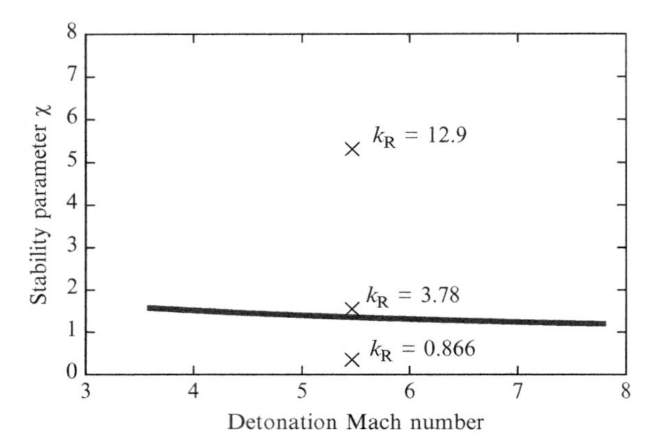

Fig. 3.35. Characterization of the two-dimensional cellular structures shown in Fig. 3.34 using the neutral stability map in the $\chi - M_{CJ}$ plane

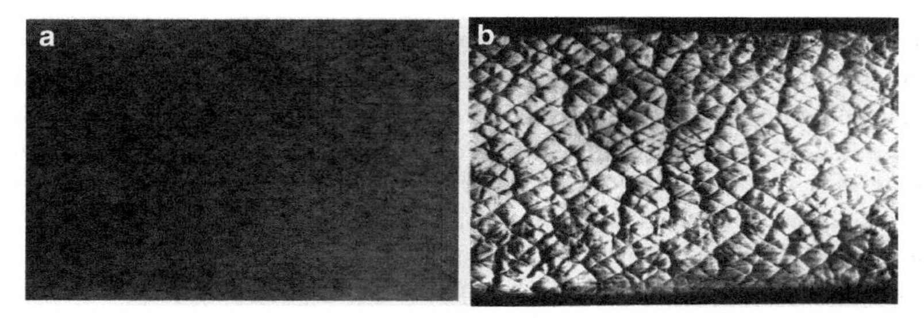

Fig. 3.36. Experimental smoked foil record for (**a**) $C_2H_2 + 2.5O_2 + 85\%Ar$ (courtesy of P. Pinard); (**b**) undiluted $C_2H_2 + 2.5O_2$ [182]

A sample experimental smoked foil obtained for stoichiometric acetylene–oxygen with 85% argon dilution is shown in Fig. 3.36a. The result shows a very regular cell pattern and the transverse waves are weak as indicated by the weak marking on the foil. The cell length measured from the smoked foil for these highly argon diluted cases and the oscillation period of the one-dimensional pulsating detonation obtained from numerical simulations, given in Table 3.6, are in good agreement. This may perhaps suggest that the regular cells observed experimentally in these highly diluted mixtures are due to the one-dimensional instability mechanism [130].

In contrast, a completely different picture can be seen from the smoked foil obtained in the experiments with low argon dilution or undiluted acetylene–oxygen mixtures, which display much stronger transverse waves with higher cellular irregularity (Fig. 3.36b). From the numerical simulation shown in Fig. 3.25, it can be seen that below a critical argon dilution of approximately 70%, the large fluctuation in reaction rates due to the increasing instabilities causes the one-dimensional detonation to quench. The self-sustenance of a

Table 3.6. Comparison of experimental cell lengths with oscillation periods of the 1D pulsating detonation

Mixture	1D pulsating length (mm)	Cell length (mm)
$C_2H_2 + 2.5O_2 + 90\%$ Ar	9	10.7
$C_2H_2 + 2.5O_2 + 85\%$ Ar	7.5	8

1D detonation wave, which relies on shock-ignition (i.e., the only propagation mechanism in the 1D case), is no longer possible. However, despite the inability of one-dimensional pulsating detonations to propagate in low-argon-diluted acetylene–oxygen mixtures as shown in the numerical simulation (Fig. 3.25), three-dimensional cellular detonations are commonly observed in experiments even for 0% argon dilution. In two- or three-dimensional detonations, instabilities can give rise to multidimensional effects such as transverse wave interactions and turbulent mixing, etc., which may play a significant role in the detonation propagation mechanism. These mechanisms can provide alternate means to cause local auto-ignition and sustain the sufficiently high burning rates necessary for the self-sustenance of real detonation waves. Consequently, these strong multidimensional effects can also result in highly irregular cellular detonation structures with strong transverse waves as observed experimentally.

Apart from characterizing the degree of stability of a detonation, a recent attempt has been made to use this stability parameter in correlations for predicting the characteristic cell size, λ [108, 111]. Numerous attempts have been made to relate experimentally measured cell sizes to some characteristic chemical length scale, Δ, in the one-dimensional ZND detonation structure. In general, a linear proportionality relationship between the cell size, λ, and the steady chemical induction length scale, Δ, has previously been proposed, i.e., $\lambda = A\,\Delta$, where A is a proportionality factor [105, 148, 149, 174, 188]. The factor A is generally determined by matching with one experimental data point for a particular combustible mixture (e.g., the value at stoichiometric composition), and the relationship is then used to extend the cell size prediction over any desired range of initial conditions. However, cell sizes predicted by this technique are usually valid only for mixtures with conditions that are similar to that of the matching point. The factor A significantly varies for different mixture compositions (especially off-stoichiometric and diluted mixtures) and initial conditions.

Since the parameter χ appears to be an intrinsic detonation property of different gaseous combustible mixtures and essentially includes all the important elements controlling instability, i.e., energetic, temperature sensitivity, induction, and chemical energy release zone length, it is suggested that the functional behavior of the factor A can be modeled by incorporating this stability parameter. A simple polynomial function has been proposed:

Table 3.7. Coefficients for the detonation cell correlation by Ng et al. [111] with $N = 3$

Coefficients	Values
$A_o = a_0 + b_0$	30.466
a_1	89.554
a_2	−130.793
a_3	42.025
b_1	−0.02929
b_2	1.02633×10^{-5}
b_3	-1.0319×10^{-9}

$$\lambda = A(\chi) \cdot \Delta_I = \sum_{k=0}^{N} \left(a_k \chi^{-k} + b_k \chi^k \right) \cdot \Delta_I, \tag{3.89}$$

where the necessary coefficients are predetermined from a set of available cell size data, as given in Table 3.7 with $N = 3$. It is shown that such a simple correlation can provide a good estimate over a wide range of different mixtures and initial conditions [100, 102, 108, 111]. Despite its simplicity, this relationship illustrates the significance of the stability parameter χ in cell size prediction. However, the validity should not be overstated. It may be expected that even better results could have been obtained by using other functional expressions derived for the correlation. Nevertheless, this simple study strongly suggests that an improved correlation of detonation cell size with chemical kinetics can be obtained by taking into account the effect of detonation instability in the formulation.

From all these analyses, there is strong evidence that the detonation chemistry behind the front controls the dynamics of the cellular structure. Several recent studies have also shown that the effect of chemistry is significant when the heat release rate exhibits two peaks of thermicity, resulting in some interesting double cellular structures. For instance, two-cell detonation structures have been observed experimentally in gaseous nitromethane and nitromethane-oxygen detonations, hydrogen, methane, ethane mixed with NO_2/N_2O_4 by Presles, Desbordes and co-workers [33, 71, 91, 128, 167] and corresponding numerical simulations were also conducted by the same group. (e.g., see [59] and more references in Chap. 5.) There are studies which also suggest that in very lean H_2–N_2O mixtures [75] and lean fuel-N_2O mixtures [179], the ZND structure is characterized by a double peak thermicity profile and a cellular detonation front resembling a double cellular structure is also observed [101]. Details of the double cellular detonation and the related two-peak thermicity heat release mechanism can be found in Chap. 5.

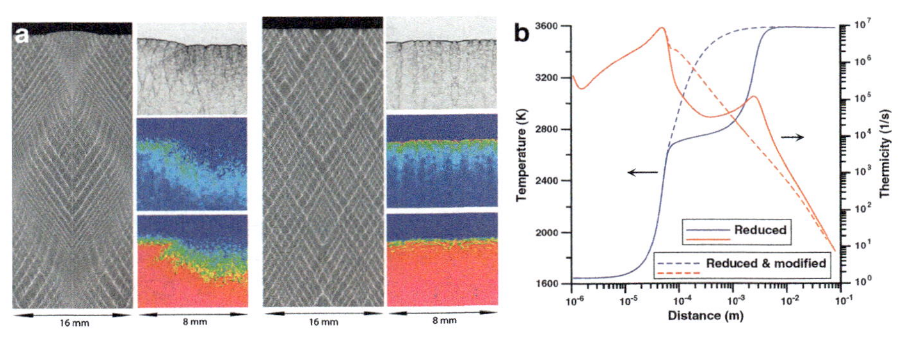

Fig. 3.37. (**a**) Numerical smoked foil records, density Schlieren plots, N, and N_2 mass fractions for detonations in a rich H_2–NO_2/N_2O_4 mixture. *Left*: Two-step energy release. *Right*: Modified single-step energy release. (**b**) Thermicity and temperature profiles: comparison between the two-step energy release with the modified model of one-step energy release for a rich H_2–NO_2/N_2O_4 mixture [30]

Some interesting computational work on the effect of non-monotonic energy release on the detonation structure was performed by Davidenko et al. [30]. Using detailed chemical kinetics and one-dimensional ZND analysis, they systematically described the origin of the non-monotonic two-step energy release from chemical kinetics considerations, observed in rich H_2–NO_2/N_2O_4 and in very lean H_2–N_2O mixtures. They also used reduced kinetics schemes to investigate numerically the two-dimensional cellular structures. To explain the effect of two-step energy release on the cellular structure, they modified some reaction parameters in the reduced kinetics model to suppress one of the two thermicity peaks.

What they found is that in rich H_2–NO_2/N_2O_4 mixtures, the double cellular structure, as observed experimentally, exists in the numerical simulations and is characterized by the superposition of a large cellular structure on one of a smaller size (Fig. 3.37). Comparing the results for both single-step and two-step energy release, there is a distinct relationship that can be observed between the double cellular structure and the evolution of N and N_2 mass fractions which confirms the kinetic origin of the double cellular structure of the detonation. On the contrary, classical substructures (due to high degree of detonation front instability) were only observed in very lean H_2–N_2O mixtures (Fig. 3.38). Calculations made with the detailed kinetic model of Mével et al. [100,102], supported by experimental observation, indeed show that detonations propagating in H_2–N_2O(-diluent) mixtures range from unstable to highly unstable ones. The rather random, small-scale instabilities that result in the formation of small cells can be interpreted based on the results of stability analysis. Although having the characteristics of a two-step heat release profile, the critical conditions for the formation of the double-cellular structure, if it exists, are rather difficult to distinguish, and it appears that the cellular dynamics and substructure stems from the instability effect, which

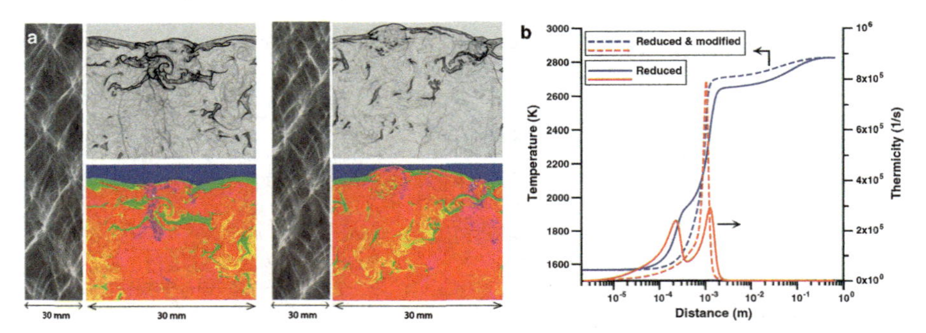

Fig. 3.38. (**a**) Numerical smoked foil records, density Schlieren plots, and temperature fields for detonations in a very lean H_2–N_2O mixture. *Left*: Two-step energy release. *Right*: Modified single-step energy release. (**b**) Thermicity and temperature profiles: comparison between the two-step energy release with the modified model of one-step energy release for a very lean H_2–N_2O mixture [30]

dominates over the effect of non-monotonic heat release. It is worth noting that the experimental observation [112] for a dimethyl-ether mixture at very low initial pressure and lean conditions, which also has a double-peak in thermicity and a high instability parameter χ, supports these numerical results, reinforcing the concluding remarks of Davidenko et al. [30].

3.5 Instability of Nonideal Detonations

In many situations real detonations are nonideal, in the sense that they deviate from the Chapman–Jouguet and the classic one-dimensional ZND model. Nonideal detonations can arise from interaction with the boundaries, such as detonation front curvature of a condensed explosive in a tube (will be further discussed in Chap. 7), diverging flow in an expanding nozzle, or friction and heat loss to a rough tube wall. They can also result from the chemistry of the mixture with both exothermic and endothermic reactions, or with mole decrement during reaction progress [50], or with multiphase mixtures consisting of an explosive gas and condensed particles with momentum and heat transfer from the explosive to the particles [196].

3.5.1 Nonideal Detonation with Losses

In a quasi-one-dimensional detonation, nonideal effects are represented by the use of source terms in the mass, momentum, and energy conservation equations in a laboratory coordinate system:

$$\begin{aligned}
&\frac{\partial \tilde{\rho}}{\partial \tilde{t}} + \frac{\partial (\tilde{\rho}\tilde{u})}{\partial \tilde{x}} = m \\
&\frac{\partial \left(\tilde{\rho}\tilde{u}\right)}{\partial \tilde{t}} + \frac{\partial (\tilde{\rho}\tilde{u}^2 + \tilde{p})}{\partial \tilde{x}} = f \\
&\frac{\partial \left(\tilde{\rho}\tilde{E}\right)}{\partial \tilde{t}} + \frac{\partial \tilde{u}(\tilde{\rho}\tilde{E} + \tilde{p})}{\partial \tilde{x}} = q,
\end{aligned} \tag{3.90}$$

where the source terms m, f, q represent mass, momentum, and heat losses, respectively. For example, the effect of area divergence, curvature of a wave front, friction, heat transfer, and body forces can be analyzed by introducing the appropriate mass, momentum, and heat loss terms in the conservation equations.

The above conservation equations can be transformed to a coordinate system attached to the detonation shock front propagating at velocity D. In the shock-attached coordinate system ($t = \tilde{t}$, $x = \int D(t)\mathrm{d}t - \tilde{x}$), the steady, quasi-one-dimensional ZND structure equation is then obtained by setting all of the time derivatives equal to zero. For the case of nonideal detonations where these source terms are present, the steady detonation solution can only be determined from the integration of the quasi-one-dimensional ZND structure equations. For instance, the change of the fluid velocity u along the propagation distance x is given by

$$\frac{\mathrm{d}u}{\mathrm{d}x} = \frac{\Phi}{1 - u^2/c^2} = \frac{\Phi^+ - \Phi^-}{1 - u^2/c^2}, \tag{3.91}$$

where c denotes the local sound speed. The quantity Φ represents the thermicity, which is a measure of the rate of net energy release resulting from the chemical heat release rate, Φ^+, and the energy dissipation rate, Φ^-, from the source terms. The detailed expression of the thermicity depends on the specified forms of source terms, chemical reaction laws, and equations of state. From the above equation, a "generalized CJ criterion" is derived as the rear boundary condition at the sonic locus embedded in the reaction zone:

$$\Phi = \Phi^+ - \Phi^- = 0, \text{ at } u = c. \tag{3.92}$$

The generalized CJ point determined by the above equation is a mathematical saddle point and integration of the one-dimensional ZND equations becomes an eigenvalue problem. Readers are referred to the details in Chap. 2.

Through the saddle point, the subsonic flow relative to the shock front becomes supersonic, where, without any support behind, the rest of the reaction after the sonic locus will not influence the upstream detonation structure in a steady solution. In an ideal CJ detonation, all negative source terms disappear; thus $\Phi^- \equiv 0$, (3.91) becomes the classic CJ criterion. Unlike the CJ detonation, nonideal detonations cannot be determined from the global conservation laws for a control volume bounded by two-equilibrium planes, since the energy gain and loss rate competition takes place in the reaction zone. Furthermore, the presence of source terms can also lead to a multiplicity of possible steady-state solutions, requiring additional stability analyses to arrive at the appropriate choice for a steady solution.

Alternatively, the problem of the existence of a stable, steady solution can be obtained and verified by a transient analysis. In a transient analysis, that is, numerical solution of the time-dependent reactive quasi-one-dimensional Euler equations (3.90), no criterion has to be imposed. If the detonation is stable, then the transient solution will asymptotically approach the steady-state solution as determined from the integration of the ZND equations for the structure. Furthermore, the selection of an appropriate solution, in the case of a multivalued solution as well as the stability of the solution, can be determined in a transient analysis.

Indeed, the need to consider the nonsteady flow behind the detonation front and seeking a detonation solution that can be matched to the transient flow behind the detonation was first recognized by Taylor [171]. Equivalently the existence of a steady quasi-one-dimensional detonation wave is determined by matching the unsteady downstream flow to the generalized CJ criterion, which determines the matching conditions of the upstream and downstream flow. However, even for a one-dimensional detonation without losses, the steady concept of sonic matching (i.e., the CJ condition) may not be satisfied, and the ideal steady ZND model fails as seen in the one-dimensional pulsating detonation for large enough activation energy values. Therefore, the steady quasi-one-dimensional detonation solution including nonideal effects is not guaranteed to be the correct solution, and it must be determined if the addition of the source terms results in the transient downstream flow being compatible with the saddle point condition (generalized CJ criterion) or not.

Due to the similar form of all negative source terms in the generalized CJ criterion, friction is chosen to elucidate the competing effect of the chemical energy release and the negative source terms, and to explore the common properties of the instability of nonideal detonations. For the case of nonideal detonation with friction, Zhang and Lee [197], and later Dionne et al. [34,35], carried out a numerical solution of the time-dependent conservation equations. What they found is that the generalized CJ criterion can be satisfied in the presence of low frictional drag in mixtures with small activation energy. For large frictional drag, the competing effect of the chemical energy release and the momentum loss results in an oscillatory detonation and a steady wave solution does not exist.

For example, Figs. 3.39 and 3.40 show the transient simulation results obtained using one-step Arrhenius kinetics with small $E_a = 22$ and frictional drag. (Note that the stability limit is $E_a = 25.26$ for a frictionless ideal detonation as described in Sect. 3.3.2.) Without friction (corresponding to the limit of an infinite hydraulic diameter tube, $D_\mathrm{h} = \infty$), the detonation is therefore ideal and propagates at the CJ velocity. As friction increases, the wave velocity evolves to a steady asymptotic value below the normal CJ value ($D_\mathrm{h} = 40$), and then displays an unsteady oscillatory behavior with a decrease in average

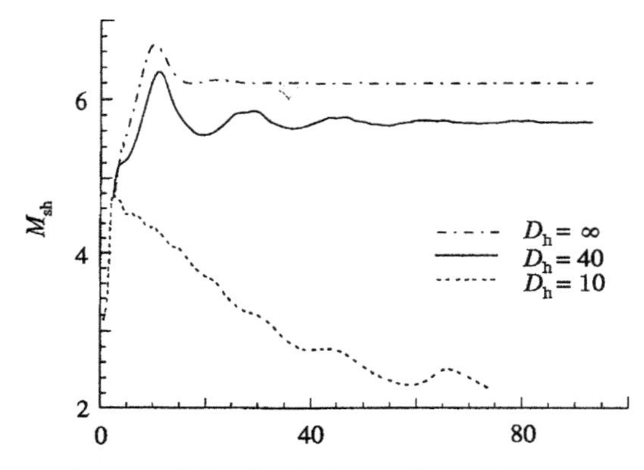

Fig. 3.39. Time evolution of shock Mach number of detonations in an unsteady wave solution for different hydraulic diameters; $\gamma = 1.2$, $Q = 50$ and $E_a = 22$ [197]

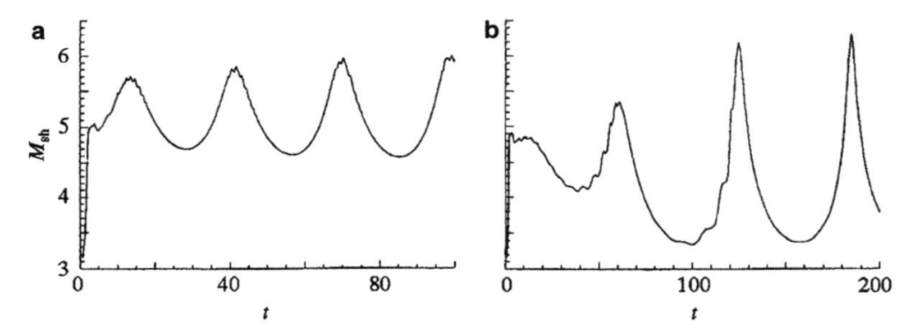

Fig. 3.40. Time evolution of shock Mach number of detonation in an unsteady wave solution; $\gamma = 1.2$, $Q = 50$ and $E_a = 22$. (**a**) $D_h = 20$; (**b**) $D_h = 14$ [197]

velocity ($D_h = 20$ and 14). Finally with very large friction ($D_h = 10$), the wave decays after initiation and eventually fails. For activation energy above the stability limit of an ideal CJ detonation ($E_a = 27$), Fig. 3.41 shows that the addition of friction drives the instability toward larger amplitude oscillations.

As shown above, an increase in friction causes a decrease in detonation velocity and therefore in shock temperature, leading to an increase in E_a/RT_s. The route to instability with an increase in friction is therefore similar in nature to the pulsating detonation for the frictionless case when the activation energy increases. Thus, it can be expected that the same trend of an increase in cellular irregularity of the detonation structure occurs in two- or three-dimensional numerical calculations as friction increases. In summary, the introduction of an energy dissipation rate in the reaction zone from a negative source term such as friction results in a detonation velocity deficit and drives a detonation toward instability. The velocity deficit and instability level increase with increasing energy dissipation, eventually leading to failure of the detonation.

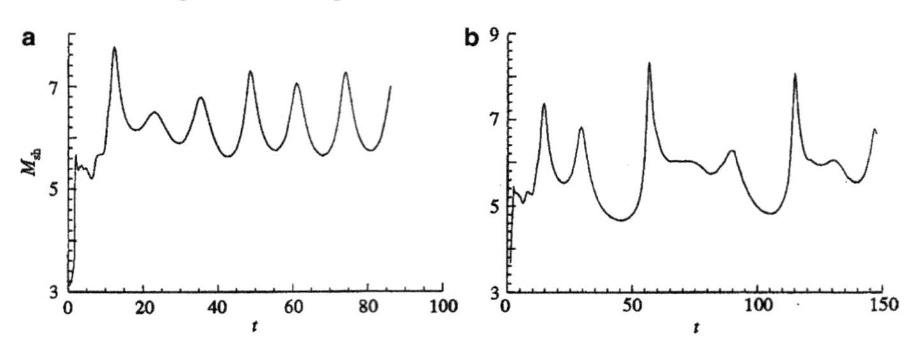

Fig. 3.41. Time evolution of the shock Mach number of detonations in an unsteady wave solution; $\gamma = 1.2$, $Q = 50$ and $E_a = 27$. (**a**) $D_{\mathrm{h}} = \infty$ (i.e., frictionless); (**b**) $D_{\mathrm{h}} = 60$ [197]

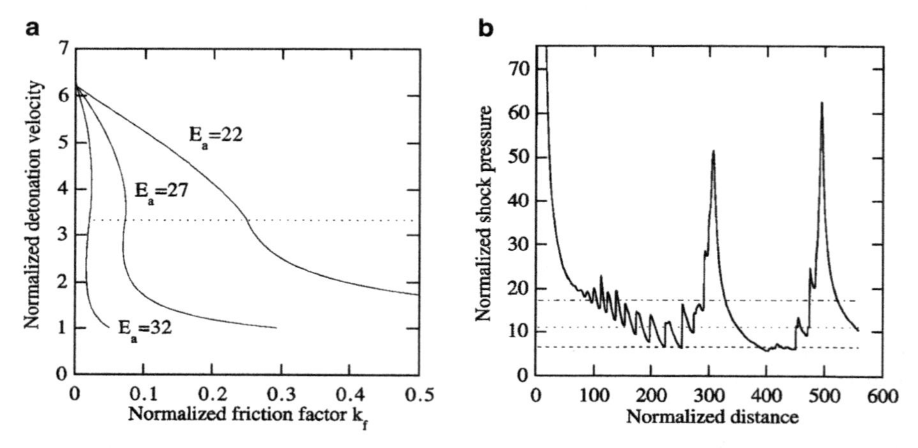

Fig. 3.42. (**a**) Steady solution for detonation with friction and for three different activation energies E_a. (**b**) The shock pressure versus distance for an unstable low velocity detonation wave with friction. The three possible steady solutions are shown by *dotted lines* [35]

Finally, Fig. 3.42a illustrates the multiplicity of solutions for quasi-steady ZND detonations with friction and sufficiently large activation energy. Indeed, for certain values of friction factor k_{f}, there are three possible detonation velocities when $E_a = 27$. Figure 3.42b shows the unstable pulsating detonation obtained from the transient analysis as the initiating blast decays asymptotically, together with the three possible steady-state solutions obtained from the quasi-steady ZND analysis. The correct solution is obtained from the long-time transient analysis for given initial conditions, particularly when conditions are outside the stability limits. On the other hand, the analysis of the steady ZND structure does not always lead to a unique solution or indicate its stability, in other words, the true solution.

For reference, examples of the instability of nonideal detonations with area divergence $A(x)$ and moving engine nozzles can be found in Zhang et al. [198] and those in gas-particle flow in Zhang [196].

3.5.2 Pathological Detonation

Nonideal detonations can arise from the chemistry of the mixture itself such as an exothermic reaction followed by an endothermic reaction, or mole decrement during reaction progress. In this case, the detonation velocity is actually greater than the equilibrium CJ velocity. For example, for the H_2–Cl_2 system [195], the exothermic reaction of $H_2 + Cl_2 \rightarrow 2HCl$ via the Nernst chain is competing against a much slower endothermic dissociation reaction of chlorine ($Cl_2 + M \rightarrow 2Cl + M$). As discussed in Chap. 2, a qualitative demonstration of the existence of pathological detonations can be obtained using the simple two-step irreversible reaction model suggested by Fickett and Davis [50], that is, A $\rightarrow$ B exothermic and B $\rightarrow$ C endothermic with an Arrhenius law for both reactions:

$$\begin{aligned} \frac{\mathrm{d}\lambda_1}{\mathrm{d}t} &= k_1 (1 - \lambda_1) \exp\left(\frac{-E_{a1}}{RT}\right) \\ \frac{\mathrm{d}\lambda_2}{\mathrm{d}t} &= k_2 (\lambda_1 - \lambda_2) \exp\left(\frac{-E_{a2}}{RT}\right), \end{aligned} \tag{3.93}$$

where λ_1 and λ_2 can be related to the mass fraction of reactant A and product C. The net chemical energy release is given by $Q = \lambda_1 Q_1 + \lambda_2 Q_2$, where $Q_1 > 0$ and $Q_2 < 0$ denote the heat release in the exothermic and endothermic reactions.

The eigenvalue solution is obtained through the integration of the steady one-dimensional ZND equations, e.g.,

$$\frac{\mathrm{d}u}{\mathrm{d}x} = \frac{\gamma - 1}{c^2} \frac{(\dot{\lambda}_1 Q_1 + \dot{\lambda}_2 Q_2)}{1 - u^2/c^2} \tag{3.94}$$

by satisfying the generalized CJ criterion at the sonic point,

$$\dot{\lambda}_1 Q_1 + \dot{\lambda}_2 Q_2 = 0 \text{ at } u = c. \tag{3.95}$$

For a very slow endothermic reaction, that is, $E_{a2} >> E_{a1}$, $Q^* \rightarrow Q_1$ when $u \rightarrow c$, the detonation velocity will then be governed by Q_1, an energy release higher than the equilibrium value, resulting in a higher detonation wave velocity. For this reason, this kind of detonation has been called a pathological detonation, providing a good demonstration that the classical equilibrium CJ solution does not always correspond to the correct steady-state detonation solution of a given explosive mixture.

The steady pathological detonation solution shown above is also referred to as a weak detonation, because it meets the necessary conditions. Namely, (1) within the reaction zone, there is at least one sonic locus embedded at which the generalized CJ condition is satisfied; and (2) the final equilibrium Hugoniot is not the upper bound of all partial equilibrium Hugoniot curves.

To verify if the one-dimensional pathological detonation is stable or can be obtained from arbitrary initial conditions, numerical simulation of transient reactive Euler equations using the two-step kinetic model were performed by Dionne [34], and Sharpe and Falle [144]. In the transient simulation where the entire nonsteady flow field is computed, the generalized CJ criterion does not need to be imposed. The detonation is initiated directly by a strong blast wave, and the existence of pathological detonation is determined from the asymptotic behavior with a detonation velocity greater than the equilibrium CJ value.

Figure 3.43 shows cases of stable and unstable pathological detonations initiated directly by a strong blast wave. It shows the asymptotic decay of

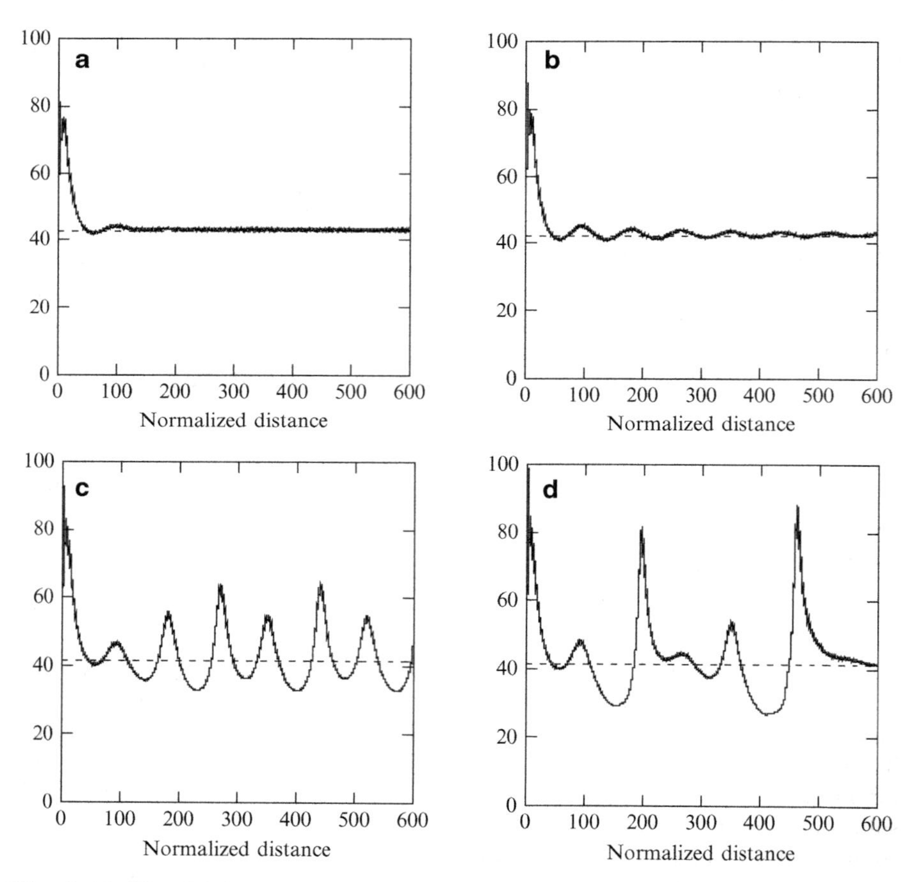

Fig. 3.43. The shock pressure profiles for a transient pathological detonation ($Q_1 = 50$, $Q_2 = -10$, $\gamma = 1.2$, $E_{a2} = 32$, and $k_1 = k_2$) with (**a**) $E_{a1} = 22$, (**b**) 24, (**c**) 26, and (**d**) 27 [34]

the strong blast to a steady solution that corresponds identically to the solution from the integration of the steady ZND equations (represented by the dashed lines). As the activation energy for the exothermic reaction (i.e., E_{a1}) increases, a nonsteady pulsating detonation is obtained and the instability level increases with E_{a1}. From the stability analysis of pathological detonations carried out by Sharpe [140], it was also found that the stability limit is determined by the value of the exothermic activation energy in accord with the transient analysis. Some other interesting observations of the nonlinear behavior of pathological detonations were also investigated by Sharpe, and his results showed that the rear/downstream boundary condition can have a significant influence on the instability and that nonlinear evolution of supported and unsupported pathological detonations can be very different. The flow downstream of this sonic point is supersonic if the detonation is unsupported but can become subsonic if the detonation is supported, the two cases having very different detonation wave structures. These effects cannot be understood from the steady ZND solutions without looking at the complete nonsteady flow field and the matching condition of the upstream and downstream flow.

The pathological detonation can indeed be considered as a nonideal detonation in a general sense. In this regard, a velocity deficit can be viewed with respect to the detonation velocity that is governed by Q_1. Hence, it can be expected that an increase in instability results from increasing the endothermic rate, that is, $\dot{\lambda}_2|Q_2|$ [140]. Figure 3.44 shows the numerical simulation of unsupported pathological detonation with increasing Q_2, and it can be clearly seen that the detonation becomes more unstable (eventually fails) as the degree of endothermicity increases [144].

3.5.3 Diverging Detonation

A cylindrically or spherically expanding detonation has a curved shock front, which the fluid particles upon crossing undergo subsequent lateral expansion. The flow divergence due to the lateral expansion can be modeled by

$$m = -\frac{j}{r}\rho u, \; f = -\frac{j}{r}\rho u^2, \; q = -\frac{j}{r}(\rho u E + pu) \tag{3.96}$$

with $j = 1$ for the cylindrical geometry or $j = 2$ for the spherical geometry. These negative source terms will lead to an energy rate that competes with the chemical energy release rate. The rate decreases as the radius increases and asymptotically disappears as the radius becomes sufficiently large, at which time the detonation approaches to the CJ velocity of an ideal detonation on average.

While an expanding detonation is an unsteady detonation, it may be assumed that the front evolves on a timescale that is long compared to the particle transit time through the reaction zone. In this case, the instantaneous speed and structure of the detonation at a given shock radius can be obtained

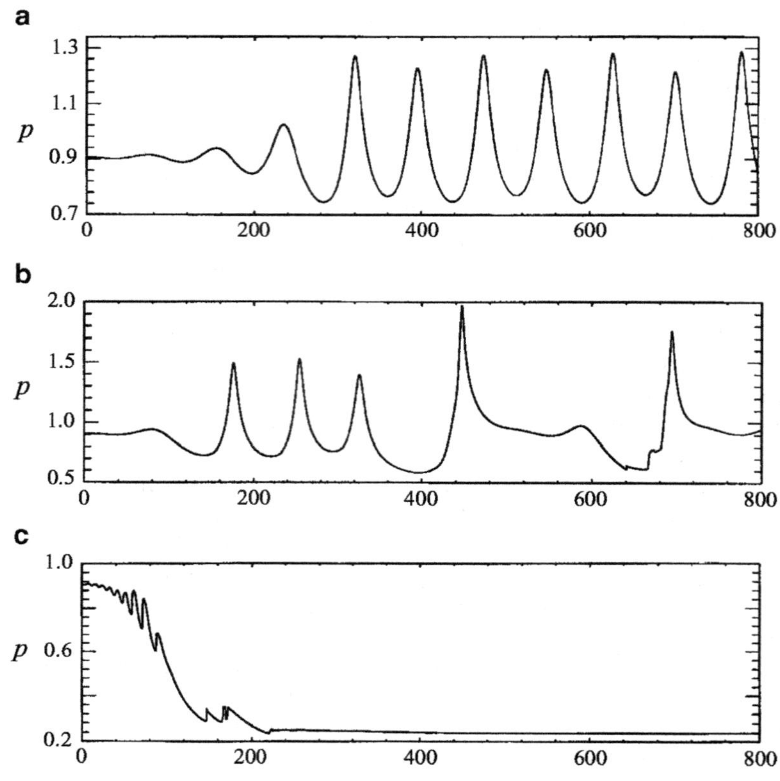

Fig. 3.44. Shock pressure history for (**a**) $Q_2 = -5$, (**b**) $Q_2 = -10$, and (**c**) $Q_2 = -20$ (all with $Q_1 = 50$, $E_1 = 25.26$ and $E_2 = 30$) for the unsupported pathological detonation [144]

by a quasi-steady analysis, which follows the same treatment used for the nonideal detonations (i.e., integration of the quasi-steady one-dimensional ZND structure equations with the generalized CJ criterion at the sonic locus as boundary condition). The quasi-steady analysis provides an eigenvalue relationship between the detonation speed and radius of curvature of the front (also see Chaps. 2 and 7), and there exists a critical radius of curvature below which there is no quasi-steady solution [63, 186, 191]. The expanding detonation therefore has the same nature as the nonideal detonations with respect to the competition of positive and negative energy rates, leading to shock velocity reduction and detonation instability.

A systematic study of the linear and nonlinear instability of radially expanding detonation waves has been carried out by Watt and Sharpe [186, 187]. In their study, a one-step Arrhenius reaction law and the ideal gas EOS were used with $Q = 50$ and $\gamma = 1.2$ and varying activation energies E_a. Figure 3.45 shows the neutral stability boundary as determined by the

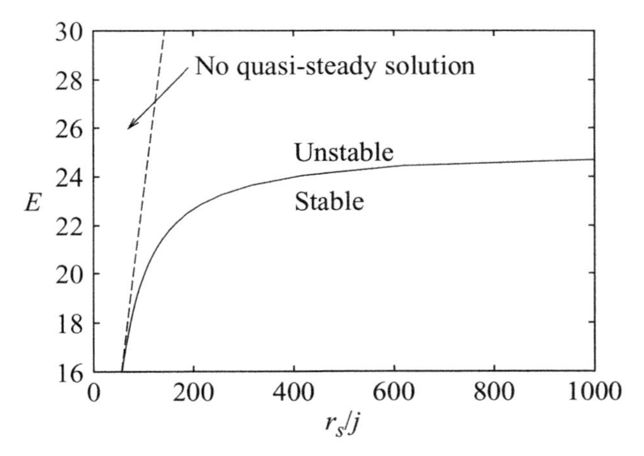

Fig. 3.45. The neutral stability boundary (*solid curve*) and the critical shock radius r_s for the quasi-steady solutions (*dashed line*). $j = 1, 2$ for cylindrical and spherical geometry, respectively [186]

leading-order linear stability analysis of the underlying quasi-steady waves in an activation energy-shock radius diagram, together with the locus of the critical radius of curvature, below which there are no quasi-steady solutions, i.e., detonations cannot propagate in a quasi-steady fashion. From Fig. 3.45, one clear observation from the linear stability analysis is the exponentially changing behavior of the neutral stability boundary as the radius of curvature decreases. At sufficiently small shock radius, even a small increase in front curvature can have a rapid destabilizing effect on detonation. As the neutral stability boundary appears to asymptote to the quasi-steady critical shock radius locus, the detonation is therefore always unstable sufficiently near the quasi-steady critical radius of curvature. At the other limit, as the shock radius approaches infinity, the critical activation energy asymptotically reaches the planar wave stability limit ($E_a = 25.26$ in this case).

Watt and Sharpe [187] compared the one-dimensional linear stability predictions for weakly curved detonation waves with the numerical simulations of the nonlinear unsteady governing equations (3.90) with source terms in (3.96) for a radially expanding detonation and focused on direct initiation. They performed numerical analysis of both cylindrically and spherically expanding detonations to study the effect of curvature on the stability and propagation of detonations. Apart from the fact that the stability of the wave changes more slowly, the main conclusions are the same for spherical geometry as for cylindrical geometry. Hence, the discussions here concentrate on the cylindrical geometry only.

For detonations with low activation energy, from Sect. 3.3.3 it is discussed that the planar wave (i.e., corresponding to infinite shock radius) is inherently stable. As already shown in Fig. 3.45, for the diverging detonation to

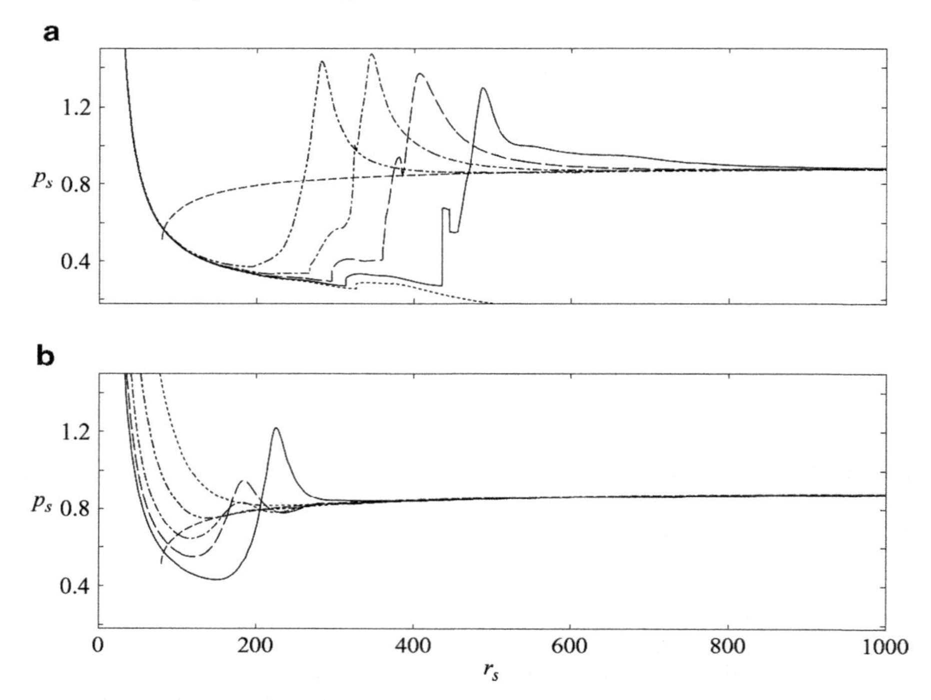

Fig. 3.46. Shock pressure versus shock radius for a cylindrical detonation with $E_a = 20$ and (**a**) Initiation energy $E_s = 572.3$ (*dotted line*), 572.4 (*solid line*), 573 (*long-dashed line*), 574 (*dot-dashed line*) and 580 (*double-dot-dashed line*); (**b**) $E_s = 600$ (*solid line*), 700 (*long-dashed line*), 900 (*dot-dashed line*), 1,500 (*double-dot dashed line*) and 3,000 (*dotted line*). The short-dashed curve is the quasi-steady $p_s - r_s$ relation [187]

be one-dimensionally unstable, the radius of front curvature has to be sufficiently near the critical quasi-steady radius. For the case with low activation energy such as $E_a = 20$, Fig. 3.46 plots the shock pressure variation versus shock radius ($p_s - r_s$) curves obtained from the unsteady computations using different source energies for direct initiation, E_s. Above the critical initiation energy ($E_s \approx 572.4$), the blast wave pressure first drops and then a transient initiation process causes an overshoot in pressure or an overdriven detonation. Afterward the detonation becomes self-sustaining and approaches eventually the quasi-steady $p_s - r_s$ curve. In the numerical simulations, the oscillations in the very early transient phase are referred by Watt and Sharpe [187] to as the ignition or initiation pulse, which depends on the initial condition and initiation method. It is important to note that care must be taken, as pointed out by Watt and Sharpe, to distinguish the initiation pulse produced by the transient initiation process from the pulsation due to the instability of the expanding detonation which is subsequently formed. In other words, a propagating detonation is said to be formed after this initiation pulse has taken place, and only the subsequent wave behavior when the self-propagating detonation approaches again the quasi-steady state after the initiation pulse shall be related to the inherent instability of the diverging detonation. In any cases, for detonations with sufficiently low activation energy, any initial oscillation is rapidly damped out and the subsequent evolution returns to being quasi-steady or stable.

In the stable cases with low activation energy, another interesting result is that the decaying critical blast wave crosses the quasi-steady $p_s - r_s$ curve very close to its turning point obtained in the $D - \kappa$ space plot, which has been used as a criterion to predict the critical energy for direct initiation [63]. The shock radius at this turning point may be linked to Zel'dovich's criterion for direct blast initiation, which requires that the blast wave should engulf a critical kernel size of radius $r_s = R^*$ for successful initiation as it decays to a certain strength of shock Mach number M_s. The turning point of the quasi-steady detonation speed D at the critical radius of curvature thus provides a critical condition for detonation initiation, since failure occurs due to the loss of the quasi-steady solution below the turning point. Nevertheless, this criterion is necessary but not sufficient, particularly, for unstable detonations where instability plays a more prominent role in controlling the initiation process. More generally, rational theories for estimating the critical initiation energy which account for unsteadiness (such as the shock acceleration $\dot{D}$ resulting in a $\dot{D} - D - \kappa$ relation for the detonation evolution) have been proposed by Eckett et al. [39] and Kasimov and Stewart [78,79]. By including consideration of the shock acceleration with shock velocity and front curvature, the model provides better agreement with experiments.

For higher activation energy, the neutrally stable radius of curvature predicted from linear stability analysis moves further away from the critical quasi-steady radius (Fig. 3.45). Therefore, decreasing the radius of curvature toward the critical quasi-steady radius rapidly destabilizes the wave. It is thus expected that the oscillatory behavior (amplitude and growth rate) due to instability is sensitive to the radius at which the detonation first forms after the initiation pulse. The smaller this initial radius, the more quickly the initial oscillation amplitude grows.

For example, Fig. 3.47 shows the results for activation energy $E_a = 25$ which is just below the planar wave stability limit. For this activation energy, the neutrally stable shock radius is significantly increased (i.e., to $r_s/j = 1{,}933$ as determined from the linear stability analysis of Watt and Sharpe [186]). For initiation energies $E_s = 1{,}500$ and $2{,}000$, which are very close to the critical energy of direct initiation (Fig. 3.47a–b), initially there is again a large-amplitude transient initiation pulse. The wave then drops back near the quasi-steady condition at a much larger shock radius location. The subsequent oscillations appear to have a smaller amplitude and a slower growth rate as well. These oscillations attain a maximum and then decay. However, as the initiation energy is increased to a more supercritical value,

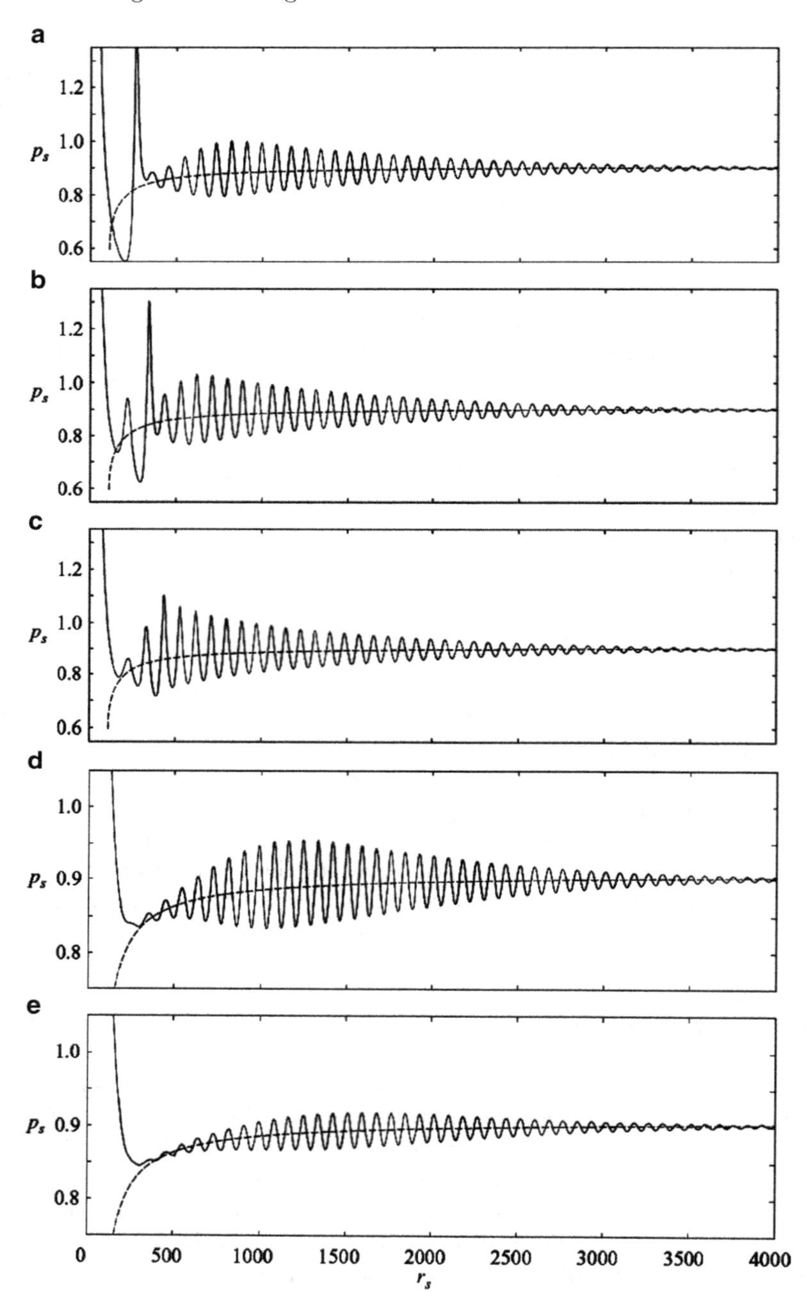

Fig. 3.47. Shock pressure versus shock radius for a cylindrically diverging detonation with $E_a = 25$ and (**a**) initiation energy $E_s = 1{,}500$, (**b**) $E_s = 2{,}000$, (**c**) $E_s = 2{,}500$, (**d**) $E_s = 4{,}000$, and (**e**) $E_s = 5{,}000$. The *dashed curve* is the quasi-steady $p_s - r_s$ relation [187]

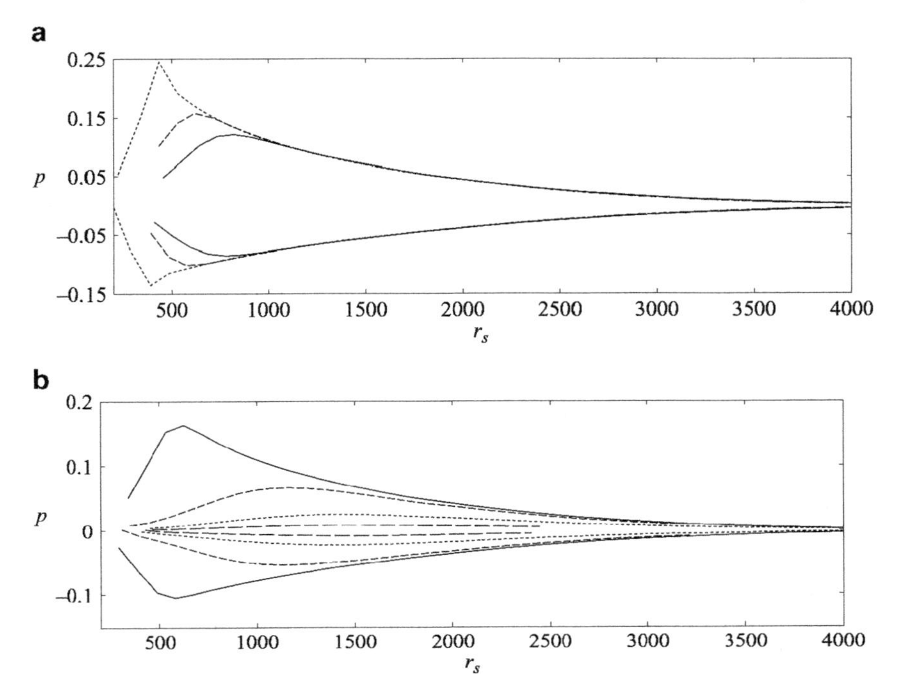

Fig. 3.48. Envelope of oscillation amplitude for $E_a = 25$. *Upper*: $E_s = 1{,}500$ (*solid lines*), 2,000 (*dashed lines*), and 2,500 (*dotted lines*). *Lower*: $E_s = 3{,}000$ (*solid lines*), 4,000 (*short-dashed lines*), 5,000 (*dotted lines*), and 6,000 (*long-dashed lines*) [187]

i.e., $E_s = 2{,}500$ (Fig. 3.47c) for which the influence of the initiation pulse is suppressed with a smaller amplitude, the subsequent oscillations due to the instability of the expanding detonation formed quickly reach a maximum of saturated nonlinear amplitude at a radius of $r_{s,\min} \leq 400$, significantly before the neutrally stable radius of $r_s = 1{,}933$. Subsequently the oscillation amplitude decays following what Watt and Sharpe [187] defined as a "saturated nonlinear envelope." It is interesting to note that for any E_s larger than $E_s = 2{,}500$ (Fig. 3.47d, e), after the transient initiation pulse, the envelopes of oscillation amplitude all lie within that of $E_s = 2{,}500$. However, the oscillations in these cases reach their maximum at a radius larger than $r_{s,\min}$. This saturated nonlinear envelope and oscillatory behavior are shown more clearly in Fig. 3.48, where the quasi-steady $p_s - r_s$ relation has been subtracted from the numerical shock pressure once the detonation is formed after the initiation transient pulse. This plot thus gives a clearer picture of the subsequent growth and decay of the amplitude of oscillations around the quasi-steady values for different initiation energies.

From the numerical observations, it is interesting as suggested by Watt and Sharpe [187] that the unstable one-dimensional diverging detonation evolves in two stages: i.e., after the transient initiation pulse that leads to the detonation formation, an initial stage of growth of the oscillation occurs toward the saturated nonlinear envelope, followed by a stage in which the amplitude saturates and subsequently follows the nonlinear envelope. In the first stage, the location of the maximum amplitude depends on the radius at which the growth rate begins. As shown in the linear stability result, the detonation is more unstable at smaller radius, and hence the more rapid the growth rate of the amplitude oscillation. The subsequent evolution that follows the envelope is also found to be independent of the initial conditions (e.g., the initiation energy for direct initiation). Furthermore, it also appears that it is the radius after the initiation pulse where the detonation first reaches the saturated nonlinear envelope that governs the subsequent stability of the detonation, rather than the radius where the blast wave decay first intersects the quasi-steady p_s–r_s curve [187].

For a further increase of activation energy where the planar case becomes inherently unstable (e.g., for $E_a = 27$), if the diverging detonation is successfully initiated, then the pulsations will always eventually reach a saturated nonlinear behavior, regardless of the amount of initiation energy E_s, as there is no stability boundary in this case (Fig. 3.49). Nevertheless, as the wave front expands toward larger radii, the saturated nonlinear dynamics evolves to that of more stable behavior. For instance, as shown in Fig. 3.49, the amplitude of the saturated nonlinear oscillation decreases and the oscillations change from multimode to a period-doubled limit cycle as the nonlinear behavior approaches to that of a planar detonation and as the front curvature continues to decrease.

Finally, it is worth noting from numerical simulations that the critical energy of direct initiation of a cylindrically or spherically expanding detonation can be a multiple value problem for an unstable mixture. Mazaheri [98] and later Eckett et al. [39] showed that a detonation was successfully initiated for initiation energy above the minimum critical energy. However, for a range of initiation energy larger than the minimum critical energy, detonation can once again fail to initiate. As the initiation energy increased beyond that range, detonation was again initiated successfully (e.g., Fig. 3.50). According to the study of Watt and Sharpe [187], this may suggest the existence of a limiting failure shock radius associated with instability. For instance, at and near the minimum critical energy, due to the transient large-amplitude initiation pulse, the detonation is established after this limiting failure radius and therefore continues to propagate. Then, for a range of initiation energy near $E_s{}^*$ that corresponds to a minimum radius of detonation formation, detonation could form before this failure radius but the rapid growth rate of the instability quenches the detonation. Thus, failure not only takes place below the minimum critical initiation energy (or below the critical quasi-steady

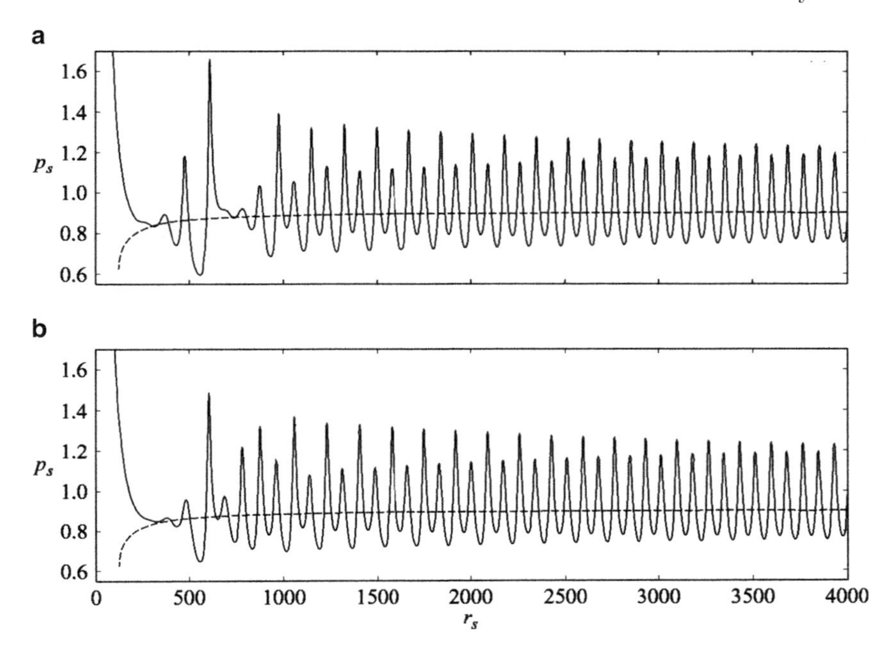

Fig. 3.49. Shock pressure versus shock radius for a cylindrically diverging detonation with $E_a = 27$ and (**a**) initiation energy $E_s = 5,000$, (**b**) $E_s = 6,000$. The *dashed curve* is the quasi-steady $p_s - r_s$ relation [187]

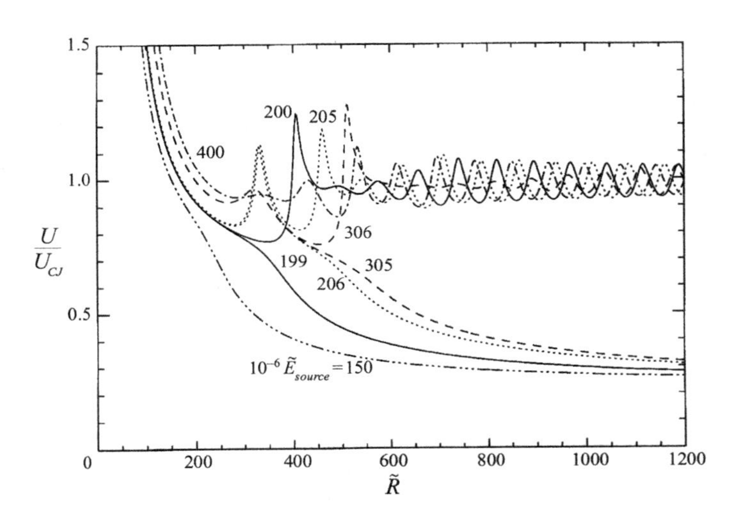

Fig. 3.50. Leading shock velocity versus shock radius for a spherical diverging detonation with several different source energies and $\gamma = 1.4$, $Q = 12$ and $E_a = 25$ [39]

radius), but also possibly for a range of initiation energy near $E_s{}^*$ depending on the instability of the mixture.

It is perhaps important to also point out that detonation failure due to the 1D instability described above may not happen in reality, since real detonation waves exhibit multidimensional instabilities such as transverse wave interactions and turbulent mixing [187]. The multidimensional instability effects are critical in determining the failure limits and propagation of diverging detonation waves, and may play a more prominent role than the one-dimensional instability effects just presented.

3.5.4 Transverse Wave Generation of Diverging Detonation

From two-dimensional simulations, it was shown that detonation instabilities manifest themselves by a characteristic cellular structure comprised of transverse waves sweeping across the detonation front. Many experimental and theoretical studies have shown that these transverse wave instabilities are essential for the self-sustained propagation of a detonation wave.

Soloukhin [163] obtained an open shutter photograph (Fig. 3.51) of a diverging cylindrical detonation initiated in a circular tube and radially expanded in an adjoining cylindrical disk. This photograph shows a detonation front structure with transverse wave spacing or cell size asymptoting toward a constant size while the front surface continuously increases. Hence, to maintain a self-propagating diverging detonation wave, the growth in the number of transverse waves needs to match the rate of increase in surface area as the diverging detonation wave continuously expands.

Jiang et al. [70] and Asahara et al. [6] carried out detailed numerical studies to investigate this problem, and concentrated on the origin and self-generation of transverse waves in diverging detonations. Their numerical solutions were obtained from the two-dimensional Euler equations with a source energy in the cylindrical center for direct initiation. In the work of Jiang et al. [70], a two-step chemical kinetic model was applied to a 30° annular region covered by 3,000 grid points radially and 1,000 grid points in the circumferential direction, equivalent to 15 points per reaction zone length (Fig. 3.52). They observed four types of transverse wave generation and self-adjustment, namely, through concave front focusing (Mode I), kinked front evolution (Mode II), wrinkled front evolution (Mode III), and weak wave engulfment (Mode IV).

In the mechanism of concave front focusing, the concave front on the weak Mach stem between two triple points is developed from the flow expansion induced by both transverse waves moving away and detonation front expansion. As the detonation expands, local-decoupling occurs between the chemical reaction and the shock at the weakest wave front location where the shock becomes slower than the adjacent part of the detonation front, as illustrated in Fig. 3.53. As the concave wave front develops, local focusing and therefore explosion occur in the unburned flow region between the reaction zone and the leading shock, thus resulting in a pair of new transverse waves at the

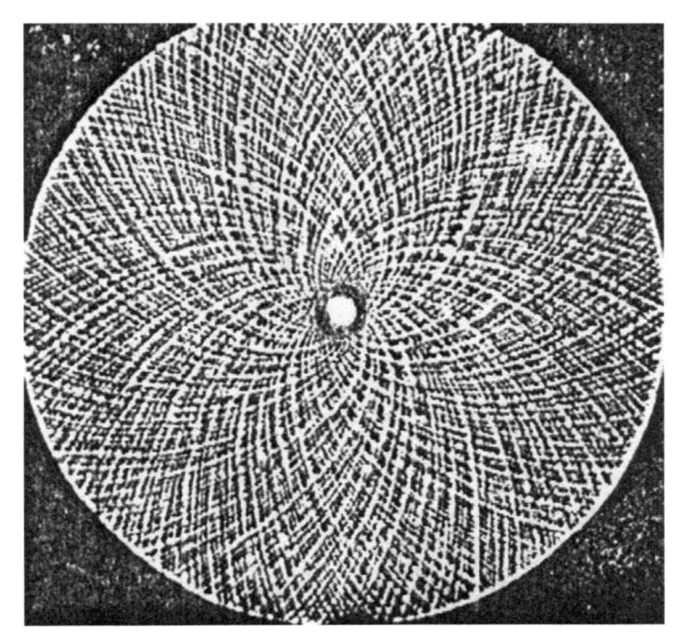

Fig. 3.51. Open-shutter photograph showing end-on view of a diverging cylindrical detonation in a C_2H_2–O_2 mixture at an initial pressure of 60 mmHg [163]

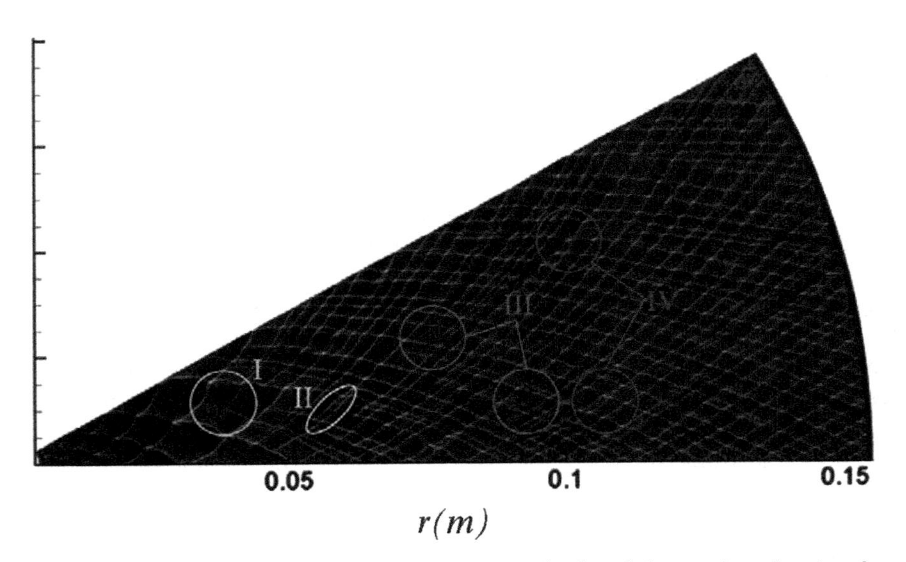

Fig. 3.52. Numerical cell pattern of diverging cylindrical detonation showing four types of self-organized transverse wave generation [70]

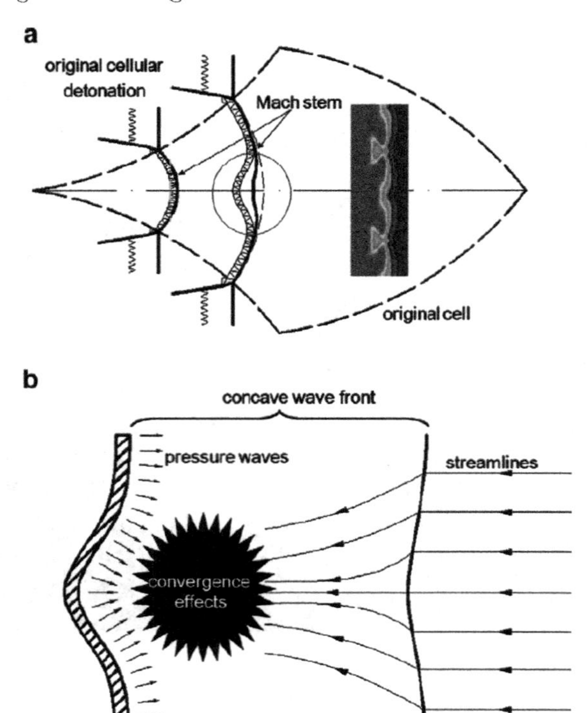

Fig. 3.53. Schematic of the mechanism for transverse wave generation resulting from concave wave front focusing. (**a**) concave wave front structure; (**b**) details of the circle in (**a**) [70]

front. The concave front focusing usually takes place at the smaller radii of curvature where a large rate of increase in front surface area occurs.

The kinked front evolution takes place near a transverse wave as the diverging detonation expands. Flow instabilities in the transverse direction develop between the decoupled reaction zone and the leading shock and cause the front to be deformed or kinked behind the traveling transverse wave. This results in the development of instability at the contact interface which mixes the burned and unburned mixture, and the subsequent combustion in turn amplifies the front kinking progressively to form a new triple point with a transverse wave propagating into the unburned mixture [70].

Another mode, the wrinkled front evolution, originates from the kinked transverse instabilities at multiple locations on the detonation front within a detonation cell. The evolution of the weak instabilities of the detonation front is predominantly due to the sensitivity of the chemistry to the smaller

Fig. 3.54. Numerical cell pattern from maximum pressure history of a cylindrically diverging detonation in a 21×21 mm region, with a central direct initiation energy just above the critical value [6]

perturbations and therefore happens preferentially at a larger shock radius of curvature.

The first three modes described above are mainly mechanisms to generate new transverse waves that lead to cell bifurcation. In contrast, it was also observed from the numerical simulation of Jiang et al. [70] that a self-adjustment process occurs in order to maintain a constant transverse wave spacing on the diverging detonation front. For instance, extra weak transverse waves could be eliminated by the interaction of two transverse waves. After the interaction, the weaker transverse wave propagates in the combustion products and subsequently attenuates without any further energy release support.

Similar to the study of Jiang et al. [70], Asahara et al. [6] computed the same problem using the Petersen and Hanson's reaction scheme [122] that contains 18 elementary reactions and eight species (H_2, O_2, H, O, OH, HO_2, H_2O_2 and H_2O) (see Fig. 3.54). The computation region shown in Fig. 3.54 is a 21-mm-long square region covered by a Cartesian mesh of $5{,}001 \times 5{,}001$ grid points with a resolution of ten points per half reaction zone length. The generation of transverse waves near a shock radius of 10 mm can be related to the reaction of the unburned hydrogen pockets near the wave front displayed in Fig. 3.55.

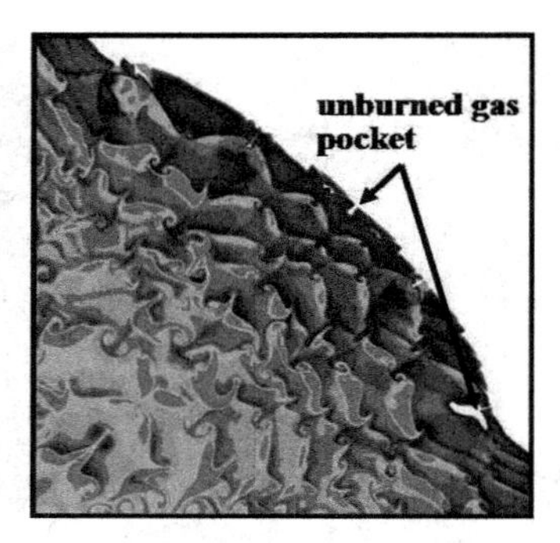

Fig. 3.55. Enlargement of the hydrogen mass fraction contour near the front between 25 and 65° at a shock radius of $r_s = 12.2\,\text{mm}$, from the same computation shown in Fig. 3.54 [6]

The two-dimensional numerical simulations suggest that the transverse wave instabilities play an important role in the propagation of diverging detonations. The birth rate of transverse waves, intrinsic to the chemistry of a reactive mixture, competes with the rate of increase in the front surface area in order to maintain a constant transverse wave spacing at the expanding detonation front. The reaction or explosion of unburned gas regions or pockets, found in simulations of Jiang et al. [70] and Asahara et al. [6], is a consequence of the flow expansion, particularly at smaller radii, and appears to be one of the main origins of new transverse waves. This phenomenon is similar to that observed in highly unstable planar detonation waves. The birth rate of transverse waves can be qualitatively linked (i.e., the level of the transverse wave instability) to the one-dimensional instability route with increasing shock radius. In the case of the diverging expanding detonation for a given reactive system, the instability route would be an inverse bifurcation diagram, in the opposite direction to that in Fig. 3.14 of Sect. 3.3.3, with the shock radius as the control parameter. Such as in the simple system shown in Fig. 3.49, the large oscillation period with high levels of instability decreases to that of period-two, and then to a single period limit-cycle as the shock radius increases. Obviously, the quantitative description of the birth rate of transverse wave must consider the multidimensional effects and their interaction with the chemistry, which remain a challenging area of research.

3.6 Concluding Remarks

Since all observed detonations in gaseous explosive mixtures exhibit a cellular structure characteristic of instability, a considerable amount of research in the last 50 years has been devoted to understanding and modeling detonation instability. This chapter presented a general outline of analytic approaches to detonation stability theory and methodologies that can be applied to numerical simulations of the full nonlinear dynamics of unstable detonations. As in most classical studies of unstable fluid systems, the linear hydrodynamic stability analysis of the steady ZND solution is the first step in investigating the instability of detonation waves. While the theory of linear stability for detonations is well established, there remain some gaps in the application of this methodology, such as the inclusion of detailed chemical kinetics and the consideration of other equations of state. It should also be kept in mind that linear stability analysis only deals with small perturbations in the linear spectrum, and the results are still far from describing the true nature of detonation dynamics. The nonlinear dynamics of detonations can perhaps be better investigated via the application of asymptotic analysis and thus the development of nonlinear evolution equations. Although one- and two-dimensional asymptotic analyses lead to some important information regarding the nature of detonation instability, mathematical complications generally restrict study to very idealized detonation models with simple chemistry and at the limit of very large activation energy and weak heat release. Hence, investigations that rely more on direct numerical simulations for a complete description of the detonation instability problem have become common. Nevertheless, linear stability and asymptotic analysis will continue to play a role in validating the numerical algorithms before their application to the full nonlinear dynamics.

There is no doubt that the central role of nonlinear dynamics will remain the key to understanding the origin and mechanism of multidimensional detonation instability. Throughout this chapter, analyses demonstrate that detonations possess some universal nonlinear dynamics features of simple dynamical systems such as those observed in a nonlinear oscillator. From a nonlinear dynamics consideration, the identification of symmetry breaking and bifurcation sequence to chaos may provide a possible mechanism for the development of pulsating and cellular instabilities. It is worth noting that a considerable number of analytical techniques and nonlinear concepts have been developed in the field of nonlinear science. Many of these methodologies have been extended to deal with more complicated systems such as those found in reactive flows. These analytical tools in nonlinear dynamics theory should be further explored in analyzing detonation models and their analogues. The nonlinear theories shall help to further reveal the intrinsic nature embedded in the reactive Euler or Navier–Stokes fluid dynamics equations, where the coupled chemistry generally dominates the instability transitions, and other parameters controlling instability and instability routes that can explain the existence of different cellular detonation structures. New nonlinear dynamics concepts may ultimately lead toward a rigorous theory of multidimensional detonation instability hierarchy.

Many higher order numerical schemes are now available that are capable of accurately integrating the inviscid Euler equations in order to capture the full nonlinearity of unstable detonation dynamics. Presently, such simulations with high resolution, large domain size, and the examination of long-duration dynamics are still computationally very expensive. Furthermore, it is becoming recognized that turbulence plays an important role in the complex interactions that comprise the cellular structure of highly unstable detonations (e.g., shock–shock, shock–vortex, and shock–interfacial turbulence interactions), and the effect of these interactions on chemical reactions cannot yet be adequately resolved with current computational capacity. While continuous effort is being expended to resolve these issues with increasingly powerful computational facilities, it is also necessary to focus on the development of nonlinear dynamics models and statistical methods to analyze the tremendous amount of flow field detail resulting from these simulations, such as the recent study of Sharpe and Radulescu [147], rather than relying solely on computational power. Once developed, these diagnostic tools can then be applied to the various chemico-physical mechanisms present when it becomes feasible to routinely simulate detonation propagation with full chemistry and with high resolution in three dimensions.

Another future challenge in detonation theory and simulation lies in the development of sub-grid models to handle small-scale fluctuations, as is the case of turbulence. Such difficulties lie in the accurate calculation of the Navier–Stokes equations to fully describe the compressible turbulent flow structures that occur within the detonation front. Having a better understanding of the role of these small-scale instabilities and turbulence in detonation would then permit a better modeling of propagation dynamics and boundary interactions, which could in turn lead to further development of models for the appropriate source terms in the Euler equations accounting for non-ideal effects in the macroscopic propagation of detonation waves. These source terms will offer another set of control parameters for future analysis to better understand the nonlinear dynamics of detonation instability.

Acknowledgement

The authors would like to extend full gratitude to many colleagues, particularly Gary Sharpe, Matei Radulescu, Mark Short, Charles Kiyanda, Andrew Higgins, Ashwani K. Kapila, Nikos Nikiforakis, and John Lee, for their valuable advice, assistance, and numerous fruitful discussions on the field of detonation physics and scientific computing.

This chapter contains figures reprinted from Radulescu et al. (2007), Short and Stewart (1998), Eckett et al. (2000), Austin (2003), Henrick et al. (2006), Sharpe and Radulescu (2011), Voitsekhovskii et al. (1958), Davidenko et al. (2011), Sharpe and Falle (2000), Watt and Sharpe (2004, 2005), Soloukhin (1966), Jiang et al. (2009), Asahara et al. (2010). The authors gratefully acknowledge Matei Radulescu, Mark Short, Joanna Austin, Gary Sharpe, Ann Karagozian, Joseph Powers, Tariq Aslam, Rmy Mvel and Jean-Philippe Dionne for sharing their figures and data in this chapter.

References

1. Abarbanel, H.D.: The analysis of observed chaotic data in physical systems. Rev. Mod. Phys. **65**, 1331–1392 (1993)
2. Abouseif, G.E., Toong, T.Y.: Theory of unstable detonations. Combust. Flame **45**, 67–94 (1982)
3. Ait Abderrahmane, H., Paquet, F., Ng, H.D.: Applying nonlinear dynamic theory to one-dimensional pulsating detonations. Combust. Theor. Model. **15**, 205 (2011)
4. Albano, A.M., Muench, J., Schwartz, C., Mees, A.I., Rap, P.E.: Singular-value decomposition and the Grassberger-Procaccia algorithm. Phys. Rev. A. **38**, 3017–3026 (1988)
5. Alpert, R.L., Toong, T.Y.: Periodicity in exothermic hypersonic flows about blunt projectiles. Astro. Acta. **17**, 539–560 (1972)
6. Asahara, M., Tsuboi, N., Hayashi, A.K.: Two-dimensional simulation on propagation mechanism of H_2/O_2 cylindrical detonation with a detailed reaction model. Combust. Sci. Technol. **182**(11–12), 1884–1900 (2010)
7. Austin, J.M.: The role of instabilities in gaseous detonation. PhD thesis, California Institute of Technology, CA (2003)
8. Barthel, H.O.: Predicted spacings in hydrogen-oxygen-argon detonations. Phys. Fluids **17**(8), 1547–1553 (1974)
9. Bates, K.R., Nikiforakis, N., Holder, D.: Richtmyer-Meshkov instability induced by the interaction of a shock wave with a rectangular block of SF_6. Phys. Fluids **19**, 036101 (2007)
10. Ben-Dor, G.: Shock Reflection Phenomena. Springer, New York (1992)
11. Bergé, P., Pomeau, Y., Vidal, C.: Order Within Chaos. Wiley, New York (1986)
12. Bourlioux, A.: Numerical study of unstable detonation. PhD Thesis, Princeton University, New Jersey (1991)
13. Bourlioux, A., Majda, A.J.: Theoretical and numerical structure for unstable two-dimensional detonations. Combust. Flame **90**, 211–229 (1992)
14. Bourlioux, A., Majda, A.J., Roytburd, V.: Nonlinear development of low frequency one-dimensional instabilities for reacting shocks. In: Fife, P., Linan, A., Williams, F. (eds.) Dynamical Issues in Combustion Theory. IMA Volumes in Mathematics and Its Applications, vol. 35, pp. 63–83. Springer, New York (1991)
15. Bourlioux, A., Majda, A.J., Roytburd, V.: Theoretical and numerical structure for unstable one-dimensional detonations. SIAM J. Appl. Math. **51**, 303–343 (1991)
16. Broomhead, D.S., King, G.P.: Extracting qualitative dynamics from experimental data. Physica D **20**, 217–236 (1986)
17. Browne, S., Shepherd, J.E.: Linear stability of detonations with reversible chemical reactions. In: Proceedings of the Western States Section/Combustion Institute, Sandia National Laboratories, Livermore (2007)
18. Buckmaster, J.D., Ludford, G.S.S.: The effect of structure on the stability of detonations. I – Role of the induction zone. Proc. Combust. Inst. **21**, 1669–1676 (1987)
19. Buckmaster, J.D., Neves, J.: One-dimensional detonation stability: The spectrum for infinite activation energy. Phys. Fluids **31**(12), 3571–3576 (1988)

20. Cael, G., Ng, H.D., Bates, K.R., Nikiforakis, N., Short, M.: Numerical simulation of detonation structures using a thermodynamically consistent and fully conservative (TCFC) reactive flow model for multi-component computations. Proc. R. Soc. Lond. A. **465**, 2135–2153 (2009)
21. Choi, J.Y., Ma, F.H., Yang, V.: Some numerical issues on simulation of detonation cell structures. Combust. Expl. Shock Waves **44**(5), 560–578 (2008)
22. Clavin, P., Daou, R.: Instability threshold of gaseous detonations. J. Fluid Mech. **482**, 181–206 (2003)
23. Clavin, P., Denet, B.: Diamond patterns in the cellular front of an overdriven detonation. Phys. Rev. Lett. **88**, 044502 (2002)
24. Clavin, P., He, L.: Stability and nonlinear dynamics of one-dimensional overdriven detonations in gases. J. Fluid Mech. **306**, 353–378 (1996)
25. Clavin, P., William, F.A.: Dynamics of planar gaseous detonations near Chapman-Jouguet conditions for small heat release. Combust. Theor. Model. **6**, 127–139 (2002)
26. Clifford, L.J., Milne, A.M., Turanyi, T., Boulton, D.: An induction parameter model for shock-induced hydrogen combustion simulations. Combust. Flame **113**, 106–118 (1998)
27. Courant, R., Friedrichs, K.O.: Supersonic Flow and Shock Waves. Springer, New York (1948)
28. Daimon, Y., Matsuo, A.: Detailed features of one-dimensional detonations. Phys. Fluids **15**(1), 112–122 (2003)
29. Daimon, Y., Matsuo, A.: Unsteady features on one-dimensional hydrogen-air detonations. Phys. Fluids **19**, 116101 (2007)
30. Davidenko, D., Mével, R., Dupré, G.: Numerical study of the detonation structure in rich H_2-NO_2/N_2O_4 and very lean H_2-N_2O mixtures. Shock Waves **21**(2), 85–99 (2011)
31. Deiterding, R., Bader, G.: High-resolution simulation of detonations with detailed chemistry. In: Warnecke, G. (ed.) Analysis and Numerics for Conservation Laws, pp. 69–91. Springer, Berlin (2005)
32. Deledicque, V., Papalexandris, M.V.: Computational study of three-dimensional gaseous detonation structures. Combust. Flame **144**, 821–837 (2006)
33. Desbordes, D., Presles, H.N., Joubert, F., Douala, C.G.: Etude de la détonation de mélanges pauvres H_2–NO_2/N_2O_4. CR Mécanique **332**, 993–999 (2004)
34. Dionne, J.-P.: Numerical study of the propagation of non-ideal detonations. PhD Thesis, McGill University, Montréal (2000)
35. Dionne, J.-P., Ng, H.D., Lee, J.H.S.: Transient development of friction induced low velocity detonations. Proc. Combust. Inst. **28**, 645–651 (2000)
36. Döring, W.: On detonation processes in gases. Ann. Phys. **43**, 421–436 (1943)
37. Dou, H.-S., Tsai, H.M., Khoo, B.C., Qiu, J.: Simulations of detonation wave propagation in rectangular ducts using a three-dimensional WENO scheme. Combust. Flame **154**, 644–659 (2008)
38. Eckett, C.A.: Numerical and analytical studies of the dynamics of gaseous detonations. PhD thesis, California Institute of Technology, Pasadena (2000)
39. Eckett, C.A., Quirk, J.J., Shepherd, J.E.: The role of unsteadiness in direct initiation of gaseous detonations J. Fluid Mech. **421**, 147–183 (2000)
40. Eckmann, J.P., Kamphorst, S.O., Ruelle, D., Ciliberto, S.: Lyapunov exponents from time series. Phys. Rev. A. **34**(6), 4971–4979 (1986)

41. Edwards, D.H., Jones, A.T., Phillips, D.E.: The location of the Chapman-Jouguet surface in multi-headed detonations. J. Phys. D Appl. Phys. **9**, 1331–1342 (1976)
42. Erpenbeck, J.J.: Stability of steady-state equilibrium detonations. Phys. Fluids **5**, 604–614 (1962)
43. Erpenbeck, J.J.: Stability of idealized one-reaction detonations. Phys. Fluids **7**, 684–696 (1964)
44. Erpenbeck, J.J.: Nonlinear theory of unstable one-dimensional detonations. Phys. Fluids **10**, 274–288 (1967)
45. Erpenbeck, J.J.: Nonlinear theory of unstable two-dimensional detonation. Phys. Fluids **13**, 2007–2026 (1970)
46. Eto, K., Tsuboi, N., Hayashi, A.K.: Numerical study on three-dimensional CJ detonation waves: Detailed propagating mechanism and existence of OH radical. Proc. Combust. Inst. **30**, 1907–1913 (2005)
47. Feigenbaum, M.J.: Quantitative universality for a class of nonlinear transformations. J. Stat. Phys. **19**(1), 25–52 (1978)
48. Feigenbaum, M.J.: Low dimensional dynamics and the period doubling scenario. Lect. Notes Phys. **179**, 131–148 (1983)
49. Feigenbaum, M.J.: Universal behavior in nonlinear systems. Phys. D **7**, 16–39 (1983)
50. Fickett, W., Davis, W.C.: Detonation. University of California Press, Berkeley (1979)
51. Fickett, W., Wood, W.W.: Flow calculations for pulsating one-dimensional detonations. Phys. Fluids **9**, 903–916 (1966)
52. Fujiwara, T., Reddy, K.V.: Propagation mechanism of detonation-three-dimensional phenomena. Mem. Facul. Eng. Nagoya Univ. **41**, 1–18 (1989)
53. Gamezo, V.N., Desbordes, D., Oran, E.S.: Formation and evolution of two-dimensional cellular detonations. Combust. Flame **116**, 154–165 (1999)
54. Gamezo, V.N., Desbordes, D., Oran, E.S.: Two-dimensional reactive dynamics in cellular detonation waves. Shock Waves **9**, 11–17 (1999)
55. Gavrikov, A.I., Efimenko, A.A., Dorofeev, S.B.: A model for detonation cell size prediction from chemical kinetics. Combust. Flame **120**, 19–33 (2000)
56. Gorchkov, V., Kiyanda, C.B., Short, M., Quirk, J.J.: A detonation stability formulation for arbitrary equations of state and multi-step reaction mechanisms. Proc. Combust. Inst. **31**(2), 2397–2405 (2006)
57. Grassberger, P., Procaccia, I.: Measuring the strangeness of strange attractors. Phys. D **9**, 189–208 (1983)
58. Grassberger, P., Procaccia, I.: Dimensions and entropies of strange attractors from a fluctuating dynamics approach. Phys. D **13**, 34–54 (1984)
59. Guilly, V., Khasainov, B., Presles, H.N., Desbordes, D.: Simulation numérique des détonation à double structure cellulaire. CR Mecanique **334**, 679–685 (2006)
60. Guirguis, R., Oran, E.S., Kailasanath, K.: The effect of energy release on the regularity of detonation cells in liquid nitromethane. Proc. Combust. Inst. **21**, 1639–1668 (1986)
61. Hao, B.L.: Chaos, An Introduction and Reprints Volume. World Scientific, Singapore (1984)
62. He, L.: Theory of weakly unstable multi-dimensional detonation. Combust. Sci. Technol. **160**, 65–101 (2000)

63. He, L., Clavin, P.: On the direct initiation of gaseous detonations by an energy source. J. Fluid Mech. **277**, 227–248 (1994)
64. He, L., Lee, J.H.S. The dynamical limit of one-dimensional detonations. Phys. Fluids **7**(5), 1151–1158 (1995)
65. Helzel, C.: Numerical approximation of conservation laws with stiff source term for the modelling of detonation waves. Thesis, Otto-von-Guericke Universität Magdeburg (2000)
66. Henrick, A.K., Aslam, T.D., Powers, J.M.: Simulations of pulsating one-dimensional detonations with true fifth order accuracy. J. Comp. Phys. **213**, 311–329 (2006)
67. Hu, X.Y., Khoo, B.C., Zhang, D.L., Jiang, Z.L.: The cellular structure of a two-dimensional $H_2/O_2/Ar$ detonation wave. Combust. Theor. Model. **8**, 339–359 (2004)
68. Hwang, P., Fedkiw, R.P., Merriman, B., Aslam, T.D., Karagozian, A.R., Osher, S.J.: Numerical resolution of pulsating detonation waves. Combust. Theor. Model. **4**, 217–240 (2000)
69. Inaba, K., Matsuo, A., Shepherd, J.E.: Soot track formation by shock waves and detonations. In: Proceedings of the 20th International Colloquium of Dynamics of Explosion and Reactive Systems, Montreal (2005)
70. Jiang, Z.L., Han, G., Wang, C., Zhang, F.: Self-organized generation of transverse waves in diverging cylindrical detonations. Combust. Flame **156**, 1653–1661 (2009)
71. Joubert, F., Desbordes, D., Presles, H.N.: Structure de la détonation des mélanges H_2–NO_2/N_2O_4. CR Mécanique **331**, 365–372 (2003)
72. Kailasanath, K., Oran, E.S., Boris, J.P., Young, T.R.: Determination of detonation cell size and the role of transverse waves in two-dimensional detonations. Combust. Flame **61**, 199–209 (1985)
73. Kantz, H.: A robust method to estimate the maximal Lyapunov exponent of a time series. Phys. Lett. A **185**, 77–87 (1994)
74. Kaplan, D., Glass, L.: Understanding Nonlinear Dynamics. Springer, New York (1995)
75. Karnesky, J., Shepherd, J.E.: Detonation in nitrated hydrocarbons. In: 32nd International Symposium on Combustion, Work in progress poster W4P023, Montréal, Canada, Aug. 3–8 (2008)
76. Kasimov, A.R., Stewart, D.S.: Spinning instability of gaseous detonations. J. Fluid Mech. **466**, 179–203 (2002)
77. Kasimov, A.R., Stewart, D.S.: On the dynamics of self-sustained one-dimensional detonations: A numerical study in the shock-attached frame. Phys. Fluids **16**, 3566–3578 (2004)
78. Kasimov, A.R., Stewart, D.S.: Theory of detonation initiation and comparison with experiment. Report #1035, Theoretical and Applied Mechanics, UIUC (2004)
79. Kasimov, A.R., Stewart, D.S.: Asymptotic theory of evolution and failure of self-sustained detonations. J. Fluid Mech. **525**, 161–192 (2005)
80. Khokhlov, A.M., Austin, J.M., Pintgen, F., Shepherd, J.E.: Numerical study of the detonation wave structure in ethylene-oxygen mixtures. In: 42th AIAA Aerospace Science Meeting and Exhibit, AIAA 2004-0792, Reno (2004)
81. Lee, H.I., Stewart, D.S.: Calculation of linear detonation instability: One dimensional instability of planar detonations. J. Fluid Mech. **216**, 103–132 (1990)

82. Lee, J.H.S.: The Detonation Phenomenon. Cambridge University Press, Cambridge (2008)
83. Lee, J.H.S., Guirao, C.M.: Gasdynamic effects of fast exothermic reactions. In: Capellos, C., Walker, R.F. (eds.) Fast Reactions in Energetic Systems, pp. 245–313. D. Reidel Publishing Company, Dordrecht (1981)
84. Lee, J.H.S., Radulescu, M.I.: On the hydrodynamic thickness of cellular detonations. Combust. Expl. Shock Waves **41**, 745–765 (2005)
85. Lee, J.H.S., Soloukhin, R.I., Oppenheim, A.K.: Current views on gaseous detonation. Astro. Acta **14**, 565–584 (1969)
86. Leung, C., Radulescu, M.I., Sharpe, G.J.: Characteristics analysis of the one-dimensional dynamics of chain-branching detonations. Phys. Fluids **22**, 126101 (2010)
87. Li, T.Y., Yorke, J.A.: Period three implies chaos. Am. Math. Mon. **82**, 985–992 (1975)
88. Liang, Z., Bauwens, L.: Detonation structure with pressure-dependent chain-branching kinetics. Proc. Combust. Inst. **30**(2), 1879–1887 (2005)
89. Liang, Z., Khastoo, B., Bauwens, L.: Effect of reaction order on stability of planar detonation waves. Int. J. Comput. Fluid Dynam. **19**(2), 131–142 (2005)
90. Lorenz, E.N.: Deterministic nonperiodic flow. J. Atmos. Sci. **20**, 130–141 (1963)
91. Luche, J., Desbordes, D., Presles, H.N.: Détonation de mélanges H_2–NO_2/N_2O_4–Ar. CR Mecanique **334**, 323–327 (2006)
92. Lutz, A.E., Kee, R.J., Miller, J.A., Dwyer, H.A., Oppenheim, A.K.: Dynamic effects of autoignition centers for hydrogen and C1,2-hydrocarbon fuels. Proc. Combust. Inst. **22**, 1683–1693 (1988)
93. Mach, P., Radulescu, M.I.: Mach reflection bifurcations as a mechanism of cell multiplication in gaseous detonations. Proc. Combust. Inst. **33**(2), 2279–2285 (2011)
94. Majda, A.: Criteria for regular spacing of reacting Mach stems. Proc. Natl. Acad. Sci. U.S.A. **84**(17), 6011–6014 (1987)
95. Majda, A., Roytburd, V.: Low-frequency multidimensional instabilities for reacting shock waves. Stud. Appl. Math. **87**, 135–174 (1992)
96. Mañé, R: On the dimension of the compact invariant sets of certain nonlinear maps. Lect. Notes Math. **898**, 230–242 (1981)
97. Massa, L., Austin, J.M., Jackson, T.L.: Triple-point shear layers in gaseous detonation waves. J. Fluid Mech. **586**, 205–248 (2007)
98. Mazaheri, B.K.: Mechanism of the onset of detonation in blast initiation. PhD thesis, McGill University, Montreal (1997)
99. McVey, J.B., Toong, T.Y.: Mechanism of instabilities of exothermic hypersonic blunt body flows. Combust. Sci. Tech. **3**, 63–76 (1971)
100. Mével, R., Lafosse, F., Catoire, L., Chaumeix, N., Dupré, G., Paillard, C.E.: Induction delay times and detonation cell size prediction of hydrogen–nitrous oxide–argon mixtures. Combust. Sci. Technol. **180**, 1858–1875 (2008)
101. Mével, R., Davidenko, D., Lafosse, F., Dupré, G., Paillard, C.E.: Prediction of detonation cell size in hydrogen-nitrous oxide-argon mixtures using chemical kinetics correlations and 2-D numerical simulation code. In: Proceedings of the 7th International Symposium on Hazards, Prevention, and Mitigation of Industrial Explosions (ISHPMIE), vol. 2, pp. 41–53, St. Petersburg, Russia (2008)

102. Mével, R., Javoy, S., Lafosse, F., Chaumeix, N., Dupré, G., Paillard, C.E.: Hydrogen-nitrous oxide delay time: Shock tube experimental study and kinetic modeling. Proc. Combust. Inst. **32**, 359–366 (2009)
103. Meyer, J.W., Oppenheim, A.K.: Coherence theory of the strong ignition limit. Combust. Flame **17**, 65–68 (1971)
104. Meyer, J.W., Oppenheim, A.K.: On the shock-induced ignition of explosive gases. Proc. Combust. Inst. **13**, 1153–1164 (1971)
105. Moen, I.O., Funk, J.W., Ward, S.A., Rude, G.M., Thibault, P.A.: Detonation length scales for fuel-air explosives. Prog. Astronaut. Aeronaut. **94**, 55–79 (1984)
106. Moen, I.O., Sulmistras, A., Thomas, G.O., Bjerketvedt, D., Thibault, P.A.: Influence of cellular regularity on the behavior of gaseous detonations. Prog. Astronaut. Aeronaut. **106**, 220–243 (1986)
107. Ng, H.D., Lee, J.H.S.: Direct initiation of detonation with a multi-step reaction scheme. J. Fluid Mech. **476**, 179–211 (2003)
108. Ng, H.D., Lee, J.H.S.: Comments on explosion problems for hydrogen safety. J. Loss Preven. Proc. Ind. **21**(2), 136–146 (2008)
109. Ng, H.D., Higgins, A.J., Kiyanda, C.B., Radulescu, M.I., Lee, J.H.S., Bates, K.R., Nikiforakis, N.: Nonlinear dynamics and chaos analysis of one-dimensional pulsating detonations. Combust. Theor. Model. **9**, 159–170 (2005)
110. Ng, H.D., Radulescu, M.I., Higgins, A.J., Nikiforakis, N., Lee, J.H.S.: Numerical Investigation of the instability for one-dimensional Chapman-Jouguet detonations with chain-branching kinetics. Combust. Theor. Model. **9**, 385–401 (2005)
111. Ng, H.D., Ju, Y., Lee, J.H.S.: Assessment of detonation hazards in high-pressure hydrogen storage from chemical sensitivity analysis. Int. J. Hydrogen Energ. **32**(1), 93–99 (2007)
112. Ng, H.D., Chao, J., Yatsufusa, T., Lee, J.H.S.: Measurement and chemical kinetic prediction of detonation sensitivity and cellular structure characteristics in dimethyl ether-oxygen mixtures. Fuel **88**(1), 124–131 (2008)
113. Ng, H.D., Ait Abderrahmane, H., Bates, K.R., Nikiforakis, N.: Geometrical characterization of cellular irregularity of gaseous detonation front using a fractal approach. In: 23rd International Colloquium on the Dynamics of Explosions and Reactive Systems, Irvine (2011)
114. Ng, H.D., Ait Abderrahmane, H., Bates, K.R., Nikiforakis, N.: The growth of fractal dimension of a scalar interface evolution from the interaction of a shock wave with a rectangular block of SF6. Commun. Nonlinear Sci. Numer. Simul. **16**(11), 4158–4162 (2011)
115. Oppenheim, A.K.: Dynamic features of combustion. Phil. Trans. R. Soc. A **315**, 471–508 (1985)
116. Oran, E.S., Boris, J.P.: Numerical Simulation of Reactive Flow. Elsevier, New York (1987)
117. Oran, E.S., Young, T.R., Boris, J.P., Picone, J.M., Edwards, D.H.: A study of detonation structure: The formation of unreacted gas pockets. Proc. Combust. Inst. **19**, 573–582 (1982)
118. Oran, E.S., Kailasanath, K., Guirguis, R.H.: Numerical simulations of the development and structure of detonations. Prog. Astron. Aero. **114**, 155–169 (1988)

119. Oran, E.S., Weber, J.W., Stefaniw, E.I., Lefebvre, M.H., Anderson, J.D.: A numerical study of a two-dimensional H_2-O_2-Ar detonation using a detailed chemical reaction model. Combust. Flame **113**, 147–163 (1998)
120. Packard, N.H., Crutchfield, J.P., Farmer, J.D., Shaw, R.S.: Geometry from a time series. Phys. Rev. Lett. **45**, 712–716 (1980)
121. Papalexandris, M.V., Leonard, A., Dimotakis, P.E.: Unsplit schemes for hyperbolic conservation laws with source terms in one space dimension. J. Comput. Phys. **134**, 31–61 (1997)
122. Petersen, E.L., Hanson, R.K.: Reduced kinetics mechanisms for ram accelerator combustion. J. Prop. Power. **15**, 591 (1999)
123. Pintgen, F., Shepherd, J.E.: Quantitative analysis of reaction front geometry in detonation. In: Roy, G.D., Berlin, A.A., Frolov, S.M., Shepherd, J.E., Tsyganov, S.A. (eds.) International Colloquium on Application of Detonation for Propulsion, pp. 23–28. Torus Press, Moscow (2004)
124. Pintgen, F., Eckett, C.A., Austin, J.M., Shepherd, J.E.: Direct observations of reaction zone structure in propagating detonations. Combust. Flame **133**, 211–229 (2003)
125. Powers, J.M.: Review of multiscale modeling of detonation. J. Propul. Power **22**, 1217–1229 (2006)
126. Powers, J.M., Aslam, T.D.: Exact solution for multidimensional compressible reactive flow for verifying numerical algorithms. AIAA J. **44**(2), 337–344 (2006)
127. Powers, J.M., Paolucci, S.: Accurate spatial resolution estimates for reactive supersonic flow with detailed chemistry. AIAA J. **43**(5), 1088–1099 (2005)
128. Presles, H.N., Desbordes, D., Guirard, M., Guerraud, G.: Gaseous nitromethane and nitromethane–oxygen mixtures: A new detonation structure. Shock Waves **6**, 111–114 (1996)
129. Quirk, J.J.: Godunov-type schemes applied to detonation flows. In: Buckmaster, J., et al. (eds.) Combustion in High-Speed Flows, pp. 575–596. Kluwer, Dordrecht (1994)
130. Radulescu, M.I.: The propagation and failure mechanism of gaseous detonations: Experiments in porous-walled tubes. PhD thesis, McGill University, Montreal (2003)
131. Radulescu, M.I., Lee, J.H.S.: The failure mechanism of gaseous detonations – Experiments in porous wall tubes. Combust. Flame **131**, 29–46 (2002)
132. Radulescu, M.I., Ng, H.D., Lee, J.H.S., Varatharajan, B.: The effect of argon dilution on the stability of acetylene-oxygen detonations. Proc. Combust. Inst. **29**, 2825–2831 (2002)
133. Radulescu, M.I., Sharpe, G.J., Lee, J.H.S., Kiyanda, C., Higgins, A.J., Hanson, R.K.: The ignition mechanism in irregular structure gaseous detonations. Proc. Combust. Inst. **30**, 1859–1867 (2005)
134. Radulescu, M.I., Sharpe, G.J., Law, C.K., Lee, J.H.S.: The hydrodynamic structure of unstable cellular detonations J. Fluid Mech. **580**, 31–81 (2007)
135. Rosenstein, M.T., Collins, J.J., De Luca, C.J.: A practical method for calculating largest Lyapunov exponents from small data sets. Phys. D **65**, 117–134 (1993)
136. Sánchez, A.L., Carretero, M., Clavin, P., Williams, F.A.: One-dimensional overdriven detonations with branched-chain kinetics. Phys. Fluids **13**, 776–792 (2001)

137. Schultz, E., Shepherd, J.E.: Validation of detailed reaction mechanisms for detonation simulation. GALCIT Technical Report, FM99-05, California Institute of Technology, USA (2000)
138. Sello, S.: Time series forecasting: A nonlinear dynamics approach. LANL 2726 preprint archive Physics, arXiv:physics/9906035v2 [physics.data-an] (1999)
139. Sharpe, G.J.: Linear stability of idealized detonations. Proc. R. Soc. Lond. A **453**, 2603–2625 (1997)
140. Sharpe, G.J.: Linear stability of pathological detonations. J. Fluid Mech. **401**, 311–338 (1999)
141. Sharpe, G.J.: Numerical simulations of pulsating detonations: I. nonlinear stability of steady detonations. Combust. Theor. Model. **4**, 557–574 (2000)
142. Sharpe, G.J.: Transverse waves in numerical simulations of cellular detonations. J. Fluid Mech. **447**, 31–51 (2001)
143. Sharpe, G.J., Falle, S.A.E.G.: One-dimensional numerical simulations of idealized detonations. Proc. R. Soc. Lond. A **455**, 1203–1214 (1999)
144. Sharpe, G.J., Falle, S.A.E.G.: One-dimensional nonlinear stability of pathological detonations. J. Fluid Mech. **414**, 339–366 (2000)
145. Sharpe, G.J., Falle, S.A.E.G.: Two-dimensional numerical simulations of idealized detonations. Proc. R. Soc. Lond. A **456**, 2081–2100 (2000)
146. Sharpe, G.J., Quirk, J.J.: Nonlinear cellular dynamics of the idealized detonation model: Regular cells. Combust. Theor. Model. **12**(1), 1–21 (2008)
147. Sharpe, G.J., Radulescu, M.I.: Statistical analysis of cellular detonation dynamics from numerical simulations: One-step chemistry. Combust. Theor. Model. **15**(6), 691–723 (2011)
148. Shchelkin, K.I., Troshin, Y.K.: Gasdynamics of Combustion. Mono Book Co., Baltimore (1965)
149. Shepherd, J.E.: Chemical kinetics of hydrogen-air-diluent mixtures. Prog. Astro. Aeronaut. **106**, 263–293 (1986)
150. Short, M.: An asymptotic derivation of the linear stability of the square-wave detonation using the newtonian limit. Proc. R. Soc. Lond. A **452**, 2203–2224 (1996)
151. Short, M.: A nonlinear evolution equation for pulsating Chapman–Jouguet detonations with chain-branching kinetics. J. Fluid Mech. **430**, 381–400 (2001)
152. Short, M.: Theory and modeling of detonation wave stability: A brief look at the past and toward the future. In: Proceedings of the 20th International Colloquium on the Dynamics of Explosions and Reactive Systems, Montreal (2005)
153. Short, M., Blythe, P.A.: Structure and stability of weak-heat-release detonations for finite Mach numbers. Proc. R. Soc. Lond. A **458**, 1795–1807 (2002)
154. Short, M., Dold, J.W.: Linear stability of a detonation wave with a model three-step chain-branching reaction. Math. Comput. Model. **24**, 115–123 (1996)
155. Short, M., Quirk, J.J.: On the nonlinear stability and detonability limit of a detonation wave for a model three-step chain-branching reaction. J. Fluid Mech. **339**, 89–119 (1997)
156. Short, M., Sharpe, G.J.: Pulsating instability of detonations with a two-step chain-branching reaction model: Theory and numerics. Combust. Theor. Model. **7**, 401–416 (2003)
157. Short, M., Stewart, D.S.: Cellular detonation stability: A normal mode linear analysis. J. Fluid Mech. **368**, 229–262 (1998)
158. Short, M., Stewart, D.S.: The multi-dimensional stability of weak-heat-release detonations. J. Fluid Mech. **382**, 109–135 (1999)

159. Short, M., Kapila, A.K., Quirk, J.J.: The chemical-gas dynamic mechanisms of pulsating detonation wave instability. Phil. Trans. R. Soc. Lond. A **357**, 3621–3637 (1999)
160. Short, M., Bdzil, J.B., Anguelova, I.I.: Stability of Chapman–Jouguet detonations for a stiffened-gas model of condensed-phase explosives. J. Fluid Mech. **552**, 299–309 (2006)
161. Short, M., Anguelova, I.I., Aslam, T.D., Bdzil, J.B., Henrick, A.K., Sharpe, G.J.: Stability of detonations for an idealized condensed-phase model. J. Fluid Mech. **595**, 45–82 (2008)
162. Sichel, M., Tonello, N.A., Oran, E.S., Jones, D.A.: A two-step kinetics model for numerical simulation of explosions and detonations in H_2–O_2 mixtures. Proc. R. Soc. Lond. A **458**, 49–82 (2002)
163. Soloukhin, R.I.: Shock Waves and Detonations in Gases, State Publishing House, Moscow; English Translation, Mono Book Corporation, Baltimore (1966)
164. Stewart, D.S., Kasimov, A.R.: State of detonation stability theory and its application to propulsion. J. Propul. Power. **22**(6), 1230–1244 (2006)
165. Stewart, D.S., Aslam, T.D., Yao, J.: On the evolution of cellular detonation. Proc. Combust. Inst. **26**, 2981–2989 (1996)
166. Strehlow, R.A., Fernandes, F.D.: Transverse waves in detonations. Combust. Flame **9**, 109–119 (1965)
167. Sturtzer, M.O., Lamoureux, N., Matignon, C., Desbordes, D., Presles, H.N.: On the origin of the double cellular structure of the detonation in gaseous nitromethane and its mixture with oxygen. Shock Waves **14**(1–2), 45–51 (2004)
168. Takai, R., Yoneda, K., Hikita, T.: Study of detonation wave structure. Proc. Combust. Inst. **15**, 69–78 (1974)
169. Takens, F.: Detecting strange attractors in turbulence. Lect. Notes Math. **898**, 366–381 (1981)
170. Taki, S., Fujiwara, T.: Numerical analysis of two-dimensional non-steady detonations. AIAA J. **16**, 73–77 (1978)
171. Taylor, G.I.: The dynamics of the combustion products behind plane and spherical detonation fronts in explosives. Proc. R. Soc. Lond. A **200**, 235–247 (1950)
172. Taylor, B.D., Kasimov, A.R., Stewart, D.S.: Mode selection in weakly unstable two-dimensional detonations. Combust. Theor. Model. **13**, 973–992 (2009)
173. Theiler, J.: Estimating fractal dimension. J. Opt. Soc. Am. A **7**, 1055–1073 (1990)
174. Tieszen, S.R., Sherman, M.P., Benedick, W.B., Shepherd, J.E., Knystautas, R., Lee, J.H.S.: Detonation cell size measurements in hydrogen-air- steam mixtures. Prog. Astronaut. Aeronaut. **106**, 205–219 (1986)
175. Toro, E.F.: Riemann Solvers and Numerical Methods for Fluids Dynamics, 1st edn. Springer, Berlin (1997)
176. Tsuboi, N., Hayashi, A.K., Matsumoto, Y.: Three-dimensional parallel simulation of cornstarch-oxygen two-phase detonation. Shock Waves **10**, 277–285 (2000)
177. Tsuboi, N., Katoh, S., Hayashi, A.K.: Three-dimensional numerical simulation for hydrogen/air detonation: Rectangular and diagonal structures. Proc. Combust. Inst. **29**, 2783–2788 (2002)
178. Varatharajan, B., Williams, F.A.: Chemical-kinetic descriptions of high-temperature ignition and detonation of acetylene-oxygen-diluent systems. Combust. Flame **124**, 624–645 (2001)

179. Vasil'ev, A.A., Trotsyuk, A.V.: Multiscaled cellular structure of gaseous detonation. In: Proceedings of the 5th International Seminar on Flame Structure, Novosibirsk, Russia (2005)
180. Vasil'ev, A.A., Gavrilenko, T.P., Topchian, M.E.: On the Chapman-Jouguet surface in multi-headed detonations. Astro. Acta **17**, 499–502 (1972)
181. Vermeer, D.J., Meyer, J.W., Oppenheim, A.K.: Auto-ignition of hydrocarbons behind reflected shock waves. Combust. Flame **18**, 327–336 (1972)
182. Voitsekhovskii, B.V., Mitrofanov, V.V., Topchian, M.E.: Optical studies of transverse detonation waves. Izv. Sibirsk. Otd. Acad. Nauk SSSR **9**, 44 (1958)
183. Von Neumann, J.: Theory of Detonation Wave. John von Neumann, Collected Works, vol. 6. Macmillan, New York (1963)
184. Voyevodsky, V.V., Soloukhin, R.I.: On the mechanism and explosion limits of hydrogen-oxygen chain self-ignition in shock waves. Proc. Combust. Inst. **10**, 279–283 (1965)
185. Wang, S.P., Anderson, M.H., Oakley, J.G., Corradini, M.L., Bonazza, R.: A thermodynamically consistent and fully conservative treatment of contact discontinuities for compressible multi-component flows. J. Comput. Phys. **195**, 528–559 (2004)
186. Watt, S.D., Sharpe, G.J.: One-dimensional linear stability of curved detonations. Proc. Roy. Soc. Lond. A **460**, 2551–2568 (2004)
187. Watt, S.D., Sharpe, G.J.: Linear and nonlinear dynamics of cylindrically and spherically expanding detonation waves. J. Fluid Mech. **522**, 329–356 (2005)
188. Westbrook, C.K., Urtiew, P.A.: Chemical-kinetic prediction of critical parameters in gaseous detonations. Proc. Combust. Inst. **19**, 615–623 (1982)
189. Williams, D.N., Bauwens, L., Oran, E.S.: Detailed structure and propagation of three-dimensional detonations. Proc. Combust. Inst. **26**, 2991–2998 (1997)
190. Wolf, A., Swift, J.B., Swinney, L.H., Vastano, J.A.: Determining Lyapunov exponent from a time series. Phys. D **16**, 285–317 (1985)
191. Yao, J., Stewart, D.S.: On the dynamics of multi-dimensional detonation. J. Fluid Mech. **309**, 225–275 (1996)
192. Yungster, S., Radhakrishnan, K.: Structure and stability of one-dimensional detonations in ethylene-air mixtures. Shock Waves **14**, 61–72 (2005)
193. Zaidel, R.M.: Stability of detonation waves in gaseous mixtures. Dokl. Akad. Nauk. SSSR. **136**, 1142–1145 (1961)
194. Zel'dovich, Y.B.: On the theory of the propagation of detonation in gaseous systems. Zh. Eksp. Teor. Fiz. 10, 542–568 (1940)
195. Zel'dovich, Y.B., Ratner, S.B.: Calculation of the detonation velocity in gases. Acta Physciochimica USSR. XIV(5), 587–612 (1941)
196. Zhang F.: Detonation of gas-particle flow. In: Zhang, F. (ed.) Heterogeneous Detonation, pp. 87–168. Springer, Berlin (2009)
197. Zhang, F., Lee, J.H.S.: Friction-induced oscillatory behavior of one dimensional detonations. Proc. R. Soc. Lond. A **446**, 87–105 (1994)
198. Zhang, F., Chue, R.S., Frost, D.L., Lee, J.H.S., Thibault, P., Yee, C.: Effects of area change and friction on detonation stability in supersonic ducts. Proc. R. Soc. Lond. A 449(1935), 31–49 (1995)
199. Zhang, F., Chue, R.S., Lee, J.H.S., Klein, R.: A nonlinear oscillator concept for one-dimensional pulsating detonations. Shock Waves **8**, 351–359 (1998)

4

Dynamic Parameters of Detonation

Anatoly A. Vasil'ev

4.1 Introduction

A chemically reactive material or mixture can undergo various combustion modes from low-speed flame (cm/s to m/s) to high-speed detonation (km/s) (e.g., [46, 77, 128, 191]).

The initiation of a flame or detonation has a threshold character, where the minimum energy ensuring 100% initiation is a fundamental dynamic parameter called the critical energy. While a weak initiation is commonly used for a flame ignition, direct initiation of an unconfined detonation requires a strong ignition source. While many principles of detonation dynamics are applicable to both gas and condensed-phase reactive materials, this chapter will focus on the gas mixtures. For example, in a spherical hydrogen–air mixture, the critical energy for a flame is $E_{\text{flame}} = 1.7 \cdot 10^{-5}$ J [10] and reaches $E_c = 3.35 \cdot 10^3$ J [183] for the direct initiation of detonation, thus showing an energy difference of seven to eight orders of magnitude (Fig. 4.1). Such a diagram is typical for any mixture. The threshold phenomenon between failure and direct initiation of cylindrical expanding detonation is illustrated in Fig. 4.2.

The direct initiation of an unconfined detonation can also be achieved as a detonation wave exits a tube and expands through diffraction into a space. Figure 4.3 shows an example of the threshold phenomenon of diffraction reinitiation through hot detonation products. The diffraction reinitiation criterion is characterized through a critical tube diameter, d_c, or channel width, l_c, which is equivalent to the critical energy of direct initiation of the same diverging, expanding detonation. When the diameter of the tube $d \geq d_c$ or channel width $l \geq l_c$, detonation reinitiation occurs [11, 17, 35–37, 40, 41, 68, 69, 76, 77, 100, 103, 106, 124, 151, 157, 158, 161, 165, 170, 186, 191, 196].

The cellular structure in the photographs in Figs. 4.2 and 4.3 exhibits the main character of a successful detonation and is intrinsically determined from the instability of a reactive system (e.g., [128]). The cellular nature arises from transverse waves that sweep across the leading front, in the direction perpendicular to the direction of detonation propagation. An instantaneous schlieren

F. Zhang (ed.), *Shock Wave Science and Technology Reference Library, Vol. 6*: Detonation Dynamics, DOI 10.1007/978-3-642-22967-1_4,

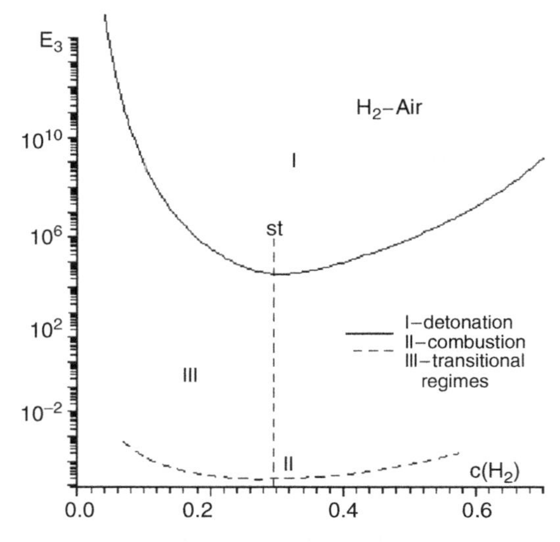

Fig. 4.1. Critical energy of flame ignition and initiation of detonation versus fuel molar concentration in hydrogen–air mixtures: I, detonation; II, flame; III, transitional regimes; including deflagration to detonation transition

Fig. 4.2. Soot imprints of subcritical (*left*, $E < E_c$) and critical (*right*, $E \geq E_c$) initiation of a cylindrical detonation in a stoichiometric acetylene–oxygen mixture (a channel of 1.0 cm thickness between two parallel plates with an exploding wire at the *center*). Fine cellular structure appears initially at the highly overdrive wave phase

photograph in Fig. 4.4 illustrates an element of the detailed wave structure of the cellular detonation front, that is, a triple point intersected with a Mach stem followed immediately by reaction, and a transverse wave combined with a shear layer behind. The transverse wave sweeps across the unburned mixture to react behind the incident shock of the detonation front. The motion of the triple points with the transverse waves forms a diamond-like shape known as the detonation cell (Fig. 4.5). Hence, the transverse waves play an important role not only during the propagation of a cellular detonation wave but also in unsteady transitional regimes such as initiation. The multidimensional and

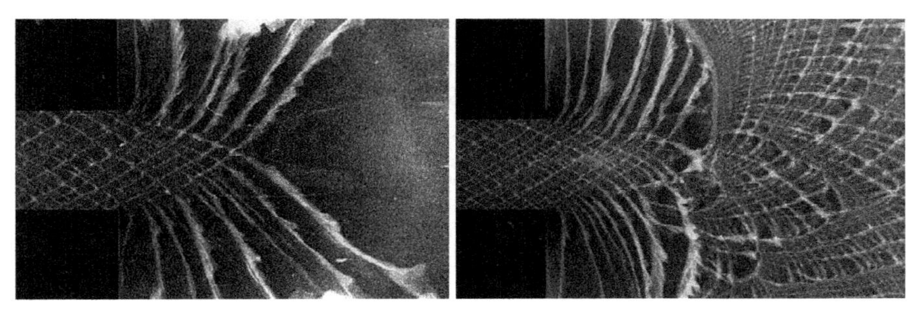

Fig. 4.3. Open-shutter photograph of failing (**a**) and reinitiating (**b**) 2D detonation through diffraction into a wide channel in an acetylene–oxygen mixture

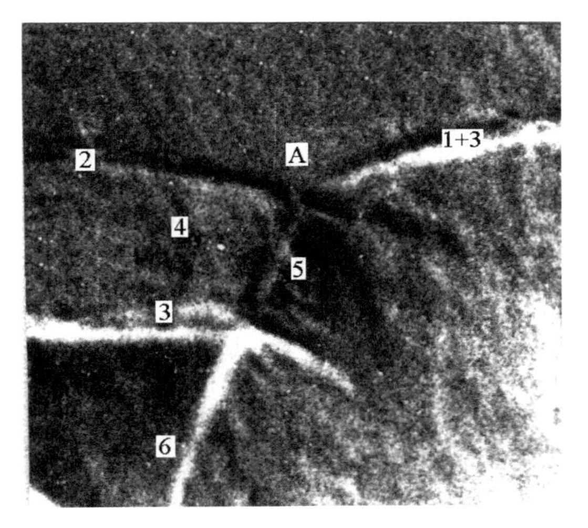

Fig. 4.4. Instantaneous schlieren photograph of a complex structure element at a detonation front: 1, shock front (Mach stem); 2, shock front (incident shock); 3, flame front; 4, induction zone; 5, transverse wave (reflected shock); and 6, tail of transverse wave; A, triple point

unsteady structure of the detonation front shown in Fig. 4.4 is observed not only in gaseous mixtures but also in some liquid, solid, and heterogeneous explosive systems.

While the detonation velocity, pressure, and temperature are regarded as detonation equilibrium parameters to be determined with thermochemical equilibrium theory, detonation cell size, critical initiation energy, and critical diameter are classified as detonation dynamic parameters since they characterize the dynamic behavior of the detonation front during the unsteady, nonequilibrium process. These dynamic parameters and relevant transient detonation problems will be the main subject of this chapter.

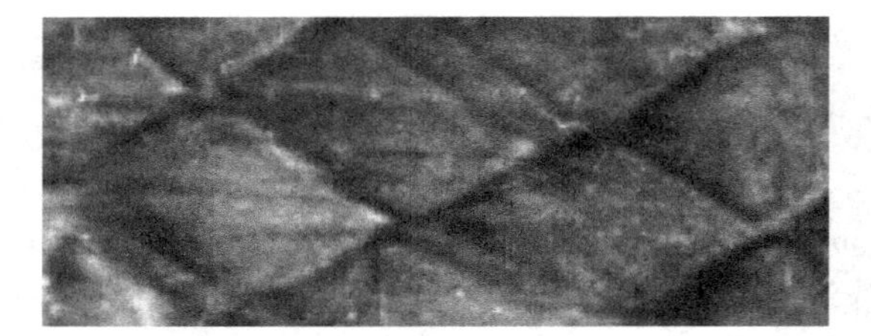

Fig. 4.5. A soot imprint of classical cellular detonation structure

4.2 Detonation Cell Size

The complicated multidimensional and transient structure of a detonation front is explained theoretically by instability of a gaseous reactive system in both chemistry and gas dynamics. As described in Chap. 3, planar detonation wave structure models are unstable, since any small perturbation (harmonics) of an idealized planar reaction front will grow with time and will, sooner or later, break the homogeneous and stationary structure of the detonation wave. Instabilities destroy the one-dimensional (1D) structure of the detonation front (i.e., planar shock wave followed by a planar induction zone and reaction zone) and transform it into a multidimensional, periodic cellular structure. To derive the detonation cell size and subsequently other dynamic parameters, correct determination of the local induction period τ becomes necessary.

4.2.1 Local Induction Time

The induction time can be derived from the 1D steady ZND detonation theory as described in Chap. 2. For a one-step global Arrhenius reaction model, the induction time can be written by

$$\tau = \frac{A \cdot \exp(E_a/RT)}{[f]^{k_1}[o]^{k_2}[in]^{k_3}}, \tag{4.1}$$

or in its logarithmic analogue

$$\ln\{[f]^{k_1}[o]^{k_2}[in]^{k_3} \cdot \tau\} = \ln A + E_a/RT, \tag{4.2}$$

where E_a is the activation energy, R is the universal gas constant and T is the mixture temperature in the induction zone. The square brackets indicate the concentration of each mixture component (f, fuel; o, oxidizer; in, inert additive), and A and k_i are coefficients determined, for example, by means of appropriate shock tube experimental data of ignition delays. The induction time can also be determined from numerically solving the 1D steady ZND equations using a detailed kinetics scheme of a combustible mixture (e.g., for heavy hydrocarbon fuels a kinetic scheme can contain as many as 1,000 elementary reactions); readers can find details of this method in Chaps. 2 and 3.

For unsteady problems, the induction time τ can be defined in an integral form:

$$1 = \int_{t_*}^{t_*+\tau} \frac{dt}{\tau_{st}} = \int_{r_1}^{r_2} \frac{dr}{D\tau_{st}} = \int_{r_1}^{r_2-\Delta_I} \frac{dr}{u\tau_{st}} = \varphi(E, r_1) - \varphi(E, r_2), \tag{4.3}$$

where r_1 is the shock radius at moment t_* when a particle crosses the shock front, and r_2 is the shock radius at which the same particle passes the induction period or a corresponding induction length, Δ_I. D and u are the shock velocity and particle velocity, respectively, and E is the energy of initiator that generates the shock wave. The induction time, τ_{st}, is introduced where subscript "st" refers to the state behind the shock wave at a given shock velocity.

Because of spatial and temporal inhomogeneities in local properties of a real multidimensional detonation front (Fig. 4.4), it is impossible to perform reliable experimental measurements of chemical kinetics under detonation conditions. The most relevant kinetic experiments have therefore been conducted under a uniform state of the mixture subjected to a shock wave. Results measured behind the shock wave are extrapolated to detonation conditions with a degree of uncertainty. There is large scatter in the global Arrhenius kinetic data on the ignition delay obtained by various authors when extrapolated to the range of detonation temperature, pressure, density, and species concentration (see [193]). Such discrepancy is observed for many fuels. Ranges of the detonation parameters in the equilibrium CJ state for various reactive gas mixtures can be found in Appendix at the end of the chapter.

4.2.2 Cell Size Models

Despite the unsteady propagation of a real detonation wave, its structure exhibits some periodicity: the propagation in opposite directions of triple points along the leading front results in self-organization and development of a cellular pattern of triple point trajectories. The individual rhombic imprint is referred to as the detonation cell, whose idealized pattern is shown in Fig. 4.6. A detonation cell is characterized by a longitudinal length, L, and a transverse width, λ, along the detonation propagation and its perpendicular direction, respectively. The velocity of the leading front varies along the cell axis by more than a factor of 2.

Numerous investigations over the past 50 years confirmed that the transverse waves are the main elements of the multidimensional structure of a detonation front and that cell size as a basic dynamic parameter is an intrinsic property of a reactive mixture. It was also shown that the transverse waves play an important role not only in detonation propagation but also in the

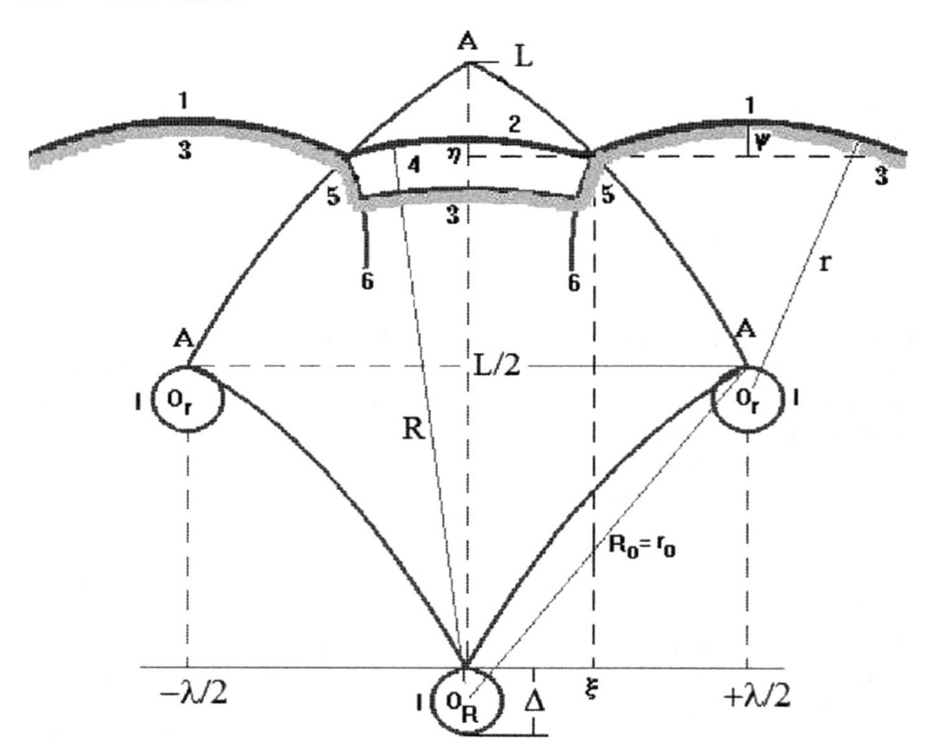

Fig. 4.6. Ideal formation of a detonation cell: I, region of transverse wave collision; 1 and 2, shock fronts; 3, flame fronts; 4, induction zone; 5, transverse waves; 6, tail of transverse wave; R and r, radius of shock waves, produced by neighboring I-regions

initiation process. The cell size depends on the fuel and oxidizer types, fuel–oxidizer ratio, phase state, dilution with promoter or inhibitor (including radicals and ions, or inert gas), mean detonation velocity, initial pressure and temperature, etc. [1, 5, 7, 12, 25, 31, 34, 69, 73, 93, 101–104, 113, 119, 120, 129–131, 133, 137, 138, 140, 142, 143, 152, 153, 161, 167, 168, 170, 172, 176, 177, 181–185, 189, 192, 193, 198]. For a given mixture in a fixed-diameter tube, the transverse cell size λ usually increases with decrease of initial pressure P_0.

Historically, analytical models for the cell size may be classified into four classes in terms of their basic postulates [161, 167, 172]:

1. Simple 1D models based on the ideal 1D detonation structure
2. "Acoustic" models based on the small perturbations in the detonation front
3. Parametric models based on certain parameters that influence the cell
4. Closed 2D models based on transverse wave collisions of the detonation front

The main idea of class-1 models is to determine the cell size λ of a cellular detonation from some characteristic length scale of the ideal 1D detonation model as the sum of the induction and reaction zone length. The uncertainty in the magnitude of this sum in real cellular detonations limits theoretical estimation to an induction zone length Δ_I:

$$\Delta_I = (D_{CJ} - u_{CJ})\tau, \tag{4.4}$$

where subscript "CJ" corresponds to the ideal planar steady CJ detonation propagating at velocity D_{CJ}, τ is the ignition delay behind the leading shock, and u is the particle velocity in the laboratory coordinate system.

For near-stoichiometric mixtures of hydrogen, acetylene, or ethylene with oxygen and air, a comparison of the CJ detonation induction length (Δ_I) calculated using detailed chemical kinetics with the experimental detonation cell size (λ) results in a linear relation [192, 193]:

$$\lambda = K\Delta_I \approx 29\Delta_I. \tag{4.5}$$

The analysis of a wider range of fuels, however, shows that the values of the factor K are not constant and differ for various fuels and equivalence ratios (for example, [35–37, 40, 41, 102, 104, 161]). Thus, the predictive capability of such estimation is questionable. It is notable that the linear relation between λ and Δ_I is confirmed only when the induction zone length considerably exceeds the recombination zone (i.e., the reaction zone). If their lengths are comparable, the corresponding relation becomes much more complicated. A more advanced model considering the ratio of the induction to reaction length has been recently introduced (see Chap. 3). Relations such as (4.5) for estimating λ are still used in studies dealing with kinetic modeling of chemical reactions at various pressures and temperatures. While the physical meaning of the relation of Δ_I with λ is rather questionable, this approach is attractive due to the "rather simple" calculation of the induction zone length Δ_I.

Class-2 models use an "acoustic" length scale in the ideal 1D detonation to determine the cell size λ of a cellular detonation wave. Historically, after development of the 1D ideal detonation theory, much of the attention was paid to the spinning detonation, including a description of the gas dynamic structure of the spinning wave with a complex shock configuration and a theoretical explanation of the rotary motion of the spinning wave. The hypothesis about the close correlation between the driving transverse wave in the spinning configuration and the acoustic oscillations of the detonation products was the most widely accepted explanation. The base frequency of the axial oscillations of the detonation products coincided with the frequency of the transverse wave rotation. At first glance, attempts to extend the acoustic treatment to transverse waves in the cellular detonation seemed natural, but were unsuccessful.

The common deficiency of the acoustic concept is the neglect of shock wave interactions in the cellular detonation front. Moreover, the concept of a "cell" in the acoustic field is vague. As a matter of fact, in this class of models, some characteristic distance is calculated based on the propagation of an acoustic wave during a certain time interval. Since the spatial distributions of sound sources in the models are not stipulated, the models fail to predict the periodic structure, and the induction zone and reaction zone are completely covered by "sound noise." The identification of some ordered structure in the sound chaos is a typical research topic on the instability theory of a dynamic system (see Chap. 3).

The cell models of class-3 are mostly physically justified by some parameters that influence the cell size such as the dependence of the cell size on the initial pressure, dilution ratio, and the decaying profile of the wave velocity along the cell axis. The essential physics and comprehensive characteristic parameters of the cell are not included in these models, and therefore they can only be used for a qualitative description.

A closed model of class-4 provides a description of a 2D regular cell and its characteristic parameters [189]. This class of models is based on a concept of the microexplosions arising from the collisions of transverse wave in a cellular detonation front. It takes into account the basic physical process of detonation propagation in the individual cell in the following sequence of events (refer to Fig. 4.6): collision of transverse waves (I-region); unsteady propagation of a cylindrical blast wave upward with instantaneous chemical reaction at its front; termination of chemical energy release; separation of the shock wave and combustion front; accumulation of the compressed and heated explosive mixture in the induction zone; and finally, the next collision of transverse waves arriving from the neighboring cells and thus the start of a new cycle. The propagation of a cellular detonation is supported by such periodically repeated transverse wave collisions. This model was named the cell model (CM).

In the 2D mathematical description, a local microexplosion due to transverse wave collision follows a self-similar solution for the linear blast explosion in a 2D flow:

$$D = \sqrt{E/\rho_0}/(2r), \tag{4.6}$$

where D is the blast wave velocity, ρ_0 is the mixture initial density, and r is the blast wave radius. The energy E was represented as a linear combination of the internal energy E_0 in the region I of transverse wave collision (see Fig. 4.6) and the chemical energy $E_Q(r)$ released behind the blast wave as it propagates:

$$E = E_0/\alpha + \beta E_Q(r), \tag{4.7}$$

with

$$E_0 = (\pi\Delta^2/4)P_3/(\gamma_\mathrm{I} - 1), \tag{4.8}$$

$$E_Q = \rho_0 q\pi(r^2 - \Delta^2/4), \tag{4.9}$$

where Δ is the diameter of the circular region I.

Due to the high degree of dissociation in detonation products during the collision of transverse waves in region I, chemical energy release is negligible ($q = 0$ hence $E_Q = 0$) and a strong explosion model can be used there. This leads to the first term of (4.7) where E_0 is from (4.8). The coefficient α is a function of the specific heat ratio γ_{I}, for instance, $\alpha = 0.983$ for $\gamma_{\text{I}} = 1.4$. P_3 is the pressure of the detonation products in domain I and is determined using the inert reflected shock relations that depend on the transverse wave velocity immediately before collision and the gas state in the induction zone in front of the two opposite transverse waves. To determine the coefficient β, the asymptotic solution for a strong shock wave is used. At $\Delta/2 \leq r \leq r_*$ (where r_* is the upper-limit radius for a strong shock wave with instantaneous chemical reaction behind it), the strong shock wave propagates as a cylindrical CJ detonation wave with constant velocity D_{CJ} where $E_0 = 0$ is assumed. Thus, $\beta = 4D_{\text{CJ}}^2/(q\pi)$. For $r_* \leq r \leq L + \Delta/2$ (L is the cell length), the reaction is considered frozen because of the rapid increase in ignition delay of the decaying wave and therefore $E = \text{const}$, thus leading to $r = r_*$ in (4.9) for $r_* \leq r \leq L + \Delta/2$. Finally, the transverse wave size at the instant of collision equals the diameter of region I, that is, Δ, which is the induction zone length immediately before the collision. Hence, it can be obtained by

$$\Delta = \int_{t_*}^{t_L} (D - u) \cdot \mathrm{d}t, \tag{4.10}$$

where u is the flow particle velocity in the induction zone and can be determined from the inert shock relations, time instant t_* corresponds to $r = r_*$, and t_L is the total time of detonation propagation in an individual cell and corresponds to $r = L + \Delta/2$.

The average detonation velocity at the cell axis along the entire cell length is defined as

$$\overline{D} = L/t_L. \tag{4.11}$$

It characterizes the losses if it is less than the CJ detonation velocity or overdrive when it is larger than the CJ value.

The wave velocity function, $D(r)$, can now be obtained by inserting (4.7)–(4.10) into (4.6). By integrating $\mathrm{d}r = D(r)\mathrm{d}t$, the blast propagation radius, $r(t)$, in an individual cell can be derived. From a combined solution of radii $r_i(t)$ calculated for each neighboring cell i (taking into account the repetition of all events with a time shift of $t_L/2$), the coordinates of the transverse wave, $\eta(t)$ and $\zeta(t)$, can be determined as the intersection point of two cylindrical waves with radii $R(t)$ and $r(t)$ as illustrated in Fig. 4.6. The axial and perpendicular components of the transverse wave velocities in the $\zeta(t)$ and $\eta(t)$ direction, D_{II} and $D_{\perp}$, are determined by differentiation of $\zeta(t)$ and $\eta(t)$. They are used to form the transverse wave velocity, whose value before the collision determines the pressure P_3. Noting the upper limit of the integration (4.10) depends on the cell length L, the problem of the transient wave structure of the periodically cellular detonation at any propagation distance requires for closure an additional equation of cell length.

The cell length equation can be obtained through the induction time of a flow particle that initially crosses the wave front (r_*, t_*) at the instant of the subsequent collision of transverse waves. This is determined from the integral relationship (4.3)

$$1 = \int_{t_*}^{t_L} \frac{\mathrm{d}t}{\overline{\tau}} = \int_{r_*}^{L+\delta\Delta} \frac{\mathrm{d}r}{D \cdot \overline{\tau}}. \tag{4.12}$$

with the integral taken along the trajectory of this particle. Here, $\overline{\tau} = \tau_{\mathrm{st}}$ is assumed as the ignition delay at constant temperature and density and is calculated from (4.1). Using the dimensionless variables $\zeta = r/L, x = r_*/L, y = \delta\Delta/L$ (theoretically $\delta = 1/2$), the formula for the cell length L is derived as a function of the physicochemical parameters of the explosive mixture:

$$L^{-1} = \int_{x}^{1+y} \frac{\mathrm{d}\zeta}{D(\zeta) \cdot \tau_{\mathrm{st}}(\zeta)}. \tag{4.13}$$

This relation can also be used as the reverse function to determine τ_{st} and average kinetic constants for the induction zone.

Equations (4.6)–(4.13) and (4.1) elucidate the basic assumptions and provide a closed model for a regular detonation cell, the details of the solution procedure can be found in [189]. The cell model is based on the concept of microexplosions from collisions of transverse waves. To facilitate the analytical derivation, a single irreversible Arrhenius kinetic model is used. The key for the application of the cell model is the availability of kinetic data for a wide range of compositions of fuel–oxygen and fuel–air mixtures.

In an approximate integration of (4.13), the particle variables along its integration path are assumed to be constant values, taken as those at the postshock state. Using the shock relations and the induction time (4.1) where $k_1 = k_3 = 0$ and $k_2 = 1$, integral (4.13) can be solved to give

$$L = (1.6D_*/x)(E_a/RT_*)A\exp(E_a/RT_*)/[\mathrm{O}_2]_*, \tag{4.14}$$

where D_*, T_*, and $[\mathrm{O}_2]_*$ are the detonation velocity, postshock temperature, and oxygen concentration at radius r_*, respectively.

Apart from the analytical models for the cell size, numerical 2D and 3D modeling of the detonation cellular structure has become more and more popular. In Chaps. 3 and 5, readers can find detailed descriptions of the cellular structure simulations and techniques for evaluating numerical data.

4.2.3 Influence Factors of Cell Size

Figures 4.7–4.17 present experimental data as well as calculated results from the class-4 cell model described above to illustrate how cell size varies as a function of initial conditions and mixture compositions. More data can be found in a review by Vasil'ev [167].

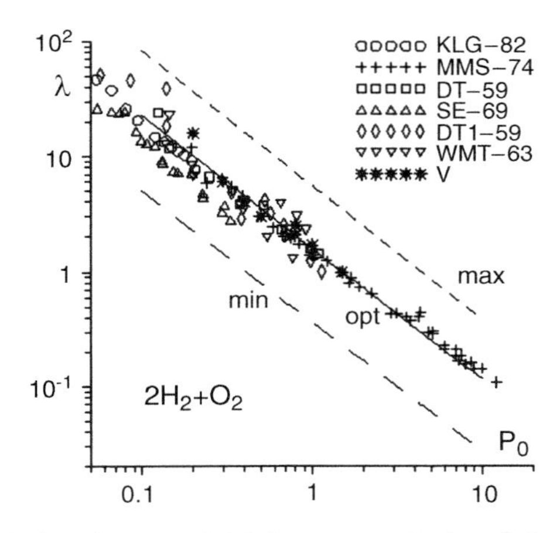

Fig. 4.7. Cell size λ (mm) versus initial pressure P_0 (atm) for mixture $2H_2 + O_2$. Experimental data: KLG, [68]; MMS, [93]; DT, [34]; SE, [138]; WMT, [191], and V, author. The *solid line* is calculated data using the cell model with optimized Arrhenius coefficients; *dotted lines* are calculated data with the minimum and maximum ignition delay from the literature for hydrogen–oxygen mixture [167]

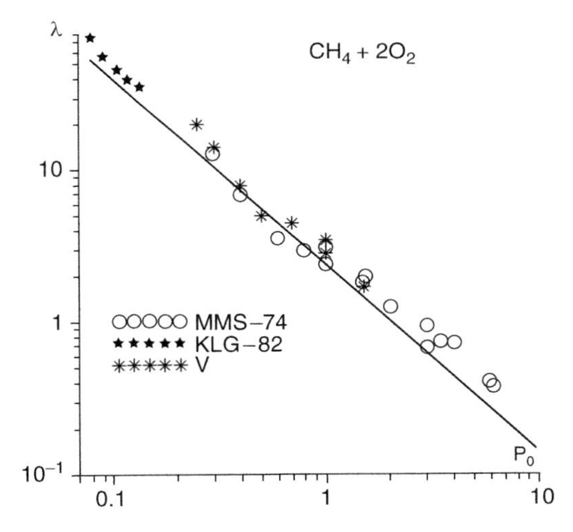

Fig. 4.8. Cell size λ (mm) versus initial pressure P_0 (atm) for mixture $CH_4 + 2O_2$. Experimental data: MMS, [93]; KLG, [68]; and V, author. The *solid line* is calculated data using the cell model with optimized Arrhenius coefficients [167]

The dependences of cell size λ on initial pressure P_0 are presented in Figs. 4.7–4.10 for stoichiometric hydrogen–oxygen and methane–oxygen mixtures as well as for acetylene–oxygen and ethane–oxygen mixtures diluted by argon. Figures 4.11 and 4.12 show cell size for stoichiometric hydrogen–oxygen

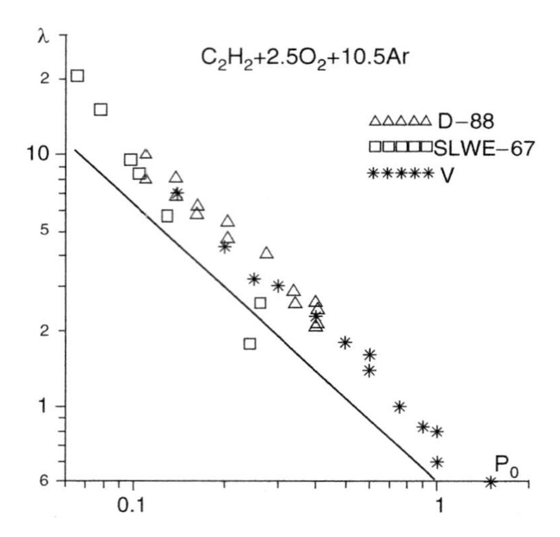

Fig. 4.9. Cell size λ (mm) versus initial pressure P_0 (atm) for mixture C_2H_2 + $2.5O_2$ + 10.5Ar. Experimental data: D, [35]; SLWE, [139]; and V, author. The *solid line* is calculated data using the cell model with optimized Arrhenius coefficients [167]

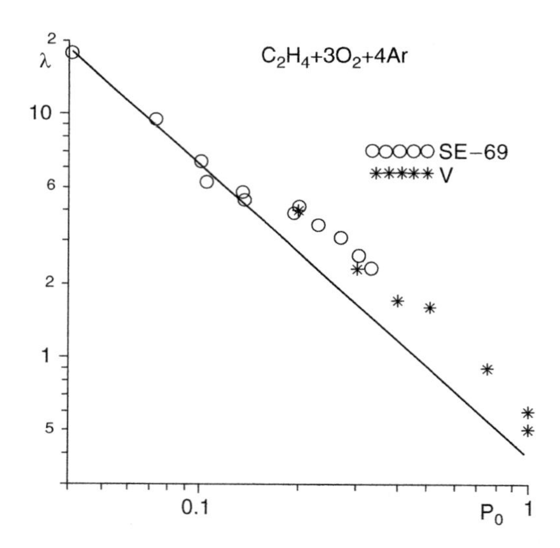

Fig. 4.10. Cell size λ (mm) versus initial pressure P_0 (atm) for mixture C_2H_4 + $3O_2$ + 4Ar. Experimental data: SE, [138] and V, author. The *solid line* is calculated data using the cell model with optimized Arrhenius coefficients [167]

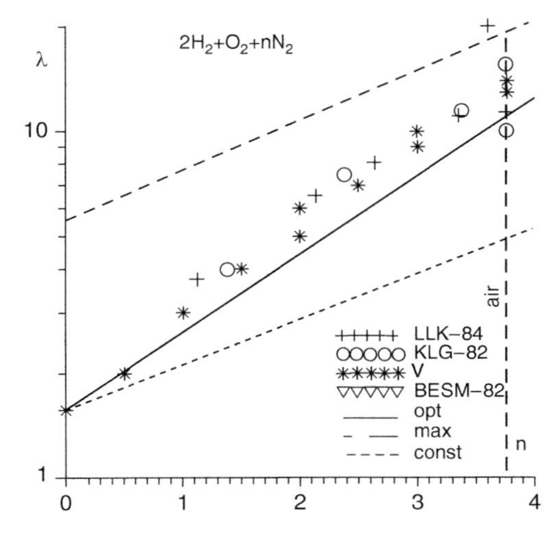

Fig. 4.11. Cell size λ (mm) versus nitrogen dilution ratio n for mixture $2H_2 + O_2 + nN_2$. Experimental data: LLK, [87]; KLG, [68]; BESM, [25]; and V, author. The *solid line* is calculated data using the cell model with optimized variable Arrhenius coefficients; *dashed lines* correspond to the same calculations but with constant coefficients from maximal induction time (*upper*) and to optimized constant Arrhenius coefficients for the undiluted mixture case ($n = 0$, *lower line*) [167]

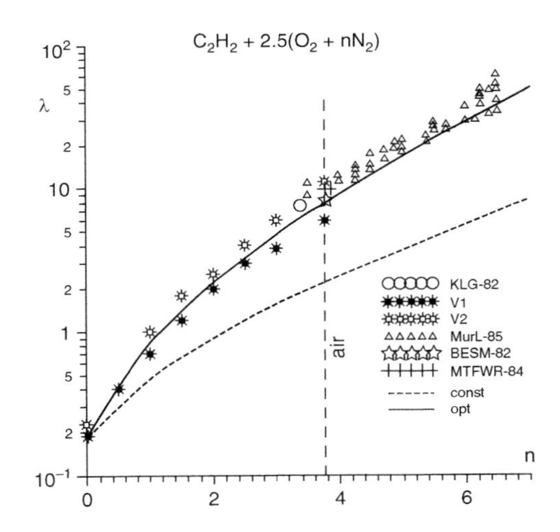

Fig. 4.12. Cell size λ (mm) versus nitrogen dilution ratio n for mixture $C_2H_2 + 2.5(O_2 + nN_2)$. Experimental data: KLG, [68]; MurL, [106]; BESM, [25]; MTFWR, [101]; and V1, V2, author. The *solid line* is calculated data using the cell model with optimized variable Arrhenius coefficients; the *dotted line* corresponds to the same calculations but with optimized constant Arrhenius coefficients for $n = 0$ [167]

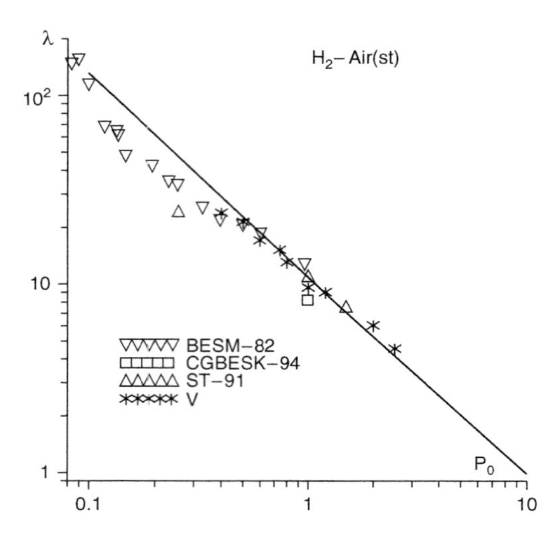

Fig. 4.13. Cell size λ (mm) versus initial pressure P_0 (atm) for stoichiometric mixture hydrogen–air. Experimental data: BESM, [25]; CGBESK, [31]; ST, [133]; and V, author. The *solid line* is calculated data using the cell model with optimized Arrhenius coefficients [167]

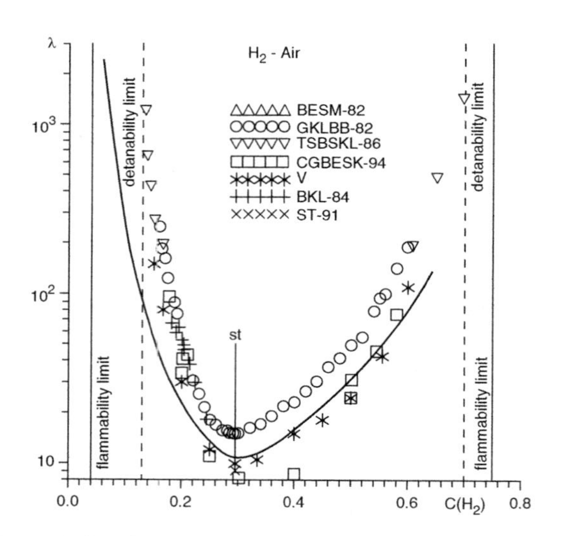

Fig. 4.14. Cell size λ (mm) versus fuel concentration (molar) for H_2–air mixtures at $P_0 = 1\,\text{atm}$. Experimental data: BESM, [25]; GKLBB, [57]; TSBSKL, [142]; CGBESK, [31]; V, author; BKL, [17]; and ST, [133]. The *solid line* is calculated data using the cell model with optimized variable Arrhenius coefficients; *vertical lines* correspond to the concentration limits (*lower* and *upper*) for detonation (*dotted*) and flame (*solid*), and to stoichiometric (st) value [167]

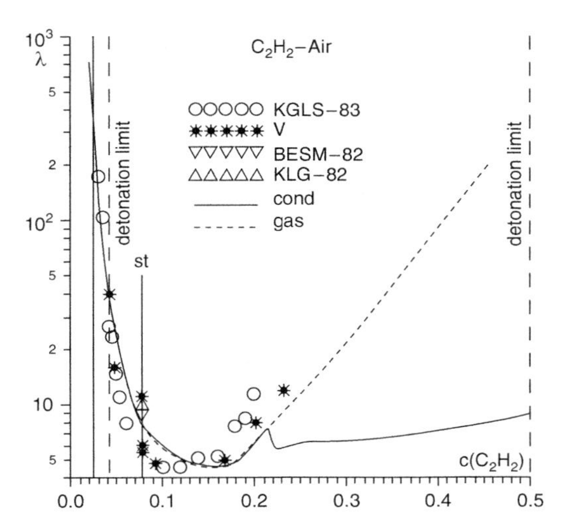

Fig. 4.15. Cell size λ (mm) versus fuel concentration (molar) for C_2H_2–air mixtures at $P_0 = 1$ atm. Experimental data: KGLS, [69]; KLG, [68]; BESM, [31]; and V, author. The *solid line* is calculated data using the cell model with optimized variable Arrhenius coefficients considering condensed carbon in equilibrium detonation products; the *dotted curve* corresponds to the same calculations but with gaseous carbon in detonation products. Vertical *dotted lines* correspond to the concentration limits for detonation (*dotted*) and flame (*solid*), and to stoichiometric (st) value [167]

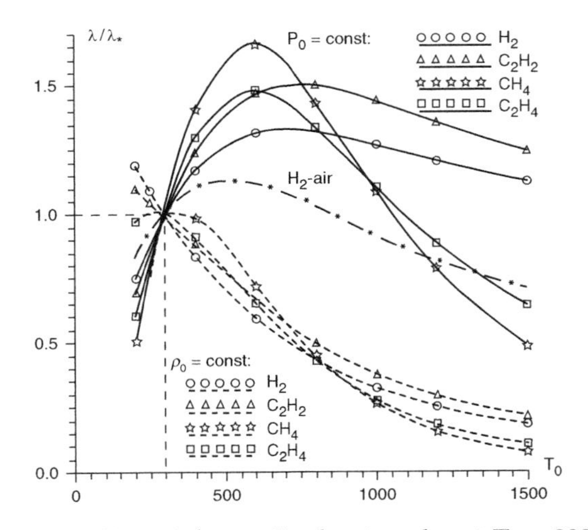

Fig. 4.16. Calculated cell size λ (normalized to its value at $T_0 = 298$ K) versus initial temperature T_0 (K) for $P_0 = \text{const} = 1.0$ atm (*upper solid lines*) and for $\rho_0 = \text{const}$ (*lower dotted lines*), for stoichiometric fuel–oxygen mixtures of hydrogen, acetylene, methane and ethylene. The long *dashed line* in the middle is for hydrogen–air mixture at $P_0 = \text{const} = 1$ atm

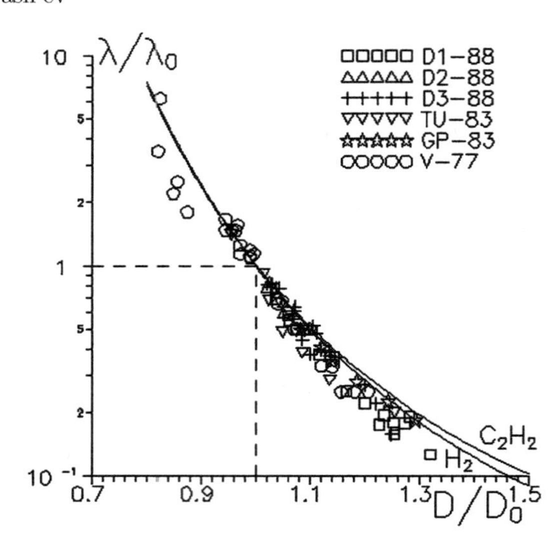

Fig. 4.17. Cell size λ in overdriven detonation (normalized to that of the CJ detonation) versus overdrive ratio D/D_0 (D_0: CJ detonation velocity) for stoichiometric hydrogen–oxygen and acetylene–oxygen mixtures. Experimental data: D1, D2, D3, [35]; TU, [145]; GP, [53]; and V, author. The *solid line* is calculated data using the cell model

and acetylene–oxygen mixtures as a function of the nitrogen dilution, where n is the ratio of the molar fractions of nitrogen to oxygen in the mixture and $n = 3.76$ for air. Figure 4.13 shows the dependence of cell size on initial pressure for the hydrogen–air mixture. Figures 4.14 and 4.15 illustrate the dependence of cell size on fuel molar concentration for different fuel–air mixtures, together with the lower and upper detonability limits.

The variation in cell size with initial temperature is shown in Fig. 4.16 for the cases of $P_0 = \text{const}$ and $\rho_0 = \text{const}$. Figure 4.17 gives the dependence of cell size on the overdrive ratio of detonation, where the cell size is normalized with respect to the cell size for a CJ detonation and D_{CJ} is the CJ detonation velocity. Similar dependences are observed for many other mixtures.

4.3 Direct Initiation of Detonation

Determination of the critical conditions for detonation initiation is a fundamental scientific and practical problem. The critical energy E_c for direct initiation of detonation has been considered as another important dynamic parameter to characterize a reactive mixture and to evaluate its detonation hazard [48, 70, 72, 74, 75, 78–80, 83–85, 154–156, 162, 164, 177, 180–185, 192, 193, 196].

Each combustible mixture is characterized by some spatial and temporal scales, r_* and t_*, at the given initial conditions of pressure, temperature, and composition (for example, the length and duration of induction or reaction zone). During initiation, the explosive mixture absorbs some quantity of energy E_ν from an initiator in some finite time interval t_0 and a finite volume V_0. From the efficiency point of view, E_ν is only a part of the energy originally stored in the initiator, E_{00}:

$$E_\nu = \int_0^{t_0} \int_0^{V_0} \varepsilon(t, V) \cdot \mathrm{d}t \cdot \mathrm{d}V = \eta E_{00}, \qquad (4.15)$$

where $\varepsilon(t, V)$ is the energy release rate of the initiator and η is the efficiency factor.

From formal mathematics, this equation represents a two-variable functional form in optimization of energy E_ν within the variable space and time limits of integration. Under influence between the spatial and temporal energy release, minimization of initiation energy to the value $E_{\min}$ will lead to a minimum power density $\varepsilon_{\min} = \varepsilon(r_*, t_*)$. If $V_0 = \text{const}$, to reach $E_{\min}$ will require optimizing the initiator power to $\varepsilon_{\min} = \varepsilon(t_*)$. Similarly, $t_0 = \text{const}$ will require optimization of the initiator power to $\varepsilon_{\min} = \varepsilon(r_*)$ corresponding to $E_{\min}$.

In general, the energy E_ν is a function of length- or timescales of both the mixture and the initiator, or possibly the ratios of their length- or timescales. The conditions $t_0 < t_*$ and $r_0 < r_*$ represent the conditions for an ideal initiator. The control of criteria of idealized initiation is the necessary condition for correct experimental comparison of various types of initiators characterized through their spatial and temporal energy release (e.g., electrical discharges, explosive charges, exploding wires or foils, high-speed bullets, laser sparks, hot combustion products). To establish a critical initiation criterion, the equivalence of the energy and power of various initiators must be established. Determination of the critical initiation energy, E_c, must account for the time and spatial characteristics of the input energy.

4.3.1 Measurement of Critical Initiation Energy

The degree of reactivity or detonability of various fuel–oxygen and fuel–air mixtures is so broad that one type of initiator cannot cover the energy range required for direct initiation of detonation in all mixtures. Experimentally, each type of initiator has specific features of energy release, and therefore correct determination of the critical initiation energy E_c for each individual initiator must consider the characteristics of its spatial and temporal energy release, E_t and E_r. Without information about E_t and E_r, the error in the estimated or measured E_c can be as large as several orders of magnitude. For example, the energy for flame self-ignition is still frequently defined as the difference between the initial and final energy of a capacitor, $E_{c,\text{flame}} = 0.5 \cdot C(U_0^2 - U_f^2)$ (C, capacitance; U, voltage), which does not account for losses in the elements of the electrical circuit.

The measurements can be standardized by defining the energy absorbed by a mixture based on the initial trajectory of an explosion wave, $r(t)$, excited by the initiator, from a strong explosion model citebib:62,bib:113:

$$r(t) = (E_0/\alpha_\nu \rho_0)^{1/(\nu+2)} t^{2/(\nu+2)}, \tag{4.16}$$

where E_0 is the explosion energy available to the blast and ρ_0 is the initial density. The index of symmetry, $\nu = 1, 2, 3$, is for the planar, cylindrical, and spherical cases respectively, and α_ν is a parameter of the strong point explosion model, depending on ν and the adiabatic index γ (for example, for $\gamma_0 = 1.4$, $\alpha_1 = 1.077$, $\alpha_2 = 0.983$, $\alpha_3 = 0.851$). This results in

$$E_0 = \alpha_\nu \rho_0 \left[\left(r_i^{(\nu+2)/2} - r_j^{(\nu+2)/2} \right) / \Delta t_{ij} \right]^2, \tag{4.17}$$

where r_i and r_j are the shock distances at times t_i and t_j, corresponding to the time interval Δt_{ij}. If critical initiation of detonation in an explosive mixture (e.g., $2H_2+O_2$) is observed at a pressure P_*, then the trajectory $r(t)$ of a shock wave in an inert mixture $2H_2 + N_2$ with a shock pressure equalling P_* can be used for determination of the effective energy of the initiator. Such replacement provides the similar initial density, $(\rho_0)_i$, and adiabatic index, $(\gamma_0)_i$, as well as similar flow parameters and their profiles in the area of detonation formation. The point explosion energy, E_0, derived from this blast trajectory $r(t)$ in such an inert mixture represents the critical initiation energy, E_c, of a detonation wave.

The validity of this technique was successfully tested using various initiators including exploding wires, high explosive charges and electrical discharges [155, 161, 170]. For example, in Fig. 4.18, the trajectory of a blast wave produced by an exploding wire is in good agreement with that of an explosion blast wave model in cylindrical symmetry ($\nu = 2$), thus indicating the blast behavior from the exploding wire with $E =$ const has a linear dependence in $r^2 - t$ coordinates at $t > t_*$ and $r > r_*$. Values of t_* and $r > r_*$ are small in comparison with the time and radius of detonation formation, t_{form} and r_{form}.

This technique provides a generic approach to determine the effective energy of various initiators and to establish the power equivalence between them. Its reliability was verified for different types of initiators.

Note that the strong explosion model technique is valid only when the length- or timescale of the initiation is below r_* and t_*. Through investigation of the influence of temporal aspects of initiator energy, E_t, for example, in initiation of a reactive mixture while only the electrical discharge duration is varied, one found that there exists a time parameter t_*. The experimental data of Lee [75] gave $(E_t)_{\min} = 0.12\,\text{J cm}^{-1}$ and $(t_0)_{\text{cr}} = t_* = 1\,\mu\text{s}$ for $C_2H_2+2.5O_2$ at $P_0 = 100\,\text{mm Hg}$, and $(E_t)_{\min} = 3.0\,\text{J cm}^{-1}$ and $(t_0)_{\text{cr}} = t_* = 3\,\mu\text{s}$ for

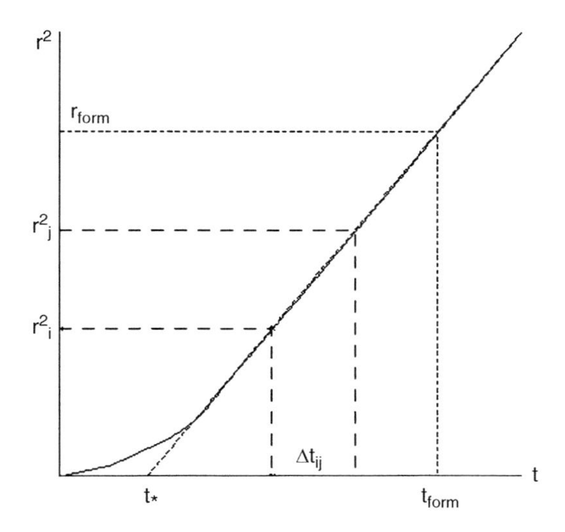

Fig. 4.18. The trajectory of a cylindrical blast wave from the strong explosion model (*solid line*) agrees well with the exploding-wire experimental data (*dotted line*) at $t > t^*$. The subscript "form" indicates the detonation formed

$2H_2 + O_2$ at $P_0 = 0.5\,\text{atm}$. In Fig. 4.19, the data of [75] are plotted in a nondimensional form and one can see a similarity between the nondimensional parameters for different mixtures and initial pressures. A sharp turning point defines t_*. If the duration of discharge $t_0 \leq t_*$, the energy required for initiation $E_t = \text{const} \cong (E_t)_{\min}$; if $t_0 > t_*$ ("stretching" discharge), E_t exceeds $(E_t)_{\min}$ and increases with t_0. The value $(E_t)_{\min}$ is accepted as the critical initiation energy of cylindrical detonation, $E_c = E_2$.

Experimental data in Fig. 4.19 were conducted with a spark duration in the order of a microsecond. Noting that modern lasers are characterized by a pulse duration in the order of a nanosecond or smaller, the transformation of energy from an initiator to the mixture using spark and laser initiation is different in nature. $E_t = \text{const} \cong (E_t)_{\min}$ for $t_0 < t_*$ may therefore not be valid at ultrashort impulse duration; instead, E_t may increase as the duration of discharge t_0 approaches zero.

The condition $t_0 < t_*$ represents a criterion of energy release rate of a given initiator into a given mixture. Only when this condition is satisfied, the successful direct initiation of a cellular detonation wave is determined by energy alone. For the critical energy for direct initiation of a diverging expanding detonation wave, the discussion below will assume the condition $t_0 < t_*$ is satisfied. Moreover, the discussion will also be limited first to the concentrated initiation source satisfying the condition $r_0 < r_*$ and will then be extended to the nonideal initiation of space- or time-distributed initiation sources.

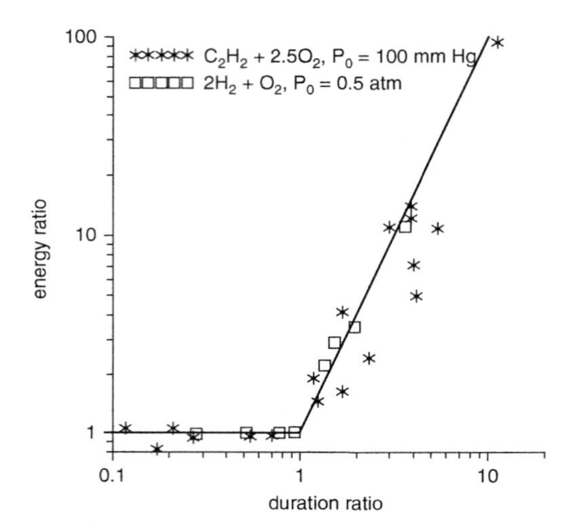

Fig. 4.19. Critical initiation energy versus duration of discharge normalized to values for an "ideal" initiator with conditions of $t_0 < t_*$ and $(E_t)_{\min} = \text{const}$

4.3.2 Peculiar Properties of Initiation Process

In Fig. 4.2, two typical soot imprints for a cylindrical detonation initiation illustrate the threshold character of the critical energy E_c. In a subcritical regime where $E < E_c$ (initiator energy less than critical energy), detonation fails (Fig. 4.2a), while in critical and supercritical regimes with $E \geq E_c$ a self-sustaining detonation is formed (Fig. 4.2b). For the detonation transmission shown in Fig. 4.3, the detonation cellular structure is destroyed in the subcritical regime and is preserved in the supercritical regime.

The mixture's chemical instability determines when and where the cellular structure appears in direct initiation of a cylindrically or spherically expanding detonation. The experimental investigations of Manzhalei and Subbotin [94] demonstrated that for the mixtures $CH_4 + 2O_2 + 2Ar$ and $2H_2 + O_2 + 3Ar$, the cellular structure appears in the decaying overdriven wave at a velocity D, corresponding to the point of intersection of the equilibrium adiabatic curve of the detonation products and the frozen shock adiabatic curve (heat release $Q = 0$) of the initial mixture. But for other mixtures ($C_2H_2 + 2.5O_2$, $2C_2H_2 + 3O_2 + 15Ar$, etc.) this criterion has not been demonstrated yet. In Table 4.1 the calculated results for the critical overdrive ratio M/M_0 (shock Mach number M at $Q = 0$ and M_0 for a CJ detonation) are presented for some mixtures. As can be seen, the critical Mach number corresponds to a high supersonic value. In Fig. 4.20 the number of transverse waves on the detonation front in the critical regime of detonation initiation is shown as a function of wave radius. Point I corresponds to the first appearance of cellular structure in the previously smooth expanding wave front and the second point II to the sharp increase in the number of transverse waves in the detonation front. While the problem of development of small disturbances up to

Table 4.1. The overdrive ratio where the equilibrium adiabat of detonation products intersects with the shock adiabat of the initial mixture

Mixture	$(M/M_0)_{Q=0}$
$2H_2 + O_2$	1.83
$CH_4 + 2O_2$	1.85
$C_2H_2 + 2.5O_2$	1.85
$2H_2 + O_2 + 3Ar$	1.56
$2H_2 + O_2 + 7Ar$	1.48
$C_2H_2 + 2.5O_2 + 10.5Ar$	1.47
H_2 + air (st)	1.76
CH_4 + air (st)	1.89
C_2H_2 + air (st)	1.88

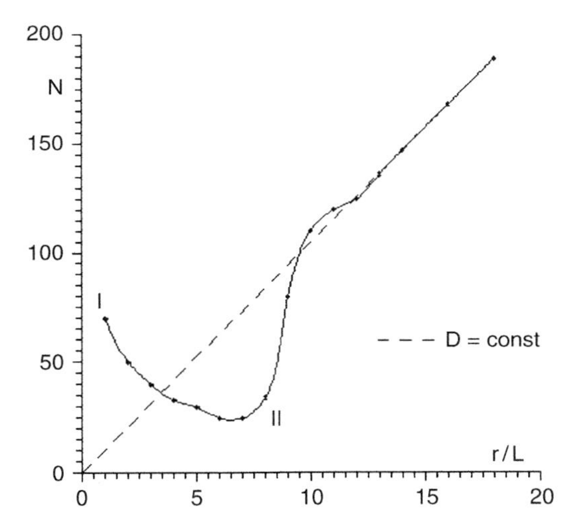

Fig. 4.20. Number of cells N in an initiating (*cylindrical*) expanding detonation wave versus wave front radius r (normalized to cell length L). *Solid line* corresponds to real unsteady initiation wave, *dotted line* to an ideal wave with constant velocity

high nonlinear values has been studied theoretically and numerically by many researchers, the precise mechanism for the appearance of a cellular structure in a decaying, expanding wave (initially overdriven) remains unresolved.

More advanced research on near-critical direct initiation ($E \cong E_c$) has revealed the behavior of the initiation process of an expanding detonation wave as shown in Fig. 4.20. The process starts at a strong initiation at point I with a high overpressure of the wave at a propagation velocity $D > D_{\mathrm{CJ}}$. "Failure" of the wave results in a continuous decrease in wave velocity below the CJ detonation value ($D < D_{\mathrm{CJ}}$) and down to a minimum $D_{\min}$ at a shock radius $r_{\min}$ at point II in Fig. 4.20. Afterward, rapid duplication of the number of transverse waves in the wave front leads to an increase in wave velocity and final formation of the self-sustaining detonation wave near the radius of

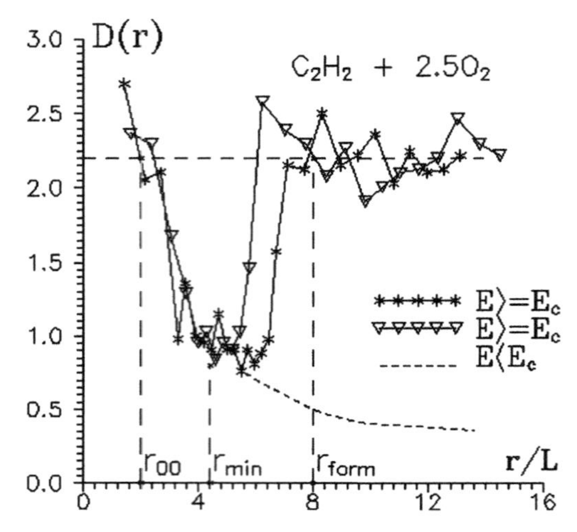

Fig. 4.21. Experimental velocity of initiating cylindrical expanding wave versus wave front radius (normalized to cell length)

Fig. 4.22. A schlieren trajectory of critical direct initiation of cylindrically expanding detonation

r_{form}. Figure 4.21 plots the velocity D of the initiating wave as a function of wave front radius: at $E < E_c$ the detonation fails and only high-velocity combustion is observed. Figure 4.22 shows the schlieren streak record of the critical initiation regime.

In accordance with experimental results similar to Figs. 4.20–4.22 in many mixtures with spherical, cylindrical, and planar initiation, formulas can be proposed for $r_{\min}$ and r_{form} at the critical initiation regime as follows:

$$r_{\min} \approx 2\nu L, \tag{4.18}$$

$$r_{\text{form}} \approx 2r_{\min}, \tag{4.19}$$

where L is the longitudinal size of the detonation cell and ν is the symmetry index. The minimum value of the $D(r)$ profile does not coincide with the maximum of the cell size profile $L(r)$, which indicates the importance of unsteadiness during an initiation process.

The radius of formation $r_{\rm form}$ characterizes the minimum size of a gaseous charge necessary for the correct definition of critical initiation energy E_c. At a charge diameter $d_0 < 2r_{\rm form}$, the obtained initiation energy value of E is underestimated compared to E_c.

4.3.3 Critical Initiation Energy Models

Theoretical models to predict the critical energy for direct initiation of detonation can be divided into two classes: models based on the 1D detonation structure and models based on the cellular detonation structure.

A number of critical initiation energy models are based on the 1D detonation structure model of Zeldovich et al. [196] for direct initiation of a spherically expanding detonation:

$$E_c \sim (\Delta_{\rm I})^3, \tag{4.20}$$

where $\Delta_{\rm I}$ is the induction zone length of a steady quasi-1D spherical detonation. Qualitatively, the idea of Zeldovich et al. [196] was based on two aspects. First, the blast wave from an initiator must be of "sufficient" intensity (with pressure $P > P_{\rm CJ}$, where $P_{\rm CJ}$ is the CJ detonation pressure) to trigger the chemical reaction behind the expanding blast wave. Second, the pressure of the decaying blast wave must remain greater than $P_{\rm CJ}$ for a duration greater than the induction time.

To apply (4.20), the value of $\Delta_{\rm I}$ and the proportionality coefficient must be determined. The ideas of Zeldovich et al. [196] and the blast wave model of Sedov [126] and Korobeinikov [71] were widely used to form approximate initiation models.

Since the cell width, λ, is assumed to be proportional to the induction zone length, as shown in (4.5), (4.20) was also extended to [193]:

$$E_c = E_3 = B\lambda^3. \tag{4.21}$$

The constant coefficient B is determined from experimental data.

Zhdan and Mitrofanov [199] proposed a critical blast radius model to estimate the critical energy for detonation initiation in gaseous and heterogeneous mixtures:

$$R_{\rm cr} \cong 4(\nu - 1)k^2(\sigma_{\rm CJ} + 1/\sigma_{\rm CJ} - 2)\Delta_{\rm I} \cdot E_a/RT_{\rm I}, \tag{4.22}$$

$$E_\nu = \alpha_\nu P_0(8R_{\rm cr}/\nu)^\nu. \tag{4.23}$$

Here, $R_{\rm cr}$ is the critical blast radius and E_ν is the critical energy for detonation initiation, $\nu = 1$, 2, and 3 is for the planar, cylindrical and spherical symmetry, respectively. $k = I/U$ is the ratio of the enthalpy I to the internal energy U, σ is the density ratio across the CJ shock wave and $\Delta_{\rm I}$ is the induction zone length calculated from (4.4). $T_{\rm I}$, E_a and α_ν are the mixture temperature behind the leading shock, global activation energy and the blast coefficient, respectively. Similar to (4.21), the critical blast radius was then correlated with the detonation cell length L to obtain:

$$E_2 \cong 250 P_0 L^2, \tag{4.24}$$

$$E_3 \cong 6,000 P_0 L^3, \tag{4.25}$$

for cylindrical and spherical symmetry, respectively, where P_0 is in $\mathrm{N\,m^{-2}}$ and L in m.

In summary, 1D models are based on a detonation induction zone length. Additional models based on the one-dimensional detonation structure can be found in the reviews of Benedick et al. [18] and Vasil'ev [156, 162].

Lee et al. have proposed a "surface energy" model that relates the point blast initiation and detonation diffraction reinitiation [76]. In this concept, the surface energy contained in the point blast-initiated spherical detonation wave at the time when the wave has decayed from its overdriven to CJ state is assumed to equal the surface energy of the detonation wave in the critical tube diameter, d_c, for diffraction reinitiation. Equating the two surface areas leads to

$$4\pi R_{\mathrm{CJ}}^2 = \pi d_c^2/4, \tag{4.26}$$

where R_{CJ} is the radius of the spherical detonation wave front at the CJ state. From point blast theory, the blast energy is

$$E_c = 4\pi\gamma_0 P_0 J M_S^2 \lambda^3, \tag{4.27}$$

where the subscript S refers to the shock front and 0 for the initial state. For $R_S = R_{\mathrm{CJ}}$ at $M_S = M_{\mathrm{CJ}}$ and $d_c = 13\lambda$ (see Sect. 4.4), the above two equations result in

$$E_c = \frac{2{,}197}{16}\pi\gamma_0 P_0 J M_{\mathrm{CJ}}^2 \lambda^3, \tag{4.28}$$

where P_0 is the initial pressure, γ_0 the adiabatic index of the initial mixture, M_{CJ} the shock Mach number of the CJ detonation (velocity D_{CJ}), and J is the blast integral that can be approximated by $J \cong Q/(\nu D_{\mathrm{CJ}})$.

In a cellular detonation front, the induction zone length varies by orders of magnitude between various locations along the front. The assumption of a uniform ignition delay after a planar shock front in the 1D theory may not reflect the multidimensional initiation condition. The large local temperature at the "hot spots" where transverse waves collide can induce initiation and therefore lower the critical initiation energy, as compared to 1D models. A critical initiation model based on the collision of transverse waves in the detonation front was therefore proposed by Vasil'ev based on the detonation cell model described in Sect. 4.2.2 [154, 161].

It was assumed by Vasil'ev [154, 161] that the critical energy for direct initiation of a cellular detonation wave, $E_c = E_\nu$, is proportional to the energy, $(E_0)_\nu$, in the area of a transverse wave collision:

$$E_\nu = n_{c\nu} E_{0\nu}, \tag{4.29}$$

where $n_{c\nu}$ is the coefficient accounting for the collective effect of a number of transverse wave collisions in the front, and $\nu = 1, 2$, and 3 are the symmetry indices for planar, cylindrical, and spherical geometries, respectively. The model is hence called the multiple-points initiation (MPI) model. Following the detonation cell model of transverse wave collisions in Sect. 4.2.2, the $E_{0\nu}$ value can be derived to be proportional to the detonation cell size. For the 2D cell model [189],

$$E_{02} = 4\varepsilon^2 \alpha_2 \rho_0 D_{\mathrm{CJ}}^2 L^2, \tag{4.30}$$

where ε is the undimensional parameter of the cell model and α_2 is the parameter in the blast explosion model (for $\nu = 2$ and $\gamma_0 = 1.4$, $\alpha_2 = 0.983$).

The MPI model was developed not only for 2D symmetry but also for spherical and planar symmetries [161, 170, 180]. The final formulas for the critical initiation energy can be written as

$$E_3 = \frac{512\varepsilon^2 \alpha_3 \cdot tg\varphi}{(\gamma_0 - 1)^2} \cdot \left(\frac{E_a}{Q}\right)^2 \cdot \rho_0 D_{\mathrm{CJ}}^2 L^3 = \mathrm{A}_3 \cdot \rho_0 D_{\mathrm{CJ}}^2 L^3, \tag{4.31}$$

$$E_2 = \frac{16\varepsilon^2 \alpha_2 \sqrt{\pi}}{\gamma_0 - 1} \cdot \frac{E_a}{Q} \cdot \rho_0 D_{\mathrm{CJ}}^2 L^2 = \mathrm{A}_2 \cdot \rho_0 D_{\mathrm{CJ}}^2 L^2, \tag{4.32}$$

$$E_1 = \frac{16\varepsilon^2 \alpha_1 \sqrt{\pi}}{\gamma_0 - 1} \cdot \frac{E_a}{Q} \cdot \rho_0 D_{\mathrm{CJ}}^2 L = \mathrm{A}_1 \cdot \rho_0 D_{\mathrm{CJ}}^2 L, \tag{4.33}$$

where E_a is an effective activation energy within the framework of the global Arrhenius reaction law, Q is the chemical energy release in the detonation front, $tg\varphi = \lambda/L$, where λ and L are the detonation cell width and length, and α is the parameter in the blast explosion model.

Although the essential assumption of multidimensional transverse wave collisions of this model cannot be satisfied in a one-dimensional scenario, the equation for E_1 is given to complete the mathematics and can be used to estimate the critical initiation energy for a planar explosive or a planar exploding foil in a long tube.

The critical energy for direct initiation of detonation can also be obtained through diffraction reinitiation of a detonation wave emerging from a critical-diameter tube and expanding into a space (the so-called diffraction reinitiation of multifront detonation (DRMD) model developed in the work of Vasil'ev [164]). In this concept, the critical initiation energy equals the work performed by the expanding detonation products along a path length that equals the cell length L of the cellular detonation front. The integration path starts at a critical diffraction shock radius, R_ν, measured from the tube exit. This critical radius corresponds to the distance traveled by the shock at the moment of collision of the inward propagating rarefaction waves at the central axis of the reactive mixture (i.e., the origin of detonation diffraction reinitiation as shown in Fig. 4.3). The critical diffraction shock radius can be expressed by

$$R_3 = 0.9 d_c, \tag{4.34}$$

$$R_2 = 0.9 l_c, \tag{4.35}$$

$$R_1 = 0.45 \delta_c, \tag{4.36}$$

where d_c and l_c are the critical tube diameter and channel width of diffraction reinitiation of spherical and cylindrical detonation waves, respectively (see Sect. 4.4.1), and δ_c is the critical distance for diffraction reinitiation from a planar detonation wave.

The critical initiation energy can therefore be derived:

$$\begin{aligned} E_3 &= \int_{R_3}^{R_3+L} P_{\mathrm{CJ}} 4\pi r^2 \mathrm{d}r = 4/3\pi P_{\mathrm{CJ}}[3R_3^2 L + 3R_3 L^2 + L^3] \\ &= 4/3\pi \rho_0 D_{\mathrm{CJ}}^2 L^3 [2.4 tg^2\varphi (d_c/\lambda)^2 + 2.7 tg\varphi (d_c/\lambda) + 1]/(\gamma+1) \\ &= B_3 \rho_0 D_{\mathrm{CJ}}^2 L^3, \end{aligned} \tag{4.37}$$

$$\begin{aligned} E_2 &= \int_{R_2}^{R_2+L} P_{\mathrm{CJ}} 2\pi r \mathrm{d}r = \pi P_{\mathrm{CJ}}[2R_2 L + L^2] \\ &= \pi \rho_0 D_{\mathrm{CJ}}^2 L^2 [1.8 tg\varphi (l_c/\lambda) + 1]/(\gamma+1) = B_2 \rho_0 D_{\mathrm{CJ}}^2 L^2, \end{aligned} \tag{4.38}$$

$$E_1 = \int_{R_1}^{R_1+L} P_{\mathrm{CJ}} \mathrm{d}r = P_{\mathrm{CJ}} L = \rho_0 D_{\mathrm{CJ}}^2 L/(\gamma+1) = B_1 \rho_0 D_{\mathrm{CJ}}^2 L, \tag{4.39}$$

where D_{CJ} is the CJ detonation velocity, ρ_0 and γ are the initial density and the adiabatic index of the mixture, $P_{\mathrm{CJ}} = \rho_0 D_{\mathrm{CJ}}^2/(\gamma+1)$ is used for the CJ detonation pressure, and $\lambda = L \cdot tg\varphi$ is the detonation cell width. Note that E_1 does not depend on R_1.

It is remarkable that the correlation

$$E_\nu = B_\nu \rho_0 D_{\mathrm{CJ}}^2 L^\nu \tag{4.40}$$

is identical in the DRMD model, the MPI model, and the surface energy model (4.22) except for the coefficients B_ν. The fact that the same form derived from various approaches may indicate the generality of this functional dependence for the critical initiation energy.

With improvements in computing power, direct numerical simulation has become available in computing the critical initiation problem using one- and multidimensional unsteady Euler equations with appropriate chemical kinetics models [38, 59, 60, 72, 83–85, 109]. Recent numerical studies have provided more mathematically rigorous treatments and greater insight into the physics of critical initiation (e.g., see Chap. 3), but their predictive ability for the critical initiation energy relies on numerically solving at least the quasi-one-dimensional unsteady Euler equations.

It is notable that the concept of a critical layer is useful not only for detonation initiation but also for spark ignition of a flame. Using the reinitiation of a flame emerging from a critical quenching diameter, d_q, the critical burning energy is assumed to equal the work performed by the expanded combustion products along the path length that equals the characteristic thickness of the flame front. The starting point of the integration path is chosen to be equal to one half of the critical quenching diameter d_q. This point plays a role in determining the critical point of self-reinitiation of a combustion wave. The Peclet criteria for the limiting combustion regime $Pe_* = d_q/b_{\text{th}} = \text{const} = 65$ can be used for estimation, where b_{th} is the thermal characteristic length of the reaction zone of the flame front. Hence, formulas for the critical ignition energy for planar, cylindrical, and spherical symmetry can be obtained:

$$E_{3*} = \int_{R_{3*}}^{R_{3*}+b_{\text{th}}} P_0 4\pi r^2 \mathrm{d}r = 4/3 \cdot \pi P_0 [3/4 \cdot d_{q3}^2 b_{\text{th}} + 3/2 \cdot d_{q3} b_{\text{th}}^2 + b_{\text{th}}^3] = C_3 P_0 (Pe_*)^2 b_{\text{th}}^3, \quad (4.41)$$

$$E_{2*} = \int_{R_{2*}}^{R_{2*}+b_{\text{th}}} P_0 2\pi r \mathrm{d}r = \pi P_0 [d_{q2} b_{\text{th}} + b_{\text{th}}^2] = C_2 P_0 \cdot Pe_* \cdot b_{\text{th}}^2 \quad (4.42)$$

$$E_{1*} = \int_{R_{1*}}^{R_{1*}+b_{\text{th}}} P_0 \mathrm{d}r = P_0 b_{\text{th}}. \quad (4.43)$$

Equations (4.41)–(4.43) bear a common form:

$$E_{\nu*} = C_\nu P_0 (Pe_*)^{\nu-1} b_{\text{th}}^\nu = C_\nu P_0 d_q^\nu / Pe_*. \quad (4.44)$$

4.3.4 Data of Critical Initiation Energy

Since the critical initiation energy is defined as the minimum energy required for successful initiation of a detonation wave, it characterizes the sensitivity of a reactive mixture to detonation. The lower the critical initiation energy, the more detonable the mixture is. There exist numerous experimental results of critical initiation energies in various gaseous fuels [3, 6, 8, 15, 18, 23–27, 48–50, 54, 57, 70, 74–80, 86, 88, 89, 95–97, 107, 111, 112, 114, 123, 148, 155, 170, 196, 200].

These data have been used to examine a variety of models mainly based on the one-dimensional detonation structure, some of which showed deviations by several orders of magnitude, even in simple hydrogen–air mixtures [162], for example, as shown in Fig. 4.23. Despite some qualitative similarity between calculated curves, quantitative differences are clearly marked.

Figure 4.24 presents the calculated and experimental critical energy for direct initiation of a cylindrically expanding detonation as a function of initial pressure for stoichiometric mixtures of hydrogen and acetylene with oxygen. Figures 4.25 and 4.26 give the critical initiation energy, E_1 for planar

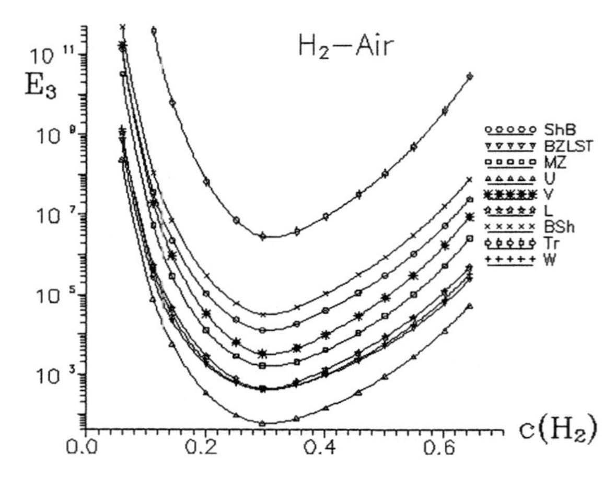

Fig. 4.23. Calculated critical initiation energy for spherical detonation versus fuel molar concentration using different initiation models but identical kinetic coefficients. The models presented are ShB, [132]; BZLST, [20]; MZ, [99]; U, [149]; L, surface energy model [18]; BSh, [19]; Tr, [144], and W, [193]

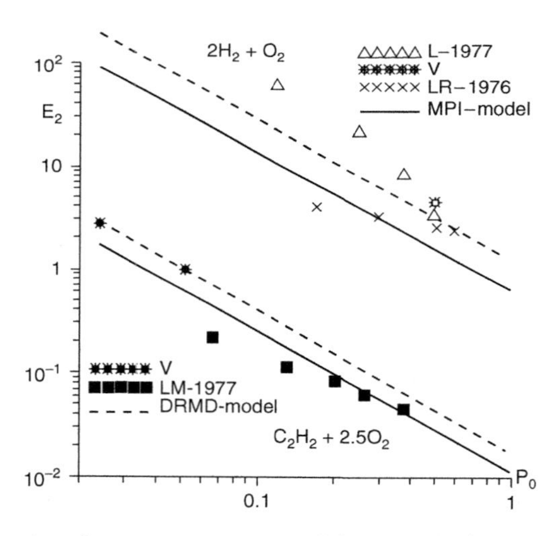

Fig. 4.24. Comparison between experimental (*data points*) and calculated (*curves*) critical initiation energy in $\mathrm{J\,cm^{-1}}$ versus initial pressure in atm for cylindrically expanding detonations. Experimental data: L, [75]; LR, [79]; LM, [78]; and V, author. The *solid lines* correspond to calculated data using MPI model and the *dotted lines* using DRMD model

initiation and E_2 for cylindrically expanding detonations, as a function of molar ratio of nitrogen dilution in stoichiometric methane–oxygen mixtures, while Fig. 4.27 provides the critical charge mass of TNT, m_{3*}, for direct initiation of spherically expanding detonations for the same mixtures. Note that

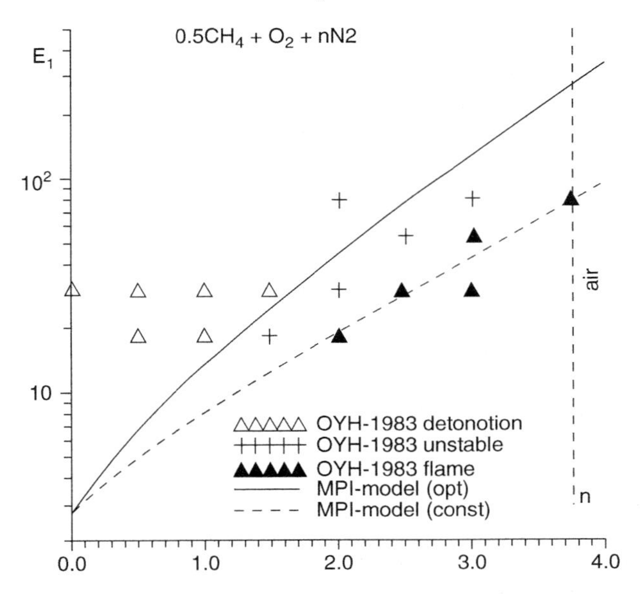

Fig. 4.25. Correlation between experimental (*data points*) and calculated (*curves*) critical initiation energy in $\mathrm{J\,cm^{-2}}$ for a planar detonation versus nitrogen dilution ratio of mixture $0.5\mathrm{CH_4} + \mathrm{O_2} + n\mathrm{N_2}$. Experimental data: OYH, [114]. The *solid* and *dotted lines* correspond to calculated data using MPI model with optimized and constant Arrhenius coefficients, respectively [162]

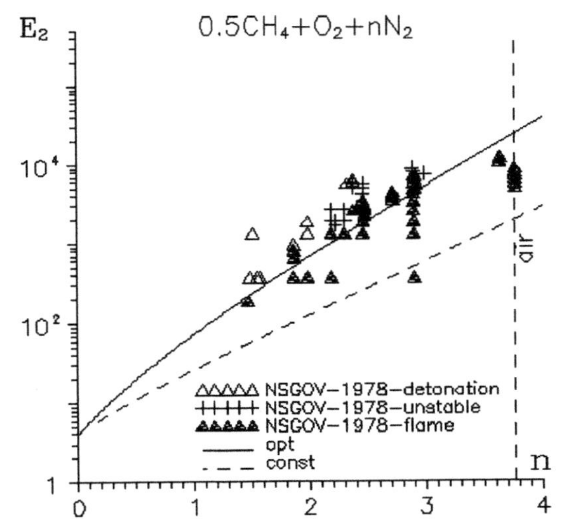

Fig. 4.26. Comparison between experimental (*data points*) and calculated (*curves*) critical initiation energy in $\mathrm{J\,cm^{-1}}$ versus nitrogen dilution ratio for cylindrically expanding detonation in $0.5\mathrm{CH_4} + \mathrm{O_2} + n\mathrm{N_2}$ ($n = 3.76$ for air). Experimental data: NSGOV, [112]. The *solid* and *dotted lines* correspond to calculated data using MPI model with optimized and constant Arrhenius coefficients, respectively [162]

the methane–oxygen–nitrogen mixtures in Figs. 4.25–4.27 were the only mixtures for which experimental data on initiation energy are available for all types of symmetry. Figures 4.28–4.30 display the critical charge mass of TNT as a function of fuel concentration in air for direct initiation of a spherically expanding detonation of hydrogen, acetylene, and ethylene. Readers can also find the critical energy values for direct initiation of spherical detonation in various reactive gas mixtures in the Appendix at the end of the chapter.

Among many models, the MPI and the DRMD models have provided the closest agreement with experiments in various fuel–oxygen and fuel–air mixtures over a range of initial pressures, equivalence ratios, and dilutions (a similar conclusion was drawn in the review of Benedick et al. [18]). The one-dimensional model of Zhdan and Mitrofanov, (4.22) and (4.23), [199] and the surface energy model of Lee et al. (4.28) [76] also provide reasonable agreement with experimental data. The good agreement with experiments demonstrates the correctness of the functional dependence of these models (i.e., $E_\nu \sim \rho_0 D_{\mathrm{CJ}}^2 L^\nu$) and therefore their predictive ability.

Finally, Fig. 4.31 provides an example of a hazard graph for some fuels ($NH_3, H_2O_2, O_3, H_2, C_2H_2, CH_4$, and N_2H_4), illustrating the dependence of E_3 on molar fuel concentration. Such a graph allows the comparison of the hazards posed by various mixtures. Alternatively, E_3 can be plotted versus initial pressure or other parameters.

4.3.5 Supersonic Bullet Initiation

Direct initiation of unconfined detonation can also be established based on an aerodynamic hypothesis using a "planar cross section" of a supersonic bullet [71]. In this direct initiation model, the work of an aerodynamic drag force along the unit length of the trajectory of a supersonic bullet that travels in an explosive mixture should exceed the minimum initiation energy for a cylindrically expanding detonation:

$$c_x \cdot \rho_0 w^2 \cdot \pi d^2/8 \geq \beta E_2 \equiv \beta \cdot A_2 \rho_0 D_{\mathrm{CJ}}^2 L^2, \tag{4.45}$$

where c_x is the aerodynamic drag coefficient of the bullet, ρ_0 is the mixture density, w is the relative velocity between the bullet and the mixture, d is the diameter of the maximum cross section of the bullet normal to the flight direction, and β is an effectiveness factor. The formula of the critical energy of cylindrical initiation E_2 in expression (4.45) is taken from (4.32) in Sect. 4.3.3. Upon rearrangement one obtains

$$d/L \geq \sqrt{8A_2\beta/(\pi c_x)} \cdot D_{\mathrm{CJ}}/w. \tag{4.46}$$

Equation (4.46) gives a relationship between the aerodynamic characteristics of the bullet and the explosive mixture properties. For a given mixture and initial pressure, the values of D_{CJ}, A_2, and L are constant. Therefore, if the bullet does not change its shape and orientation (c_x = const) during flight, the criterion of detonation initiation has a simple form:

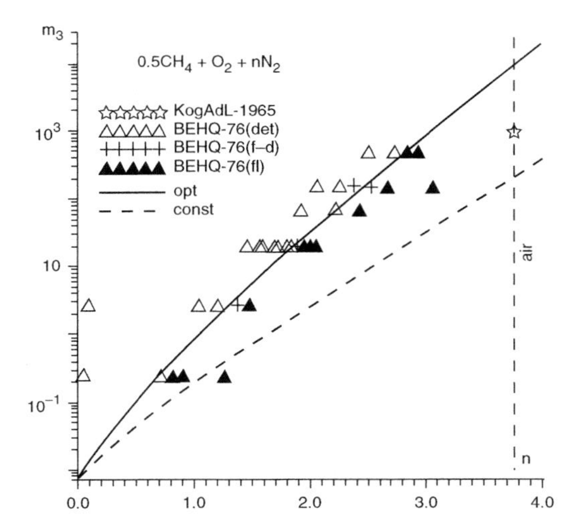

Fig. 4.27. Comparison between experimental (*data points*) and calculated (*curves*) critical initiating charge mass m_3 of TNT in grams versus nitrogen dilution ratio for spherically expanding detonations in $0.5CH_4 + O_2 + nN_2$ ($n = 3.76$ for air). Experimental data: KogAdL, [70] and BEHQ, [23]. The *solid* and *dotted lines* correspond to calculated data using MPI model with optimized and constant Arrhenius coefficients, respectively [162]

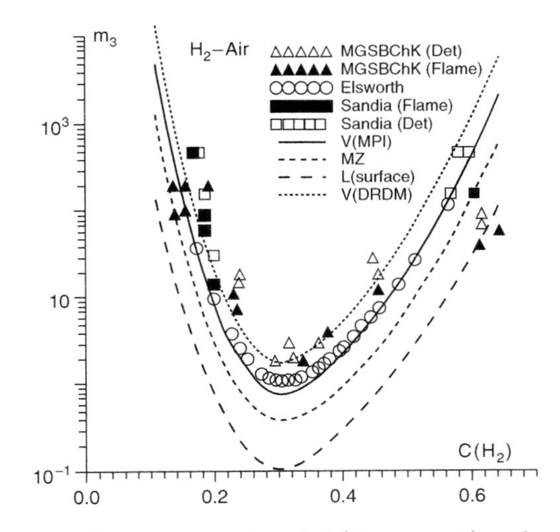

Fig. 4.28. Comparison between experimental (*data points*) and calculated (*curves*) critical initiating charge mass m_3 of TNT in grams versus fuel molar concentration for spherically expanding detonations in H_2–air mixtures. Experimental data: MGSBChK, [89]; Elsworth and Sandia, data from paper [18]; indexes "det" and "flame" refer to initiated regimes. Lines are calculated data: *solid line*, MPI model; *dotted line*, DRMD model [162]; *dashed line*, Zhdan and Mitrofanov [199], and *long dashed line*, surface energy model [18]

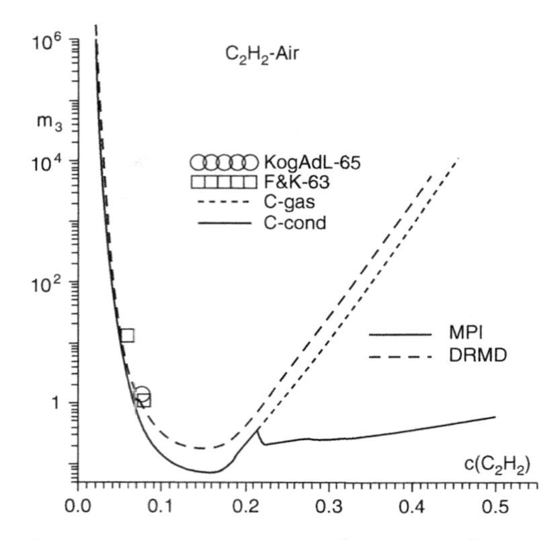

Fig. 4.29. Comparison between experimental (*data points*) and calculated (*curves*) critical initiating charge mass m_3 of TNT in grams versus fuel molar concentration for spherically expanding detonations in C_2H_2–air mixtures. Experimental data: KogAdL, [70] and F&K, [48]. The *solid line* is calculated data using MPI model with optimized variable Arrhenius coefficients including condensed and gaseous carbon in equilibrium detonation products and the *dotted line* corresponds to the same calculations but with gaseous carbon in detonation products. The *dashed line* is calculated using DRMD model [182]

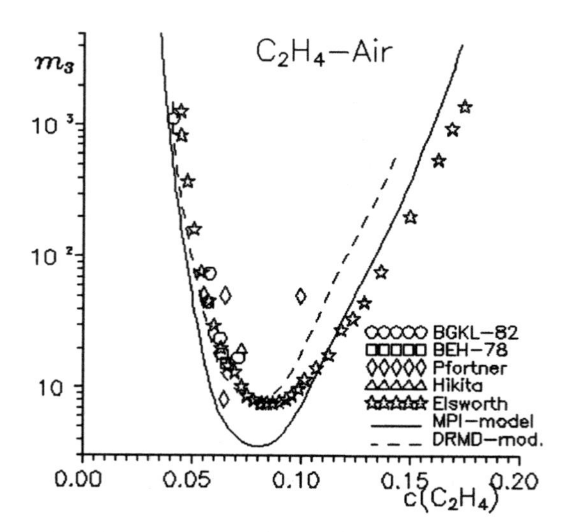

Fig. 4.30. Comparison between experimental (*data points*) and calculated (*curves*) critical initiation charge mass m_3 of TNT in grams versus fuel molar concentration for spherically expanding detonations in C_2H_4–air mixtures. Experiment data: BGKL, [18]; BEH, [23]; Pfortner, Hikita, Elsworth, data from paper [18]. The *solid line* is calculated from MPI and *dashed line* from DRMD models

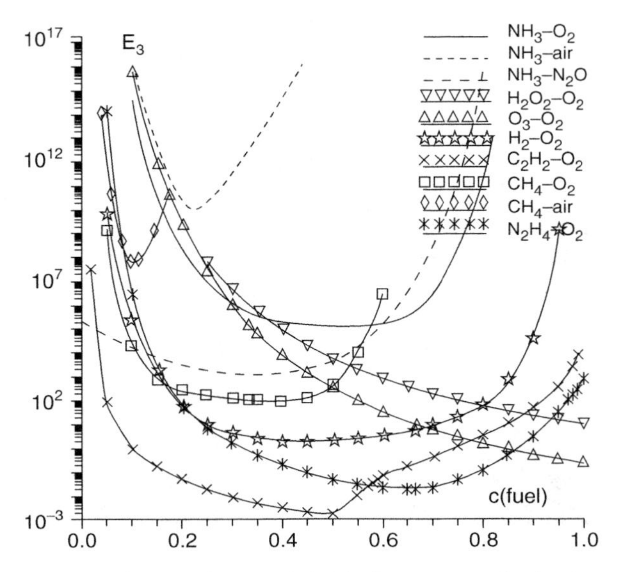

Fig. 4.31. The dependence of critical initiation energy in joules on fuel molar concentration for spherically expanding detonations in various mixtures

$$d \cdot w \geq \text{const.} \tag{4.47}$$

Figure 4.32 displays schlieren streak records of supersonic bullet initiation of combustion (upper photo) and detonation (lower). Experiments demonstrate that over a range of bullet diameters (5–250 mm) and speeds (800–3,500 m s^{-1}), the above criterion reliably predicts direct initiation of detonation in various mixtures of fuel–oxygen and fuel–air mixtures including gasoline [66, 77, 161, 163, 171, 188]. In Figs. 4.33 and 4.34 the experimental data for hydrogen–oxygen and hydrogen–air mixtures are presented. The critical line obtained from (4.46) well divides the regime of detonation initiation (above the line) from that of combustion only.

4.4 Diffraction Initiation of Detonation

4.4.1 Critical Scales for Diffraction Initiation

As a self-sustained cellular detonation wave transmits from a constant cross-section tube into a volume containing the same reactive mixture, the wave decays due to the diverging expansion. The exiting hot detonation products may then cause diffraction initiation of an expanding detonation wave in the diverging flow (Fig. 4.3). Successful diffraction initiation defines a critical diameter d_c for a round detonation tube connected to a spherical diffraction geometry ($\nu = 3$) or a critical width l_c for a rectangular detonation

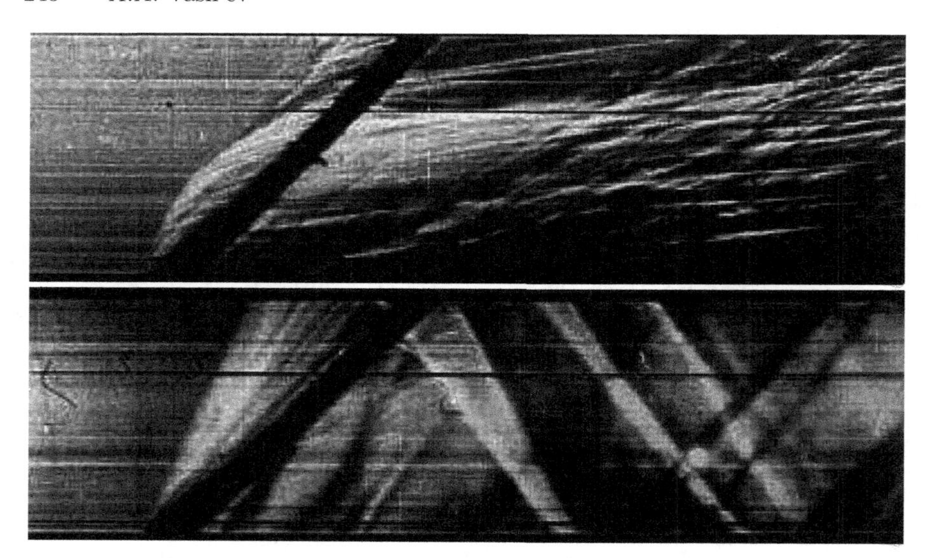

Fig. 4.32. Schellieren streak record of supersonic bullet initiation of combustion (*upper*) and self-sustaining detonation (*lower*) in an acetylene–oxygen mixture

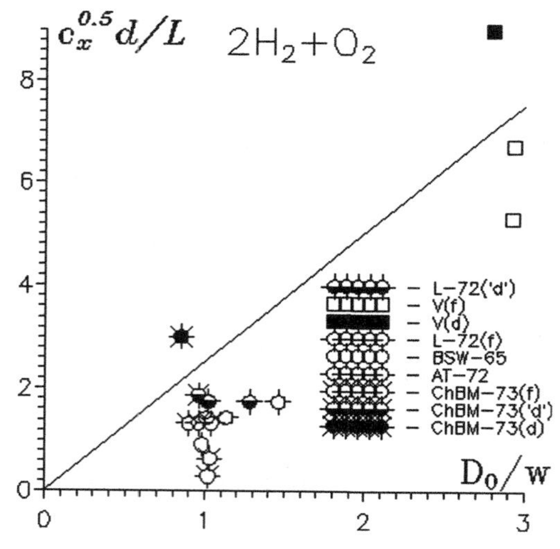

Fig. 4.33. Data of supersonic bullet initiation of combustion and detonation in hydrogen–oxygen mixture. Experimental data: L, [82]; V, author; BSW, [16]; AT, [4]; and ChBM, [28]. The legends d and f refer to detonation and flame, respectively. The *solid line* is calculated from criterion (4.46)

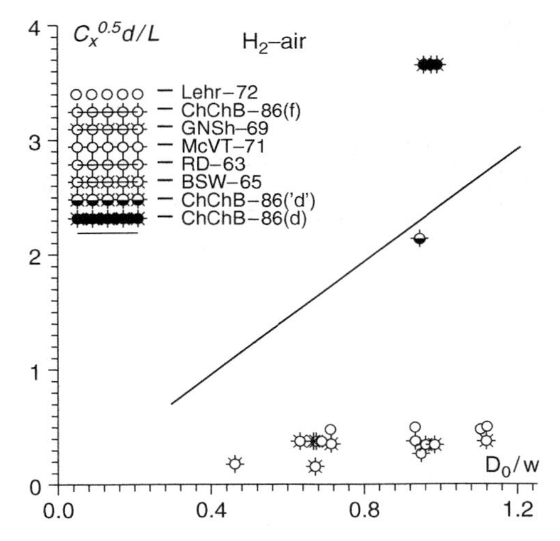

Fig. 4.34. Data of supersonic bullet initiation of combustion and detonation in hydrogen–air mixture. Experimental data: L, [82]; ChChB, [30]; GNSh, data from paper [55]; McVT, [98]; RD, [125], and BSW, [16]. The legends d and f refer to detonation and flame, respectively. The *solid line* is calculated from criterion (4.46)

channel connected to a cylindrical diffraction geometry ($\nu = 2$). An empirical correlation between detonation cell width, λ, and the critical tube diameter, d_c, or channel width l_c, was found:

$$d_c/\lambda = 13 \tag{4.48}$$

$$l_c/\lambda = 10 \tag{4.49}$$

(e.g., [100, 196]). The assumption of d_c/λ = const has been widely used for various mixtures to calculate their dynamic parameters of detonation [35,68,69,76,77,192,193]. However, this assumption may not be generally valid and the resulting errors can be considerable, in particular, for fuel–air mixtures, or mixtures near the concentration limits, or diffraction of overdriven detonations [36,37,40,41,157,161]. The photograph in Fig. 4.35 illustrates an example of failure of detonation reinitiation where the ratio of channel width to detonation cell width greatly exceeds the critical value in formula (4.49). Hence, the value of d_c/λ or l_c/λ for various mixtures is not always constant and can significantly vary.

Figure 4.36 provides the experimental critical diffraction diameter as a function of initial pressure for a stoichiometric hydrogen–oxygen mixture. Figures 4.37–4.39 give the experimental values of d_c as a function of nitrogen dilution fraction for stoichiometric fuel–oxygen mixtures of hydrogen, ethylene, and propane, respectively.

Both the critical energy for direct initiation of detonation and the critical diameter for diffraction initiation have been used to characterize reactive

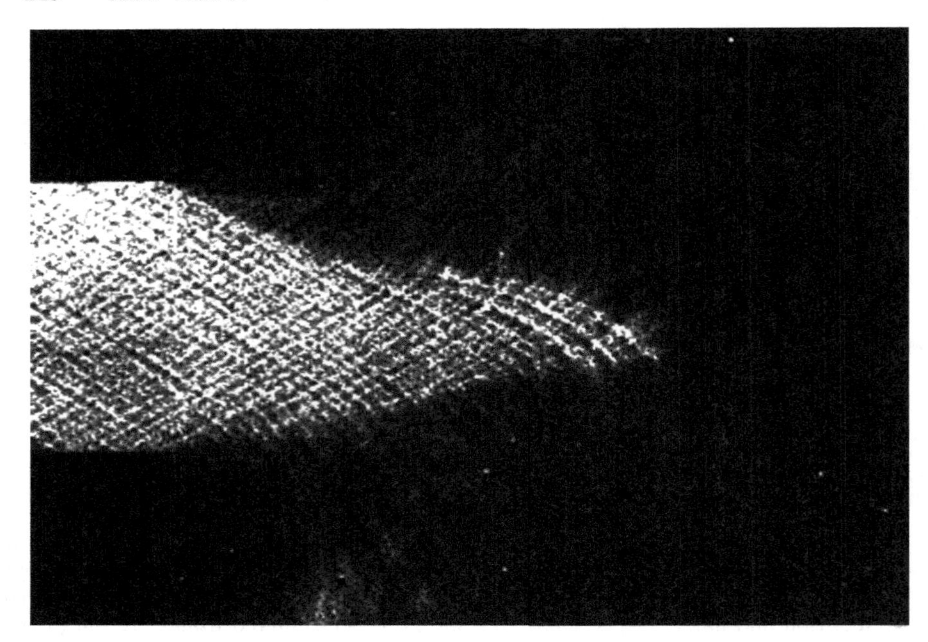

Fig. 4.35. Self-luminosity photograph for the subcritical regime of detonation diffraction in a highly diluted mixture, $C_2H_2 + 2.5O_2 + 10.5Ar$, where $l_c/\lambda >> 10$

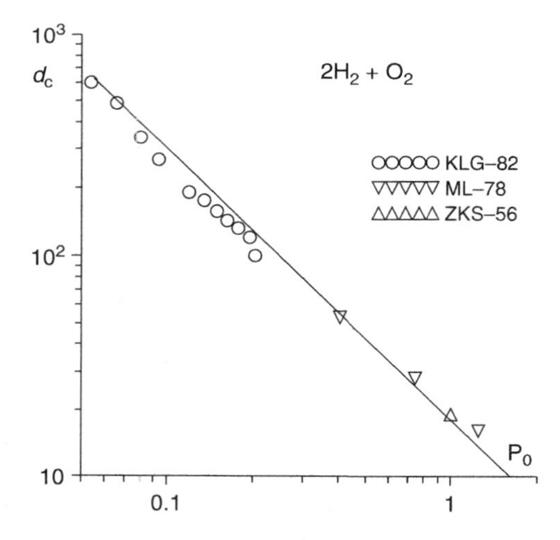

Fig. 4.36. Critical diameter of diffraction reinitiation in mm versus initial pressure in atm for $2H_2 + O_2$. Experimental data: KLG, [68]; ML, [99]; ZKS, [198]. *Solid line*, calculated using the cell model as in Fig. 4.7 and (4.48)

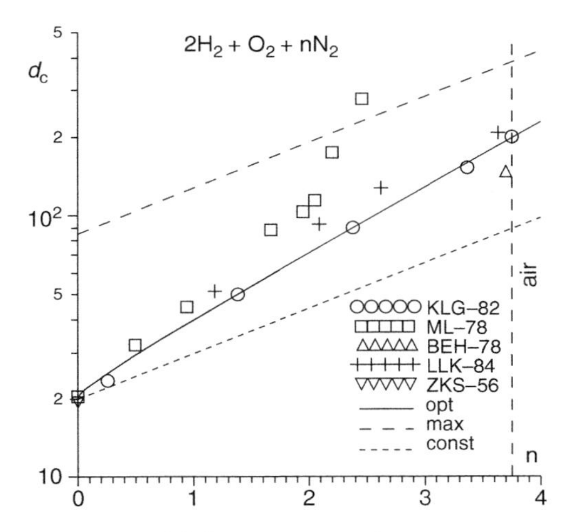

Fig. 4.37. Critical diameter of diffraction re-initiation in mm versus nitrogen dilution fraction for $2H_2 + O_2 + nN_2$ ($n = 3.76$ is for air). Experimental data: KLG, [68]; ML, [99]; BEH, [24]; LLK, [87]; ZKS, [198]. *Lines*, calculated using the cell model as in Fig. 4.11 and (4.48)

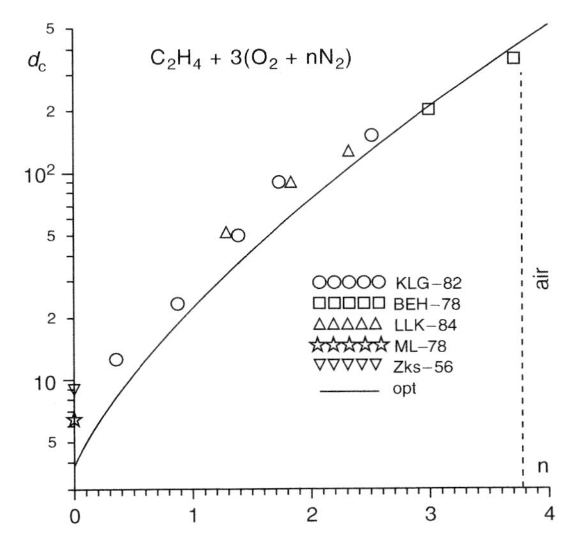

Fig. 4.38. Critical diameter of diffraction re-initiation in mm versus nitrogen dilution fraction for $C_2H_4 + 3(O_2 + nN_2)$ ($n = 3.76$ is for air). Experimental data: KLG, [68]; BEH, [24]; LLK, [87]; ML, [99]; ZKS, [198]. *Solid line*, calculated using the cell model and (4.48)

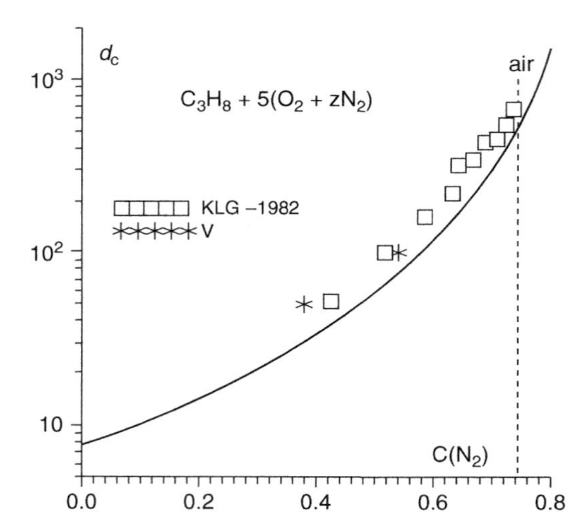

Fig. 4.39. Critical diameter of diffraction re-initiation in mm versus nitrogen dilution fraction for $C_3H_8 + 5(O_2 + nN_2)$ ($n = 3.76$ is for air). Experimental data: KLG, [68] and V, author. *Solid line*, calculated using the cell model and (4.48)

mixtures reactive mixtures for cylindrical or spherical expanding detonations. Hence, they can be considered as two fundamental detonation dynamic parameters and their correlation has been studied carefully.

Liu et al. have investigated the critical diffraction initiation of spherically expanding detonations through a number of detonation tube exit nozzle geometries (ellipse, square, rectangle, and triangle) [87]. They defined an effective critical diameter as

$$d_c = \sqrt{d_1 \cdot d_2}, \tag{4.50}$$

where d_1 and d_2 are the two semiaxes of an ellipse, the side of a square or an equal-sided triangle, and the length and width of a rectangular exit nozzle, respectively. The experimental results of the effective critical diameter for the four shapes are presented in Fig. 4.40 as a function of initial pressure. Correlation (4.48) applies to the effective critical diameters as well as to the diameter of a round detonation tube.

In an experiment, the chamber, in which the diffraction initiation takes place, must have a sufficient dimension to avoid the influence of wave reflection on the determination of d_c (or l_c). Hence, the diffraction chamber diameter, d_{chamb}, must satisfy $d_{\text{chamb}} \geq 5d_{\text{tube}}$ where d_{tube} is the detonation tube diameter. In the diffraction chamber, one must also avoid any small disturbances, since the diffraction initiation of detonation is sensitive to small obstacles, gaps, crevices, screw heads, steps, etc. (e.g., see Fig. 4.52).

The influence of overdriven detonation on the diffraction initiation has been studied by Vasil'ev [157, 189] and Desborders et al. [35–37]. In [157] a

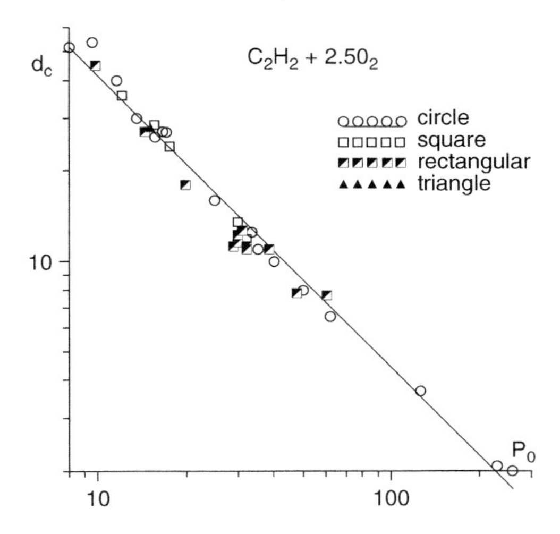

Fig. 4.40. Effective critical diameter for diffraction initiation in mm versus initial pressure in atm for different exit nozzle shapes [87]. *Solid line*, calculated using the cell model and (4.48) and (4.50)

detonation wave is overdriven through transmission from a constant diameter tube (d_{tube} =100 mm) into a convergent cone, with a half cone angle α and a variety of exit diameters (d_{exit} = 8–20 mm), and then enters an unconfined volume filled with the same mixture. In this case, the diffraction initiation is governed by the parameters α and the exit diameter d_{exit} (or the overdrive degree). Figure 4.41 provides the experimental critical initial pressure, P_{odd}, as a function of α, where P_{odd} is normalized to the critical initial pressure, P_{ssd}, of the self-sustained CJ detonation. The ratio $P_{\mathrm{odd}}/P_{\mathrm{ssd}} \leq 1$ indicates an efficiency in initiation. By varying α at a given d_{exit}, the efficiency of the diffraction initiation by overdriven waves is found to be greatest at $\alpha \approx 20°$ as shown in Fig. 4.41. At $\alpha \geq 60°$, the critical initial pressure ratio equals 1 and the overdriven wave no longer provides an enhanced efficiency.

Figure 4.42 shows the experimental results of the critical diameter in the ratio of $n_{\mathrm{odd}} = d_{\mathrm{exit}}/\lambda$ using the overdriven detonation, normalized by the value of the self-sustained CJ detonation, $n_{\mathrm{ssd}} = d_c/\lambda$ versus the overdrive degree defined by D/D_{CJ} where D is the overdriven detonation velocity. At small overdrive degrees (less than 10%), n_{odd} is nearly the same as that self-sustained detonation. Thereafter, n_{odd} linearly increases with the overdrive degree.

4.4.2 Diffraction Initiation in Other Divergent Geometries

For the case where transition of detonation from a straight tube or channel into a diverging tube or channel (with the half expanding angle, $\alpha = 0 - 90°$), the expanding detonation decays with increasing α up to a limit angle α_*, above

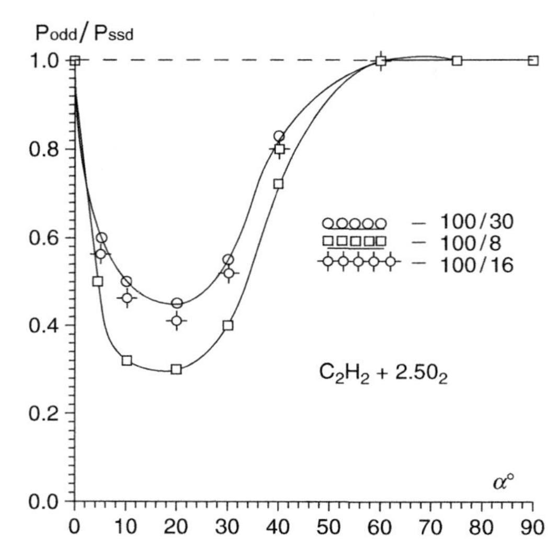

Fig. 4.41. Critical initial pressure, P_{odd}, for diffraction initiation by overdriven detonation (normalized to the value of the self sustained CJ detonation, P_{ssd}) in $C_2H_2 + 2.5O_2$ versus convergent cone angle with $d_{\mathrm{tube}} = 100\,\mathrm{mm}$ and $d_{\mathrm{exit}} = 30$, 16, and 8 mm, respectively

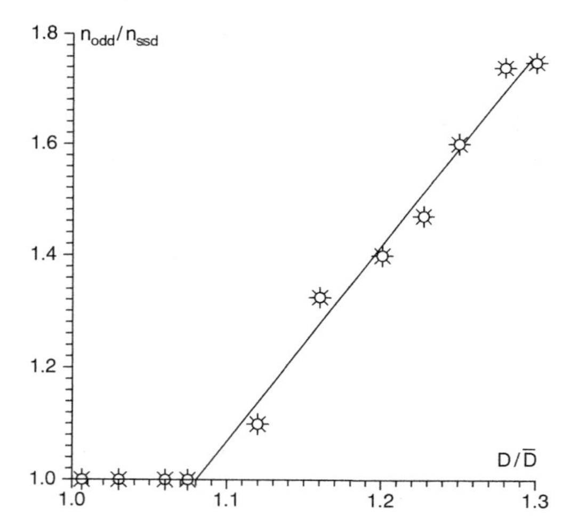

Fig. 4.42. Critical number of detonation cells, n_{odd}, across the channel width for diffraction initiation using overdriven detonation (normalized to the value of the self-sustained CJ detonation, n_{ssd}) versus overdrive degree D/D_{CJ} in $C_2H_2 + 2.5O_2$

which detonation fails. For example, in a $C_2H_2 + 2.5O_2$ mixture, $\alpha_* \approx 30°$ for $\nu = 3$ and $\alpha_* \approx 45°$ for $\nu = 2$. Figure 4.43 shows the critical diffracting angle as a function of initial pressure for $C_2H_2 + 2.5O_2$ in a rectangular cross-section

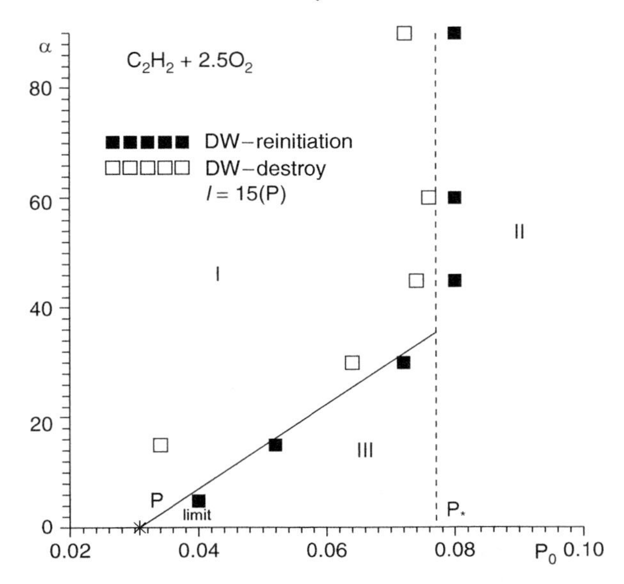

Fig. 4.43. Correlation of half expanding angle, in degree, and critical initial pressure, in atm, for detonation diffraction in $C_2H_2+2.5O_2$ in a rectangular cross-section channel with a width of 30 mm. Regions II and III: reinitiation; region I: failure

channel. For $\alpha \geq \alpha_*$ detonation failure is virtually independent of the divergent cone angle.

Diffraction failure or reinitiation can also take place in the propagation of a cellular detonation wave along a curvilinear boundary. Rotating detonation on a cylinder is used as an example to elucidate the relevant critical diffraction phenomena. Continuous rotating detonation in a coaxial annular cylinder has received attention due to its possible application in pulse detonation engines (for example, [26, 194]). The rotating detonation concept can be traced to the early experimental and analytical work by Voitsekhovsky [190, 191] and Nicholls et al. [110]. The rotating detonation can take place in a gas mixture confined in an annular gap between two coaxial cylinders or as a gas layer around a cylindrical surface as sketched in Fig. 4.44.

For the rotating detonation confined in a cylindrical annular tube, the outer concave circular wall provides continuous compression that turns the flow streamlines and results in a converging flow to promote the detonation, while the inner convex circular wall causes continuous diffraction that leads to the failure of detonation. In contrast, detonation in a gas layer on a cylindrical surface has a free exterior boundary. In the limiting case, the propagation of a rotating detonation wave must therefore rely on the diffraction reinitiation in the diverging flow caused by the convex circular surface. Hence, there must be a minimum radius of curvature of the convex circular surface,

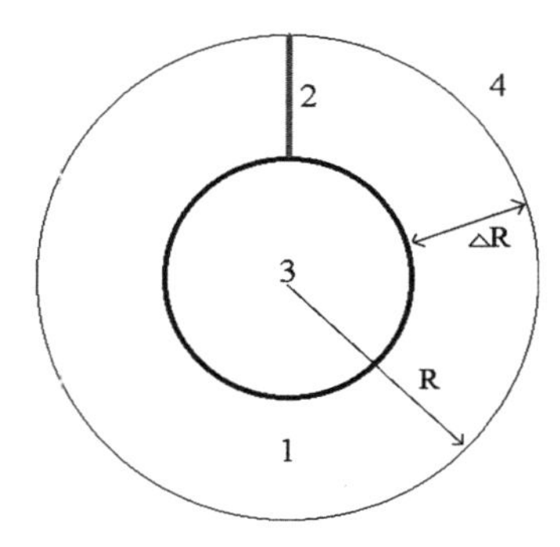

Fig. 4.44. Schematic of rotating detonation in a gas layer around a cylinder surface: 1, gas layer; 2, rotating detonation wave; 3, solid cylindrical surface boundary; 4, air

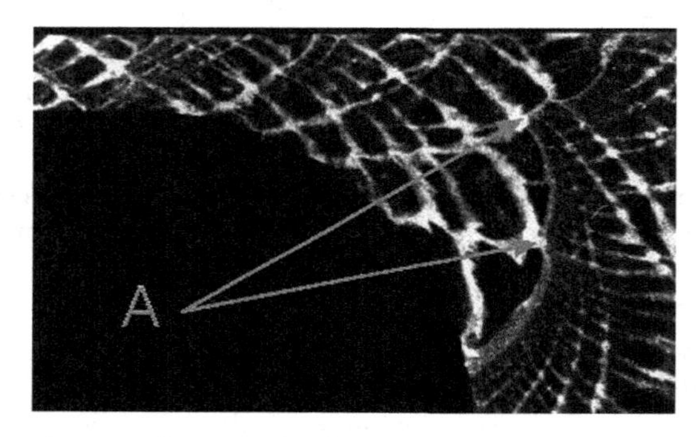

Fig. 4.45. Open-shutter photograph of critical detonation diffraction from a 15 × 1.5 mm channel onto a $R = 60$ mm cylindrical surface (A, reinitiation centers)

$R_{\min}$, and a minimum thickness of the gas layer, ΔR, below which the rotating detonation fails. These critical parameters also provide an upper limit of the geometry for the propagation of a rotating detonation in a cylindrical annular tube. The excitation and quasisteady propagation of a cellular detonation wave around a cylindrical surface were investigated for various cylinder radii, gas layer thicknesses, and initial mixture pressures by Vasil'ev [165].

The photos in Figs. 4.45 and 4.46 illustrate the mechanism for the propagation of rotating cellular detonation around a cylindrical surface, that is, new centers of reinitiation in the diverging flow appear nearly periodically some distance from the convex surface. The thickness of the gas layer around the cylindrical surface must be sufficiently large to contain these microcenters of reinitiation for a given radius of the surface curvature that is above the minimum radius. The minimum radius of the surface curvature and the minimum thickness of the gas layer were derived to be [165]:

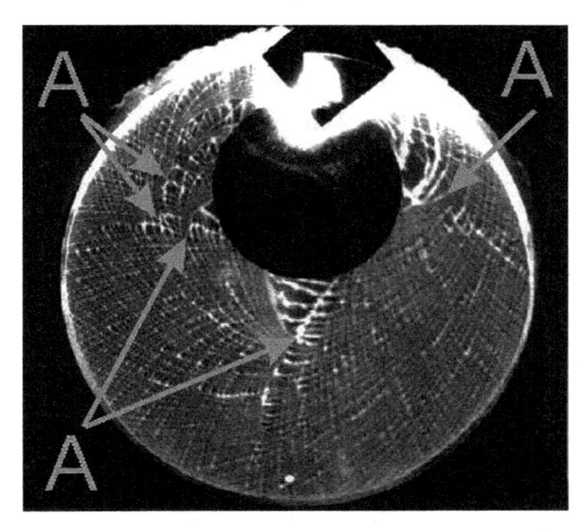

Fig. 4.46. Open-shutter photograph of a rotating detonation (*clockwise*) around a cylindrical surface of radius $R = 40$ mm in $C_2H_2 + 2.5O_2$ (A, reinitiation centers)

$$R_{\min} >> 0.4l_c \tag{4.51}$$

and

$$\Delta R > 0.9l_c, \tag{4.52}$$

where l_c is the critical channel width for diffraction initiation defined in Sect. 4.4.1.

4.5 Advanced Initiations

More efficient detonation initiation schemes can be considered through: (1) spatial distribution of input energy; (2) a time series of impulses; (3) reacting particles; (4) gradients of parameters (e.g., density, temperature, composition); (5) shock wave reflection and focusing; and (6) initiator geometry.

The change of a concentrated initiator into multiple ones provides spatial distribution of input energy. In detonation initiation of an unconfined mixture ($\nu = 3$), initiator charges can be modeled in the following geometric forms as shown in Fig. 4.47: (a) a circular disk with radius R; (b) an annulus with radii r and R; (c) a multipoint scheme represented by n charges of diameter d spatially distributed, for example, on a circle of radius R; (d) a single rectangular planar charge with $H \times 2w$; (e) a system of parallel line charges $H \times w$, separated by a distance $2z$, or (f) a system of nonparallel line charges, for example, shaped with an angle α or closed triangle.

The idea for the distributed initiation models listed above lies in the shock wave collisions. The detonation initiation occurs as the total energy of transverse wave collisions cumulated on the same surface or smaller than that

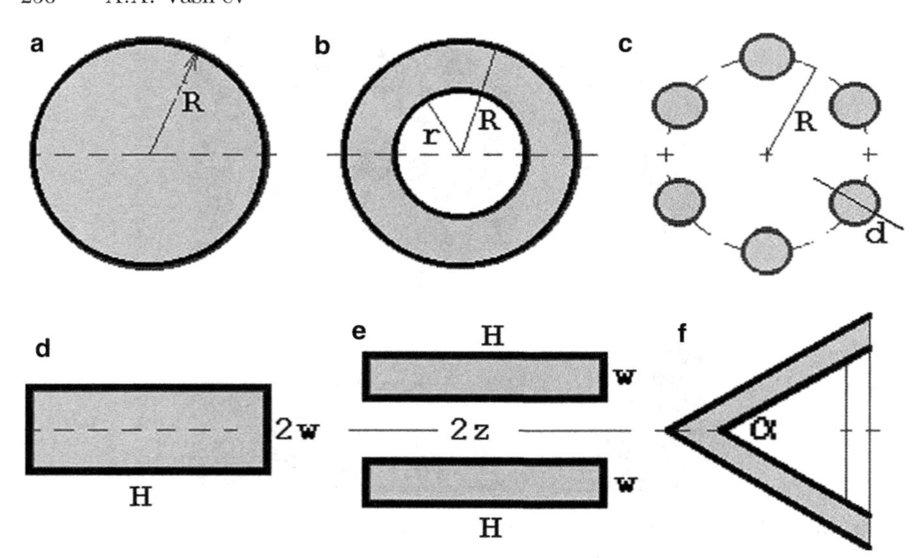

Fig. 4.47. Simple models of the spatial distribution of input energy

facing a reference concentrated initiator. The ratio of the maximum pressure or temperature with respect to that of the reference concentrated initiator is therefore greater than one. Hence the critical initiation energy E_c can be reduced; in other words, the efficiency of initiation is increased.

Figures 4.48 and 4.49 present some critical initiation experimental data for quasispherical expanding detonations using annular initiator charges (Fig. 4.47b) and parallel line initiator charges (Fig. 4.47e). The results are plotted as critical initial pressure for successful initiation versus charge geometry, where the critical initial pressure is normalized with respect to the baseline initiator, the circular disk charge (Fig. 4.47a), and the planar charge (Fig. 4.47d), respectively. Since $E_c \sim 1/P_0$, the distributed initiator scheme in the regions below a normalized pressure of 1 has higher efficiency.

The photograph in Fig. 4.50 illustrates the central initiation of a cylindrical detonation wave by a multilines scheme (Fig. 4.47c) at lower mixture initial pressure than the critical pressure for initiation using a homogeneous circular disk initiator (Fig. 4.47a), thus indicating higher efficiency of initiation due to cumulative interactions of shock waves from the distributed initiators.

A combination of the first two initiation schemes listed at the beginning of this section results in sequential triggering of distributed initiators, for example, a scheme with a central charge surrounded by an annular charge. The central charge is initiated first generating a blast wave and then the annular charge is initiated with some optimal delay relative to the position of the decaying blast wave. Mathematical modeling reveals that the two-dimensional circular initiation causes energy cumulation on the charge center to form a new blast wave which enhances the initiation of the material subjected to the

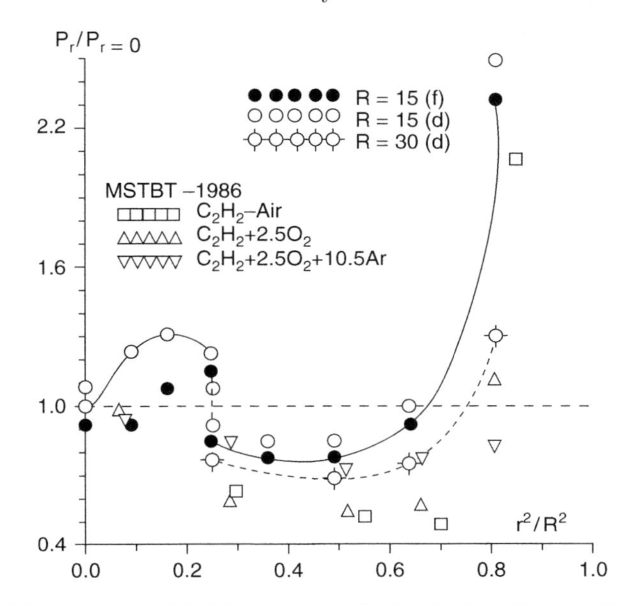

Fig. 4.48. Mixture critical initial pressure for initiation of a quasispherical detonation by an annular charge (Fig. 4.47b), normalized to the critical initial pressure of a circular disk charge of the same diameter (Fig. 4.47a) versus the square of the ratio of inner to outer radius of the annulus. Experimental data from $R = 15$ and 30 mm, with index "*f*" for decaying processes and index "*d*" for reinitiation of detonation (*upper legend*); the MSTBT legend from data of [104]. The region below the horizontal *dashed line* indicates higher efficiency of annular charges

primary central explosion. Under certain conditions, this two-stage initiation scheme appears more effective than a single stage explosion of a homogeneous disk charge.

The results of these investigations demonstrate that the behavior of the spatial factor of initiator energy E_r is not the same as the temporal factor of the initiator E_t. While $E_t = \text{const} = E_c$ in a region of $t_0 \le t_*$ (see Fig. 4.19 and Sect. 4.3.1), the dependence of E_r on the initiator charge distribution radius shows a U-shaped curve with a minimal value $E_{\min} < E_c$ at some optimized distribution of the initiator for a given mixture. The greatest efficiency (in comparison with the concentrated charge) typically occurs for the annular initiators, triangle liners, and multipointed initiators. At an optimized distribution, the efficiency can be increased as much as up to one order of magnitude.

The efficiency of detonation initiation can also be increased by injection of hot or reactive substances including ionized substances into the induction zone of the decaying blast wave generated by a single initiator. Such particles serve to promote initiation. It must be noted that in any posttriggering scheme the energy of the first initiator is much lower than the critical initiation energy.

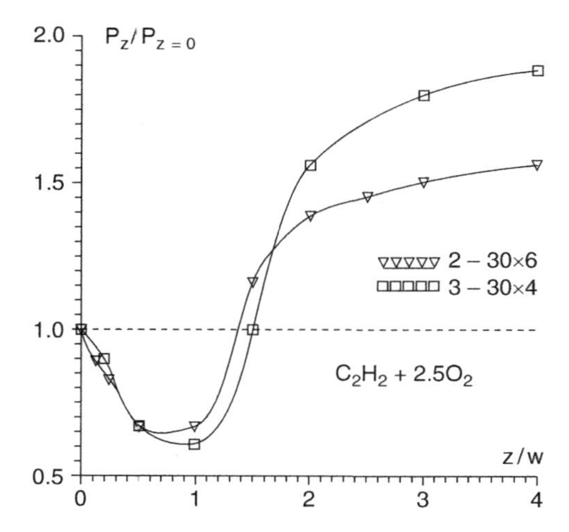

Fig. 4.49. Mixture critical initial pressure for initiation of a quasicylindrical detonation by parallel line charges (Fig. 4.47e), normalized to the initial pressure of the rectangular charge (Fig. 4.47d), versus the distance between the charges normalized to the width of the line charge. The two curves correspond to the data for two ($H \times w = 30 \times 6$ mm) and three ($H \times w = 30 \times 4$ mm) parallel line charges but with equal area of the total cross sections. The region below the horizontal *dashed line* indicates higher efficiency of parallel line charges

Fig. 4.50. Self-luminosity photo of cylindrical detonation initiation by a multipoint scheme (Fig. 4.47c): mixture, $C_2H_2/H_2/O_2$

Detonation initiation in a mixture with gradients of parameters is interesting both from a fundamental point of view and for its potential to optimize applications; however, it remains a challenging area of research.

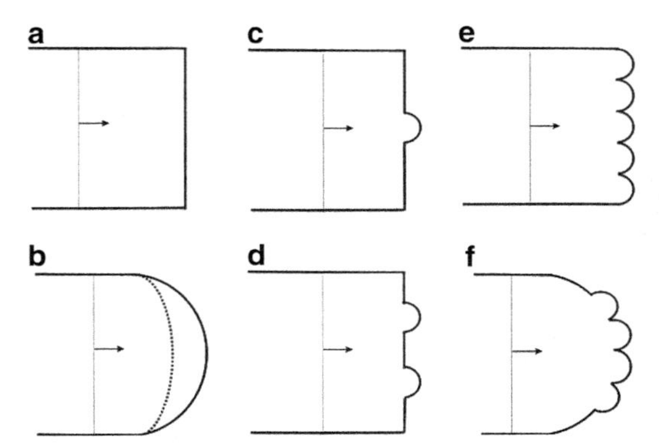

Fig. 4.51. Concave surface configurations for the optimization of mixture initiation at shock reflection: (**a**) a baseline planar reflector; (**b**) an elliptical reflector over the cross section; (**c**) a localized elliptical reflector in the cross section; (**d**) two localized elliptical reflectors; (**e**) a series of elliptical reflectors; (**f**) a series of elliptical reflectors on a concave elliptical line

Relatively weak shocks can be made stronger when they are reflected off a planar surface. They can be made even stronger, locally, when they are reflected off a concave surface, due to wave focussing effects. This shock enhancement through reflection and focusing prompts the self-ignition processes. Some multifocusing configurations are shown in Fig. 4.51. The critical incident shock Mach number for successful initiation can be significantly decreased in a multifocusing system.

A sharp increase in temperature, pressure, and density in a local region of a decaying wave front is possible through reflection from obstacles. These obstacles can be solid (with a characteristic size much smaller than the channel size) or perforated (e.g., metallic mesh). Such local obstacles are particularly effective near the critical initiation conditions. Figure 4.52 is an example of a successful reinitiation via reflection of the decaying expanding wave from a localized small obstacle.

The influence of the initiator geometry is shown in Fig. 4.53, where the critical initiation energy from an ideal blast ($\nu = 1, 2$, and 3 for planar, cylindrical, and spherical geometries, respectively) in the stoichiometric hydrogen–oxygen mixtures is calculated as a function of initial pressure using (4.29)–(4.31). At lower initial pressures, the critical initiation energy for the planar detonation wave is the lowest, while at larger initial pressures the critical initiation energy for the cylindrical and spherical detonation waves becomes less than the planar case [183]. This effect of geometry is generally valid for any mixture.

In the case of a long bar initiator charge to initiate a cylindrical detonation, the effect of varying bar length, H, and width, l, at constant charge mass is illustrated in Fig. 4.54. In the figure, the experimental critical initiation charge

Fig. 4.52. Self-luminous photograph of the reinitiation of a subcritical diffraction wave via reflection on a localized obstacle near the expanding cone

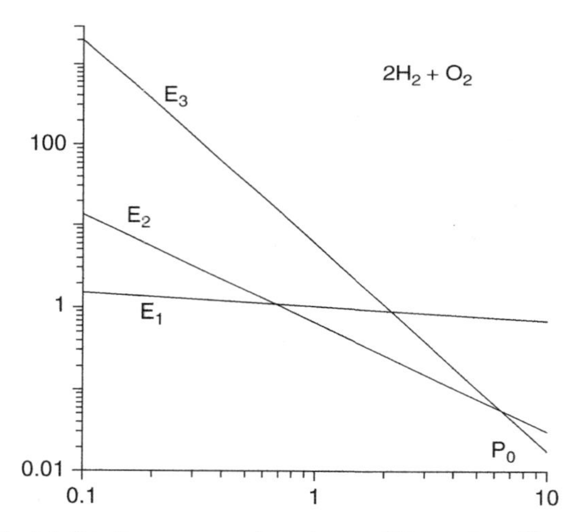

Fig. 4.53. Critical initiation energy for planar (E_1 in J m^{-2}), cylindrical (E_2 in J cm^{-1}) and spherical (E_3 in J) detonation wave in $2H_2 + O_2$ mixtures versus initial pressure (in atm)

width normalized by the cell width, λ, is plotted versus the ratio of length to width, $k = H/l$. From the above examples, it is clear that the geometry of the actual initiator will influence the critical initiation of detonation. Optimization of the critical initiation energy will depend on all aspects of the spatial and time characteristics of the initiator and the explosive mixture.

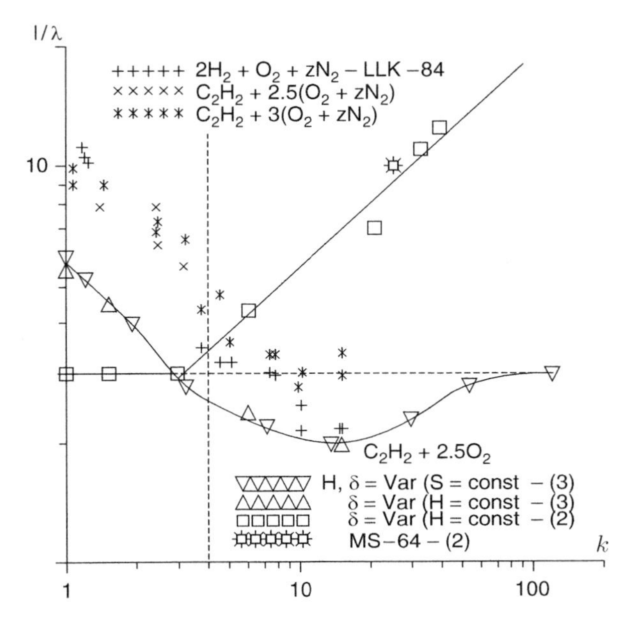

Fig. 4.54. Critical width of a long bar initiation charge normalized to cell width (l/λ) versus the ratio of initiation charge length to width (k) for successful detonation in various mixtures. Experimental data: LLK, [87]; MS, [101]; *other symbols*, author's data using different DRMD schemes through diffraction into space (3) or into wider channel (2) by varying charge length H, width l, or area $S = H \times l$

4.6 Critical Sizes for Limiting Detonations

The propagation regime of cellular detonations in mixtures well within the concentration limits can become critical if the characteristic size of the detonation tube approaches the characteristic length scale of the detonation. This device size limit is discussed below.

The cross section-averaged properties of a detonation wave propagate in a steady manner, but only when this average is calculated for a large enough number of cell widths. Thus, a steady detonation wave can exist only if the cross-sectional dimension of a detonation tube or channel is greater or equal to the detonation cell width, λ (e.g., [54, 159, 160, 169]). As a rule, the cell width increases as initial pressure decreases (inverse proportionality). Hence, the number of transverse waves in the cross section of a tube decreases with decreasing P_0.

The *spinning* detonation with a single transverse wave in the front structure represents the limiting regime for the propagation of steady detonation in a circular tube; therefore,

$$d_s \cong \lambda/\pi, \tag{4.53}$$

where d_s is defined as the spinning detonation tube diameter.

For a rectangular channel with a cross-sectional length and width, $l \times \delta$, the limiting (or marginal) regime of steady detonation (e.g., [105, 137, 138, 140]) is defined by the relationship

$$l_s = (k+1)\lambda/2\pi, \qquad (4.54)$$

with l_s being the limiting channel size, where $k = l/\delta$ [159].

When the tube dimension is below that for the limiting regime of steady detonation, there exits a *galloping* detonation which manifests itself in a longitudinal pulsation cycle, observed in long tubes [42, 91, 92, 150, 169]. The cycle begins with a strongly overdriven detonation that continuously decays. Consequently, the combustion front lags behind the leading shock front as the induction zone length grows. At a certain time, a local explosion occurs within the reaction zone and develops into a detonation wave in the compressed and still unreacted gas downstream of the shock front. This secondary detonation catches up with the leading shock front, forming a new overdriven detonation wave. The cycle then repeats. Galloping detonation is essentially a series of deflagration-to-detonation transition (DDT) cycles. The spatial frequency of the one-dimensional pulsations spans hundreds of tube diameters. The average propagation velocity of the galloping detonation is 10–30% lower than the CJ detonation velocity.

Ulianitsky established an approximate analytical model for galloping detonations [118]. The model provides a satisfactory description by assuming that the oscillations are regular with a mean propagation velocity equal to the experimental value and that the deflagration is instantly transformed into a detonation after an induction period in the gas compressed by the decaying shock front.

The limiting tube diameter for galloping detonation can be expressed in the following form:

$$d_g \approx 2\Delta_{\mathrm{I}}, \qquad (4.55)$$

where Δ_{I} is induction length behind the CJ detonation wave. Below this diameter, the galloping mode cannot be sustained.

The so-called *low-velocity* detonation or *quasidetonation* regime is experimentally observed for tube dimensions smaller than for the *galloping* detonation regime (e.g., [159]). It is a supersonic wave regime with the average propagation velocity approximately equal to one half of the CJ detonation velocity. The low-velocity detonation can exist, for instance, in rough tubes with large obstacles or in narrow smooth tubes such as glass capillaries [92]. The wave structure of low-velocity detonation is caused by transport processes near the flame surface rather than by accumulation of active centers in the induction zone behind the shock front. The flame is located at a distance of three to eight tube diameters behind the shock front and has the shape of an almost flat disk in the flow core adjoined with an oblique flame in the boundary layer. Note that the length of a cycle of oscillation based on the

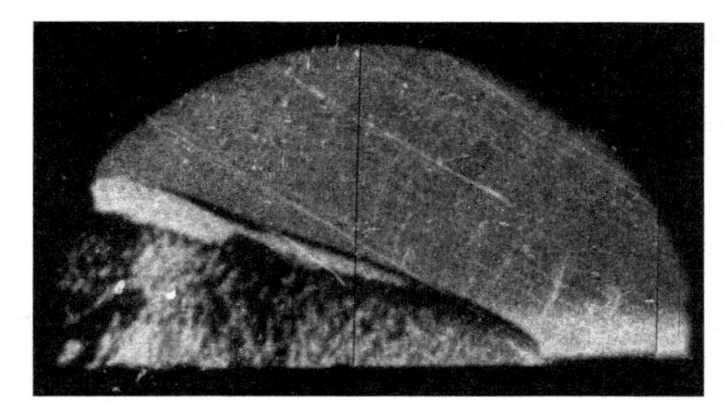

Fig. 4.55. Schlieren photograph of detonation propagation through a gas layer free at the upper boundary

flame velocity equals 160 capillary tube diameters [92], which are many times greater than the distance between the flame and the shock front.

The limiting tube diameter for low-velocity detonations in smooth tubes may be expressed in the following form:

$$d_{lv} \approx \Delta_{\mathrm{I}}, \tag{4.56}$$

where Δ_{I} is induction length behind the CJ detonation wave.

Finally, in a free cylindrical charge without any tube confinement, a critical charge diameter $d_{\min}$ is defined as the minimum charge diameter, above which $(d > d_{\min})$ a steady detonation can propagate in any length of charge once the detonation is initiated. For $d < d_{\min}$ the detonation fails and thereafter only unstable high-velocity combustion is observed for any strong initiator power. The values of $d_{\min}$ for a condensed-phase high explosives and for a gas mixture vary from millimeters to tens of centimeters.

The limiting detonation in a free gaseous charge is usually a cellular detonation wave, that is, spinning, galloping, and quasidetonations do not form in free charges. Figure 4.55 shows an instantaneous schlieren photograph of the propagation of a near-limiting detonation wave from left to right through a free reactive gas layer on a planar surface. The detonation front is attached by a lateral shock wave that propagates into the surrounding air.

For a free subsonic gaseous jet, Vasil'ev and Zak [178] established that if the jet diameter d exceeds critical value $d_{\min}$, then the detonation propagates along the charge axis in a self-sustaining regime. Based on experimental observations, an approximate relation between the critical diameter of a free gaseous charge and the gaseous charge in the steel tube for diffraction initiation is

$$d_{\min} \approx 2.5 d_c. \tag{4.57}$$

4.7 Concluding Remarks

A fundamental analysis for the most important dynamic parameters of cellular detonation has been described, including the detonation cell size, the critical energy for direct initiation of detonation, and the critical diameter (or width) for diffraction initiation of a spherically (or cylindrically) expanding detonation. Critical phenomena and related properties equivalent to critical direct initiation or critical diffraction initiation have also been introduced. Geometrical limits for maintaining the propagation of limiting detonations have been briefly presented.

The analysis and experimental evidence have shown that the most adequate models correlating the dynamic parameters have the common functional form of $E_c \sim \rho_0 D_{\mathrm{CJ}}^2 \lambda^{\nu}$. These models include the "multiple transverse wave collision point initiation" (MPI) model, the "diffraction re-initiation of multifront detonation" (DRMD) model by Vasil'ev et al. [162], the model of Zhdan and Mitrofanov [200], and the "surface energy" model by Lee et al. [18]. They capture the essential detonation phenomena of the transverse wave collisions as local explosions to form detonation cells and the energy equivalence between the point spherical explosion initiation and planar detonation diffraction initiation.

There are still many challenging problems unresolved for the detonation dynamic parameters. Figure 4.56 indicates that, as one of the diffraction initiation modes, DDT was observed due to the influence of rarefaction waves on the flame front near the center of the expansion cone. DDT due to the rarefaction wave effect seems unusual, since traditionally the rarefaction wave would weaken the flame. Figure 4.57 shows another example of a DDT event in the expanding wave after the transmission of the flame front through the orifice. Another unusual phenomenon is in the subcritical regime, where the decaying of a detonation wave at diffraction is accompanied with the complete destruction of not only the detonation but also the flame luminosity as

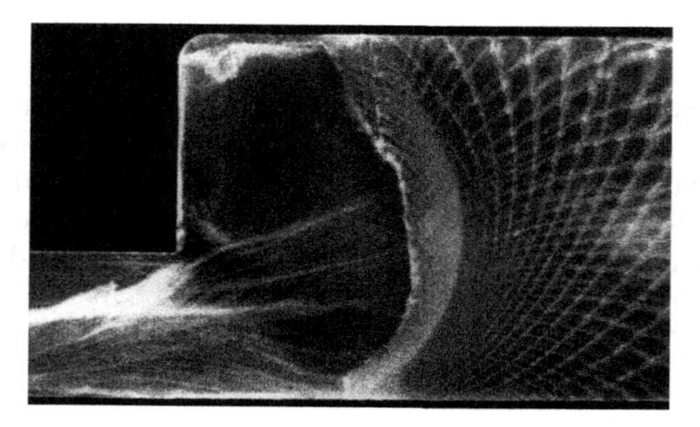

Fig. 4.56. Spontaneous initiation of cylindrical expanding detonation at a sharp expansion of the flame front

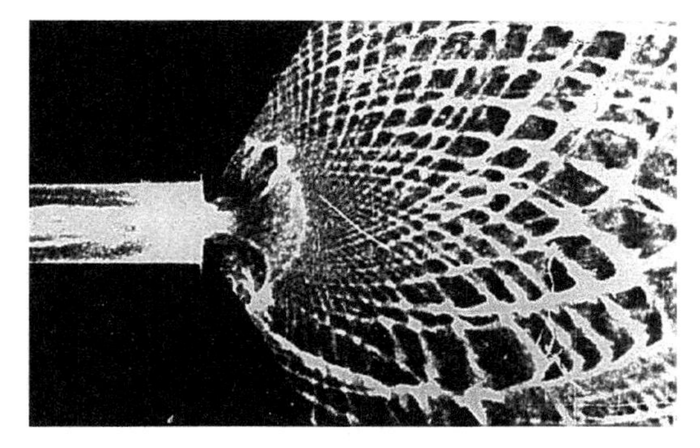

Fig. 4.57. Spontaneous initiation of cylindrical expanding detonation at a sharp expansion through an orifice after reflection–compression of flame front at the end wall of the channel

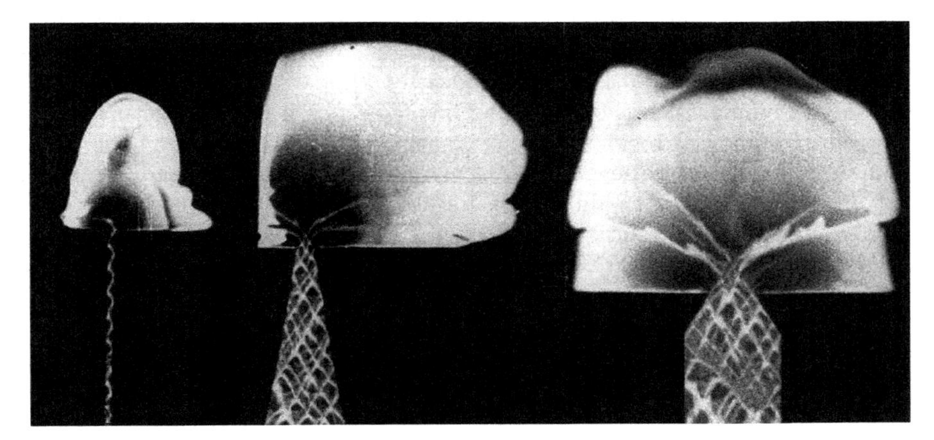

Fig. 4.58. Complete destruction of detonation and flame at diffraction for marginal detonation (*left*) and overdriven detonation (*middle* and *right*)

displayed in Fig. 4.58. The complete destruction of the flame front has been observed for both marginal and overdriven detonations. These recent critical phenomena will need further numerical and experimental investigations.

For detonation in a moving mixture, especially at a supersonic speed, it has been established that the detonation velocity increases or decreases and the cell length decreases or increases as the detonation propagates upstream or downstream, respectively (compared to the CJ detonation in a mixture at rest initially) [190]. The dynamic parameters and near-limiting behaviors in a supersonic moving mixture are interesting subjects for future investigation.

Another interesting area is the cellular detonation structure in heterogeneous detonation (for example, [197]). In particular, the detonation transforms a homogeneous mixture to a heterogeneous state during the chemical reaction, such as with carbon condensation in the products [43, 179]. How does the carbon condensation influence the detonation cell structure and other dynamic parameters? How different exotic structures (such as nanotubes, fullerenes, and clusters) are generated when condensation and coagulation of carbon atoms occur in detonation products?

Finally, the recent discovery of new types of cellular structures including diagonal structure in a rectangular channel [43, 179], and particularly a double-sized cellular detonation structure, has attracted wide attention, the details of which can be found in Chap. 5. In contrast to the traditional cellular detonation structure where a dominant cell size is dispersed to a size spectrum depending on the inherent instabilities of a given mixture, in the double-sized cellular structure large cells are superimposed on smaller cells. The existence of the double-sized cellular structure has been found in many mixtures with a nonmonotonic energy release, some of which immediately begin at the detonation front [176]. The double-sized cellular detonation might be a part of a more comprehensive detonation structure regarding the characteristic cell sizes in multifuel systems with widely different chemical kinetics of the individual fuels. For example, how does the cellular structure appear in the methane–hydrogen system by varying the two-fuel ratio in the oxidizing gas? This class of new cellular structure will motivate research involving the direct initiation and diffraction initiation of detonation, relevant critical parameters in their relations to the cell sizes, and the geometrical limits of the double-sized structure in comparison with that of a single-sized cellular detonation.

Appendix A: Explosion Parameters

Explosion parameters for various reactive gas mixtures are listed in Table 4.2, with D_{CJ}, CJ detonation velocity in m/s and P_{CJ}, CJ detonation pressure normalized to the initial pressure ($P_0 = 1.0\,\mathrm{atm}$), T_{CJ}, CJ detonation temperatures in K; P_v, constant volume explosion pressure normalized to $P_0 = 1.0\,\mathrm{atm}$; T_v, constant volume explosion temperature in K; T_p, constant pressure combustion temperature in K; and E_3, critical energy for direct initiation of spherical detonation in J. Indices used in the table are: g, gaseous carbon in the products only; c, gaseous and condensed carbon in the products; st, stoichiometric; and lq, initial fuel in liquid phase. The critical initiation energy E_3 is calculated using the MPI model; all the other parameters are calculated using the standard thermochemical equilibrium methods.

Table 4.2. Computed explosion parameters for reactive gas mixtures in various combustion processes

Mixture	D_{CJ}	P_{CJ}	T_{CJ}	P_v	T_v	T_p	E_3
$2H_2 + O_2$	2,837	18.79	3,682	9.59	3,504	3,079	6
$2CO + O_2$	1,798	18.48	3,523	9.45	3,358	2,976	3,777
$1.6CO + O_2 + 0.4H_2$	1,905	18.3	3,501	9.36	3,340	2,962	765
$1.2CO + O_2 + 0.8H_2$	2,049	18.33	3,530	9.37	3,365	2,980	162
$CO + O_2 + H_2$	2,136	18.37	3,550	9.39	3,383	2,993	80
$0.8CO + O_2 + 1.2H_2$	2,236	18.43	3,573	9.42	3,404	3,008	42
$0.4CO + O_2 + 1.6H_2$	2,485	18.59	3,624	9.5	3,453	3,042	14
$2CO + O_2 + 10\%H_2$	1,880	18.33	3,500	9.37	3,338	2,960	1,623
$2CO + O_2 + 25\%H_2$	2,042	18.26	3,503	9.34	3,334	2,950	612
$2CO + O_2 + 50\%H_2$	2,381	17.08	3,261	8.75	3,054	2,645	958
$2CO + O_2 + 75\%H_2$	2,707	12.02	2,260	6.22	2,024	1,650	1.01E + 06
$C_2H_2 + 2.5O_2$ – g	2,424	33.83	4,214	17.07	3,980	3,342	1.10E – 02
$C_2H_2 + O_2$ – g	2,934	46.01	4,507	23.18	4,161	3,324	1.76E – 03
$11C_2H_2 + 9O_2$ – g	2,730	39.96	4,117	20.12	3,827	3,133	1.05E – 02
$3C_2H_2 + 2O_2$ – c	2,542	34.54	3,772	17.65	3,552	3,113	6.95E – 02
$13C_2H_2 + 7O_2$ – c	2,440	32.76	3,632	16.59	3,480	3,073	1.63E – 01
$3C_2H_2 + O_2$ – c	2,304	28.81	3,476	14.63	3,342	2,995	1.16E + 00
$19C_2H_2 + O_2$ – c	2,042	21.97	3,194	11.22	3,094	2,848	3.69E + 02
C_2H_2 – c	1,977	20.44	3,130	10.46	3,036	2,812	
$0.4C_2H_2$ + air(st) – g	1,866	19.1	3,112	9.77	2,918	2,538	1.56E + 03
C_2H_2 + air – g	2,048	21.85	3,115	11.11	2,811	2,269	3.49E + 02
$1.2C_2H_2$ + air – c	1,978	20.42	2,924	10.48	2,662	2,265	9.31E + 02
$1.5C_2H_2$ + air – c	1,951	20.63	2,877	10.51	2,690	2,319	9.48E + 02
$2C_2H_2$ + air – c	1,953	20.71	2,897	10.56	2,727	2,387	1.05E + 03
$3C_2H_2$ + air – c	1,956	20.76	2,927	10.6	2,777	2,473	1.42E + 03
$4.76C_2H_2$ + air – c	1,959	20.74	2,960	10.6	2,829	2,556	2.47E + 03
$7C_2H_2$ + air – c	1,961	20.68	2,987	10.57	2,867	2,612	4.51E + 03
$CH_4 + 2O_2$	2,390	29.3	3,725	14.8	3,541	3,053	200
$C_2H_6 + 3.5C_2H_6 + 3.5O_2$	2,369	33.97	3,800	17.15	3,609	3,084	0.54
$C_3H_8 + 5O_2$	2,357	36.2	3,827	18.26	3,634	3,094	0.62
$C_4H_{10} + 6.5O_2$	2,350	37.5	3,841	18.9	3,646	3,099	0.78
$C_5H_{12} + 8O_2$	2,346	38.4	3,851	19.4	3,655	3,102	0.76
$C_6H_{14} + 9.5O_2$	2,343	39	3,858	19.7	3,661	3,104	0.93
$C_7H_{16} + 11O_2$	2,342	39.5	3,863	19.9	3,665	3,106	1.33
$C_8H_{18} + 12.5O_2$	2,340	39.9	3,866	20.1	3,669	3,107	1.54
$C_{10}H_{22} + 15.5O_2$	2,338	40.4	3,872	20.4	3,674	3,109	3.3
$C_{10}H_{22} + 15.5O_2$(lq)	2,302	39.2	3,814	19.7	3,622		1.5
H_2 + air(st)	1,969	15.58	2,945	8	2,749	2,380	3.36E + 03
CO + air(st)	1,671	15.49	2,864	7.96	2,699	2,383	2.60E + 07
C_2H_2 + air(st)	1,866	19.1	3,112	9.77	2,918	2,538	1.56E + 03
CH_4 – air(st)	1,803	17.2	2,778	8.8	2,585	2,224	7.06E + 07
C_2H_6 + air(st)	1,802	17.95	2,812	9.19	2,620	2,257	7.70E + 04
C_3H_8 + air(st)	1,800	18.24	2,821	9.33	2,629	2,266	1.39E + 05
C_4H_{10} + air(st)	1,798	18.39	2,823	9.4	2,632	2,268	2.23E + 05
C_7H_{16} – air(st)	1,796	18.6	2,229	9.5	2,638	2,274	5.39E + 05
C_8H_{18} – air(st)	1,796	18.6	2,830	9.5	2,639	2,274	6.65E + 05

References

1. Abid, S., Dupre, G., Paillard, C.: Oxidation of gaseous unsymmetrical dimethylhydrazine at high temperatures and detonation of UDMH/O_2 mixtures. In: Progress in Astronautics and Aeronautics, vol. 153, pp. 162–181. American Institute of Aeronautics and Astronautics, VA (1991)
2. Afanas'ev, A.N., Bortnikov, L.N.: Reduction of motor toxic by hydrogen addition. In: Proceedings of the 5th International Conference on Technologies and Combustion for a Clean Environment, Lisbon, Portugal, vol. 2, pp. 1075–1077 (1999)
3. Alekseev, V., Dorofeev, S., Sidorov, V.: Direct initiation of detonations in unconfined gasoline sprays. Shock Waves **6**, 67–71 (1996)
4. Alpert, R.L., Toong, T.: Periodicity in exothermic hypersonic flows about projectiles. Astronaut. Acta **17**(4–5), 539–560 (1972)
5. Aminallah, M., Brossard, J., Vasil'ev A.: Cylindrical detonations in methane-oxygen-nitrogen mixtures. In: Progress in Astronautics and Aeronautics, vol. 153, pp. 203–228. American Institute of Aeronautics and Astronautics, VA (1993)
6. Atkinson R., Bull D.C., Shuff R.I.: Initiation of spherical detonation in hydrogen–air. Combust. Flame **39**, 287–300 (1980)
7. Austin, J.M., Shepherd, J.E.: Detonations in hydrocarbon fuel blends. Combust. Flame **132**, 73–90 (2003)
8. Bach, G.G., Knystautas, R., Lee, J.H.: Initiation criteria for diverging gaseous detonation. In: 13th Symposium (International) on Combustion, pp. 1097–1110. The Combustion Institute, Pittsburgh (1970)
9. Bannikov, N.V., Vasil'ev, A.A.: Multipoints ignition of gaseous mixture and its influence on transition of deflagration to detonation. Combust. Explo. Shock Waves **28**(3), 65–69 (1992)
10. Baratov, A.N., Korolchenko, A.Ya. (eds.): Fire-Explosion Hazard of Substances: Both Materials and Means for Suppression (in Russian). Khimia, Moscow (1990)
11. Bauer, P., Brochet, C., Presles H.N.: The influence of initial pressure on critical diameters of gaseous explosive mixtures. In: Progress in Astronautics and Aeronautics, vol. 94, pp. 118–129. American Institute of Aeronautics and Astronautics, VA (1984)
12. Bauer, P., Presles, H.N., Heuze, O., Brochet C.: Measurement of cell lengths in the detonation front of hydrocarbon oxygen and nitrogen mixtures at elevated initial pressures. Combust. Flame **64**(1), 113–123 (1986)
13. Bauer, P.A., Dabora, E.K., Manson, N.: Chronology of early research on detonation wave. In: Progress in Astronautics and Aeronautics, vol. 133, pp. 3–18. American Institute of Aeronautics and Astronautics, VA (1991)
14. Bauwens, L., Williams, D.N., Nikolic, M.: Failure and reignition of one-dimensional detonations: The high activation energy limit. In: Twenty-seventh Symposium (International) on Combustion. The Combustion Institute, Pittsburgh (1998)
15. Beeson, H.D., McClenagan, R.D., Bishop, C.V., Benz, F.J., Pitz, W.J., Westbrook, C.K., Lee J.H.S.: Detonability of hydrocarbon fuels in air. In: Progress in Astronautics and Aeronautics, vol. 133, pp. 19–36. American Institute of Aeronautics and Astronautics, VA (1991)

16. Behrens, H., Struth, W., Wecken W.: Studies of hypervelocity firings into mixtures of hydrogen with air or with oxygen. In: 10th Symposium (International) on Combustion. Combustion Institute, Pittsburgh (1964)
17. Benedick, W.B., Knystautas, R., Lee J.H.S.: Large-scale experiments on the transmission of fuel-air detonations from two-dimensional channels. In: Progress in Astronautics and Aeronautics, vol. 94, pp. 546–555. American Institute of Aeronautics and Astronautics, VA (1984)
18. Benedick, W.B., Guirao C.M., Knystautas R., Lee J.H.: Critical charge for direct initiation of detonation in gaseous fuel-air mixtures. In: Progress in Astronautics and Aeronautics, vol. 106, pp. 181–202. American Institute of Aeronautics and Astronautics, VA (1986)
19. Bohon, Yu.A., Shulenin, Yu.V.: The minimal initiation energy for spherical detonation for some hydrogen mixtures (in Russian). Rep. USSR Acad. Sci. **245**(3), 623–626 (1979)
20. Borisov, A.A., Zamansky, V.M., Lisyansky, V.V., Skachkov, G.I., Troshin, K.Ya.: The estimation of critical initiation energy for detonable gaseous system by the ignition delays (in Russian). Chem. Phys. **5**(12), 1683–1689 (1986)
21. Borisov, A.A., Khomik, S.V., Mikhalkin, V.N., Saneev E.V.: Critical energy of direct detonation initiation in gaseous mixtures. In: Progress in Astronautics and Aeronautics, vol. 133, pp. 142–155. American Institute of Aeronautics and Astronautics, VA (1991)
22. Borissov, A.A., Sharypov, O.V.: Physical model of dynamic structure of the surface of detonation wave. In: Borrisov, A. (ed.) Dynamic Structure of Detonation in Gaseous and Dispersed Media. Fluid Mechanics and its Applications, vol. 5, pp. 27–49. Kluwer, Dordrecht (1991)
23. Bull, D.C., Elsworth, J.E., Hooper, G., Quinn, C.P.: A study of spherical detonation in mixtures of methane and oxygen diluted by nitrogen. J. Phys. D: Appl. Phys. **9**, 1191 (1976)
24. Bull, D.C., Elsworth, I.E., Hooper, G.: Initiation of spherical detonation in hydrocarbon–air mixtures. Acta Astronaut. **5**, 997–1008 (1978)
25. Bull, D.C., Elsworth, J.E., Shuff, P.J., Metcalfe, E.: Detonation cell structures in fuel-air mixtures. Combust. Flame **45**(1), 7–22 (1982)
26. Bykovskii, F.A., Zhdan, S.A., Vedernikov, E.F.: Continuous spin detonations. J. Propul. Power. **22**(6), 1204–1216 (2006)
27. Carlson, G.A.: Spherical detonations in gas-oxygen mixtures. Combust. Flame **21**(3), 383–385 (1973)
28. Chernayvsky, S.Yu., Baulin, N.N., Mkrtumov, A.S. Flowing of high-speed bullet by hydrogen-oxygen mixture. Combust. Explo. Shock Waves **9**(6), 786–791 (1973)
29. Chernyi, G.G.: Gas Flow with Large Supersonic Speed (in Russian). Fizmatgiz, Moscow (1959)
30. Chernyi, G.G., Chernayvsky, S.Yu., Baulin, N.N.: Bullet moving with high velocity in mixture of hydrogen-air. Dokladi Akademii Nauk SSSR (in Russian) **290**(1), 44–47 (1986)
31. Ciccarelli, G., Ginsberg, T., Boccio, J., Economos, C., Sato, K., Kinoshita M.: Detonation cell size measurements and predictions in hydrogen–air–steam mixtures at elevated temperatures. Combust. Flame **99**(2), 212–220 (1994)
32. Clavin, P., He, L., Williams, F.A.: Multidimensional stability analysis of overdriven gaseous detonations. Phys. Fluids **9**(12), 3764–3785 (1997)

33. Deledisque, V., Papalexandris, M.V.: Numerical analysis of the rectangular and diagonal structures in three-dimensional detonations. In: Proceedings of the European Combustion Meeting, Louvain Catolique University, Belgium (2005)
34. Denisov, Yu.N., Troshin, Ya.K.: Pulsating and spinning detonation of gaseous mixtures in tubes. Dokl. Akad. Nauk SSSR. **125**(1), 110–113 (1959)
35. Desbordes, D.: Transmission of overdriven plane detonations: Critical diameter as a function of cell regularity and size. In: Progress in Astronautics and Aeronautics, vol. 114, pp. 170–185. American Institute of Aeronautics and Astronautics, VA (1988)
36. Desbordes, D., Vachon, M.: Critical diameter of diffraction for strong plane detonations. In: Progress in Astronautics and Aeronautics, vol. 106, pp. 131–143. American Institute of Aeronautics and Astronautics, VA (1986)
37. Desbordes, D., Guerraud, C., Hamada, L., Presles, H.N.: Failure of the classical dynamic parameters relationships in highly regular cellular detonation systems. In: Progress in Astronautics and Aeronautics, vol. 153, pp. 347–359. American Institute of Aeronautics and Astronautics, VA (1993)
38. Eckett, C.A., Quirk, J.J., Shepherd, J.E.: The role of unsteadiness in direct initiation of gaseous detonation. J. Fluid Mech. **421**, 147–183 (2000)
39. Edwards, D.H., Hooper G., Morgan J.M.: An experimental investigation of the direct initiation of spherical detonation. Acta Astronaut. **3**, 117–130 (1976)
40. Edwards, D.H., Thomas, G.O., Nettleton, M.A.: The diffraction of a planar detonation wave at an abrupt area change. J. Fluid Mech. **95**(1), 79–96 (1979)
41. Edwards, D.H., Thomas, G.O., Nettleton, M.A.: Diffraction of a planar detonation in various fuel-oxygen mixtures at an area change. In: Progress in Astronautics and Aeronautics, vol. 75, pp. 341–357. American Institute of Aeronautics and Astronautics, VA (1981)
42. Elsworth, J.E., Shuff, P.J., Ungut, A.: "Galloping" gas detonations in the spherical mode. In: Progress in Astronautics and Aeronautics, vol. 94, pp. 130–150. American Institute of Aeronautics and Astronautics, VA (1984)
43. Emelianov A., Eremin A., Fortov V., Makeich A., Jander H., Deppe J.: Detonation wave driven by condensation. In: Proceedings of the 4th European Combustion Meeting. CD, Vienna University of Technology, P810055, Vienna (2009)
44. Erpenbeck, J.J.: Stability of idealized one-reaction detonations. Phys. Fluids. **7**, 684–696 (1964)
45. Erpenbeck, J.J.: Detonation stability for disturbances of small transverse wavelength. Phys. Fluids. **9**(7), 1293–1306 (1966)
46. Fickett, W., Davis, W.C.: Detonation. University of California Press, Berkeley (1979)
47. Fischer, M., Pantow, E., Kratzel, T.: Propagation, Decay and re-ignition of detonations in technical structures. In: Roy, G., Frolov, S., Kailasanath, K., Smirnov, N. (eds.) Gaseous and Heterogeneous Detonations, Science to Applications, pp. 197–212. ENAS Publishers, Moscow (1999)
48. Freiwald, H., Koch, H.W.: Spherical detonation on acetylene-oxygen-nitrogen mixtures as a function of nature and strength of initiation. In: 9th Symposium (International) on Combustion, pp. 275–281. Academic, New York (1963)
49. Frolov, S.M., Basevich, V.Y., Vasil'ev, A.A.: Dual-fuel concept for advanced propulsion. In: Roy, G., Frolov, S., Netzer, D., Borisov, A. (eds.) High-Speed

Deflagration and Detonation. Fundamental and Control, pp. 315–332. ELEX-KM Publishers, Moscow (2001)
50. Fry, R.S., Nicholls, J.A.: Blast initiation and propagation of cylindrical detonations in MAPP-Air mixtures. AIAA J. **12**(12), 1703–1708 (1974)
51. Fujiwara, T., Reddy, K.V.: Propagation mechanism of detonation: Three dimensional phenomenon. In: Memoirs of the Faculty of Engineering, vol. 41(1), pp. 93–111. Nagoya University, Nagoya (1989)
52. Gavrikov, A.I., Efimenko, A.A., Dorofeev, S.B.: A model for detonation cell size prediction from chemical kinetics. Combust. Flame **120**, 19–33 (2000)
53. Gavrilenko, T.P., Prokhorov, E.S.: Overdriven gaseous detonations. In: Progress in Astronautics and Aeronautics, vol. 87, pp. 244–250. American Institute of Aeronautics and Astronautics, VA (1983)
54. Gelfand, B.E., Frolov, S.M., Nettleton M.A.: Gaseous detonations – a selective review. Prog. Energy Combust. Sci. **17**, 327–371 (1991)
55. Gilinsky, S.N., Zapraynov, Z.D., Chernyi, G.G.: Supersonic flowing of sphere by combustible mixture. Izvestia Akademii Nauk SSSR: Mechanics of liquid and gas (in Russian) **5**, 8–13 (1966)
56. Guilly, V., Khasainov, B., Presles, H.N., Desbordes, D., Vidal P.: Numerical study of detonation cells under non-monotonous heat release. In: CD: Proceeding of the 20th International Colloquium on the Dynamics of Explosion and Reactive Systems (ICDERS), McGill University, Montreal (2005)
57. Guirao, C.M., Knystautas, R., Lee, J.H., Benedick, W., Berman M.: Hydrogen-air detonations. In: 19th Symposium (International) on Combustion, pp. 583–590. The Combustion Institute, Pittsburgh (1982)
58. Hanana, M., Lefebvre, M.H., Van Tiggelen, P.J.: On rectangular and diagonal three-dimensional structures of detonation waves. In: Roy, G., Frolov, S., Kailasanath, K., Smirnov N. (eds.) Gaseous and Heterogeneous Detonations: Science to Applications, pp. 121–130. ENAS Publishers, Moscow (1999)
59. He, L.: Theoretical determination of the critical conditions for the direct initiation of detonation in hydrogen–oxygen mixtures. Combust. Flame **104**, 401–418 (1996)
60. He, L., Clavin, P.: On the direct initiation of gaseous detonations by an energy source. J. Fluid Mech. **277**, 227–248 (1994)
61. Ichikawa, T., Matsui, A.: Study of cell width and shock pressure in directly initiated spherical detonation. In: CD: Proceedings of the 22nd International Colloquium on the Dynamics of Explosion and Reactive Systems (ICDERS), Lyakov Institute of Heat and Mass Transfer, Minsk (2009)
62. Jackson, S., Shepherd, J.: Toroidal imploding detonation wave initiator for pulse detonation engines. AIAA J. **45**(1), 257–270 (2007)
63. Jones, D.A., Kemister, G., Tonello, N.A.: Numerical simulation of detonation reignition in H2-O2 mixtures in area expansions. In: Conference Proceedings of the 16th International Colloquium on the Dynamics of Explosions and Reactive Systems, p. 102. University of Mining and Metallurgy, Kraków (1997)
64. Kailasanath, K., Oran, E.S., Boris, J.P., Young, T.R.: Determination of detonation cell size and the role of transverse waves in two-dimensional detonations. Combust. Flame **61**, 199–209 (1985)
65. Kaneshige, M., Schultz, E., Pfahl, U.J., Shepherd, J.E., Akbar, R.: Detonations in mixtures containing nitrous oxide. In: 22nd International Symposium on Shock Waves, pp. 251–256. Imperial College, London (1999)

66. Kaneshige, M.J.: Gaseous detonation initiation and stabilization by hypervelocity projectiles. Ph.D. thesis, California Institute of Technology (1999)
67. Khariton, Yu.B.: About detonability of explosives (in Russian). In: Voprosi teorii vzrivchatih veshestv, vol. 1, pp. 7–28, Moscow-Leningrad, AN USSR (1947)
68. Knystautas, R., Lee, J.H., Guirao, C.M.: The critical tube diameter for detonation failure in hydrocarbon–air mixtures. Combust. Flame **48**, 63–83 (1982)
69. Knystautas, R., Guirao, C., Lee, J.H., Sulmistras, A.: Measurement of cell size in hydrocarbon–air mixtures and predictions of critical tube diameter, critical initiation energy and detonation limits. In: Progress in Astronautics and Aeronautics, vol. 94, pp. 23–37. American Institute of Aeronautics and Astronautics, VA (1983)
70. Kogarko, S.M., Adyshkin, V.V., Ljamin, A.G.: Investigations of spherical detonation in gaseous mixtures. Combust. Explo. Shock Waves **2**, 22–34 (1965)
71. Korobeinikov, V.P.: Some Problems of the Point Explosion Theory in Gases (in Russian). Nauka, Moscow (1973)
72. Korobeinikov, V.P., Levin, V.A., Markov, V.V., Chernyi, G.G.: Propagation of blast waves in a combustible gas. Astronaut. Acta. **17**(4–5), 529–537 (1972)
73. Kumar, R.K.: Detonation cell widths in hydrogen–oxygen-diluent mixtures. Combust. Flame **80**(2), 157–169 (1990)
74. Lee, J.H., Knystautas, R., Yoshikawa, N.: Photochemical initiation of gaseous detonations. Acta Astronaut. **5**, 971–982 (1978)
75. Lee, J.H.: Initiation of gaseous detonation. Ann. Rev. Phys. Chem. **28**, 75–104 (1977)
76. Lee, J.H.S.: Dynamic parameters of gaseous detonations. Ann. Rev. Fluid Mech. **16**, 311–336 (1984)
77. Lee, J.H.S.: The Detonation Phenomenon. Cambridge University Press, Cambridge (2008)
78. Lee, J.H., Matsui, H.: A comparison of the critical energies for direct initiation of spherical detonations in acetylene-oxygen mixtures. Combust. Flame **28**, 61–66 (1977)
79. Lee, J.H., Ramamurthi, K.: On the concept of the critical size of a detonation kernel. Combust. Flame **27**, 331–340 (1976)
80. Lee, J.H., Lee, B.H.K., Knystautas, R.: Direct initiation of cylindrical gaseous detonations. Phys. Fluids **9**(1), 221–222 (1966)
81. Lefebvre, M.H., Oran, E.S., Kailasanath, K., Van Tiggelen, P.J.: Simulation of cellular structure in a detonation wave. In: Progress in Astronautics and Aeronautics, vol. 153, pp. 64–77. American Institute of Aeronautics and Astronautics, VA (1993)
82. Lehr, H.F.: Experiments on shock-induced combustion. Astronaut. Acta. **17**(4–5), 589–597 (1972)
83. Levin, V.A., Markov, V.V.: The initiation of detonation wave at concentrating energy admission. Combust. Explo. Shock Waves **11**(4), 623–633 (1975)
84. Levin, V.A., Osinkin, S.F., Markov, V.V.: Direct initiation of detonation in a hydrogen–air mixture. In: Combustion, Detonation, Shock Waves. Proceedings of the Zeldovitch Memorial, vol. 2, pp. 363–365. Russian Section of the Combustion Institute, Moscow (1994)
85. Levin, V.A., Markov, V.V., Osinkin, S.F.: Initiation of detonation in hydrogen–air mixture by spherical TNT charge. Combust. Explo. Shock Waves **31**(2), 91–95 (1995)

86. Litchfield, E.L., Hay, M.H., Forshey, D.R.: Direct electrical initiation of freely expanding gaseous detonation waves. In: 9th Symposium (International) on Combustion, pp. 282–286. The Combustion Institute, Pittsburgh (1963)
87. Liu, Y.K., Lee, J.H., Knystautas, R.: Effect of geometry on the transmission of detonation through an orifice. Combust. Flame **56**(2), 215–225 (1984)
88. Macek, A.: Effect of additives on formation of spherical detonation waves in hydrogen–oxygen-mixtures. AIAA J. **1**(8), 1915–1918 (1963)
89. Makeev, V.I., Gostintsev, Ju.A., Strogonov, V.V., Bochon, Ju.A., Chernushkin, Ju.N., Kulikov, V.N.: Burning and detonation of hydrogen–air mixes in free volumes. Combust. Explo. Shock Waves **19**(5), 16–18 (1983)
90. Manson, N., Dabora, E.K.: Chronology of research on detonation waves: 1920–1950. In: Progress in Astronautics and Aeronautics, vol. 153, pp. 3–42. American Institute of Aeronautics and Astronautics, VA (1993)
91. Manson, N., Brochet, C., Brossard, J., Pujol, Y.: Vibration phenomena and instability of self-sustained detonation in gases. In: 9th Symposium (International) on Combustion, pp. 461–469. The Combustion Institute, Pittsburgh (1962)
92. Manzhalei, V.I.: Detonation regimes of gases in capillaries. Combust. Explo. Shock Waves **28**(3), 296–301 (1992)
93. Manzhalei, V.I, Mitrofanov, V.V., Subbotin, V.A.: Measurement of inhomogeneities of a detonation front in gas mixtures at elevated pressures. Combust. Explo. Shock Waves **10**, 89–95 (1974)
94. Manzhalei, V.I, Subbotin, V.A.: Experimental Investigation of Stability of Overdriven Detonation in Gases. Combust. Explo. Shock Waves **12**(6), 935–942 (1976)
95. Matsui, H.: On the measure of the relative detonation hazards of gaseous fuel-oxygen and air mixtures. In: 17th Symposium (International) Combustion Proceedings, pp. 1269–1280. The Combustion Institute, Pittsburgh (1979)
96. Matsui, H., Lee, J.H.: Influence of electrode geometry and spacing on the critical energy for direct initiation of spherical gaseous detonations. Combust. Flame **27**, 217–220 (1976)
97. Matsui, H., Lee, J.H.: On the measure of the relative detonation hazards of gaseous fuel-oxygen and air mixtures. In: 17th Symposium (International) on Combustion, pp. 1269–1280. The Combustion Institute, Pittsburgh (1978)
98. McVey, J.B., Toong, T.: Mechanism of instabilities of exothermic hypersonic blunt-body flows. Combust. Sci. Technol. **3**, 63–76 (1971)
99. Mitrofanov, V.V.: Some critical phenomena in detonations connected to losses of an impulse. Combust. Explo. Shock Waves **19**(4), 169–174 (1983)
100. Mitrofanov, V.V., Soloukhin, R.I.: About diffraction of a multifront detonation wave (in Russian). Rep. USSR Acad. Sci. **159**(5), 1003–1006 (1964)
101. Moen, I., Funk, J., Ward, S., Rude, G., Thibault, P.: Detonation length scales for fuel–air explosives. In: Progress in Astronautics and Aeronautics, vol. 94, pp. 55–79. American Institute of Aeronautics and Astronautics, VA (1985)
102. Moen, I., Sulmistras, A., Thomas, G., Bjerketvedt, D., Thibault, P.: The influence of cellular regularity on the behaviour of gaseous detonations. In: Progress in Astronautics and Aeronautics, vol. 106, pp. 220–243. American Institute of Aeronautics and Astronautics, VA (1986)

103. Moen, I.O., Murray, S.B., Bjerketvedt, D., Rinnan, A., Knystautas, R., Lee, J.H.: Diffraction of detonation from tubes into a large fuel-air explosive cloud. In: 19th Symposium (International) on Combustion, pp. 635–644. The Combustion Institute, Pittsburgh (1982)
104. Moen, I.O., Ward, S.A., Thibault, P.A., Lee, J.H., Knystautas, R., Dean, T., Westbrook, C.K.: The influence of diluents and inhibitors on detonations. In: 20th Symposium (International) on Combustion Proceedings, pp. 1717–1726. The Combustion Institute, Pittsburgh (1985)
105. Munday, G., Ubbelohde, A.R., Wood, I.F.: Marginal detonation in cyanogen-oxygen mixtures. Proc. Roy. Soc. A **306**(1485), 179–184 (1968)
106. Murray, S.B., Lee, J.H.: On the transformation of planar detonations to cylindrical detonation. Combust. Flame **52**(3), 269–289 (1983)
107. Murray, S., Thibault, P., Moen, I., Knystautas, R., Lee, J., Sulmistras, A.: Initiation of hydrogen–air detonations by turbulent fluorine-air jets. In: Progress in Astronautics and Aeronautics, vol. 133, pp. 91–117. American Institute of Aeronautics and Astronautics, VA (1991)
108. Murray, S.B., Thibault, P.A., Zhang, F., Bjerketvedt, D., Sulmistras, A., Thomas, G.O., Jenssen, A., Moen, I.O.: The role of energy distribution on the transmission of detonation. In: Roy, G.D., Frolov, S.M., Netzer, D.W., Borisov, A.A. (eds.) Detonation and High-Speed Deflagration: Fundamentals and Control, vol. 6/18, pp. 139–162. Elex-KM Publisher, Moscow (2001)
109. Ng, H.D., Radulescu, M.I., Higgins, A.J., Nikiforakis, N., Lee, J.H.S.: Numerical investigation of the instability for one-dimensional Chapman-Jouguet detonations with chain-branching kinetics. Combust. Theor. Model. **9**(3), 385–401 (2005)
110. Nicholls, J.A., Dabora, E.K., Gealer, R.L.: Studies in connection with stabilized gaseous detonation waves. In: 7th Symposium (International) on Combustion, pp. 766–772. Combustion Institute, Pittsburgh (1958)
111. Nicholls, J., Sichel, M., Fry, R., Glass, D.: Theoretical and experimental study of cylindrical shock and heterogeneous detonation waves. Acta Astronaut. **1**, 385–404 (1974)
112. Nicholls, J.A., Sichel, M., Gabrijel, Z., Oza, R.D., Vandermolen, R.: Detonability of unconfined natural gas-air clouds. In: 17th Symposium (International) on Combustion, pp. 1223–1234. Combustion Institute, Pittsburgh (1978)
113. Nikolaev, Yu.A., Vasil'ev, A.A., Ulianitsky, V.Yu.: Gas detonation and its technological adaptation. Combust. Explo. Shock Waves **39**(4), 22–54 (2003)
114. Ohyagi, S., Yoshihashi, T., Harigaya, Y.: Direct initiation of planar detonation waves in methane/oxygen/nitrogen mixtures. In: Progress in Astronautics and Aeronautics, vol. 94, pp. 3–22. American Institute of Aeronautics and Astronautics, VA (1983)
115. Oran, E., Boris, J.: Numerical Simulation of Reactive Flow. Elsevier, New York (1987)
116. Oran, E.S., Boris, J.P., Young, T., Flanigan, M., Burks, T., Picone, M.: Numerical simulations of detonations in hydrogen–air and methane-air mixtures. In: 18th Symposium (International) on Combustion, pp. 1641–1649. The Combustion Institute, Pittsburgh (1981)
117. Oran, E.S., Young, T.R., Boris, J.P., Picone, J.M., Edwards, D.H.: A study of detonation structure: The formation of unreacted gas pockets. In: 19th Symposium (International) on Combustion, pp. 573–582. The Combustion Institute, Pittsburgh (1982)

118. Oran, E.S., Weber, J.W., Jr., Stefaniw, E.I., Lefebvre, M.H., Anderson, J.D., Jr.: A numerical study of a two-dimensional H_2-O_2-Ar detonation using a detailed chemical reaction model. Combust. Flame **113**, 147–163 (1998)
119. Papavassiliou, J., Makris, A., Knystautas, R., Lee, J.: Measurements of cellular structure in spray detonation. In: Progress in Astronautics and Aeronautics, vol. 154, pp. 148–169. American Institute of Aeronautics and Astronautics, VA (1993)
120. Pedley, M.D., Bishop, C.V., Benz, F.J., Bennett, C.A., McClenagan, R.D., Fenton, D.L., Knystautas, R., Lee, J.H., Peraldi, O., Dupre, G., Shepherd, J.E.: Hydrazine vapor detonations. In: Progress in Astronautics and Aeronautics, vol. 114, pp. 45–63. American Institute of Aeronautics and Astronautics, VA (1988)
121. Presles, H.N., Desbordes, D.: Non-ideal behavior of multi-headed self-sustained gaseous detonation. In: CD – International Conference on Combustion and Detonation: Zel'dovich Memorial II, Moscow (2004)
122. Presles, H.N., Desbordes, D., Guirard, M., Guerraud, C.: Gaseous nitromethane and nitromethane-oxygen mixture, a new detonation structure. Shock Waves **6**, 111–114 (1996)
123. Radulescu, M.I., Higgins, A.J., Murray, S.B., Lee, J.H.S.: An experimental investigation of the direct initiation of cylindrical detonations. J. Fluid Mech. **480**, 1–24 (2003)
124. Rinnan, A.: Transmission of detonation through tubes and orifices. In: Fuel-Air Explosions, pp. 553–564. University of Waterloo Press, Waterloo (1982)
125. Ruegg, F.W., Dorsey, W.W.: General discussion. In: 9th Symposium (International) on Combustion, pp. 476–477. The Combustion Institute, Pittsburgh (1962)
126. Sedov, L.I.: The Methods of Similarity and Dimensions in Mechanics (in Russian). Nauka, Moscow (1987)
127. Sichel, M.: A simple analysis of the blast initiation of detonations. Acta Astronaut. **4**(3–4), 409–424 (1977)
128. Shelkin, K.I., Troshin, Ya.K.: Gasdynamics of Combustion (in Russian). USSR Academy of Science, Moscow (1963)
129. Shepherd, J.E.: Chemical kinetics and cellular structure of detonations in hydrogen sulfide and air. In: Progress in Astronautics and Aeronautics, vol. 106, pp. 294–320. American Institute of Aeronautics and Astronautics, VA (1986)
130. Shepherd, J.E.: Chemical kinetics of hydrogen–air-diluent detonations. In: Progress in Astronautics and Aeronautics, vol. 106, pp. 263–293. American Institute of Aeronautics and Astronautics, VA (1986)
131. Shepherd, J.E., Moen, I.O., Murray, S.B., Thibault, P.A.: Analyses of the cellular structure of detonations. In: 21st Symposium (International) on Combustion Proceedings, pp. 1649–1658. The Combustion Institute, Pittsburgh (1988)
132. Shulenin, Yu.V., Bohon, Yu.A.: The minimal initiation energy of gaseous mixtures in unlimited area (in Russian). Rep. USSR Acad. Sci. **257**(3), 680–683 (1981)
133. Stamps, D.W., Tieszen, S.R.: The influence of initial pressure and temperature on hydrogen–air-diluent detonations. Combust. Flame **83**(3), 353–364 (1991)
134. Stewart, D.S., Bdzil, J.B.: The shock dynamics of stable multi-dimensional detonation. Combust. Flame **12**, 311–323 (1988)

135. Stewart, D.S., Kasimov, A.R.: State of detonation stability theory and its application to propulsion. J. Propul. Power. **22**(6), 1230–1244 (2006)
136. Stewart, D.S., Aslam, T.D., Yao, J.: On the evolution of cellular detonation. In: 26th Symposium (International) on Combustion, pp. 2981–2989. The Combustion Institute, Pittsburgh (1996)
137. Strehlow, R.A.: Multi-dimensional detonation wave structure. Astronaut. Acta. **15**(5), 345–357 (1970)
138. Strehlow, R.A., Engel, C.D.: Transverse waves in detonations: II. Structure and spacing in H_2-O_2, C_2H_2-O_2, C_2H_4-O_2 and CH_4-O_2 systems. AIAA J. **7**(3), 492–496 (1969)
139. Strehlow, R.A., Liaugminas, R., Watson, R.H., Eyman, J.R.: Transverse wave structure in detonations. In: 11th Symposium (International) on Combustion, pp. 683–692. The Combustion Institute, Pittsburgh (1967)
140. Strehlow, R.A., Maurer, R.E., Rajan, S.: Transverse waves in detonations: I. Spacings in the hydrogen–oxygen system. AIAA J. **7**(2), 323–328 (1969)
141. Taki, S., Fujiwara, T.: Numerical analysis of two-dimensional nonsteady detonations. AIAA J. **16**, 73–77 (1978)
142. Tieszen, S.R., Sherman, M.P., Benedick, W.B., Shepherd, J.E., Knystautas, R., Lee, J.H.S.: Detonation cell size measurements in hydrogen–air-steam mixtures. In: Progress in Astronautics and Aeronautics, vol. 106, pp. 205–219. American Institute of Aeronautics and Astronautics, VA (1986)
143. Tieszen, S., Stamps, D., Westbrook, C., Pitts, W.: Gaseous hydrocarbon–air detonations. Combust. Flame **84**(3), 376–390 (1991)
144. Troshin, K.Ya.: The initiation energy for divergent detonation waves (in Russian). Rep. USSR Acad. Sci. **247**(4), 887–889 (1979)
145. Trotsyuk, A.V., Ulianitsky, V.Yu.: About parameters of detonation waves in gas at concentrated energy initiation. Combust. Explo. Shock Waves **19**(6), 76–82 (1983)
146. Tsuboi, N., Hayashi, A.K.: Numerical simulation of continuous spinning detonation in a circular tube. In: Roy, G., Frolov, S., Sinibaldi, J. (eds.) Pulsed and Continuous Detonations, pp. 186–192. Torus Press Ltd, Moscow (2006)
147. Tsuboi, N., Asahara, M., Eto, K., Hayashi, A.K.: Numerical simulation of spinning detonation in square tube. Shock Waves **18**(4), 329–344 (2008)
148. Ul'yanitskii, V.Yu.: Closed model of direct initiation by gas detonation, taking account of instability II. Nonpoint initiation. Combust. Explo. Shock Waves **16**(4) 427–434 (1980)
149. Ul'yanitskii, V.Yu.: Closed model of direct initiation of gas detonation taking account of instability. I. Point initiation. Combust. Explo. Shock Waves **16**(3) 331–341 (1980)
150. Ul'yanitsky, V.Yu.: Investigation of galloping regime of gaseous detonation. Combust. Explo. Shock Waves **17**(1), 118–124 (1981)
151. Urtiew, P.A., Tarver, C.M.: Effects of cellular structure on the behaviour of gaseous detonation waves under transient conditions. In: Progress in Astronautics and Aeronautics, vol. 75, pp. 370–384. American Institute of Aeronautics and Astronautics, VA (1981)
152. Vandermeiren, M., Van Tiggelen, P.: Cellular structure in detonation of acetylene-oxygen mixtures. In: Progress in Astronautics and Aeronautics, vol. 94, pp. 104–117. American Institute of Aeronautics and Astronautics, VA (1984)

153. Vandermeiren, M., Van Tiggelen, P.: Role of an inhibitor on the onset of gas detonations in acetylene mixtures. In: Progress in Astronautics and Aeronautics, vol. 114, pp. 186–200. American Institute of Aeronautics and Astronautics, VA (1988)
154. Vasil'ev, A.A.: The estimation of initiation energy for cylindrical detonation. Combust. Explo. Shock Waves **14**(3), 154–155 (1978)
155. Vasil'ev, A.A.: Research of critical initiation of a gas detonation. Combust. Explo. Shock Waves **19**(1), 121–131 (1983)
156. Vasil'ev, A.A.: The experimental methods and calculating models for definition of the critical initiation energy of multifront detonation wave. In: Proceedings of the 16th International Colloquium on the Dynamics of Explosion and Reactive Systems (ICDERS), pp. 152–155. University of Mining and Metallurgy, AGH, Cracow (1987)
157. Vasil'ev, A.A.: Diffraction of multifront detonation. Combust. Explo. Shock Waves **24**(1), 99–107 (1988)
158. Vasil'ev, A.A.: A spatial initiation of multifront detonation. Combust. Explo. Shock Waves **25**(1), 113–119 (1989)
159. Vasil'ev, A.A.: The limits of stationary propagation of gaseous detonation. In: Borissov, A. (ed.) Dynamic structure of detonation in gaseous and dispersed media. Fluid Mechanics and its Applications, vol. 5, pp. 27–49. Kluwer, Dordrecht (1991)
160. Vasil'ev, A.A.: Near-limiting detonation in channels with porous walls. Combust. Explo. Shock Waves **30**(1), 101–106 (1994)
161. Vasil'ev, A.A.: Near-critical modes of a gas detonation (in Russian). Lavrentyev Institute of Hydrodynamics, Novosibirsk (1995)
162. Vasil'ev, A.A.: Gaseous fuels and detonation hazards. In: Eisenreih, N. (ed.) Combustion and Detonation (Proceedings of the 28th Fraunhofer ICT-Conference), Karlsruhl, Germany (1997)
163. Vasil'ev, A.A.: Modeling of detonation combustion of gas mixtures using a high-velocity projectile. Combust. Explo. Shock Waves **33**(5), 85–102 (1997)
164. Vasil'ev, A.A.: A new diffraction estimation of critical initiation energy. In: Proceeding of the Colloquim on Gas, Vapor, Hybrid and Fuel-Air Explosions, pp. 470–481. Safety Consulting Engineers, Schaumburg (1998)
165. Vasil'ev, A.A.: The character propagation regimes of multifront detonation along convex surface. Combust. Explo. Shock Waves **35**(5), 86–92 (1999)
166. Vasil'ev, A.A.: Optimization of accelerators of deflagration-to-detonation transition. In: Roy, G., Frolov, S., Santoro, R., Tsyganov, S. (eds.) Confined Detonations and Pulse Detonation Engines, pp. 41–48. Torus Press, Moscow (2003)
167. Vasil'ev, A.A.: Cell size as the main geometric parameter of a multifront detonation waves. J. Propul. Power. **22**(6), 1245–1268 (2006)
168. Vasil'ev, A.A.: Detonation characteristics of synthetic gas. Combust. Explo. Shock Waves **43**(6), 90–96 (2007)
169. Vasil'ev, A.A.: The quasi-steady regimes of wave propagation in active mixtures. Shock Waves **18**(4), 245–253 (2008)
170. Vasil'ev, A.A., Grigor'ev, V.V.: The critical conditions for detonation propagation at sharp expanded channel. Combust. Explo. Shock Waves **16**(5), 117–125 (1980)
171. Vasil'ev, A.A., Kulakov, B.I., Mitrofanov, V.V., Silvestrov, V.V., Titov, V.M.: The initiation of gaseous mixtures by high-velocity bullet (in Russian). Rep. Russ. Acad. Sci. **338**(2), 188–190 (1994)

172. Vasil'ev, A.A., Mitrofanov, V.V., Topchiyan, M.E.: Detonation waves in gases. Combust. Explo. Shock Waves **23**(5), 605–623 (1987)
173. Vasil'ev, A.A., Nikolaev, Yu.A.: The cell model for multifront detonation. Combust. Explo. Shock Waves **12**(5), 744–754 (1976)
174. Vasil'ev, A.A., Pinaev, A.V.: Formation of carbon clusters in deflagration and detonation waves in gas mixtures. Combust. Explo. Shock Waves **44**(3), 96–101 (2008)
175. Vasil'ev, A.A., Trotsyuk, A.V.: Experimental investigation and numerical modeling of expanded multifront detonation wave. Combust. Explo. Shock Waves **39**(1), 92–103 (2003)
176. Vasil'ev, A.A., Trotsyuk, A.V.: Multi-scaled cellular structure of gaseous detonation. In: Korobeinichev, O. (ed.) Proceedings of the 5th International Seminar on Flame Structure. Parallel Ltd. Novosibirsk (2005); CD ISBN 5-98901-004-4, OPr-08
177. Vasil'ev, A.A., Vasil'ev, V.A.: The steam influence on hydrogen–oxygen and hydrogen–air detonation. In: Conference Proceeding of 16th-International Colloquium on the Dynamics of Explosion and Reactive Systems (ICDERS), pp. 385–388. University of Mining and Metallurgy AGH, Cracow (1997)
178. Vasil'ev, A.A., Zak, D.V.: Detonation in gaseous jet. Combust. Explo. Shock Waves **22**(4), 82–88 (1986)
179. Vasil'ev, A.A., Topchian, M.E., Ulianitsky, V.Yu.: Influence of initial temperature on parameters of gaseous detonation. Combust. Explo. Shock Waves **15**(6), 149–152 (1979)
180. Vasil'ev, A.A., Nikolaev, Yu.A., Ulyanitsky, V.Yu.: The critical initiation energy for multifront detonation. Combust. Explo. Shock Waves **15**(6), 94–104 (1979)
181. Vasil'ev, A.A., Valishev, A.I., Vasil'ev, V.A., Panfilova, L.V., Topchian, M.E.: Detonation waves parameters at increased pressure and temperatures. Chem. Phys. Rep. **16**(9), 1659–1666 (1997)
182. Vasil'ev, A.A., Valishev, A.I., Vasil'ev, V.A.: The detonation safety of gaseous fuels. Acetylene and cyanogen. In: Conference Proceeding of the 16th-International Colloquium on the Dynamics of Explosion and Reactive Systems (ICDERS), p. 594. University of Mining and Metallurgy AGH, Cracow (1997)
183. Vasil'ev, A.A., Valishev, A.I., Vasil'ev, V.A., Panfilova, L.V., Topchian, M.E.: Detonation hazards of hydrogen mixtures. In: Proceeding of the Colloquim on Gas, Vapor, Hybrid and Fuel-Air Explosions, pp. 391–413. Safety Consulting Engineers, Schaumburg (1998)
184. Vasil'ev, A.A., Valishev, A.I., Vasil'ev, V.A., Panfilova, L.V.: Parameters of combustion and detonation of hydrazine and its methyl derivatives. Combust. Explo. Shock Waves **36**(3), 81–96 (2000)
185. Vasil'ev, A.A., Valishev, A.I., Vasil'ev, V.A.: The estimation of parameters of combustion and detonation of hydrocarbon gas-hydrates. Combust. Explo. Shock Waves **36**(6), 119–125 (2000)
186. Vasil'ev, A.A., Ttotsyuk, A.V., Fomin, P.A., Vasil'ev, V.A., Rychkov, V.N., Desbordes, D., Khasainov, B., Presles, H.N., Vidal, P., Demontis, P., Priault, C.: The basic results on reinitiation processes in diffracting multifront detonations, Part 1. Eurasian Chem-Technol. J. **5**(4), 279–289 (2003)
187. Vasil'ev, A.A., Zvegintsev, V.I., Nalivaichenko, D.G.: Detonation waves in supersonic flows of reactive mixtures. Combust. Explo. Shock Waves **42**(5), 85–100 (2006)

188. Vasiljev, A.A.: Initiation of gaseous detonation by a high speed body. Shock Waves **3**(4), 321–326 (1994)
189. Vasiljev, A.A., Nikolaev, Yu.: Closed theoretical model of a detonation cell. Acta Astronaut. **5**, 983–996 (1978)
190. Voitsekhovsky, B.V.: On spinning detonation (in Russian). Dokl. Acad. Sci. SSSR. **114**, 717–720 (1957)
191. Voitsekhovsky, B.V., Mitrofanov, V.V., Topchian, M.E.: Structure of Detonation Front in Gases (in Russian). Siberian Branch USSR Academy Science, Novosibirsk (1963)
192. Westbrook, C.K.: Chemical kinetics of hydrocarbon oxidation in gaseous detonations. Combust. Flame **46**, 191–210 (1982)
193. Westbrook, C.K., Urtiew, P.A.: Chemical kinetic prediction of critical parameters in gaseous detonations. In: 19th Symposium (International) on Combustion, pp. 615–623. The Combustion Institute, Pittsburgh (1982)
194. Wolanski, P., Kindracki, J., Fujiwara, T.: An experimental study of small rotating detonation engine. In: Roy, G.D., Frolov, S.M., Sinibali, J. (eds.) Pulsed and Continous Detonation, pp. 332–338. Torus Press, Moscow (2006)
195. Yao, J., Stewart, D.S.: On the dynamics of multi-dimensional detonation. J. Fluid Mech. **309**, 225–275 (1996)
196. Zeldovich, Ya.B., Kogarko, S.M., Simonov, N.N.: The experimental investigations of spherical gaseous detonation (in Russian). J. Tech. Phys. **26**(8), 1744–1768 (1956)
197. Zhang, F.: Heterogeneous detonation. Springer, Berlin (2009)
198. Zhang, F., Murray, S., Gerrard, K.: JP-10 vapour detonations at elevated pressures and temperatures. In: CD: Proceedings of the 18th International Colloquium on the Dynamics of Explosion and Reactive Systems (ICDERS), University of Washington, ISBN 0-9711740-0-8, Seattle (2001)
199. Zhdan, S.A., Mitrofanov, V.V.: A simple model for calculation of initiation energy for geterogeneous and gaseous detonation. Combust. Explo. Shock Waves **21**(6), 98–103 (1985)
200. Zitoun, R., Desbordes, D., Guerraud, C., Deshaies, B.: Direct initiation of detonation in cryogenic gaseous H_2-O_2 mixtures. Shock Waves **4**(6), 331–337 (1995)

5

Multi-Scaled Cellular Detonation

Daniel Desbordes and Henri-Noël Presles

5.1 Introduction

Experimental evidence of the three-dimensional features of detonation fronts can be traced back to the work of Campbell and Woodhead [9] and Bone and Fraser [6] for marginal detonations propagating in tubes. Manson [64] and Fay [36] established an acoustic theory that provided a reasonable explanation for these observations, without knowledge of the intrinsic nature of the detonation structure. In 1959, Denisov and Troshin [18] discovered the real detonation front structure (Fig. 5.1). Their results showed that the detonation reaction zone was locally transient, thus leading to three-dimensional periodic instabilities. Since then, numerous studies of modern detonation physics have been dedicated to the so-called cellular structure of the detonation front.

The cellular structure is a manifestation of the specific instabilities of detonation phenomenon. As described in Chap. 4, the basic element of the detonation front structure is the transverse wave and the related triple shock configuration, similar to that obtained in irregular oblique shock reflection on a plane surface. The triple shock configuration consists of the leading incident shock, the Mach stem and the reflected shock (i.e., the transverse wave), with non-uniform flow and reaction behind it [87, 99]. During the detonation propagation, two neighbouring transverse waves collide periodically and each collision regenerates the continuously decaying triple waves. The elementary periodic pattern drawn by two adjacent triple-point trajectories forms a diamond-like cell pattern. The cell shape and the characteristic cell width and length appear as intrinsic properties of a reactive mixture at given initial conditions.

The cellular structure can be considered universal; it exists for self-sustained and overdriven detonation, in reactive gaseous mixtures, and also in suspensions of reactive particles in oxygen or air [125], and for any propagation geometries including planar as well as expanding or converging cylindrical and spherical detonations [31]. In the propagation of expanding detonation, new triple points are continuously generated to maintain

F. Zhang (ed.), *Shock Wave Science and Technology Reference Library, Vol. 6*: Detonation Dynamics, DOI 10.1007/978-3-642-22967-1_5,

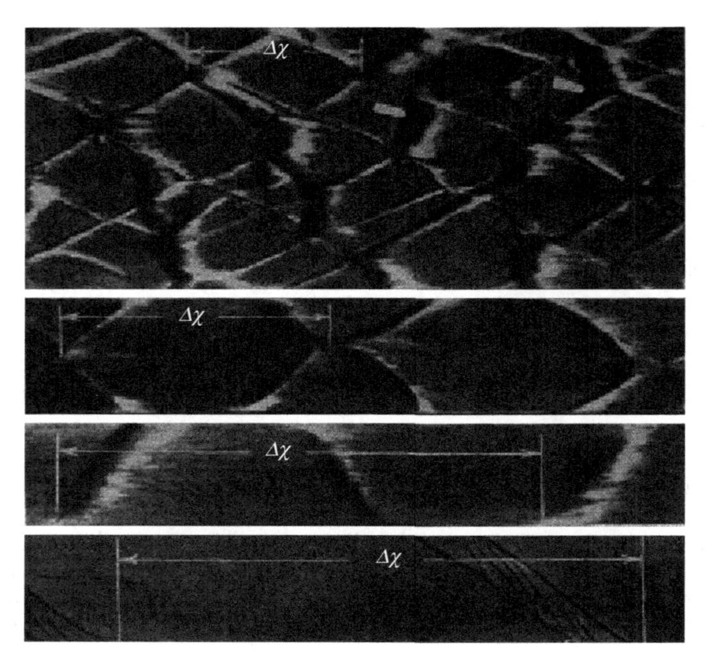

Fig. 5.1. Soot tracks on the inner surface of a $d = 16$ mm i.d. round tube for $2H_2+O_2$ mixture at initial pressure (**a**) $p_0 = 53.3$ kPa, (**b**) $p_0 = 40$ kPa, (**c**) $p_0 = 17.3$ kPa, (**d**) $p_0 = 6$ kPa. Detonations propagate from left to right [87]

a constant cell size intrinsic to a mixture when the surface area of the wave front increases (see Chap. 3). Recently, it has been found that detonation can display a double cellular structure in a class of reactive mixtures. This structure appears as a large cellular structure filled with smaller cells whose size is orders of magnitude smaller than the large cell [54, 80]. This new discovery raises questions about the heat release mechanisms that are responsible for cellular structures and for propagation of detonation waves.

This chapter attempts to link the various detonation cellular structures to different heat release mechanisms. After introduction of a variety of detonation front structures in Sect. 5.2, the intrinsic nature of different cellular structures is discussed with respect to the responsible heat release mechanisms in Sect. 5.3. This association is further confirmed by direct numerical simulations summarized in Sect. 5.4. Finally, detonation limits and related dynamic parameters are revisited in view of these heat release mechanisms.

5.2 Detonation Front Structure

5.2.1 Macroscopic Aspect

The simplest technique used to study detonation cellular structure is the soot-coated foil placed along the direction of the detonation propagation [91]. The large and abrupt variations of pressure and velocity inside the detonation cell scrub off the soot film revealing structural details as small as 0.1 mm. The foil can also be placed normally to the flow to record the head-on detonation structure. Figure 5.2 shows an example of the cellular structure imprinted in soot tracks, from which the characteristic cell length, L, in the detonation propagation direction and the cell width, λ, in the perpendicular direction can be determined.

Some other recording methods consist of:

1. A laser light through a soot-coated glass window recorded with a high speed camera (Fig. 5.3)
2. The self-luminosity of the detonation front recorded with an open shutter camera (Fig. 5.4)

These three techniques directly show that "tracks" are written by the triple points.

Head-on cellular detonation structure can also be obtained by a camera recording the deformation on a thin aluminized planar Mylar film impacted normally by the detonation as shown in Fig. 5.5 [78].

Interferometric, schlieren and planar laser-induced fluorescence (PLIF) optical methods allow recording of instantaneous shock wave positions, associated turbulent eddies and reaction zones, and evidence of unreacted pockets behind the detonation front (Figs. 5.15 and 5.16) [1,77,89,122,123]. All these techniques provide a 2D record of a 3D phenomenon.

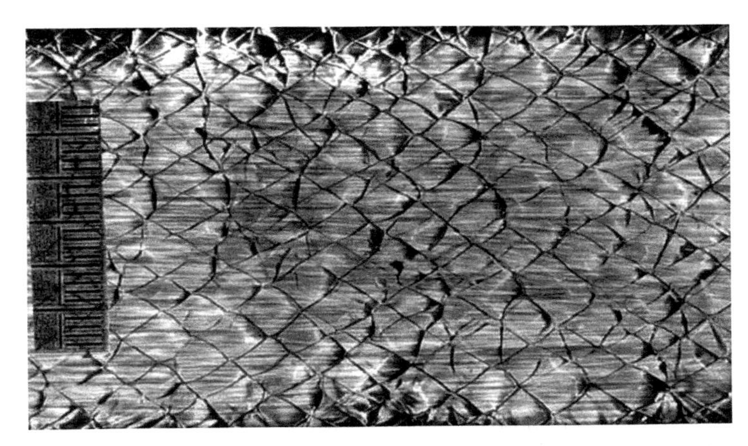

Fig. 5.2. Soot tracks in $2H_2 + O_2 + 3Ar$, $p_0 = 40$ kPa, tube diameter $d = 52$ mm. The detonation propagates from left to right [21]

5.2.1.1 Basic Cellular Structure

The basic three-shock configuration of the self-sustained detonation cellular structure, visible in Figs. 5.2–5.4, can be calculated (also see Fig. 4.4)

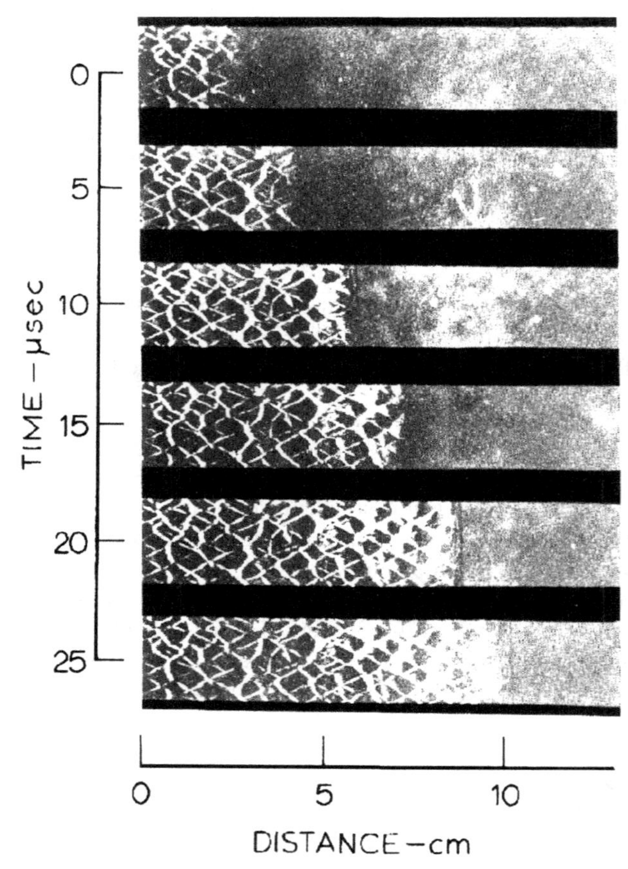

Fig. 5.3. Stroboscopic smoke-film/laser-shadow of multi-headed detonation wave of detonation in a $C_2H_2 + O_2$ mixture contained in a $1/8 \times 1$ in. tube by Urtiew [60]

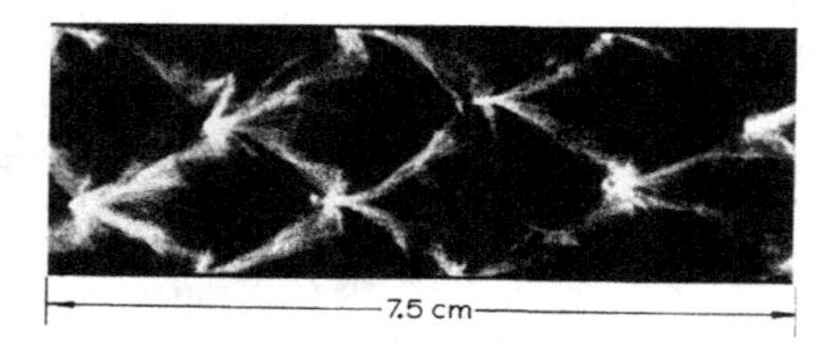

Fig. 5.4. Open shutter photograph of detonation in a $C_2H_2 + O_2$ mixture contained in a $1/8 \times 1$ in. tube by Kamel. The detonation propagates from left to right [60]

(a) $p_0 = 50\,\mathrm{kPa}$ (b) $p_0 = 40\,\mathrm{kPa}$ (c) $p_0 = 26\,\mathrm{kPa}$

Fig. 5.5. Normal interaction of a detonation wave propagating in a $d = 52\,\mathrm{mm}$ i.d. tube with an aluminized mylar film in a $2H_2 + O_2 + 7Ar$ mixture [78]

using the quasi-steady shock-polar analysis [32, 73]. Since in the three-shock configuration the Mach stem is much stronger than the incident shock, the induction time is shorter behind the Mach stem than behind the incident shock. The transverse wave moves laterally into the compressed unreacted material and causes a large part of mass flow crossing the incident wave to react. Each transverse-wave collision produces an energetic explosion and gives birth to an overdriven detonation, leading to a Mach stem in the following cell. At the apex of this new cell, two reflected transverse waves move away from the collision point along with the triple points driven by the decaying Mach stem flow. These transverse waves then propagate mostly at a near-acoustic velocity and cause the pre-compressed and unreacted material behind adjacent incident shocks inside the neighbouring cells to react. The decaying Mach stem turns into an incident decaying shock in the second half of the cell. At the same time, adjacent transverse wave collisions create new triple points and transverse waves that propagate in opposite direction and finally collide at the end of the cell, thus generating a new cycle. Collision of two transverse waves has been widely studied in the past. Assuming a triple-point trajectory entrance angle (measured at the end of the previous cell—Fig. 5.6), the exit angle (angle of the triple-point trajectory at the beginning of the cell) and other shock characteristics can be well predicted [73, 96, 100, 107].

Detonation cells can be classified on the basis of their regularity intrinsic to a reactive mixture and its initial thermodynamic state. For some specific confinements and/or reactive systems, the cell regularity also depends on boundary conditions (Sect. 5.2.1.3). As the detonation cell size and regularity are significantly dependent on the detonation propagation regime (self-sustained or overdriven), it is important to know the boundary conditions accurately. In general, to avoid the influence of the confinement on detonation cell size for self-sustained cellular detonation, the detonation tube diameter must be greater than five to ten times the detonation cell width. The velocity of such a multi-headed cellular detonation is slightly lower (1–2%) than the

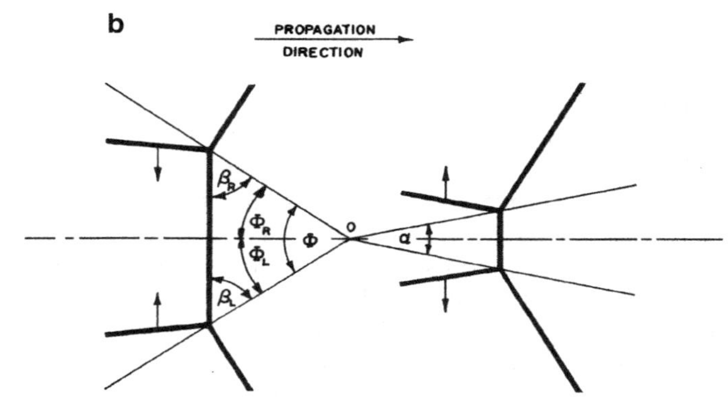

Fig. 5.6. Symmetric collision of two transverse waves with triple-point trajectory entrance angle Φ and exit angle α [96]

CJ detonation velocity; this detonation regime is called near-CJ detonation in this chapter.

5.2.1.2 Cell Regularity

The regularity of detonation cellular structure has been divided into four classes (Fig. 5.7), as proposed by Strehlow [94] and Libouton et al. [61]. They are described below:

1. Very regular structure (excellent regularity—Fig. 5.7a) where each cell can be interchanged by another, obtained in planar and orthogonal modes in tubes of a square or rectangular cross section with large height but very small width. In general, acoustic coupling (see Sect. 5.2.1.3) improves the regularity of the detonation structure of a given mixture. Such structure is observed in the detonation of light fuel–O_2 mixtures (H_2, C_2H_2, C_2H_4, CO, etc., except CH_4) highly diluted by a monatomic inert diluent, especially at low initial pressure.

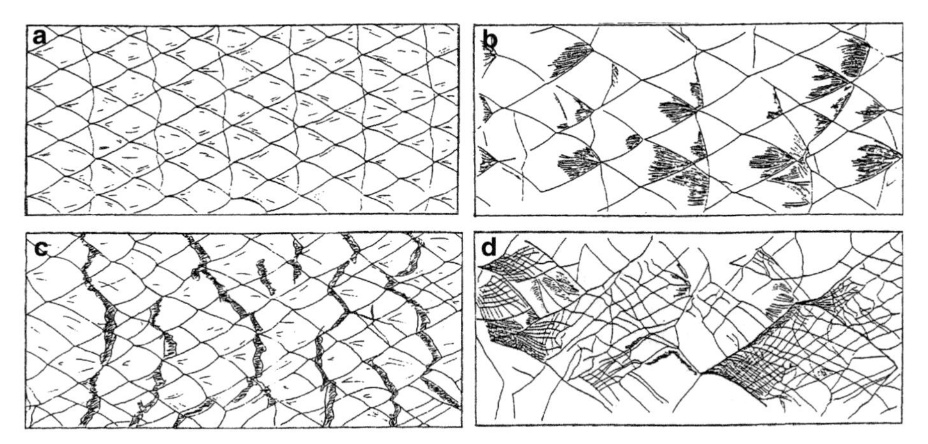

Fig. 5.7. Classification of cell regularity: (**a**) very regular, (**b**) regular, (**c**) irregular, and (**d**) very irregular [61]

2. Regular cell (good regularity—Fig. 5.7b), obtained in the same systems above but with less dilution and higher initial pressure and therefore little acoustic coupling. The cell size is fairly uniform and can be measured with reasonable accuracy.
3. Irregular cell (poor regularity—Fig. 5.7c), obtained with mixtures diluted by diatomic and triatomic molecules. Smoked foil records obtained for these mixtures typically show that a substructure begins to emerge inside the first half of the main cells. The cell shape becomes non-uniform and the accuracy of cell size measurement is not good.
4. Very irregular cell (Fig. 5.7d) obtained with fuel–O_2 mixtures heavily diluted with diatomic and triatomic molecules, or fuel–air mixtures. Reactive mixtures based on CH_4 belong to this category. Substructure is present inside the main cell whose characteristic size is often very difficult to determine.

The cell regularity and the strength of the transverse waves can be closely related to the detonation wave stability (see Chap. 3). For stable detonations with mixture conditions within the stability limits, the transverse waves are weak and the cellular pattern is regular, whereas for very unstable detonations, the transverse waves are strong and the detonation wave has a highly irregular cellular structure with the onset of a smaller cellular substructure inside.

Cell regularity also depends on the number of cells within the finite-sized confinement. For instance, as the cell width λ becomes of the same order as the detonation tube diameter, the cell shape and size vary significantly. Cell regularity is also greatly improved as the degree of overdrive is increased.

In general, early studies on detonation cellular structure were mainly performed in regular or very regular cellular systems arising in particular chemical compositions and/or from special acoustic coupling regimes such as spinning detonation and orthogonal mode detonation. This approach was useful to determine the main characteristic features and to progress towards understanding the basic mechanisms, as a necessary first step before addressing the more complex cellular structure displayed in common fuel–air mixtures.

5.2.1.3 Cell and Acoustics

In some conditions, experiments show that there is a kind of adaptation of the detonation cellular structure to the confinement. Transverse waves produce disturbances that excite and couple with the acoustic mode of vibration of the confinement, modifying the shape, regularity and size of the intrinsic detonation cellular structure. For instance, in the case of mixtures diluted by an inert monatomic gas, the inherent cells are regular (Cell Class 2). When confined in a rectangular cross-sectioned tube, cells become very regular (Cell Class 1) with orthogonal transverse waves as shown in Fig. 5.8 [50, 94, 103]. A detonation wave is called marginal when the number of half cells (or triple points or so-called modes) in the tube height equals 1. For the same initial conditions, detonation waves with different modes may be observed indicating that the detonation regime can degenerate [100]. For undiluted mixtures typically with irregular cellular structure, no such phenomenon is observed, because the intrinsically unstable nature of the cells prevents them from forming an acoustic organization.

In round cross-sectioned tubes, a decrease in number of triple points down to one leads to the propagation of a marginal detonation wave called spinning detonation. For any chemical system and cell regularity class, the triple-point trajectory is helical in this case and the spin pitch, imposed by the tube diameter, is roughly constant. For a given mixture, the spinning detonation can be observed through reducing the initial pressure or tube diameter. The spinning detonation is considered as the limiting case for a steady detonation propagation in tubes. However, depending on the reactive mixture, it is possible to observe spinning detonations in a large range of initial pressure (or a large range of dilution by an inert gas, etc.), thus demonstrating the significant role of acoustic coupling in sustaining detonation (Sect. 5.5.2).

The influence of small linear disturbances at the detonation front on acoustic vibrations in detonation products can be found in the works of Manson [64], Fay [36] and Chu [10]. Fundamental and higher transverse acoustic modes of detonations propagating in round and rectangular tubes are also shown and explained in these references.

5.2.2 Structure Details

5.2.2.1 Triple Points and Transverse Waves

According to the observations of Strehlow and Biller [96], Edwards et al. [33], Strehlow and Crooker [97], the transverse wave associated with a triple point

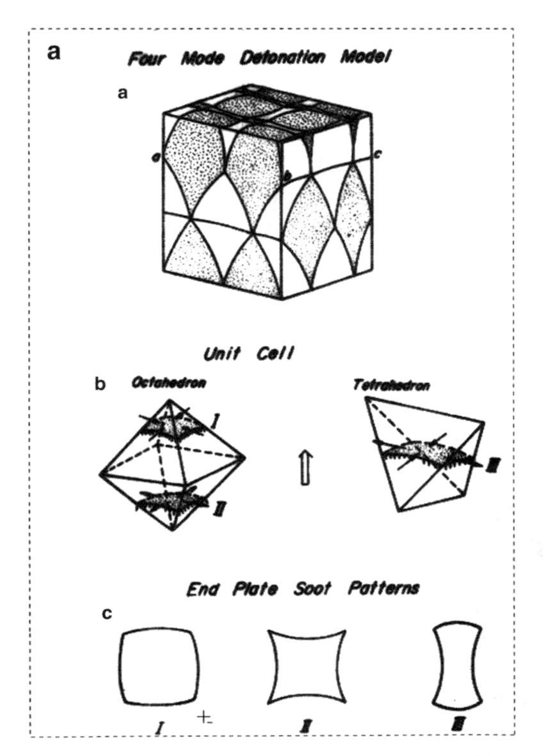

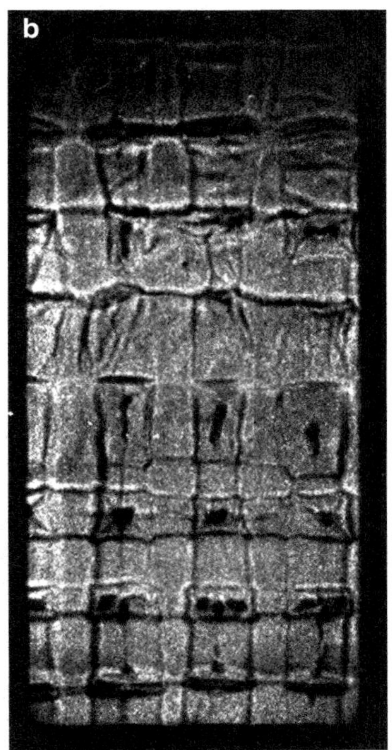

Fig. 5.8. (**a**) 3D schemes of four-mode cells in a square cross-sectioned tube [103]. (**b**) Head-on detonation interaction on a mylar aluminised film of a 7×15-mode cell detonation in a 22×47.5 mm rectangular cross-sectioned tube in $C_2H_4+3O_2+10Ar$, $p_0=20$ kPa. Presles unpublished result (1987)

(Fig. 5.9) can be either a shock wave, segment AR, followed by a reaction flame zone or a detonation, segment BC, respectively, called (1) a weak transverse wave and (2) a strong transverse wave.

In the case of a weak transverse wave, the mean track angle of the triple-point trajectory is about 30°. The weak transverse-wave structure is generally associated to near-CJ cellular detonation in most chemical systems except H_2–O_2 where periodic transitions between weak to strong structures happen. An example of such a transition is displayed Fig. 5.10 where the transverse-wave structure is weak at the beginning of the cell and turns into the strong structure at the end before the transverse-wave collisions.

The strong transverse wave occurs when detonation approaches the propagation limits. As displayed in Fig. 5.11, this transverse wave exhibits a fine cellular structure (except in the H_2–O_2 system) and its track angle is larger (30–45° range) than in the case of the weak transverse wave configuration (30°).

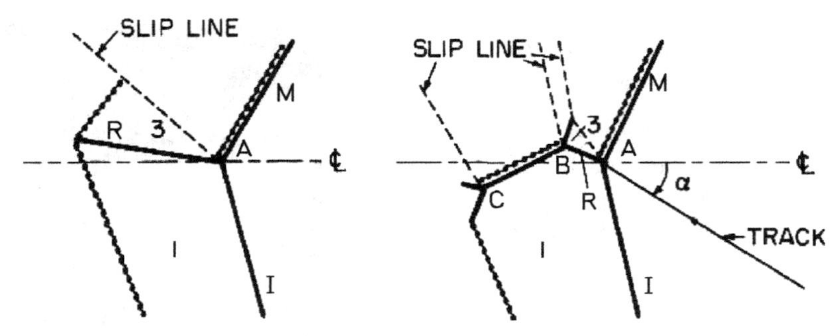

Fig. 5.9. Weak (*left*) and Strong (*right*) transverse-wave structures. The *solid lines* represent shocks and the *waved lines* are reaction fronts. A, B, C, Triple points; I, incident shock; R, reflected shock; M, Mach stem; 1 and 3, gas regions after incident and reflected shock. The wave front propagates from left to right [37]

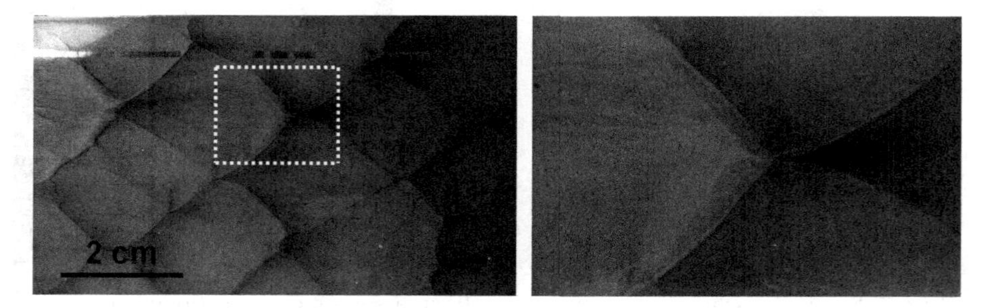

Fig. 5.10. Soot tracks in cellular detonation in $2H_2 + O_2 + 3Ar$, $p_0 = 13.3\,kPa$, tube diameter $d = 52\,mm$. Details of weak to strong transverse-wave transition and collision (*right*). The detonation propagates from left to right [21]

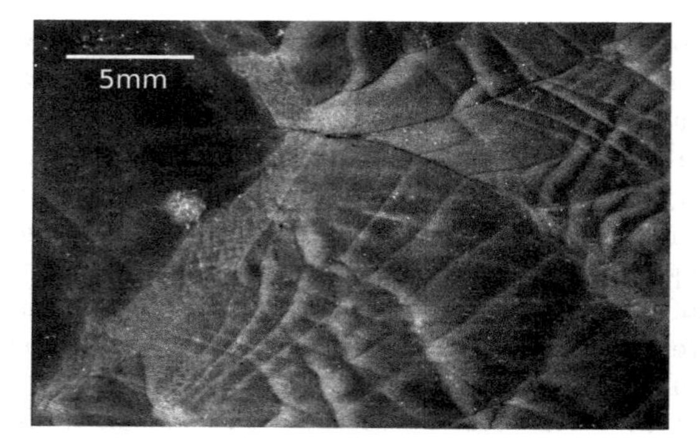

Fig. 5.11. Soot tracks in $H_2 + 0.42NO_2 + 0.35Ar$ ($\varphi = 1.2$), $p_0 = 25\,kPa$ in a tube of inner diameter $d = 52\,mm$. Details of weak to strong transverse-wave transition and collision in self-sustained cellular detonation, where strong transverse waves produce fine cells. The detonation propagates from left to right [116]

Finally, the strong transverse-wave structure is the basic structure for the single-head (i.e., one triple-point configuration) spinning detonation.

5.2.2.2 In-Cell Flow Variations

The detonation front velocity and pressure along the axis of a cell are decreasing functions of x, where x ranges between 0 and cell length L. As indicated by Steel et al. [93], Crooker [17] and Dormal et al. [29], the leading shock velocity along the centreline of the cell is continuously decreasing and varies over a range of 1.3–0.8D_{CJ} or 1.8–0.6D_{CJ} (D_{CJ} is the CJ detonation velocity), which depends on reactive mixtures and the proximity to detonation propagation limits. These measurements were generally performed in very large planar cells, more than 10 cm long as shown in Figs. 5.12 and 5.13 (i.e., for marginal detonations).

The calculated induction zone length can increase by three orders of magnitude during the shock decay inside a cell (Fig. 5.13). The large variation in induction length can cause significant fluctuation of the energy release in the reaction zone structure, thus leading to various gasdynamic instabilities in the reaction zone. As illustrated numerically using an irreversible Arrhenius law for the reaction zone (see Fig. 3.20), by increasing the activation energy E_a, the temperature sensitivity of the reaction is greater and large fluctuations of the reaction rate resulting from small gasdynamic disturbances can render the detonation unstable. As the reaction rate becomes more sensitive to local perturbation, cellular structure turns from regular to irregular, with simultaneous enlargement of the amplitude of the shock velocity variation along the cell (Fig. 5.14 where T_{ZND} refers to the von Neumann shock temperature in the ZND detonation structure). Consequently, this fluctuation leads to irregular cells containing in their first half very fine cells of increasing size [65] and the creation of unreacted pockets behind the front [38, 89], as observed experimentally (Figs. 5.15 and 5.16).

5.2.2.3 Cell Size

As detonation sensitivity of a mixture is usually expressed in terms of dynamic parameters that are related to λ, the cell width, such as critical direct initiation energy E_c ($E_c \propto \lambda^3$) and critical tube diameter d_c ($d_c \propto \lambda$) discussed in Chap. 4, it is very important to determine this intrinsic cell width for any reactive mixture.

Detonation cell size depends on the chemical system, equivalence ratio and dilution, and initial conditions of pressure and temperature. For instance, λ varies with equivalence ratio φ as a U-shaped curve with a minimum close to $\varphi = 1$. The H_2-based reactive systems with oxidizer O_2, N_2O, or NO_2 present a different behavior, where the minimum lies along a plateau $\varphi = 0.5$–1. These systems, however, return to the classical behavior when highly diluted by Ar or N_2. The dependence of cell size on initial pressure follows $\lambda \sim A p_0^{-\alpha}$ with

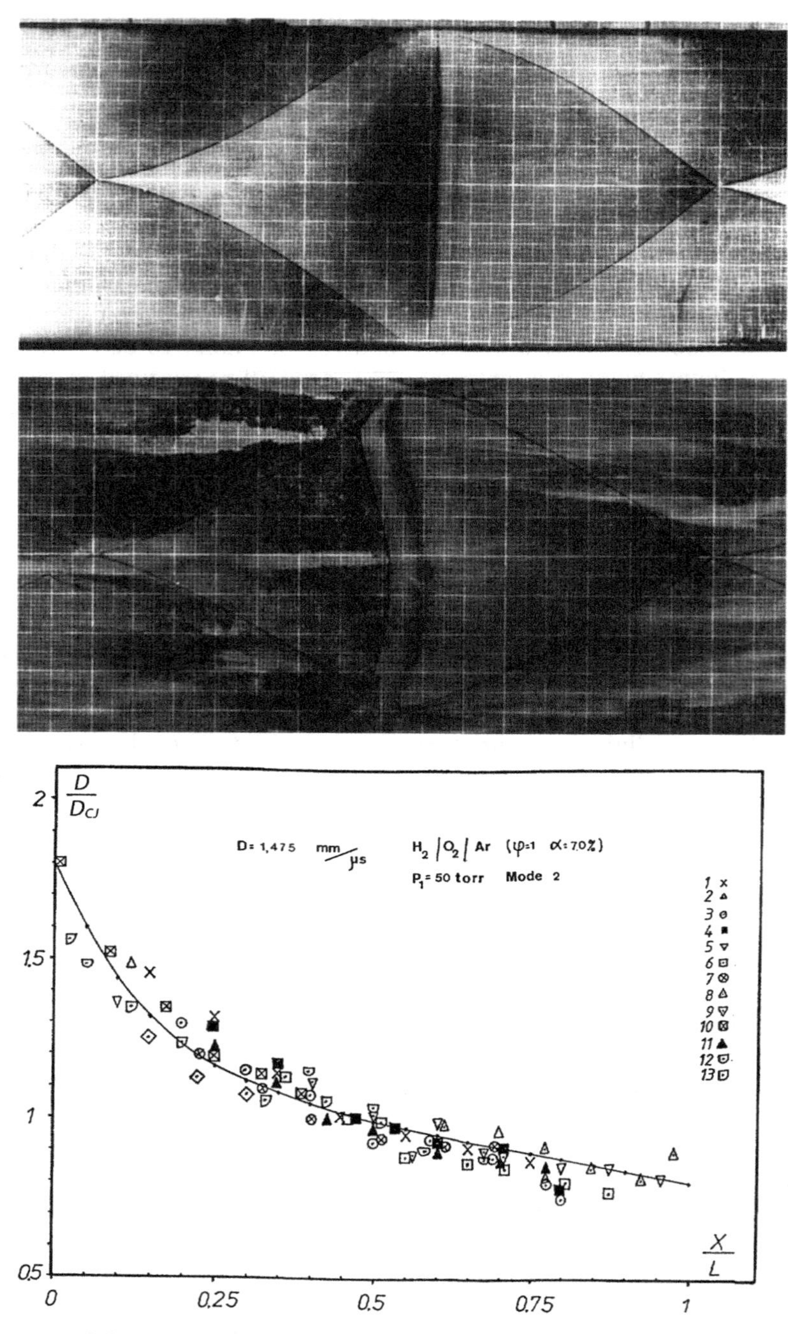

Fig. 5.12. Self-similarity of detonation cells in a rectangular tube 9.2 × 3.2 cm, $H_2/O_2/Ar$ mode 2 (*top*) and $CO/H_2/O_2/CF_3Br/Ar$ mode 5 (*middle*). *Bottom*: measured axial velocity versus normalized distance through the cell [29]

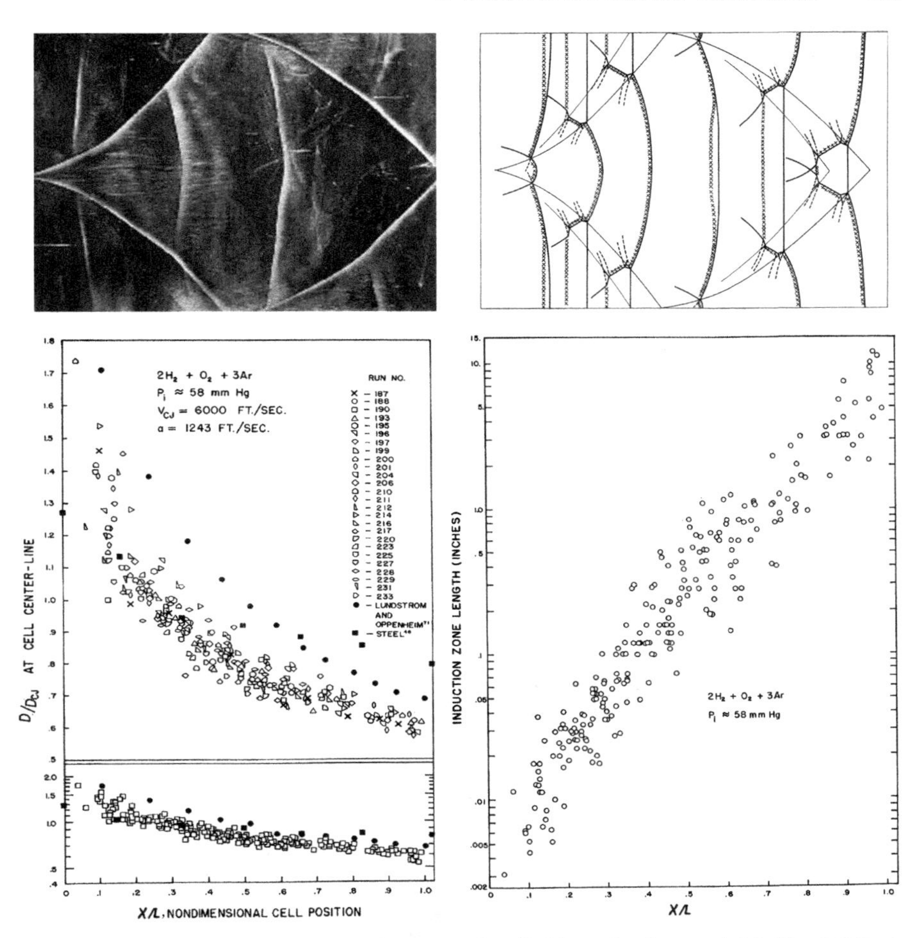

Fig. 5.13. *Top*: detonation cell (soot tracks (*left*) and scheme (*right*)) of $2H_2 + O_2 + 3Ar$ mixture ($p_0 = 7$ kPa) in a 3×0.25 in. rectangular tube. *Bottom*: measured axial velocity (*left*) and calculated induction zone length (*right*) versus normalized distance through the cell [17]

$1.1 < \alpha < 1.2$ in the range 1–1,000 kPa. Stamps and Tieszen [92] showed an exception for the H_2–O_2 system with a non-monotonous cell size variation with initial pressure. The cell size variation with the initial temperature T_0 is more complex and depends not only on the chemical system but also on the pressure and density [2,3,12–15]. A large body of data on detonation cell dependence on various parameters can be found in Chap. 4.

Atoms (He, Ne, Ar, Kr, etc.), diatomic molecules (N_2, O_2) in excess with respect to stoichiometry ($\varphi < 1$), H_2 or fuels in excess with respect to stoichiometry ($\varphi > 1$) and triatomic molecules (CO_2, H_2O, etc.) can behave as gaseous diluents. They are, however, thermodynamically and chemically

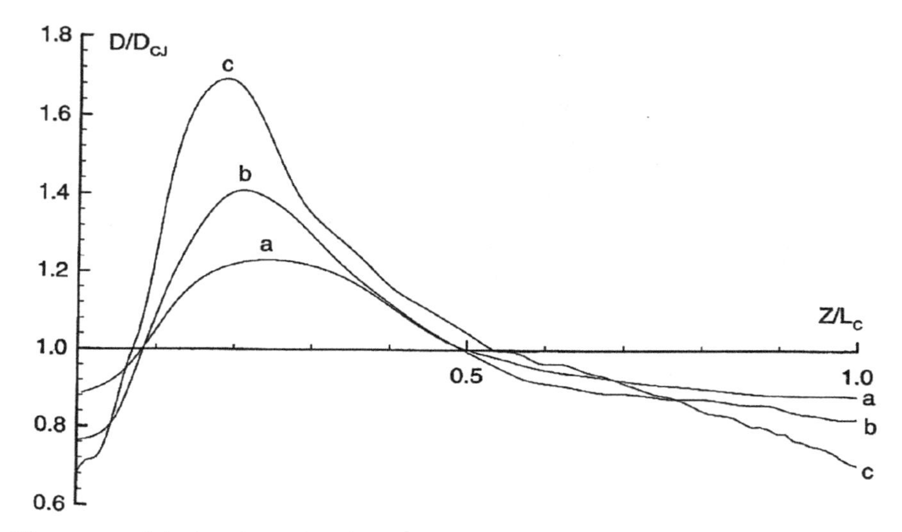

Fig. 5.14. Calculated axial scaled velocity versus normalized distance through the cell for an irreversible Arrhenius law with $E_a/RT_{\rm ZND} = 2.1$ (a), 4.9 (b) and 7.4 (c) [38]

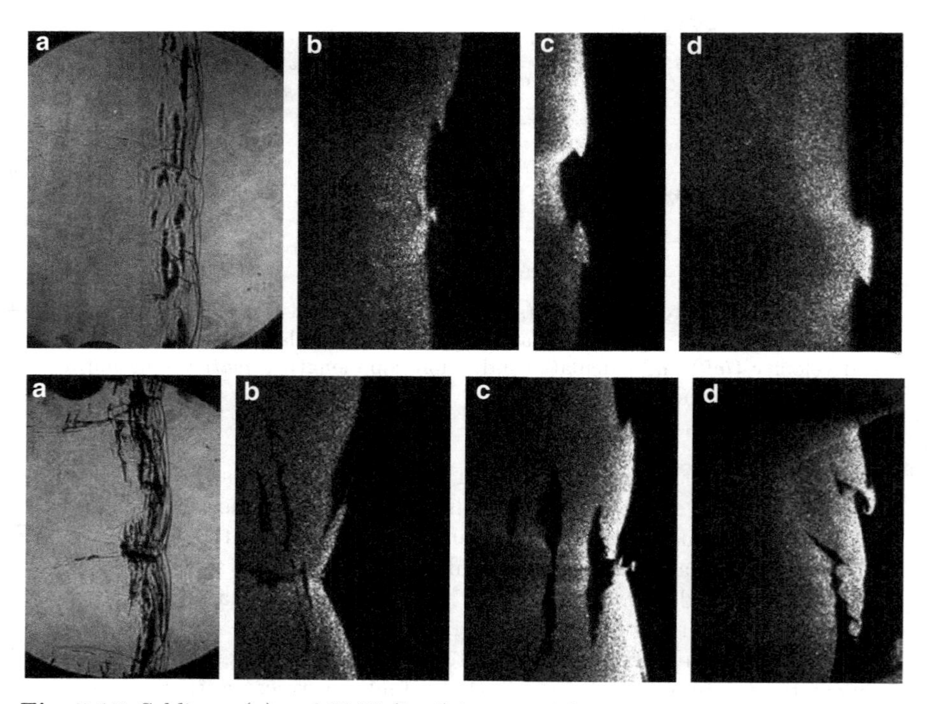

Fig. 5.15. Schlieren (**a**) and PLIF (**b**–**d**) images of detonation fronts in $H_2+0.5O_2+10Ar$ at $p_0 = 20$ Pa (*upper*), and in $H_2 + 0.5O_2 + 3N_2$ at $p_0 = 18$ Pa (*lower*) [89]

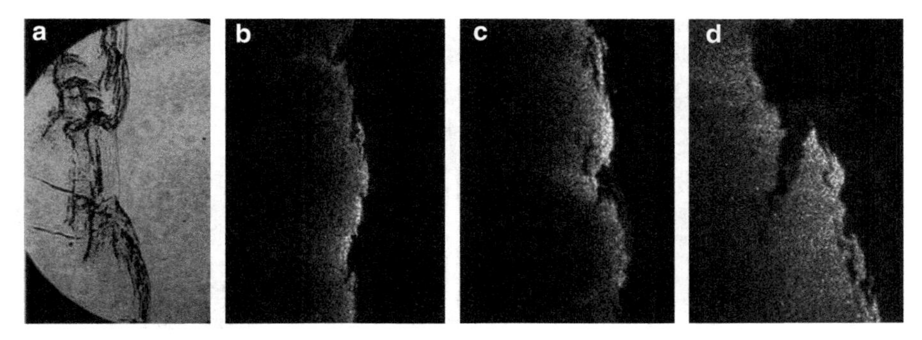

Fig. 5.16. Schlieren (**a**) and PLIF (**b**–**d**) images of detonation fronts in $C_2H_4 + 3O_2 + 9N_2$, $p_0 = 20\,\mathrm{Pa}$ [89]

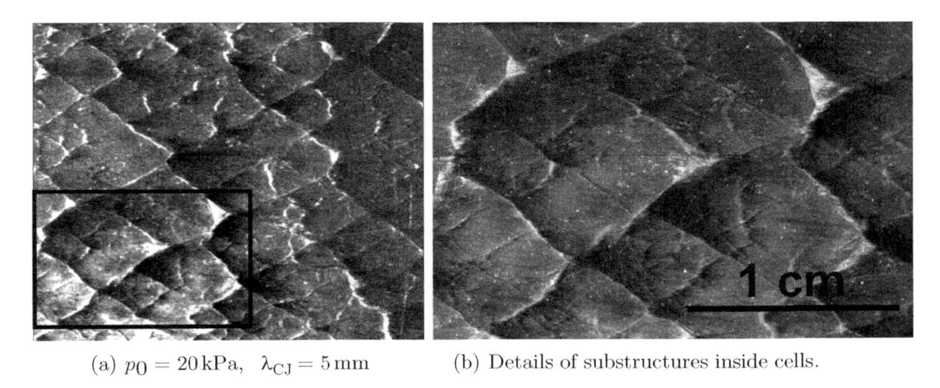

(a) $p_0 = 20\,\mathrm{kPa}$, $\lambda_{\mathrm{CJ}} = 5\,\mathrm{mm}$ (b) Details of substructures inside cells.

Fig. 5.17. Soot tracks of near-CJ self-sustained detonation in a $C_3H_8{+}5O_2$ mixture. The detonation propagates from left to right [21]

active. The cell size generally increases with increasing dilution. An exception arises again with the light fuel H_2. For instance, small and moderate dilution of H_2–O_2 mixture with a heavy monatomic diluent such as Ar or Kr decreases the cell size.

The cell size significantly depends on the detonation regime. Cell widths are usually measured for near-CJ ($D \sim D_{\mathrm{CJ}}$) self-sustained detonations and are named λ_{CJ}. For the overdriven regime ($D > D_{\mathrm{CJ}}$), λ is much smaller than λ_{CJ} as shown in Figs. 5.18 and 5.19. While the cell shape is not drastically changed (trajectory angle remains 30°), the regularity noticeably increases with the increasing degree of overdrive. In overdriven detonations, the transverse waves become weaker (towards the sonic waves) and the cell appears as a very regular lozenge. For $C_3H_8 + 5O_2$, beyond $D/D_{\mathrm{CJ}} \sim 1.8$ cells can no longer be observed, which is probably due to the complete endothermicity of the reaction, as suggested by Gordeev [43]. The variation of $\lambda/\lambda_{\mathrm{CJ}}$ as a function of D/D_{CJ} (>1) is a rapidly decreasing curve as shown in

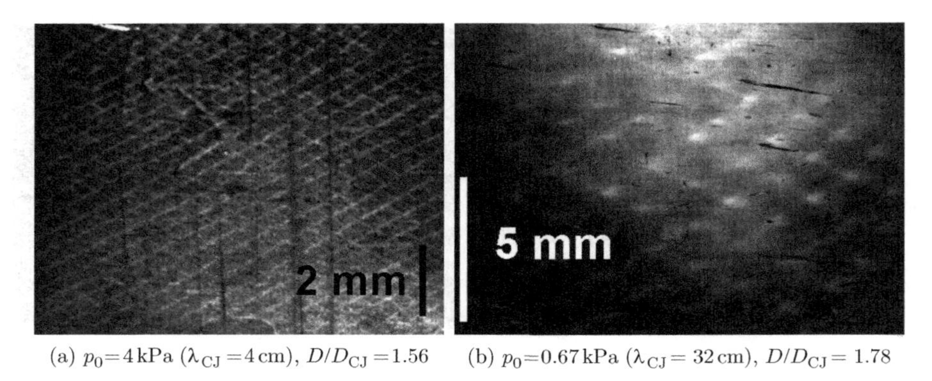

(a) p_0=4 kPa (λ_{CJ} =4 cm), D/D_{CJ} =1.56 (b) p_0=0.67 kPa (λ_{CJ}= 32 cm), D/D_{CJ}= 1.78

Fig. 5.18. Soot tracks of overdriven detonation in a $C_3H_8 + 5O_2$ mixture. The detonation propagates from left to right [21]

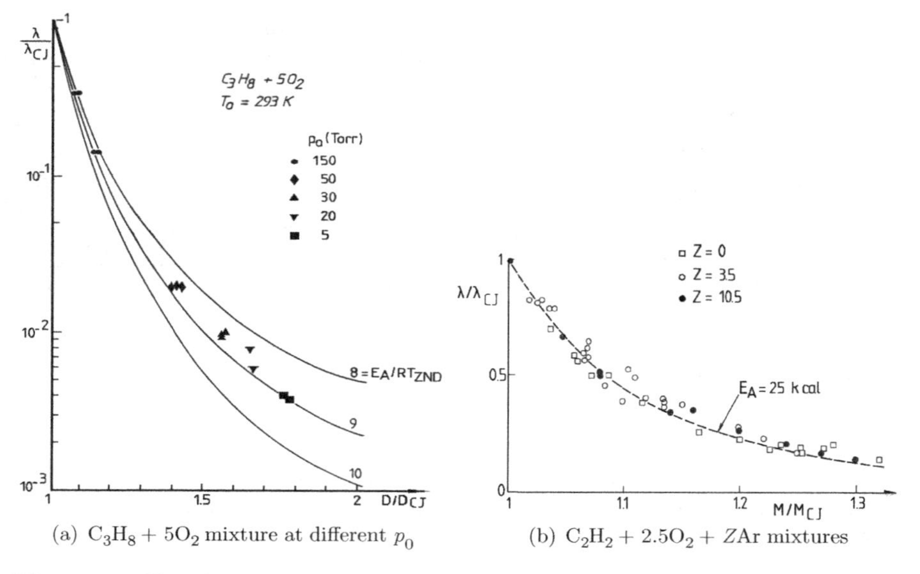

(a) $C_3H_8 + 5O_2$ mixture at different p_0 (b) $C_2H_2 + 2.5O_2 + Z$Ar mixtures

Fig. 5.19. Non-dimensional experimental cell size λ/λ_{CJ} versus overdrive factor D/D_{CJ} and theoretical curves [20, 21]

Fig. 5.19 [20, 21, 41–43, 66, 70, 105]. The cell size decreases with D/D_{CJ} for the $C_3H_8 + 5O_2$ mixture up to $D/D_{CJ} = 1.8$, where cells can no longer be observed, and for $C_2H_2 + 2.5O_2 + Z$Ar mixtures up to $D/D_{CJ} = 1.3$. Results were obtained using the detonation shock tube technique [23].

The correlation of the cell size with overdrive factor, D/D_{CJ}, follows approximately:

$$\frac{d\lambda/\lambda}{dD/D} \sim -2(E_a/RT_{ZND}), \qquad (5.1)$$

Table 5.1. Detonation cell size of stoichiometric fuel–O_2 mixtures in standard initial conditions

Fuel	C_2H_2	C_2H_4	C_2H_6	C_3H_8	H_2	CH_4	NH_3
λ (mm)	0.1	0.5	0.9	0.9	1.4	2.5	7

Table 5.2. Detonation cell size of stoichiometric fuel–air mixtures in standard initial conditions

Fuel	C_2H_2	H_2	C_2H_4	C_2H_6	JP10	C_3H_8	CH_4
λ (mm)	6	10	20	50	55	60	280

where an irreversible Arrhenius reaction law is applied. Equation (5.1) provides good agreement with the experiments, as shown in Fig. 5.19. For different mixtures, the curve slope near the CJ point ($\lambda/\lambda_{CJ} = 1$, $D/D_{CJ} = 1$) can vary from about 10 (e.g., $E_a/RT_{ZND} \sim 5$ for C_2H_2–O_2, H_2–O_2, $\varphi = 1$) to much larger values (i.e., 20–30 for fuel–air mixtures).

A database for λ is provided on the web site http://www.galcit.caltech.edu/EDL/ [90]. These internationally collected data cover common reactive mixtures. The cell size values of some stoichiometric fuel–O_2 and fuel–air mixtures at ambient conditions ($T_0 = 293\,\text{K}$, $p_0 = 100\,\text{kPa}$) are also given in Tables 5.1 and 5.2. Uncertainty of cell size measurements is less than 20% for fuel–O_2 mixtures and 20% or more for fuel–air mixtures. In the tables, fuels whose cell size is smallest (or largest) when mixed with oxygen also have the smallest (or largest) cell size when mixed with air, except for H_2 since the addition of N_2 to H_2–O_2 increases the density of the mixture and therefore reduces the cell size increase.

5.2.2.4 λ and l_i Correlation

Schelkhin and Troshin [87] first attempted to correlate ZND reaction zone length l_i to cell size λ. For a few reactive mixtures they found a linear proportionality relationship. The extensive investigations of Strehlow et al. [99], Strehlow and Engel [98] and Strehlow [95] showed that when the recombination zone is negligible compared with the induction zone, direct proportionality between the cell width, λ, and induction zone length, l_i, holds:

$$\lambda = Al_i, \tag{5.2}$$

λ is one to two orders of magnitude larger than the chemical length l_i, used as a reference scale. When the recombination zone becomes of the order of the induction zone, no simple correlation was found. Westbrook et al. [121] calculated the 1D chemical induction length for different reactive mixtures using detailed chemical kinetic mechanisms in order to predict dynamic parameters of detonation such as critical tube diameter and direct initiation energy. Comparing the measured cell sizes for stoichiometric H_2–O_2, fuel (CH_4, C_2H_2, C_2H_4, C_2H_6, C_3H_8)–O_2 and fuel–air mixtures with the same fuels to the corresponding reaction zone lengths, they found a value for $A = 29$. This value is an average between 10 (for fuel–air mixtures) and 50 (for fuel–O_2 mixtures). Some questions remain about the choice of the chemical length to be used in the correlation [88] and consequently the approximation of $A =$ constant is questionable. In practice, this approximation is often useful even in fuel-blends/air mixtures [7] since cell size measurements for real systems are not accurate, and the chemical scheme used may not always be well adapted to detonation conditions.

Different cell models based on a strong point source explosion to provide estimation of the shock decay were used to predict cell size, as described in Chap. 4. An acoustic ray trapping theory has also been used for the same goal [4].

All of the above studies were based on a one-step Arrhenius heat release law. Recent studies clearly show that the chemical reaction of most of the reactive mixtures cannot be represented by this simple heat release model. In Chap. 3, an attempt to improve the semi-empirical prediction (5.2) was made by considering the factor A as a function of a stability parameter, i.e., $A(\chi)$. Here, the stability parameter χ is the ratio of the characteristic induction length to reaction length and was introduced to quantify the detonation stability when driven by chain-branching kinetics, as a replacement for the activation energy used in the one-step Arrhenius law. ZND reaction zone calculations involving detailed chemical kinetics demonstrate that heat release may proceed by two successive exothermic steps. This will be explained in the following sections on chemical reaction mechanisms with some new arguments associated with experimental results and numerical simulations.

5.2.2.5 Spinning Detonation Structure

Since the early work of Voitsekhovskii [117] and Voitsekhovskii et al. [118], Duff [30] and Schott [85] on the structure of single-head spinning detonation, many researchers have been devoted to this particular detonation regime (e.g., Voitsekhovskii et al. [119], Huang et al. [51], and Zhang et al. [125]). The spinning detonation wave is nearly steady, but it propagates without any transverse-wave collision. The transverse-wave structure is strong and the triple-point trajectory is helical (Figs. 5.20 and 5.21). The triple point prints a band, on a soot film covering the inside of a circular tube, of constant or periodically variable width (transitions from weak transverse structure to

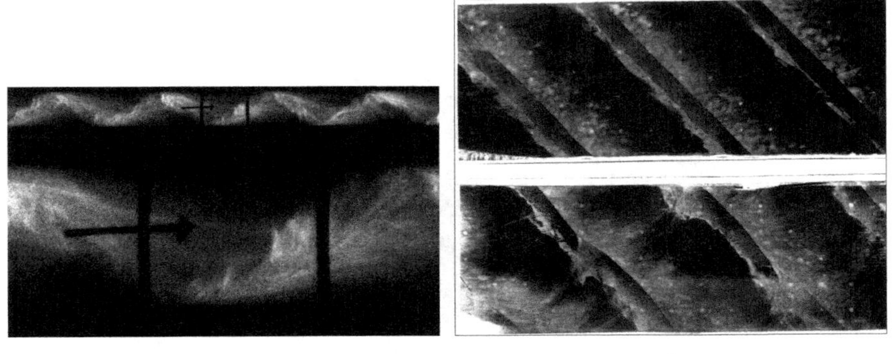

Fig. 5.20. Self-luminosity (*left*) and smoked foil records (*right*) of spinning detonation in a $d = 2.54\,\mathrm{cm}$ round tube in a $6.7C_2H_2 + 10O_2 + 83.3Ar$ mixture [85]

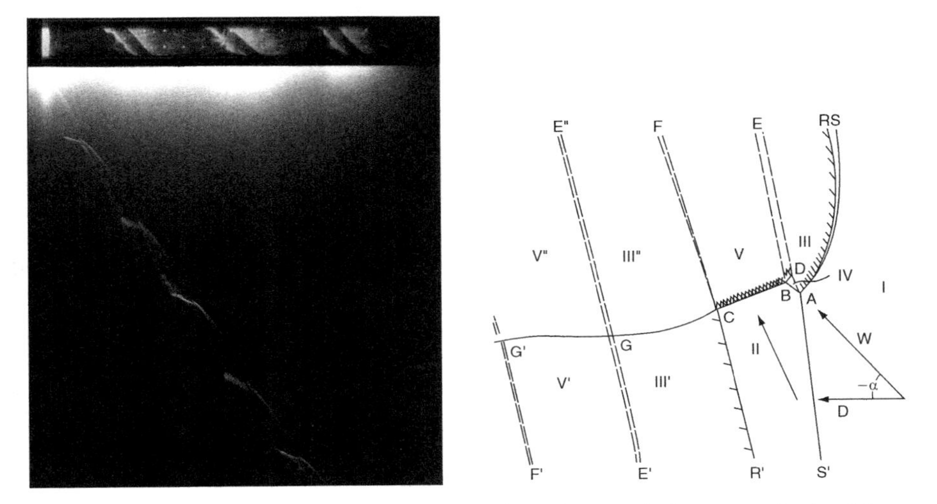

Fig. 5.21. Spinning detonation x (*horizontal*)—t (*vertical*) diagram, obtained from axial slit sweeping photograph of luminosity in a i.d. $d = 2.54\,\mathrm{cm}$ glass round tube in a $6.7C_2H_2 + 10O_2 + 83.3Ar$ mixture (*left*), and instantaneous structure of spinning detonation wave (*right*) [85]

a strong one and then to weak again, as observed in Fig. 5.20), including generally a fine cellular structure resulting from a super detonation. The angle between the band and the tube axis being about 45°, the super detonation propagates along the tube wall with a velocity equal to about $1.4D$, where D is the longitudinal mean velocity of the spinning detonation. D can be 10–15% lower than D_{CJ}. As indicated by Voitsekhovskii et al. [118], the spinning detonation wave could be slightly overdriven. The periodical spin pitch P, roughly equal to πd (d being the tube diameter), can be predicted on the

Fig. 5.22. Spinning detonation smoked foils in a square cross-sectioned tube in a $6.7C_2H_2 + 10O_2 + 83.3Ar$ mixture [60]

basis of the acoustic theories of Manson [64] and Fay [36]. For highly unstable detonation, up to the first third of the inter-band, there appears a zone of quasi-constant thickness with cellular substructure, corresponding to the Mach stem propagation.

As the spinning detonation regime is marginal, a small decrease in initial pressure or a slight variation of mixture composition can lead to its failure. The detonation limit is generally expressed by a correlation between a hydrodynamic characteristic length, for instance the pitch P ($\sim\pi d$) resulting from tube acoustics, and the cell width λ as the typical detonation length, i.e., $\lambda \sim P$. This relation is mostly valid for H_2–O_2 and fuel–O_2 systems close to stoichiometry. For the same systems but far from stoichiometric or diluted by inert gases (Ar, N_2, etc.) the existence of spinning detonations can be observed for a large range of initial pressure (i.e., for $\lambda > P$).

Variation of detonation velocity along the tube wall inside the inter-band (pitch) is in general larger than that along an ordinary detonation cell. Spinning detonation can propagate in square cross-sectioned tubes as well (Fig. 5.22).

5.2.3 Two-Cell Structure

When the oxidizer NO_2/N_2O_4 (denoted in the following only by NO_2) is used instead of O_2, a more complex detonation cellular structure appears. Indeed, the detonation of fuel–NO_2 mixtures can produce a double cellular structure called a *two-cell detonation* [52–54, 58, 79, 80, 102]. NO_2 may also be found in a monopropellant, nitromethane (CH_3NO_2) for instance. Depending mainly on the equivalence ratio, detonation of these systems can produce a double-cellular structure. Two-cell detonations, consisting of

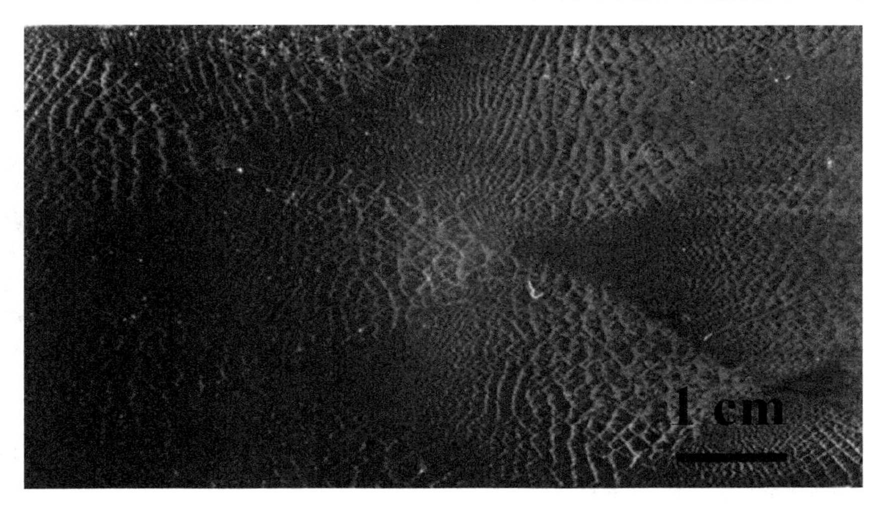

Fig. 5.23. Soot tracks of two-cell detonation in a H_2–NO_2 mixture, $\varphi = 1.1$, $p_0 =$ 50 kPa. The detonation propagates from left to right [54]

a small cellular structure within a larger one, appear mostly not only in fuel-rich mixtures but also in very lean mixtures. The size of the smaller cells increases continuously within the larger cell from its beginning to its end in the direction of detonation propagation. This is due to the fact that the smaller structure is subjected to the continuously decreasing shock wave strength of the larger cell. Two-cell structure can be clearly seen on soot records given in Figs. 5.23 and 5.24 for rich H_2–NO_2 mixtures, Fig. 5.25 for gaseous CH_3NO_2–O_2 rich mixtures ($\varphi = 1.3, 1.5$) and Fig. 5.26 for very lean gaseous nitromethane/tetranitromethane (CN_4O_8) mixtures. The double cellular structure was initially discovered in these mixtures.

One can see in Fig. 5.23 that the smaller cell size increases along the larger cell (size λ_2) axis in the direction of detonation propagation by at least an order of magnitude. In the middle of the larger cell, the smaller cell size λ_1 is roughly 20 times smaller than λ_2. The ratio λ_2/λ_1 depends on φ (Fig. 5.27). For $\varphi < 1$, the detonation of the same H_2–NO_2 mixture shows only classic one-cell-like structure (Fig. 5.28). Increasing the equivalence ratio from $\varphi = 0.8 - 1$ to $\varphi > 1$ leads to the emergence of the larger cell, superimposed on the net of single small cells, which evolve from constant size ($\varphi = 0.8 - 1$) to periodic size increase ($\varphi > 1$).

Calculations [26] in very lean H_2–NO_2 mixtures predicted the existence of a two-cell structure, which has been recently experimentally confirmed in very lean gaseous nitromethane/tetranitromethane mixtures [81]. As displayed in Figs. 5.23, 5.24 and 5.28, the H_2–NO_2 reactive system produces fairly regular single and double cells in nearly stoichiometric and fuel-rich conditions. The double detonation cellular structure of reactive systems, such as CH_4, C_2H_6 with NO_2, CH_3NO_2 and $C_3H_7NO_3$, is less regular.

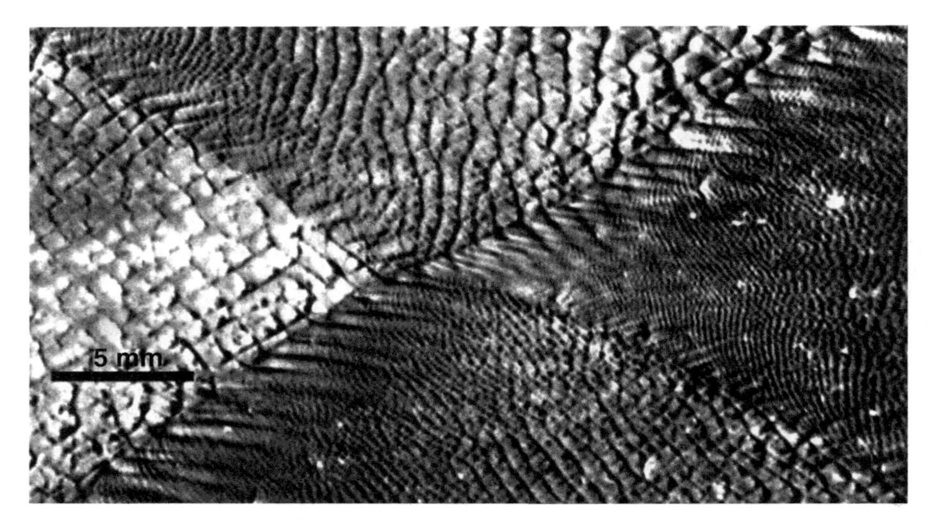

Fig. 5.24. Soot tracks of two-cell detonation in H_2–NO_2–Ar mixture (30% Ar, $\varphi = 1.2$, $p_0 = 100\,\mathrm{kPa}$) show the details of the transverse-wave collision. The detonation propagates from left to right

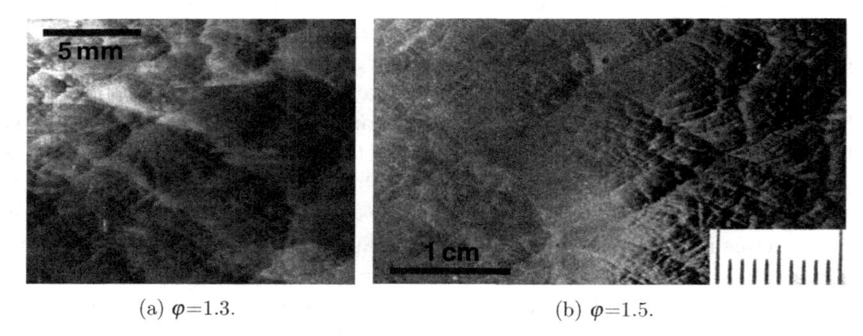

(a) φ=1.3. (b) φ=1.5.

Fig. 5.25. Soot tracks of two-cell detonation in a CH_3NO_2–O_2 mixture, $p_0 = 80\,\mathrm{kPa}$, $T_0 = 383\,\mathrm{K}$. The detonation propagates from left to right

Unexplained specific features displayed on soot track records include:

1. Accumulation of soot just after the collision of the transverse waves of the larger cells (Fig. 5.29a)
2. Particular soot tracks of an α shape at the beginning of some large cells in some records (Fig. 5.29b)

Similar to the classic one-cell detonation, the spinning two-cell detonation has a transverse cellular structure in the narrow spinning band but exhibits a cellular net of increasing size along the inter-band. The pitch P follows $P \sim \pi d$ where d is the tube internal diameter, with the band track angle θ also close to 45°. An example of a spinning detonation in pure gaseous nitromethane is

Fig. 5.26. Soot tracks of double cellular structure obtained in a very lean ($\varphi = 0.38$) gaseous nitromethane/tetranitromethane mixture at $p_0 = 10\,\text{kPa}$ and $T_0 = 360\,\text{K}$. The detonation propagates from left to right [81]

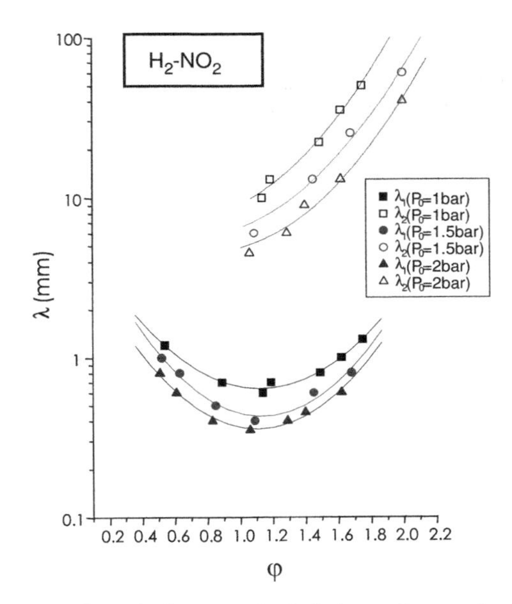

Fig. 5.27. Experimental equivalence ratio (φ) dependence of cell sizes in H_2–NO_2 mixtures for $p_0 = 100$–150 and $200\,\text{kPa}$. λ_1 is measured in the middle of the larger cells [54]

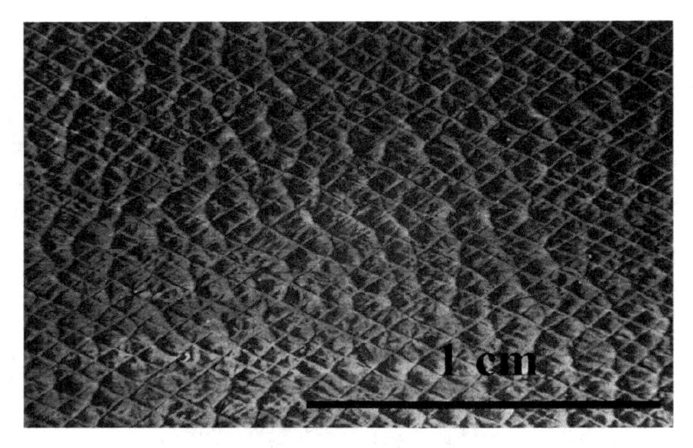

Fig. 5.28. Soot tracks of single cell detonation in a H_2–NO_2 mixture, $\varphi = 0.9$, $p_0 = 100\,\text{kPa}$. The detonation propagates from left to right [54]

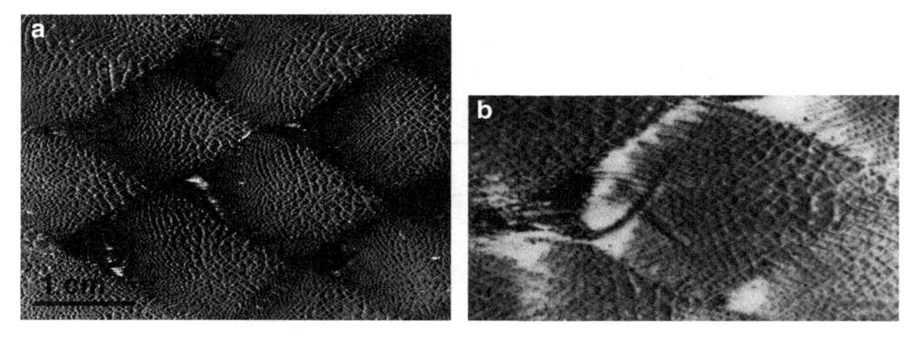

Fig. 5.29. (**a**) Soot accumulation at the beginning of cells in a H_2–NO_2 mixture, $\varphi = 1.2$, $p_0 = 100\,\text{kPa}$. (**b**) α shape observed in similar conditions

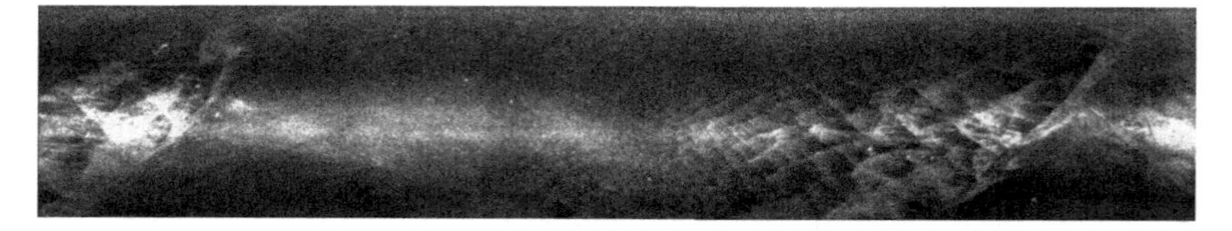

Fig. 5.30. A spinning detonation smoked foil for gaseous *nitromethane* in a round tube ($d = 52\,\text{mm}$, $P \sim 150\,\text{mm}$), $p_0 = 30\,\text{kPa}$, $T_0 = 383\,\text{K}$ [102]

given in Fig. 5.30. Figure 5.31 shows that at the end of the inter-band, the smaller cell tracks are covered obliquely by fine cells inside the band.

The smaller cell size λ_1 increases by a factor of about 100 through the inter-band and its orientation θ with respect to the tube axis varies from $-45°$ (as band angle) to $+12°$(Fig. 5.32). The λ_1 variation inside the band is similar to the variation of l_i through the classic cell given in Fig. 5.13.

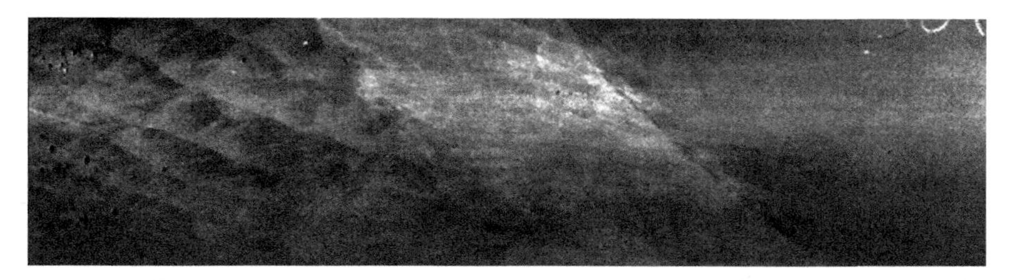

Fig. 5.31. Enlargement of the spinning track of the gaseous *nitromethane* detonation in a round tube ($d = 52\,\mathrm{mm}$), $p_0 = 30\,\mathrm{kPa}$, $T_0 = 383\,\mathrm{K}$ [102]

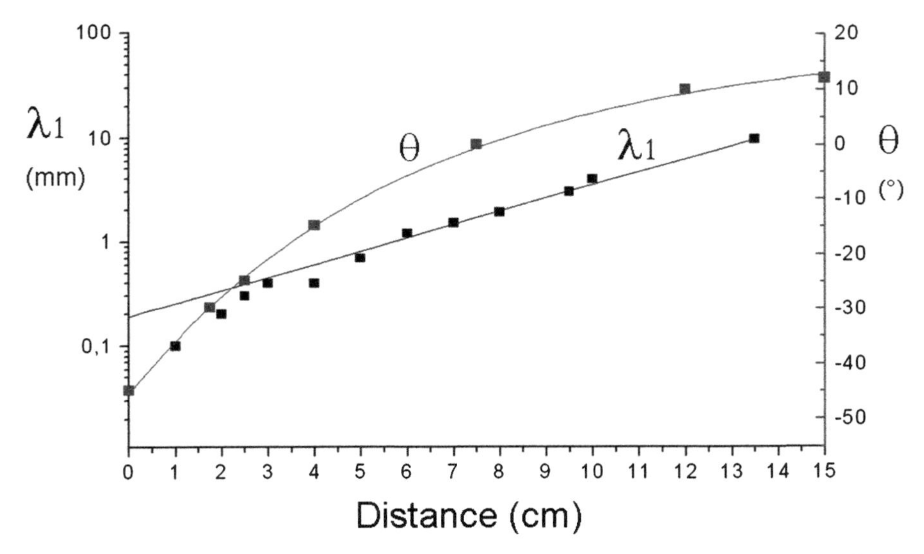

Fig. 5.32. Variation of smaller cell size and angle of inclination of small cell relative to the tube axis through the inter-band of the spinning detonation of Fig. 5.31 [101]

As it will be shown later (Sect. 5.5), the spinning detonation regime of the two-cell structure presented above is not the limiting self-sustained detonation regime of such mixtures; recent observations suggest the existence of a "low velocity one-cell detonation" (LVD) whose cellular structure is the finer cell of the two-cell detonation.

The above observations clearly show that the finer cell of the two-cell detonation differs from the substructure observed inside the classic detonation cellular structure of highly unstable reactive mixtures. Explanation of the origin of this two-cell detonation structure and its consequences for the detonation propagation regimes are presented in the next sections.

5.3 Detonation Heat Release Mechanisms

Since the detonation cellular structure finds its origin in the chemical heat release mechanism, many ZND reaction zone calculations have been performed using detailed chemical kinetic mechanisms in many reactive systems and over a large range of initial pressure and temperature conditions. These studies lead to distinguish three typical characteristic chemical heat release laws:

1. The one-step heat release law with one maximum in the reaction rate history
2. The two-step heat release law involving two successive exothermic global reaction steps, characterized by two maxima in the reaction rate
3. The two-step hybrid law involving two successive exothermic global reaction steps but only the first one being characterized by a maximum in the reaction rate

A reactive fuel-oxidizer-diluent system may exhibit these three behaviors under different compositions (equivalence ratio, dilution) and initial conditions.

5.3.1 One-Step Heat Release

Inside a detonation, the classical one step heat release law consists of a shock-induced thermally neutral induction zone followed by a rapid or stiff recombination zone of monotonic heat release, leading to the complete reaction at the equilibrium CJ state. An example of the ZND reaction zone in a $C_2H_2 + 2.5O_2$ mixture at standard initial conditions is displayed in Fig. 5.33. The time between the leading shock and the maximum reaction rate provides the induction reaction time τ_i and the corresponding induction length l_i. A global one-step kinetics or, more recently, reduced chemical kinetic mechanisms that closely reproduce the evolution of thermodynamic parameters are

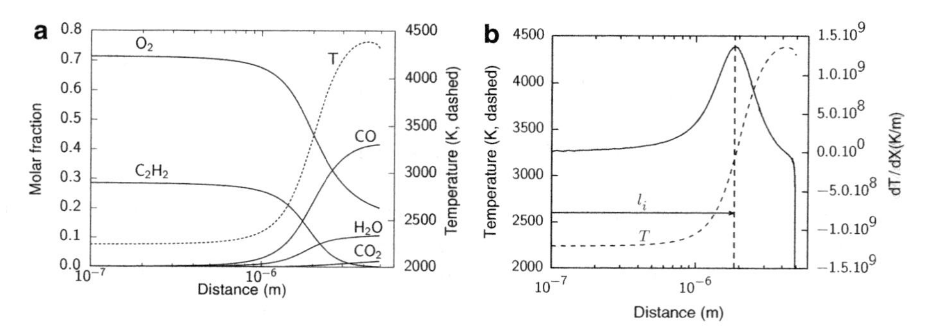

Fig. 5.33. Evolution of (**a**) chemical species (*solid curves*), (**b**) temperature (*dotted curve*) and time derivative of temperature (*solid curve*) in the ZND reaction zone in a $C_2H_2 + 2.5O_2$ mixture, $p_0 = 100\,\text{kPa}$ and $T_0 = 293\,\text{K}$ [44]

commonly used in numerical simulations of cellular detonations. The one-step heat release is fairly representative for the chemical reaction in H_2–O_2, hydrocarbons (CH_4, C_2H_6, C_3H_8, C_2H_4, C_2H_2, etc.)–O_2 and –N_2O mixtures near stoichiometry and also for these mixtures slightly diluted by monatomic, diatomic or triatomic gas in standard initial conditions.

Due to the stiffness of the recombination zone, the detonation reaction zone of these reactive mixtures has been often modelled in the past by the ZND square wave model with induction length depending on the global Arrhenius law whose activation energy E_a can vary, according to the fuel, from less than 20 kcal mol^{-1} to more than 50–55 kcal mol^{-1}. The detonation of such systems shows one cellular net with the possibility of a substructure (of the same origin) at the beginning of cells when E_a/RT_{ZND} becomes larger than 6. This one cellular net is representative of the one-step detonation energy release law.

5.3.2 Two-Step Heat Release

The two successive exothermic global reaction steps (Fig. 5.34) can be characterized by two different chemical times; each of them is composed of an induction zone and a recombination zone. Specifically,

- A first fast reaction step provides a short induction period and a first elevated maximum of the reaction rate that defines a delay τ_{i_1} and corresponding first induction length l_{i_1}.
- A second reaction step results in a secondary maximum of the reaction rate (of lower amplitude than the first one) with a delay τ_{i_2} and a corresponding induction length l_{i_2}. l_{i_2} is at least one order of magnitude larger than l_{i_1} for the systems studied up to now [52, 53, 58, 81, 102].

Stoichiometric, rich and very lean fuel–NO_2 mixtures (H_2–NO_2 (Fig. 5.34), CH_4–NO_2 and C_2H_6–NO_2) exhibit such behavior. The first exothermic reaction step corresponds to the NO_2 decomposition into NO (approximately up to the maximum of NO production) and the second one to the NO decomposition into N_2.

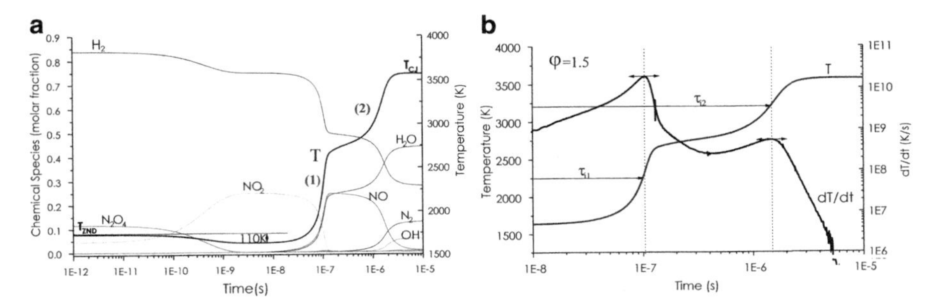

Fig. 5.34. Evolution of (**a**) chemical species, (**b**) temperature and time derivative of temperature in the ZND reaction zone of a H_2–NO_2 mixture, $\varphi = 1.5$, $p_0 = 100$ kPa and $T_0 = 293$ K

For the simple H_2–NO_2 system, the two main global reactions involved are the following:

1st Step:

- For $\varphi \leq 0.5$:
  $\varphi H_2 + \frac{1}{2}NO_2 \longrightarrow \frac{1}{2}NO + \varphi H_2O + \frac{1}{2}\left(\frac{1}{2} - \varphi\right) O_2$
- For $\varphi \geq 0.5$:
  $\varphi H_2 + \frac{1}{2}NO_2 \longrightarrow \frac{1}{2}NO + \frac{1}{2}H_2O + \left(\varphi - \frac{1}{2}\right) H_2$

2nd Step:

- For $\varphi \leq 1$:
- $\left.\begin{array}{l}\frac{1}{2}NO + \varphi H_2O + \frac{1}{2}\left(\frac{1}{2} - \varphi\right) O_2 \\ \frac{1}{2}NO + \frac{1}{2}H_2O + \left(\varphi - \frac{1}{2}\right) H_2\end{array}\right\} \longrightarrow \frac{1}{4}N_2 + \varphi H_2O + \frac{1}{2}\left(1 - \varphi\right) O_2$
- For $\varphi \geq 1$:
  $\frac{1}{2}NO + \frac{1}{2}H_2O + (\varphi - \frac{1}{2})H_2 \longrightarrow \frac{1}{4}N_2 + H_2O + (\varphi - 1)H_2$

In the two-step heat release, each maximum of the reaction rate is responsible for a specific cellular net, thus leading to two-cell detonations for stoichiometric, rich and very lean fuel–NO_2 mixtures (Figs. 5.23 and 5.24). The second maximum of the reaction rate is responsible for the larger cells while the inside smaller cells are induced by the first maximum of the reaction rate. This has been confirmed by direct numerical simulations (Sect. 5.4). Cell sizes, λ_1 and λ_2, can be linearly correlated to their respective induction lengths l_{i_1} and l_{i_2} (calculated via the detailed chemical scheme of Djebaili-Chaumeix et al. [28]). For instance, for H_2–NO_2 mixtures (Fig. 5.35), $\lambda_1 \sim 15 l_{i1}$ and $\lambda_2 \sim 20 l_{i2}$. Each reaction step corresponds to a chemical energy release, q_1 and q_2, respectively ($q_1 + q_2 = q$, total heat energy release). For the H_2–NO_2 mixture, q_1 and q_2 are of the same order.

In pure gaseous nitromethane (CH_3NO_2) or rich CH_3NO_2–O_2 mixtures, the two-cell detonation follows the two-step heat release law. Melius [69] developed a two-step reaction model to study nitromethane ignition. For nitromethane, l_{i_2} is two to three orders of magnitude larger than l_{i_1} (Fig. 5.36). For CH_3NO_2–O_2 mixtures, in the range $0.2 < \varphi < 1.75$, $\lambda_1 \sim 250 l_{i1}$ and, for $1.3 < \varphi < 1.75$, $\lambda_2 \sim 20 l_{i2}$ [102].

It is notable that this two-step heat release mechanism was already proposed and indirectly observed through a double flame structure (Fig. 5.37) in rich fuel–NO_2 mixtures [8, 76], in gaseous nitro and nitrate monopropellants [49] and in double-base propellants.

5.3.3 Two-Step Hybrid Heat Release

Between the one-step and the two-step heat release laws, a two-step hybrid law can be established. It consists of two exothermic reaction steps, but with only

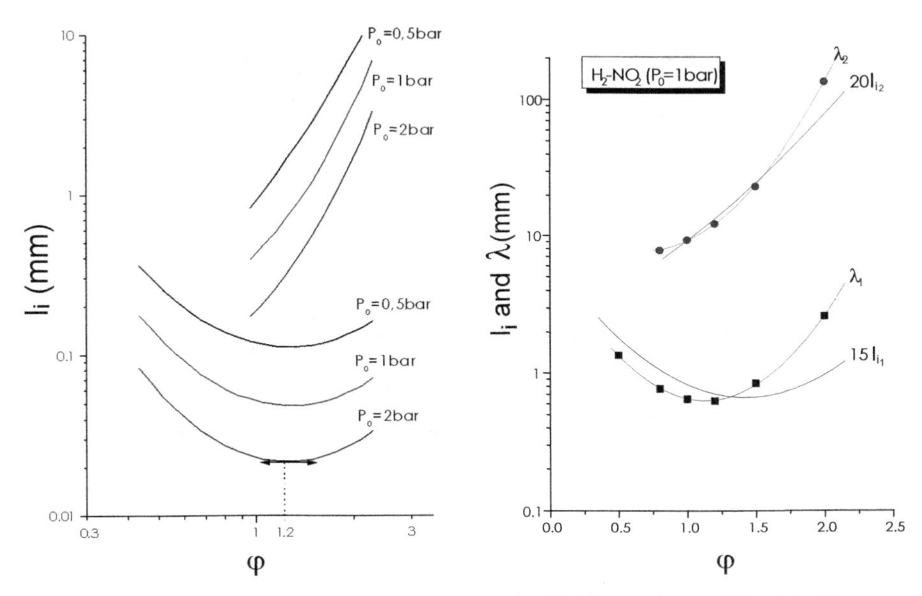

Fig. 5.35. Variation of l_{i_1} (*lower curves*) and l_{i_2} (*left*), and λ_1 and λ_2 (*right*) versus φ in a H_2–NO_2 mixture

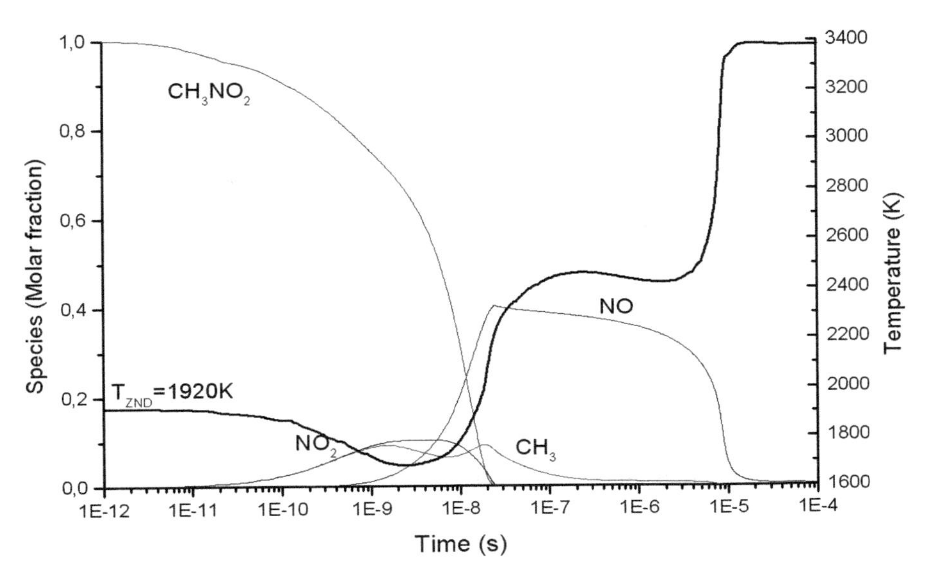

Fig. 5.36. Time evolution of chemical species and temperature in the ZND reaction zone of CH_3NO_2, $p_0 = 100\,\text{kPa}$, $T_0 = 383\,\text{K}$

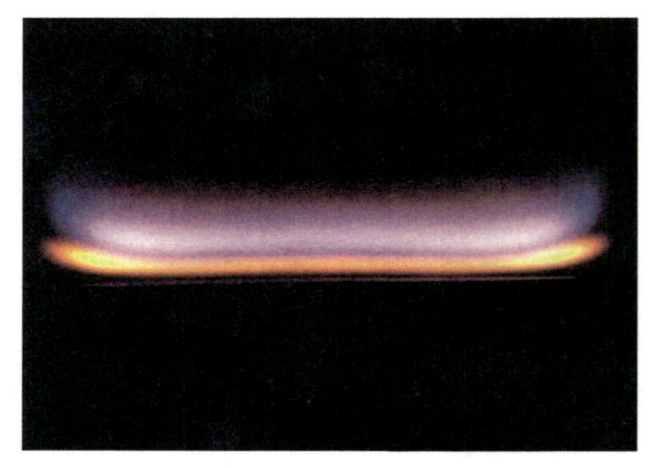

Fig. 5.37. Laminar flat double-flame in a CH_4–NO_2–O_2 mixture, $p_0 = 6.66\,\mathrm{kPa}$ [8]

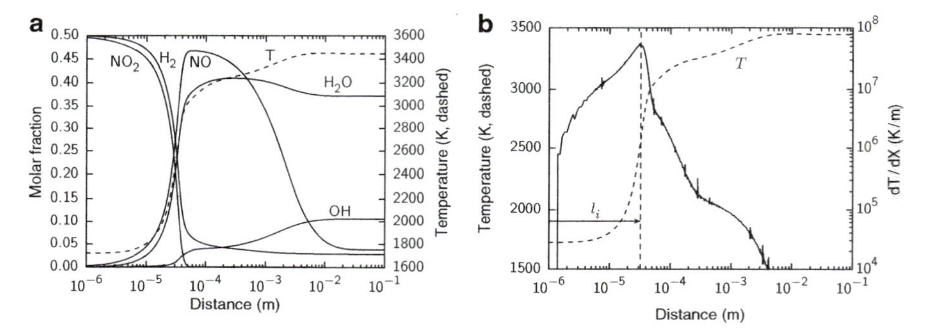

Fig. 5.38. Evolution of (**a**) chemical species, (**b**) temperature and time derivative of temperature in the ZND reaction zone of a H_2–NO_2 mixture, $\varphi = 0.5$, $p_0 = 100\,\mathrm{kPa}$ and $T_0 = 293\,\mathrm{K}$

one maximum reaction rate associated with the first fast exothermic reaction step. Therefore, only one induction length l_i associated with the first step can be determined, as the second reaction step gradually brings about the CJ state (Fig. 5.38). With this mechanism which elongates the reaction zone at least by two orders of magnitude, only one cellular net is observed (Fig. 5.28). It is the manifestation of the first exothermic step but supported by the total energy release. This class of heat release law is relevant to chemical reactive systems such as lean H_2–NO_2, CH_4–NO_2, and lean CH_3NO_2–O_2 mixtures.

It may also describe the detonation in classical chemical systems of off-stoichiometric fuel/O_2 mixtures and stoichiometric mixtures diluted by inert gas (He, Ar, etc.) and N_2, O_2, CO_2, etc. As pointed out by Varatharajan and Williams [108, 109], oxidation in many of these reactive mixtures (C_2H_2,

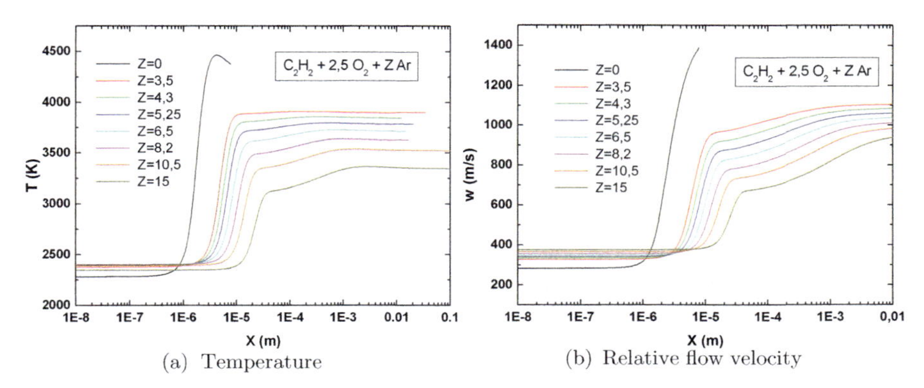

(a) Temperature (b) Relative flow velocity

Fig. 5.39. ZND reaction zone in $C_2H_2+2.5O_2+Z\mathrm{Ar}$ mixtures in function of dilution Z, $p_0 = 100\,\mathrm{kPa}$, $T_0 = 293\,\mathrm{K}$ [44]

C_2H_4, C_3H_8, $C_{10}H_{16}$ with O_2 and a diluent) can be reduced to a two-step chemical kinetic mechanism with two very different characteristic times. For example, a C_2H_2–O_2–Diluent system:

$$C_2H_2 + 1.5O_2 \longrightarrow 2CO + 0.5H_2O + 0.5OH + 0.5H \tag{5.3}$$

$$\begin{aligned} &2CO + 0.5H_2O + 0.5OH + 0.5H + cO_2 \longrightarrow \\ &\qquad (2-a)CO_2 + aCO + (1-b)H_2O + 2b(OH + rH)/(1+r), \end{aligned} \tag{5.4}$$

where a, b, c, r are parameters depending on the mixture composition and detailed chemical kinetics. The first step reaction leads to chain branching thermal explosion followed by CO oxidation and radical-recombination for the second step.

Using a detailed chemical kinetic mechanism [28], calculation of the detonation reaction zone in a $C_2H_2 + 2.5O_2$ mixture diluted by argon illustrates this chemical behavior (Figs. 5.39 and 5.40). Here, the separation between the two steps corresponds approximately to the maximum of CO production. As shown in the ZND reaction profiles of temperature T (Fig. 5.39a), relative velocity w (Fig. 5.39b) and main species molar fractions (Fig. 5.40), an increase in Ar dilution leads to a relative increase in the length of the second exothermic step.

Up to 50% Ar dilution ($Z = 3.5$), the second step has a relatively low contribution to the total heat release. Beyond 50% dilution, its contribution becomes noticeable and can reach 47% of the total heat release for very large Ar dilution ($Z = 60$ for instance, as seen in Fig. 5.40b). The length or duration of the second step is at least two orders of magnitude larger than that of the first one.

Other common fuel-oxygen diluted mixtures, namely $C_2H_2+2.5O_2+9.4N_2$ (C_2H_2/Air, $\varphi = 1$) and $C_2H_2+2.5O_2+9.4O_2$ ($\varphi = 0.21$) at ambient conditions,

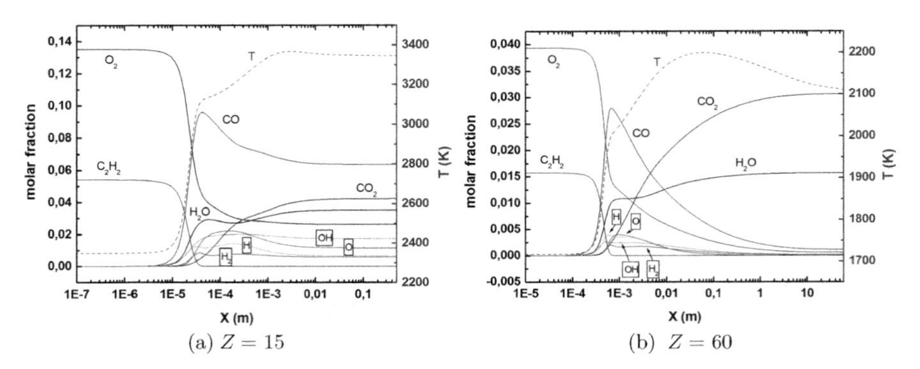

Fig. 5.40. ZND reaction zone chemical species and temperature variations in $C_2H_2 + 2.5O_2 + Z$Ar mixtures, $p_0 = 100\,\text{kPa}$, $T_0 = 293\,\text{K}$ [44]

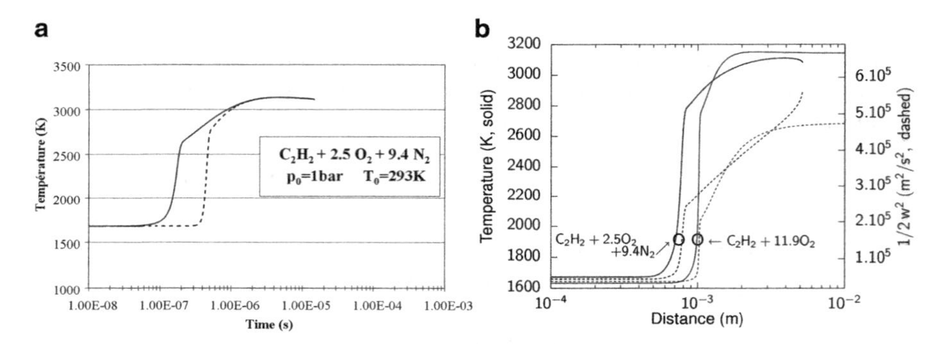

Fig. 5.41. (**a**) Temperature variation versus time in ZND reaction zone of $C_2H_2 + 2.5O_2 + 9.4N_2$ mixture, $p_0 = 100\,\text{kPa}$, $T_0 = 293\,\text{K}$. *Solid line*: detailed kinetic mechanism of Djebaili-Chaumeix et al. [28]. *Dotted line*: reduced mechanism of Varathanajan and Williams [108]. (**b**) Temperature (*solid line*) and specific kinetic energy (*dashed line*) versus distance in ZND reaction zone of $C_2H_2 + 2.5O_2 + 9.4N_2$ and $C_2H_2 + 2.5O_2 + 9.4O_2$, $p_0 = 100\,\text{kPa}$, $T_0 = 293\,\text{K}$. GRI 3.0 kinetic mechanism

are examined in Fig. 5.41. For these mixtures, calculations of ZND reaction zones from different chemical kinetic mechanisms provide results which are in qualitative agreement. The duration or length of the second reaction step is about one order of magnitude larger than that of the first step.

In summary, whatever diluent is added to the $C_2H_2 + 2.5O_2$ mixture, the detonation heat release law turns from one-step reaction for the baseline mixture to the hybrid two-step law. Thus, the most important feature of the hybrid two-step law is that it applies to an extended reaction zone length as demonstrated in all the above examples. The regularity of detonation cellular structure essentially depends on the nature of the diluent.

For example, with dilution by diatomic (N_2 or O_2) or triatomic molecules, the shocked temperature T_{ZND} decreases with the CJ detonation velocity

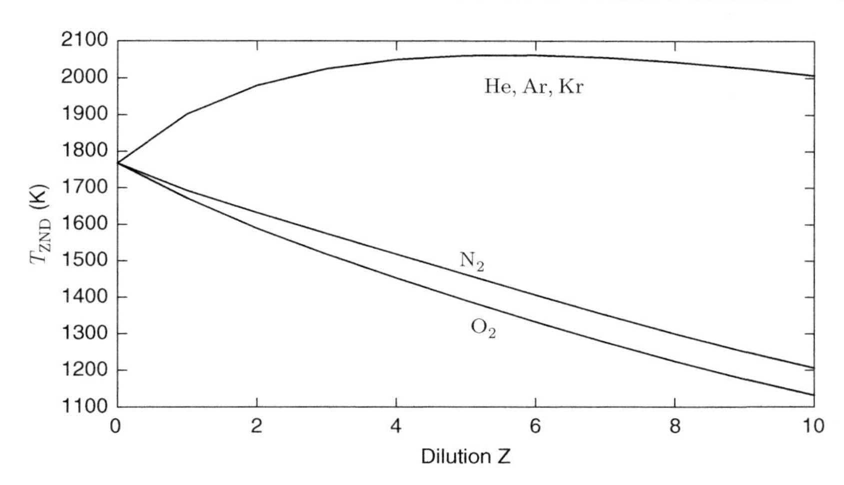

Fig. 5.42. T_{ZND} in function of Z in $2H_2 + O_2 + ZI$ mixtures, I = He, Ar, Kr, N_2 and O_2 under standard initial conditions

as the dilution increases (Fig. 5.42). The final cell regularity is determined by the competition between the decreased T_{ZND} (unstable factor) and the enlarged reaction zone length (stable factor). For monatomic gases (He, Ar, Kr, etc.), however, in spite of the CJ detonation velocity being reduced when more diluent is added, T_{ZND} increases with dilution by an increment around 200K at $Z = 5.25$ as compared with $Z = 0$ in Fig. 5.42 for $2H_2 + O_2 + ZI$ mixtures and at $Z \approx 10$ in Fig. 5.39a for $C_2H_2 + 2.5O_2 + Z$Ar mixtures. For the later mixtures towards higher dilution, T_{ZND} decreases down to 1,720 K at $Z = 60$ (Fig. 5.40b), compared to the CJ value of about 2,250 K without dilution (Fig. 5.39a). Hence, an increase in cell regularity in fuel–O_2 mixtures through high dilution of a monatomic gas cannot be explained with E_a/RT_{ZND} using a one-step Arrhenius law. The fact of a significant increase in reaction zone length with energetic contribution from the second reaction step leads to a progressive energy release. According to the description in Chap. 3 where a ratio of the characteristic induction length to reaction length was introduced to be a stability parameter controlling the detonation stability driven by chain-branching kinetics, an increase in reaction zone length results in a decrease in the length ratio of induction to reaction zone and therefore a decrease in detonation instability. Consequently small disturbances in the flow will not cause significant fluctuations of the energy release rate in the reaction zone, thus leading to a more stable system with more regular detonation structure.

A three-step reaction mechanism may be expected for certain problems such as a two-phase system containing a two-step detonable gaseous reactive mixture laden with fuel droplets or dust.

5.4 Direct Numerical Simulation

5.4.1 Simulation of One-Step/One-Cell Detonation

Numerical simulations have been performed including the early work of Taki and Fujiwara [104], Oran and coworkers [48, 74, 75], Markov [67], Schöffel and Ebert [84] and many others, in order to improve our knowledge of the detonation cellular structure beyond what can be studied using experimental diagnostics. Basically, these simulations consist of solutions to the Euler or Navier-Stokes governing equations coupled with either (1) a single Arrhenius global reaction law, (2) an induction and a recombination Arrhenius law, or (3) a reduced chemical kinetics scheme within the capability of current computer technology. The polytropic gas hypothesis is generally used. The onset of the cellular structure in the planar ZND detonation wave may originate from a variety of disturbances including local composition heterogeneity, local transverse explosion and numerical noise. The non-linearity of the governing equations, especially with the chemical kinetics, leads to the development of the entire cell structure.

2D and 3D numerical simulations of detonation structure of these systems using the single-step Arrhenius reaction law show the onset of instabilities leading to cellular detonation, whose cell regularity depends strongly on the value of E_a/RT_{ZND}. Up to now, calculated cell sizes and cellular structure globally follow the experimental trends. The calculated structure, however, is in general in the lower instability regime, cell sizes are not exactly predicted and the substructure is only qualitatively correct. Due to the stiffness of the non-linear coupling between hydrodynamics and kinetics, very fine numerical grids and high capability computers are needed to solve the 2D or 3D reaction zone with a sufficiently high resolution and accuracy. Numerical simulations and experimental results are most similar for very regular cellular systems (characterized by low values of E_a/RT_{ZND}) in conditions of favourable acoustic coupling (Fig. 5.43).

The simulated details of the three-wave interaction features [75] can be used to gain better understanding of the complex flow, particularly the formation of unburned pockets separated from the leading front by the interaction of transverse waves as experimentally observed (Figs. 5.15 and 5.44). The mean spatial distribution of these pockets is directly linked to cell-head number density in the leading front. In simulations using the single step Arrhenius law, the volume of unreacted pockets was found to increase when the instability control parameter E_a/RT_{ZND} was increased (Fig. 5.45). The existence of the unburned pockets and related instabilities could also be used to explain the experimental observation that the sonic mean locus can be located at two to ten cell widths behind the leading shock [34, 111].

Single cell spinning detonation has also been numerically simulated [40, 106, 112–114, 120]. The main characteristics of this detonation structure have been reproduced, but the band fluctuations and the cells inside the band have rarely been predicted, except for the work of Gamezo et al. [40].

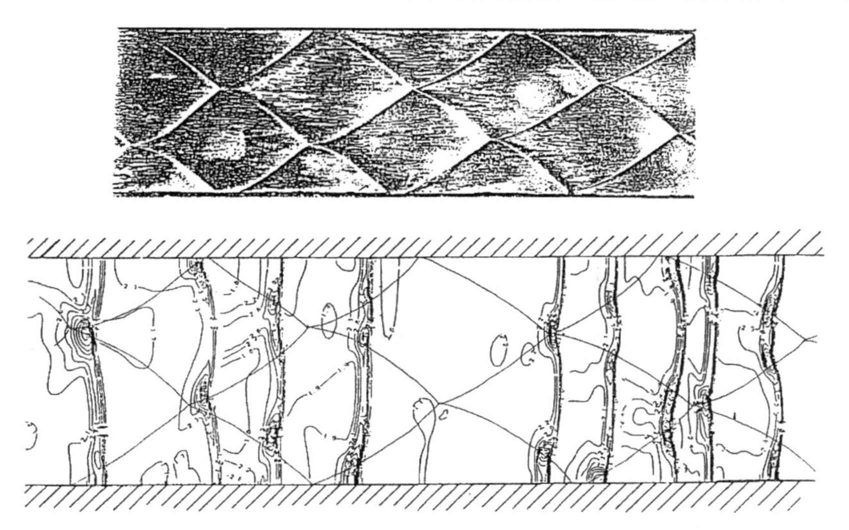

Fig. 5.43. Comparison of soot tracks [94] and 2D numerical simulation of cellular detonation [83]

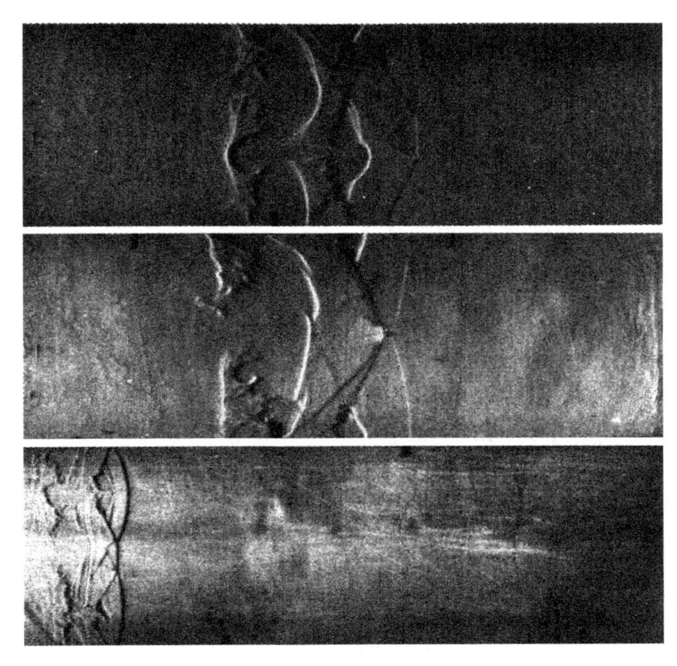

Fig. 5.44. Schlieren photographs of unreacted pockets behind detonation wave front propagating in a 3×0.25 in. tube containing a $H_2 + 0.5O_2 + 2.25Ar$ mixture for p_0 in the range of 6.7–10.7 kPa [75]

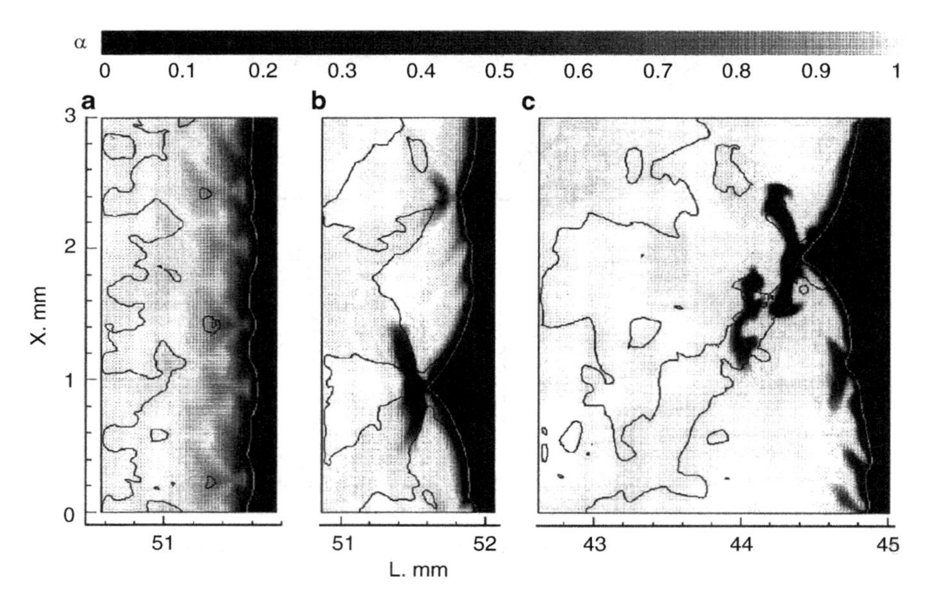

Fig. 5.45. Gray-scale contours of the reaction progress variable α calculated for $E_a/RT_{\mathrm{ZND}} = 2.1$ (**a**), 4.9 (**b**), 7.4 (**c**). The *white line* is the leading shock and the black contour is the *sonic line*. Unreacted pockets are important in (**c**) [39]

5.4.2 Simulation of Two-Step/One or Two-Cell Detonation

Guilly et al. [46] performed numerical 2D axisymmetric simulations of two-step detonations propagating in a 80 mm i.d. round tube without losses. Two irreversible successive exothermic reactions have been considered, i.e., A → B and B → C with two Arrhenius reaction rate laws, i.e.:

$$\mathrm{A} \rightarrow \mathrm{B}: \qquad \mathrm{d}\beta_1/\mathrm{d}t = K_1(1-\beta_1)^{n_1} \tag{5.5}$$

$$\mathrm{B} \rightarrow \mathrm{C}: \qquad \mathrm{d}\beta_2/\mathrm{d}t = K_2(\beta_1-\beta_2)^{n_2} \tag{5.6}$$

with $K_i = Z_i\rho^{n_i}\exp(-E_{a_i}/RT)$ and $\beta_i(0) = 0$ $(i = 1, 2)$. β_i's are reaction progress variables and they range from 0 to 1.

Here, $E_{a1} = E_{a2} = 250\,\mathrm{kJ\,mol^{-1}}$, $n_1 = n_2 = 1$, $Z_1 = 2.5 \times 10^{11}\,\mathrm{m^3\,kg^{-1}\,s^{-1}}$. The total chemical energy is taken constant and equally shared between the two steps, $q_1 = q_2 = q/2 = 2{,}200\,\mathrm{kJ\,kg^{-1}}$, for simulation of H_2–NO_2 mixtures at $\varphi = 1.2$.

By changing the Arrhenius pre-exponential coefficient Z_2 of the second chemical reaction while keeping the first Z_1 constant, the three classes of heat release laws discussed in the previous section were modelled. The results are summarized below.

1. When Z_2 is large (but less than Z_1: $Z_2 = 6 \times 10^{10}\,\mathrm{m^3\,kg^{-1}\,s^{-1}}$), the two exothermic reaction steps overlap, constituting a one-step reaction. Thus, classical one-cell detonations are numerically simulated.

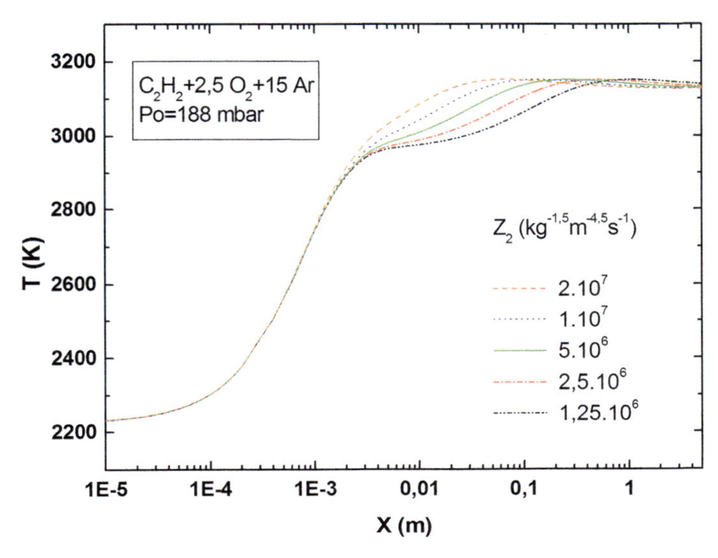

Fig. 5.46. ZND reaction zone temperature in a $C_2H_2 + 2.5O_2 + 15Ar$ mixture as a function of Z_2, Z_1(fixed)$> Z_2$, $p_0 = 18.8$ kPa, $T_0 = 293$ K [47]

2. When Z_2 is made smaller (e.g., $Z_2 = 1 \times 10^{10}\,\mathrm{m^3\,kg^{-1}\,s^{-1}}$), the first short-delay reaction step emerges, followed by the second slow reaction step without any second maximum in the reaction rate. In this case, the numerical detonation cell is induced by the first reaction step only and remains constant since the detonation is supported by the total energy of the two steps and has an elongated reaction zone.
   Particularly in $C_2H_2 + 2.5O_2 + 15Ar$ mixture (81% Ar dilution) at $p_0 = 18.8$ kPa, $T_0 = 293$ K, Guilly et al. [47] showed that the cell size remains constant at $\lambda = 10$ mm as Z_2 decreases (Fig. 5.46). This result is very important, since it shows that the separation between the two exothermic steps elongates the reaction zone while keeping the total energy release unchanged, thus leading to a constant detonation cell size with higher cell regularity.
3. With a further decrease in Z_2 (e.g., $Z_2 = 6 \times 10^8\,\mathrm{m^3\,kg^{-1}\,s^{-1}}$), the second reaction step separates from the first step and a second maximum appears in the reaction rate profile. In this case, the second reaction delay is about ten times the first one and a 2D double cellular structure is obtained (Fig. 5.47). The width of the small cells in the middle of the larger ones where $D \sim D_{CJ}$ is the same as that in case 2. This numerical two-cell detonation is qualitatively representative of the experimental one (Figs. 5.23–5.25).
   The detailed pressure field and reaction progress for the two-cell detonation are shown in Fig. 5.48. The coexistence of two nets of transverse waves is particularly evident, the first reaction zone is narrow and the second one is

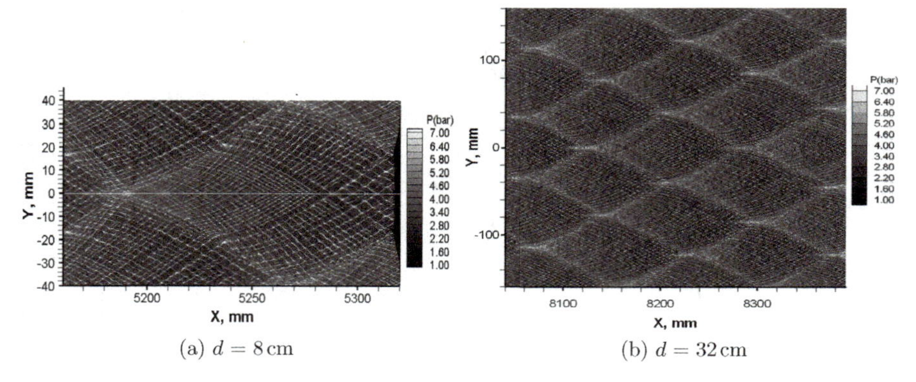

(a) $d = 8\,\text{cm}$ (b) $d = 32\,\text{cm}$

Fig. 5.47. 2D Numerical soot tracks of two-cell detonations propagating in round tubes of different inner diameters, d [46]

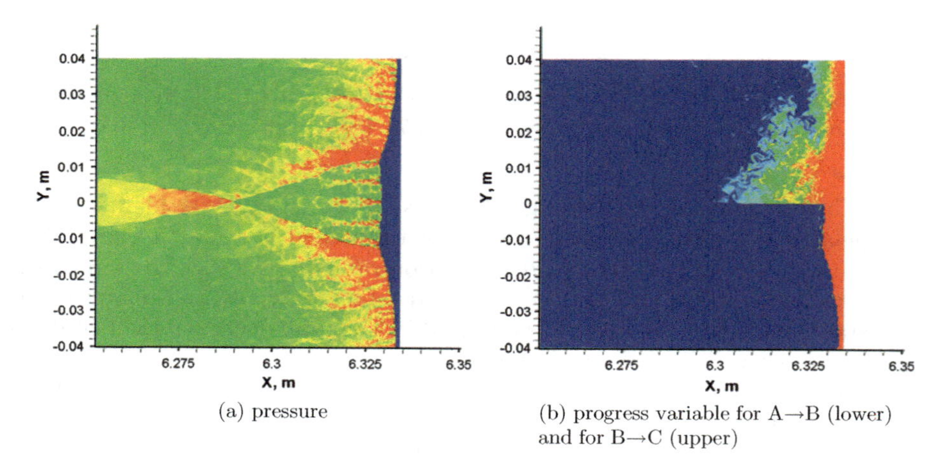

(a) pressure (b) progress variable for A→B (lower) and for B→C (upper)

Fig. 5.48. 2D Numerical two-cell detonation front [46]

extended. For each reaction step, unreacted pockets exist, fine for the first one and coarse for the second. Calculations fail to correctly resolve finer cells at the beginning of the larger ones because the mesh size is not small enough.

4. For very low Z_2 (e.g., $Z_2 = 3 \times 10^5\,\text{m}^3\,\text{kg}^{-1}\,\text{s}^{-1}$), the size of the larger cell exceeds the tube diameter. In such a way, only one cellular structure is present at the beginning of detonation propagation. This structure corresponds to the first chemical reaction step (Fig. 5.49) whose related chemical energy ensures alone the detonation propagation, while the second step is not yet activated. As the corresponding self-sustained detonation velocity is much lower than D_{CJ}, this detonation is termed a "low velocity detonation" (LVD). In this case, the cellular structure, representative of the

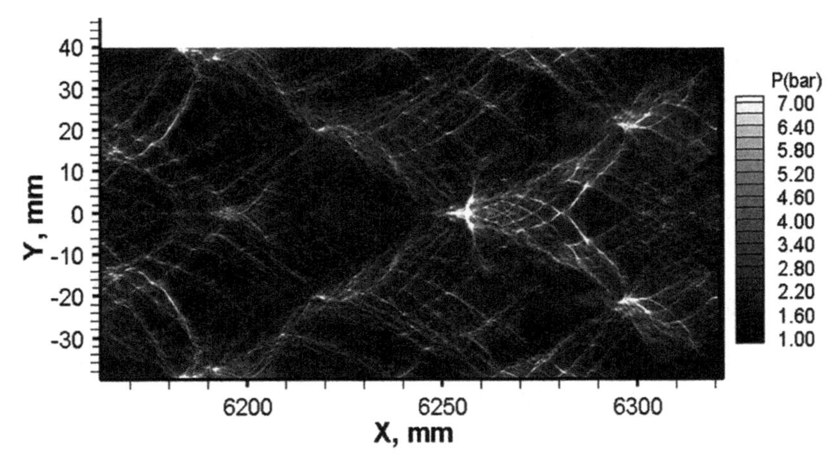

Fig. 5.49. Numerical soot tracks of self-sustained low velocity single-cell detonation propagating in a round tube of i.d. $d = 80\,\text{mm}$ [46]

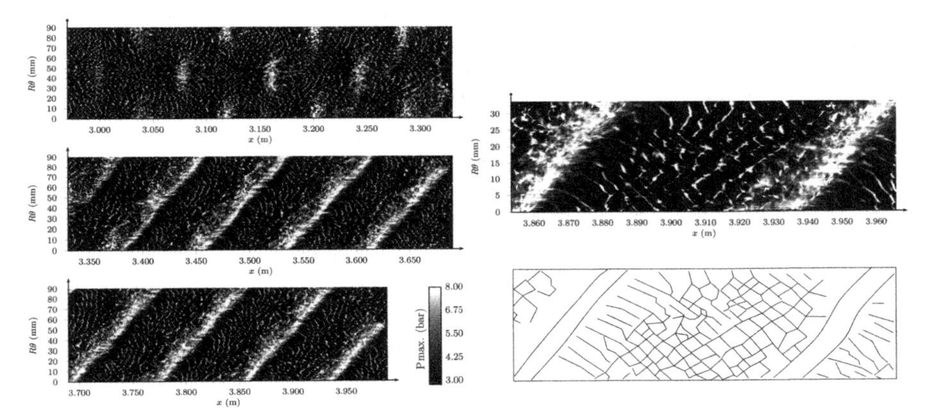

Fig. 5.50. Numerical soot tracks of onset of spinning two-cell detonation propagating in a round tube of i.d. $d = 28.7\,\text{mm}$ [116]

energy of the first reaction step only, is large and irregular with substructures in the first half of cells. This has been experimentally observed in some detonation propagation configurations (Sect. 5.5 and Fig. 5.60b).

Two-cell spinning detonations have been simulated by Virot [112] and Virot et al. [115,116], but with an insufficient resolution of the finer detonation cell structure in the band and immediately after it (Fig. 5.50).

In summary, the existence of the two-cell detonation structure was demonstrated both experimentally and numerically. This type of detonation can be reproduced numerically using a two-step heat release mechanism with two maxima in the heat release rate, characteristic reaction delay times

$\tau_{i_2}/\tau_{i_1} \sim O(10)$ and equivalent energy release for each step $q_2/q_1 \sim O(1)$. It would be interesting to verify whether this type of detonation can still be obtained experimentally and numerically for significantly different conditions, i.e., $\tau_{i_2}/\tau_{i_1} < O(10)$, or $q_2/q_1 \sim O(0.1)$ or $O(10)$.

To make progress in the numerical simulation of cellular structure in self-sustained detonations, it is necessary to check the validity of such calculations, at least by comparing calculated structures with experimental benchmarks that lead to transient detonation, such as, for instance, the sudden enlargement of the detonation tube, allowing the determination of the critical tube diameter for detonation transmission [55, 56].

5.5 Limits of Multi-Scaled Cellular Detonation

This section describes the link between the different types of heat release laws (Sect. 5.3) and the critical conditions for detonation existence as well as correlations between cell size and relevant characteristic hydrodynamic lengths.

Since the pioneering work of Zeldovich et al. [124], the induction length l_i and later the cell size λ have been considered to be fundamental length scales representing the detonability and relevant dynamic criteria, mainly for one-step/one-cell detonations (see Chap. 4). This has been the object of numerous experimental studies in the past with some unexplained departures from the classically accepted limit criteria. As seen above, cellular structures are intimately linked to detonation heat release laws which can reach great complexity for a wide range of potential reactive systems.

Detonation behavior in transient conditions is determined by a heat release law and associated cellular structure. Detonation propagation limits depend on the detonation cellular structure type (one-step/one-cell, two-step/one-cell or two-step/two-cell) as well as on the losses due to sudden expansion, wall friction or heat conduction.

The classic approach of detonation sensitivity based on a sole cell size needs to be revisited and extended because of the possibility of a second slow reaction step (as in the two-step hybrid heat release law) or because of the presence of a double cellular structure.

5.5.1 Limits of Expanding Detonation

In the case of an expanding detonation, in order to maintain periodic transverse-wave collisions and cell size, new transverse waves have to be created, otherwise the detonation fails. Expanding spherical detonation, ignited by a powerful explosion point source, becomes self-sustained if the radius of curvature R of the leading shock exceeds a critical value R_c. For one-step detonation, R_c is larger than λ [19,22] or l_i [16], respectively, by one or two to three orders of magnitude. For $R < R_c$, the detonation must be sustained by the initiation source energy. At the critical initiation of spherical detonation,

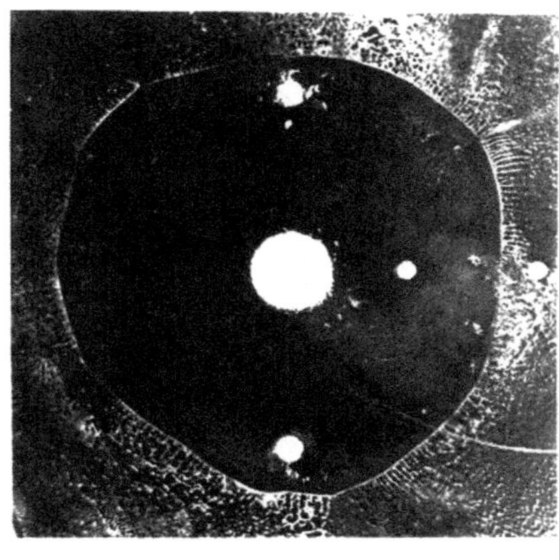

(a) 18×18 cm, $C_2H_2 + 2.5O_2$, $p_0 = 5.3$ kPa

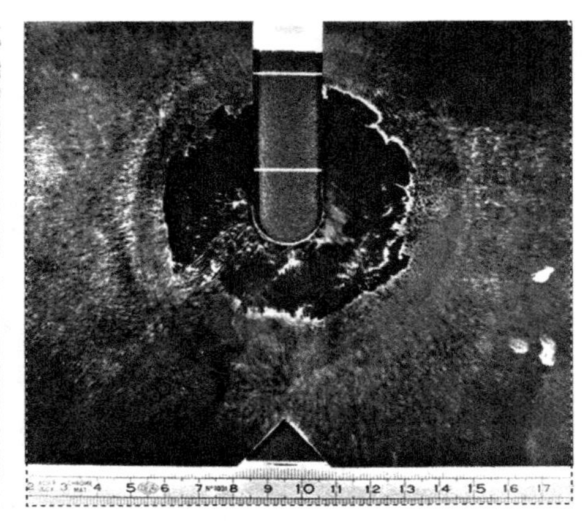

(b) 15×16.5 cm, $C_2H_2 + 2.5O_2 + 8Ar$, $p_0 = 74$ kPa

Fig. 5.51. Soot tracks of a pre-detonation zone of average radius R_c in critically initiated expanding spherical detonations. Detonation is initiated by an exploding wire [19, 21]

the energy of the point source is just sufficient for self-sustained detonation generation at $R \sim R_c$. Although reaction is ignited at different points (hot spots) in the layer of compressed unreacted material behind the curved decaying shock produced by the explosion source, it is possible to determine an average critical radius R_c at which the onset of spherical detonation takes place (Fig. 5.51). The critical radius was associated to the critical detonation initiation energy E_c by Zeldovich et al. [124], to the detonation critical kernel by Lee and Ramamurthi [59] and to the hydrodynamic thickness by Vasil'ev et al. [111], Edwards et al. [34] and Benedick et al. [5].

It has been shown by Desbordes [22] (Fig. 5.52) that R_c is close to 20λ for the $C_2H_2 + 2.5O_2$ mixture, characterized as one-step/one-cell detonation and $R_c \sim 40$–50λ for this mixture highly diluted by Ar (70–81%), characterized as a two-step/one-cell detonation [45].

Another means to initiate spherical detonation is to transmit a detonation from a tube with an internal diameter d into a large volume. When exiting the tube, the planar detonation undergoes a sudden expansion and can be quenched if the number of cells across the tube diameter is less than a threshold value. The extinction of transverse waves due to sudden expansion eliminates triple-point collisions, which leads to progressive disappearance of cellular structure and subsequent detonation failure (Figs. 5.53 and 5.54).

After the work of Zeldovich et al. [124], Mitrofanov and Soloukhin [71] and many researchers [5, 35, 68, 110], it is well known that the critical tube diameter, d_c, which permits detonation transmission from a tube to a large volume for H_2–O_2 and hydrocarbon-O_2 mixtures (not far from stoichiometry

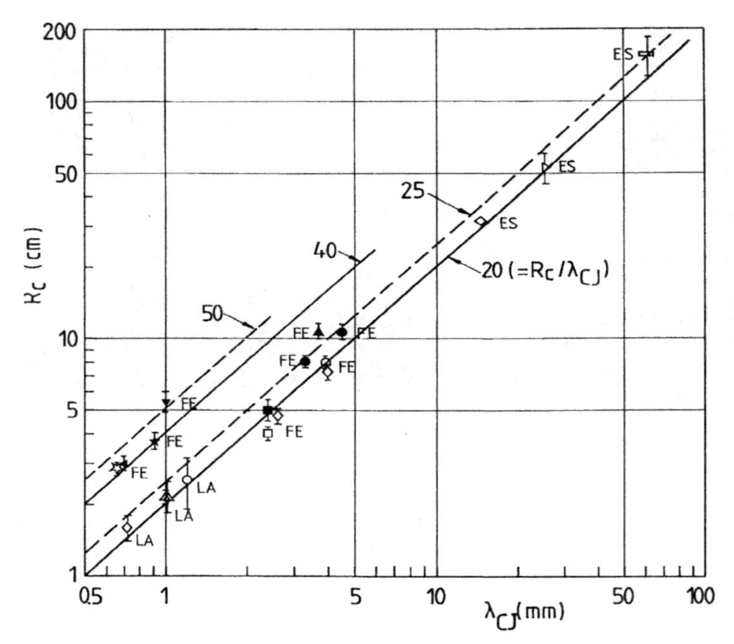

Critical predetonation radius R_c versus cell size λ_{CJ} in various reactive mixtures for different initiation sources. (FE : Exploding wire - LA : Laser spark - ES : High Explosive) ◆ C_2H_2+ O_2 - ● C_2H_2+ 2.5 O_2 - ▲ C_2H_2 + 2.5 O_2 + 3.5 Ar (50% Ar) ▲ C_2H_2 + 2.5 O_2 + 9 Ar (72% Ar) ★ C_2H_2 + 2.5 O_2 + 10 Ar (75% Ar) ▼ C_2H_2+ 2.5 O_2 + 13 Ar (79% Ar) - ■ C_2H_2+ 2.5 O_2 + 5 N_2 - H_2 + Air (ϕ=1) - C_2H_4 + Air (ϕ=1.3) - C_3H_8+ Air (ϕ = 1) (white point = spherical propagation - black point = hemispherical propagation).

Fig. 5.52. Critical detonation radius R_c in critically initiated expanding spherical detonation as a function of cell size λ [22]

and slightly diluted) obeys the empirical law $d_c \sim 13\lambda$. This is valid, for instance, for the $C_2H_2 + 2.5O_2$ mixture. It is noteworthy that the critical tube diameter of the same mixture diluted by 81% of argon, $C_2H_2 + 2.5O_2 + 15Ar$ (Ar can be equivalently replaced by another monatomic diluent such as He or Kr), has been measured to be d_c =26–30λ [25, 72].

The diffraction re-initiation of detonation takes place similarly as the point initiated detonation, characterized by a quasi-spherical pre-detonation zone with critical radius $R_c = 1.6d_c$, as shown in Fig. 5.55 [25]. From this relation, $R_c \sim 20\lambda$ since $d_c \sim 13\lambda$ for the C_2H_2+$2.5O_2$ mixture featuring one-step/one-cell detonation, and R_c ~40–50λ since d_c ~26–30λ for the same mixture highly diluted by 81% of He, Ar or Kr (with significant dilution, a two-step/one-cell detonation wave results).

The critical tube diameter of a given mixture is drastically reduced for overdriven detonation [20, 24]. An increase in degree of overdrive D/D_{CJ} up to 1.3 leads to a reduction of critical diameter by a factor of 3. However, R_c

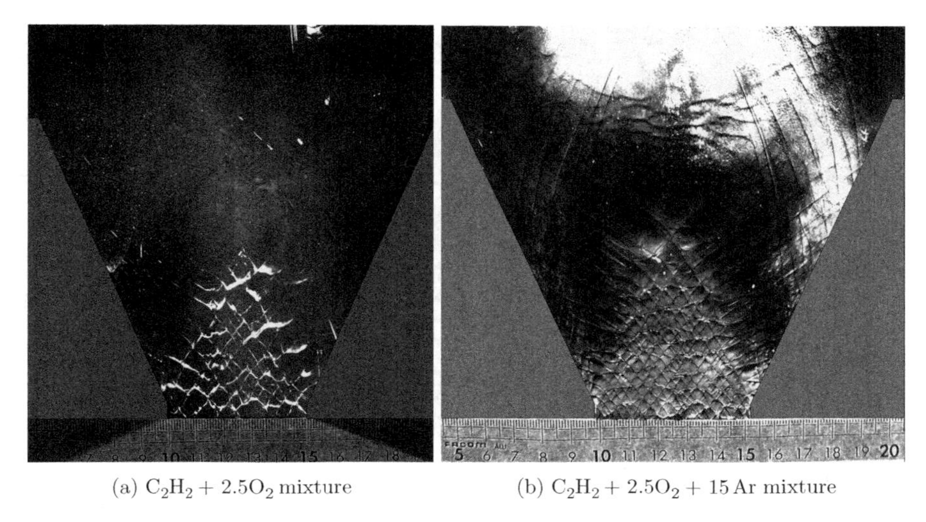

(a) $C_2H_2 + 2.5O_2$ mixture (b) $C_2H_2 + 2.5O_2 + 15\,Ar$ mixture

Fig. 5.53. Examples of detonation failure during detonation diffraction from a $d = 52\,\text{mm}$ tube to a cone. The detonation propagates from bottom to top [44]

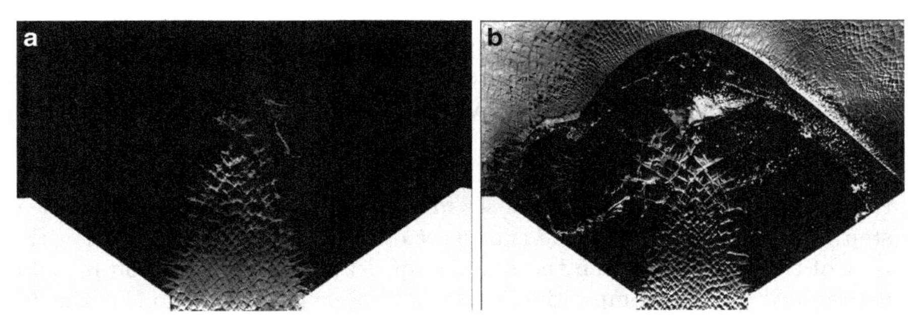

Fig. 5.54. Examples of (**a**) detonation failure ($p_0 = 17\,\text{kPa}$) and (**b**) critical reinitiation ($p_0 = 18\,\text{kPa}$) during detonation diffraction from a $d = 52\,\text{mm}$ tube to a cone in a $C_2H_2 + 2.5O_2 + 6.5Ar$ mixture [44]

remains equal to about $20\lambda_{CJ}$, and this detonation initiation source therefore approaches an ideal explosion point source.

The minimum number of cells in the critical tube diameter (and also in R_c) increases with Ar dilution, due to the evolution of the chemical heat release law from single-step to the two-step hybrid one with an elongated reaction zone. Indeed, the critical detonation diffraction from round tube to a large volume in $C_2H_2 + 2.5O_2$ mixture was found to be $d_c \sim 11 - 12\lambda$ in numerical simulations [56]. Guilly et al. [47] performed calculations of the reaction zone in the $C_2H_2 + 2.5O_2 + 15Ar$ mixture using a reduced chemical kinetic mechanism that closely follows the detailed kinetics for this two-step mixture, and they obtained d_c ~30–32λ in accordance with experiments. The

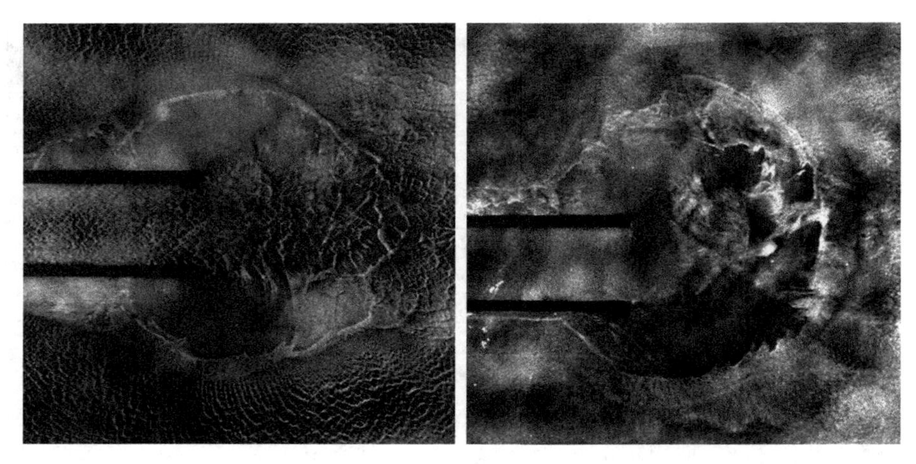

Fig. 5.55. Soot tracks of pre-detonation zones of average radius R_c in critically initiated expanding spherical detonations. Detonation is initiated by planar detonation diffraction from a round tube of diameter $d = 52\,\text{mm}$ in a $C_2H_2 + 2.5O_2$ mixture at $p_0 = 5.3\,\text{kPa}$ [21]

dependence of the $d_c - \lambda$ relationship on the reaction mechanism could be valid for many reactive mixtures for which $d_c > 13\lambda$, for example, $d_c \sim 18$–24λ for fuel–air mixtures (e.g., H_2–Air, C_2H_2–Air, etc., Ciccarelli [11], Schultz and Shepherd [86]).

During the diffraction of two-step/two-cell and two-step/one-cell detonations, the lateral expansion quenches first the second exothermic heat release step whose characteristic time is at least one order of magnitude larger than that of the first step. Then the problem of detonation re-initiation is mainly determined by the competition between the chemical reaction of the first exothermic step alone (i.e., with its corresponding energy) and expansion losses. For instance, Fig. 5.56a provides the critical initial pressure p_0 for which detonation transmission occurs, in the H_2–NO_2 mixture for a range of equivalence ratios $0.4 < \varphi < 1.3$ [27]. Thus, for two-cell detonation (stoichiometric and rich H_2–NO_2 mixtures), the critical tube diameter can be expressed in two different manners, i.e., $d_c \sim k_1\lambda_1$ or $d_c \sim k_2\lambda_2$ where always $k_1 > 13$ and $k_2 < 13$ (Fig. 5.56b), while for lean H_2–NO_2 mixtures characterized with two-step/one-cell detonation following the two-step hybrid heat release law, $d_c \sim k\lambda$ with always $k > 13$ (Fig. 5.56b).

Smoke foil examples of successful detonation transmission and failure for the H_2–NO_2 mixture at $\varphi = 1.1$ (two-cell detonation) are shown in Fig. 5.57. Thus, the critical size of the pre-detonation zone in a two-cell detonation is globally correlated to d_c with the same law $R_c \sim 1.6d_c$ as established for a one-cell detonation.

These results lead to the conclusion that k is equal to 13 for one-step/one-cell detonation. For other detonation heat release laws considered here, k (or

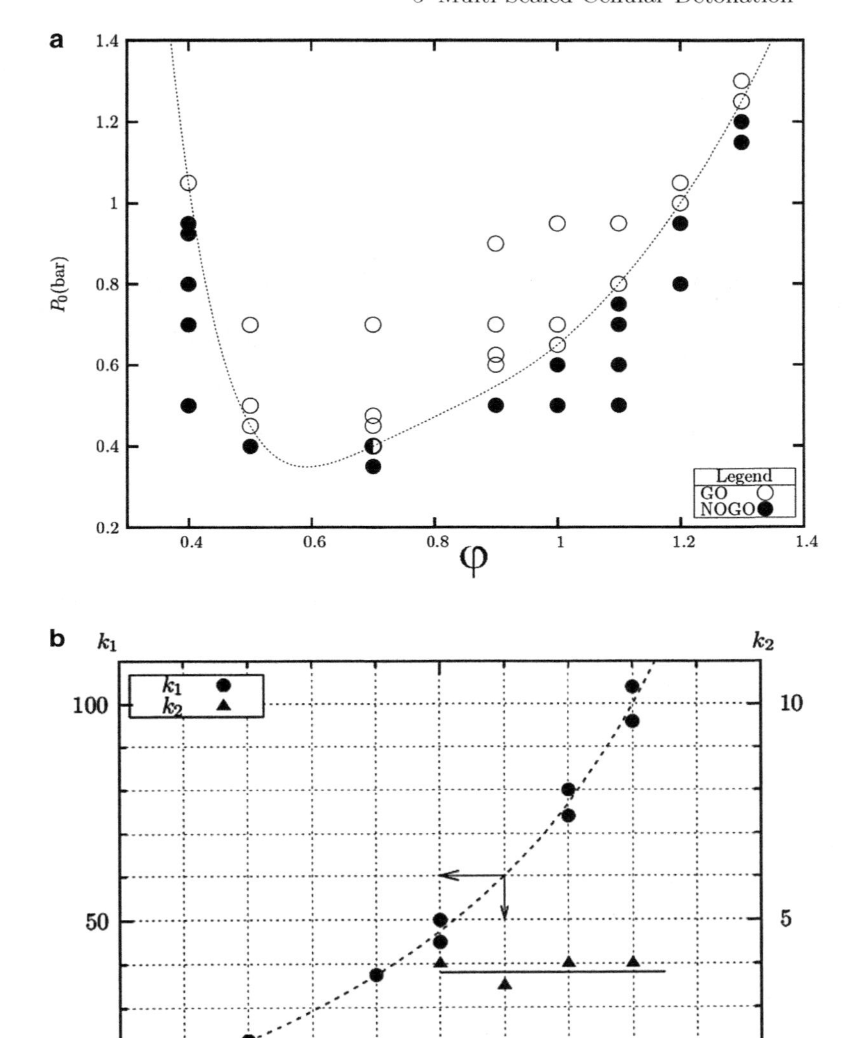

Fig. 5.56. (**a**) Critical pressure (p_0) versus equivalence ratio (φ) for successful planar detonation transmission from a tube of diameter $d = 52\,\text{mm}$ to large volume and (**b**) variation of $k_1(d_c/\lambda_1)$ and $k_2(d_c/\lambda_2)$ versus φ in a H_2–NO_2 mixture [27]

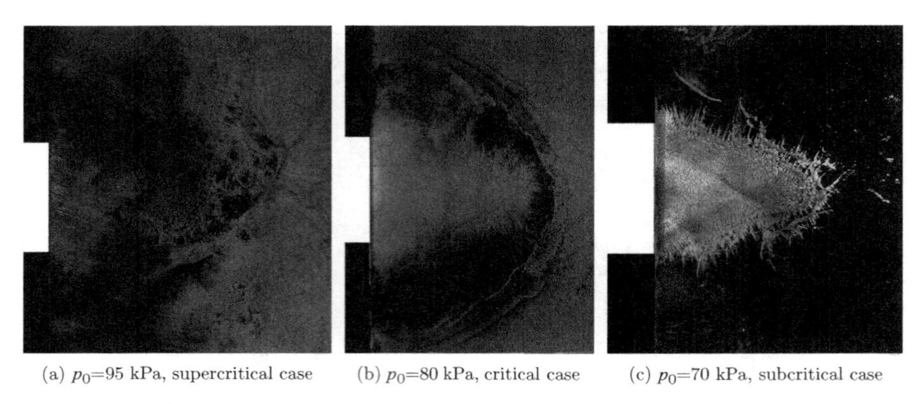

(a) p_0=95 kPa, supercritical case (b) p_0=80 kPa, critical case (c) p_0=70 kPa, subcritical case

Fig. 5.57. Soot tracks of planar detonation diffraction from a tube of diameter $d = 52\,\text{mm}$ in a H_2/NO_2 mixture for $\varphi = 1.1$ [27]

k_1) is always greater than 13, and its value depends on the specific properties of the second exothermic reaction step (duration and relative energy release).

An expanding conical tube is another setup to observe the different behavior of detonation in two reactive mixtures with similar cell sizes but obeying different heat release laws. Let us examine the self-sustained near-CJ detonation in $C_2H_2 + 2.5O_2$ (one-step/one-cell) and in $C_2H_2 + 2.5O_2 + 15Ar$ (two-step/one-cell), propagating from a 10 m long tube of diameter $d = 52\,\text{mm}$ into a slightly expanding conical tube with a 5° half angle, as shown in Figs. 5.58 and 5.59. The experimental results for the $C_2H_2 + 2.5O_2$ mixture at $p_0 = 2\,\text{kPa}$ and corresponding to $d/\lambda = 7$–8 ($\lambda = 7\,\text{mm}$) show that the detonation cell size (and velocity) are nearly unaffected by this slight change of geometry (Fig. 5.58). In the $C_2H_2 + 2.5O_2 + 15Ar$ mixture at $p_0 = 33\,\text{kPa}$, $d/\lambda = 13$ ($\lambda = 4\,\text{mm}$), results differ significantly from the previous case: the detonation entering the conical tube slows down very progressively and has a velocity deficit of about 8% at a distance of $5d$ downstream from the conical tube entrance and the cell size increases from 4 to 8–10 mm as new triple points are not created to compensate for the expansion (Fig. 5.59). If the conical tube length and d/λ are sufficiently large, the detonation may reaccelerate, leading to triple-point creations and decreasing cell sizes. In summary, for $d/\lambda \sim O(10)$, a small expansion has little effect on one-step detonation propagation, but can significantly influence two-step/one-cell detonations with initially higher cell regularity. In the latter case, the second-step reaction takes place in a time two orders of magnitude (Fig. 5.40) larger than that of the first step. As a consequence, the second step is much more sensitive to expansion cooling than the first one, thus leading to a loss, at $5d$, of about 16% of the total heat release (because $D \sim \sqrt{q}$) and a doubling of the cell size.

The same small expansion experiment has also been performed in a C_2H_2/Air ($\varphi = 1$) mixture at ambient conditions with $d/\lambda = 8$–9 ($\lambda = 6\,\text{mm}$). At $5d$ downstream from the conical tube entry, an increase in cell size by

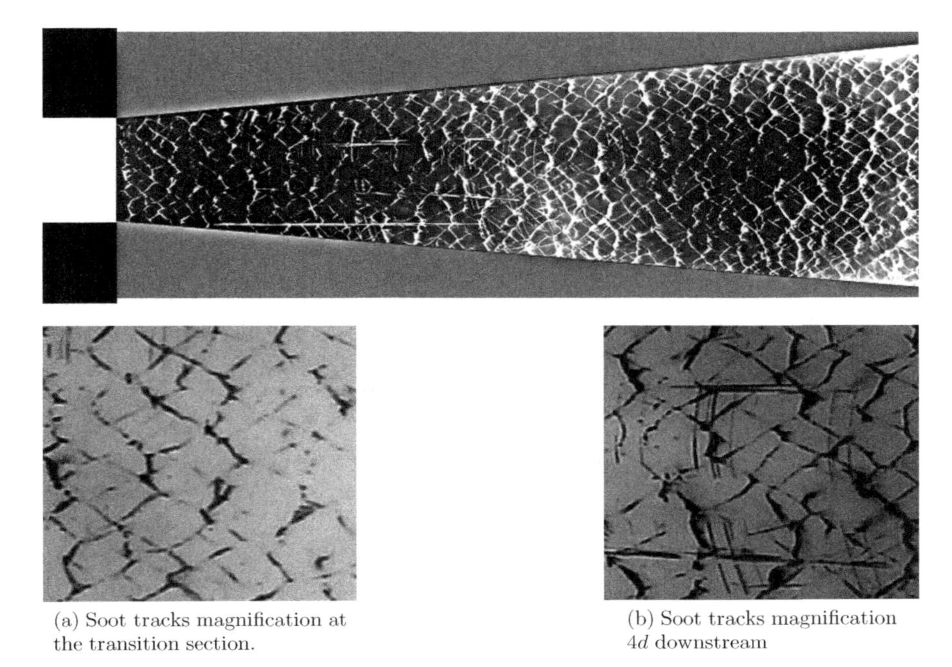

(a) Soot tracks magnification at the transition section.

(b) Soot tracks magnification $4d$ downstream

Fig. 5.58. Soot tracks on plate during detonation transition from a round tube of diameter $d = 52\,\mathrm{mm}$ to slightly expanding conical section (half angle 5°) in a $C_2H_2 + 2.5O_2$ mixture (one step/one cell) at $p_0 = 2\,\mathrm{kPa}$ corresponding to $d/\lambda = 7$–8

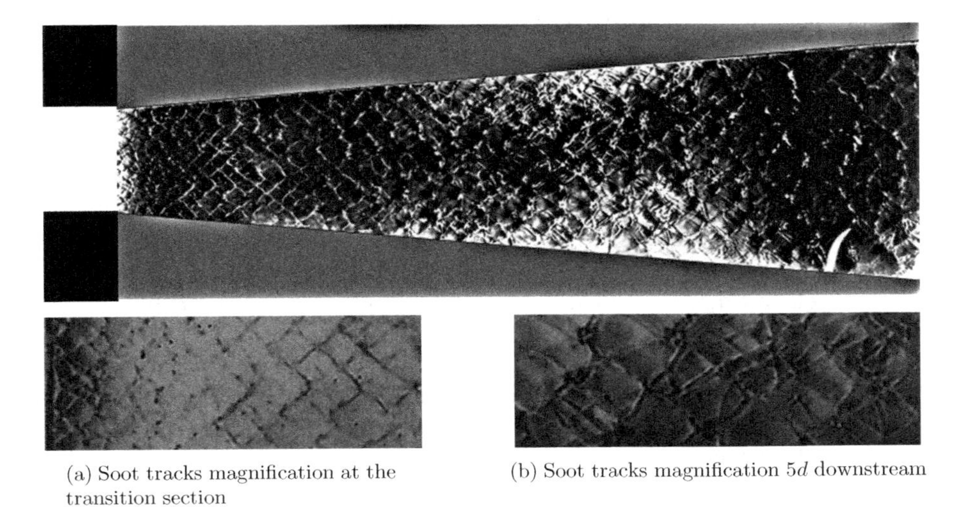

(a) Soot tracks magnification at the transition section

(b) Soot tracks magnification $5d$ downstream

Fig. 5.59. Soot tracks during detonation transition from a round tube of diameter $d = 52\,\mathrm{mm}$ to a slightly expanding conical section (half angle 5°) in a $C_2H_2 + 2.5O_2 + 15Ar$ mixture (two steps/one cell) at $p_0 = 33\,\mathrm{kPa}$ corresponding to $d/\lambda = 13$

a factor of 1.8 is noticed with a corresponding detonation velocity deficit of about 4–5%. The competition between expansion losses and chemical heat production in the small expansion experiments clearly reveals the importance of the detonation heat release laws with respect to the detonation propagation in the mixtures studied.

5.5.2 Detonation Propagation Limits in Tubes

The limiting steady detonation propagation in smooth round tubes is generally observed to be the spinning detonation, which is sustained with only part of a cell because of acoustic coupling. Taking into account the different detonation heat release laws and corresponding cellular structures, one can distinguish different trends concerning detonation limits in tubes:

1. *For one-step/one-cell detonation:* The limit of the self-sustained steady detonation propagation is the spinning detonation mode with $\lambda \sim P \sim \pi d$. This is the case in nearly stoichiometric CH_4–O_2, C_2H_6–O_2 and H_2–O_2 mixtures (following a stiff one-step heat release law). The observed spinning detonation has a small to moderate velocity deficit ($D \sim 0.95 D_{CJ}$ for CH_4–O_2) before detonation fails. Even when strongly ignited in a very long tube, detonation failure occurs as losses overcome the one-step heat release [112].
2. *For two-step/one-cell detonation:* Spinning detonation can be observed in a finite range of λ values starting from $\lambda \sim P \sim \pi d$ up to several P. In this case, while the triple point is mainly controlled by the first reaction step, the relative energy release of the second slower reaction step is important for sustaining the wave. The disappearance of the second reaction step occurs progressively and the range of the initial pressure is therefore broad for the sustenance of the spinning detonation. For example, in the $CH_4 + 2O_2 + 20Ar$ mixture, the spinning detonation is observed from $\lambda \sim P$ to $\lambda \sim$4–5P with a large velocity deficit of the detonation limit, $D \sim 0.85 D_{CJ}$.
3. *For two-step/two-cell detonation:* From the observation of the two-cell near-CJ detonation in a H_2–NO_2 mixture confined in a round tube, the following trends emerge:
   (a) The spinning mode of the larger cell (i.e., $\lambda_2 \sim P$) exhibits a small velocity deficit and can be induced by: (1) increasing Ar dilution [62], or (2) decreasing the initial pressure [116].
   (b) The above spinning mode is not the detonation limit since transition has been observed to a self-sustained one-cell LVD whose velocity is attributable to the first-step heat release. Its cellular structure is an enlargement of the smaller cell of the two-cell detonation as observed in a $\varphi = 0.3$ H_2–NO_2 mixture [26]. An example of this transition is shown in Fig. 5.60a,b, the H_2–NO_2 mixture ($\varphi = 1.2$) being diluted respectively by 30% to 60% Ar. The detonation propagates as a two-cell near-CJ detonation ($D > 0.98 D_{CJ}$) in the range of 0–50% Ar dilution and becomes a single-cell LVD with $D \sim 0.75 D_{CJ}$ when the Ar dilution

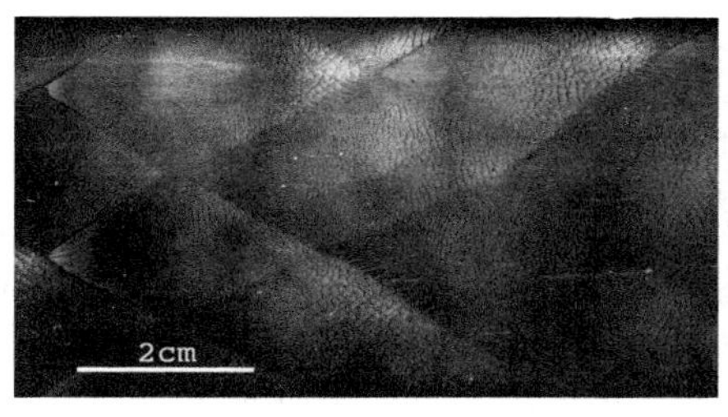

(a) $H_2 + 0.42\,NO_2 + 0.61\,Ar$ (30% Ar) near-CJ two cell detonation

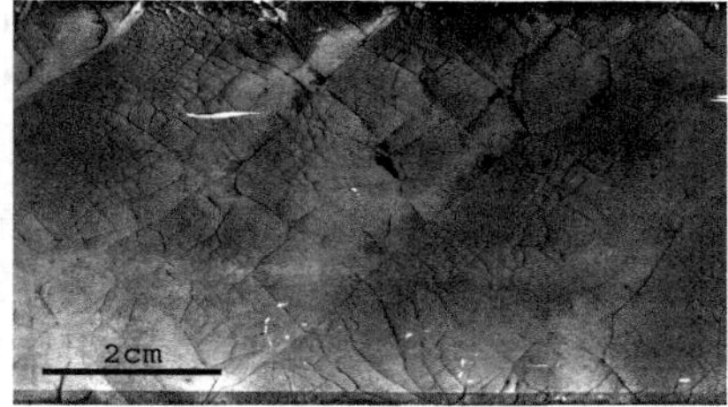

(b) $H_2 + 0.42\,NO_2 + 2.13\,Ar$ (60% Ar) low velocity single-cell detonation

Fig. 5.60. Soot tracks of self-sustained detonation in a 18 m long tube of i.d. $d = 52$ mm ($p_0 = 100$ kPa, $T_0 = 293$ K) [62]

is increased to 52–60%. Spinning detonation based on the larger cell occurs for 50–52% Ar dilution. It is remarkable that the regular two-cell detonation with very regular finer cells (Fig. 5.60a) turns into a very irregular large cell structure covered with substructures for the LVD regime (Fig. 5.60b). Detonation propagation limits in the tube were not obtained in this case due to insufficient detonation initiation energy, but one could speculate that the limit would correspond to the spinning detonation of the smaller cell. Some limiting behaviors have also been observed for two-step/one-cell systems, including cell enlargement, substantial velocity deficit and large initial pressure range of the spinning mode due to progressive quenching of the second step from the tube wall losses.

Many experiments [57, 63, 72, 82] have been performed in the past in an attempt to demonstrate that the cell regularity is at the origin of the different transient and steady detonation behaviors between fuel–O_2 mixtures and the same mixtures highly diluted by argon, classified, respectively, as group 1 and group 2 by Radulescu and Lee [82]. The reason invoked by the authors is twofold: the very different strength of the transverse waves in the two groups (strong for group 1 and weak for group 2) and the intrinsic instabilities of group 1 that enable self-regeneration of strong transverse waves to overcome losses. On the basis of this assumption, a number of measurements including the critical tube diameter of detonation propagation (tube with porous wall, tube bundles, etc.) and the critical tube diameter of detonation transmission into large volumes have been related to the cell size by $d/\lambda \sim k$ correlations. In the same experimental conditions, d/λ of group 2 is generally two to three times larger than that of group 1.

These various detonation behaviors may also be analysed in terms of the competition between different types of losses (expansion, rough wall losses, porous wall, porous media, tube bundles, etc.) and chemical energy production via the different heat release mechanisms associated with each group of reactive mixtures. Indeed, those of group 1 belong to the one-step/one-cell detonation type and those of group 2 to the two-step/one-cell hybrid detonation type. Part of the chemical energy of the mixtures of group 2 is released during the second step, i.e., far behind the detonation front. This energy is released more slowly and is easily overcome by the losses in the detonation apparatus when the detonation approaches the limit, thus explaining the large detonation velocity deficit generally observed with the mixtures of group 2 before quenching. Near detonation failure, the near CJ self-sustained cell number across the tube diameter is larger for the mixtures of group 2 since their cell size is mainly representative of the first exothermic step with only a part of the total chemical energy.

Finally, the failure of detonation in the mixtures of group 1 occurs over a shorter distance than in the mixtures of group 2 since, for the latter case, energy losses of the second step are progressive and prevent, for a moment, energy losses of the first step.

5.6 Concluding Remarks

In spite of its steady behavior in a macroscopic sense, self-sustained detonation in gaseous mixtures is locally unsteady. Its front is comprised of longitudinal and transverse shock waves connected through the triple-point configuration. Analysis of triple-point trajectories reveals that the detonation front structure is cellular and that the 2D "print" of elementary cells is of a "fish scale" type, conferring a universal character on detonation cellular structure. The extent of cell regularity (from very regular to very irregular), however, strongly depends on the reactive system and detonation regime.

The origin of this cellular structure remains not fully understood. It was assumed that reaction zone instabilities are induced by the non-linear character of the delayed global one-step heat release with the selection of a characteristic length, the cell width λ, correlated to the representative chemical length l_i associated with the maximum reaction rate. According to this assumption, the detonation cellular structure is essentially governed by the chemical reaction kinetics, but, up to now, the wide range of the ratio $\lambda/l_i \sim 10$–100 remains unexplained.

A characteristic size of detonation cellular structure has long been considered as a macroscopic representation of the global chemical induction length of reactive systems and was therefore used to evaluate their detonation sensitivity by means of correlations. It appears now that this is only true for the mixtures whose complete heat release can be modelled by a single Arrhenius law, like near stoichiometric and slightly diluted fuel–O_2 mixtures (namely the one-step/one-cell detonation). Indeed, recent observations, involving oxidation of a fuel by NO_2, have revealed a more complex double-scaled detonation cellular structure. In fact, the detonation heat release law of most reactive mixtures is not of a single Arrhenius type but more complex and can be modelled by two successive exothermic steps. When these two steps are characterized by very different induction times (defined by maximum temperature gradients associated with two maxima in the reaction rate) a double detonation cellular structure is obtained, larger cells being completely filled with a finer cellular net (namely the two-step/two-cell detonation). The finer cell is induced by the faster, first exothermic step and the large cellular structure by the slower, second step. Rich fuel–NO_2 mixtures and detonable gas-phase nitro and nitrate compounds typically display such behavior. Because of their specific detonation front structure, these mixtures can lead to the observation of the important LVD regime.

The two-step hybrid heat release law, which may possibly be applicable to a large number of mixtures including diluted fuels–O_2 mixtures (whatever the diluent), is the intermediate between the one-step and the two-step heat release laws. In such mixtures, a fraction of chemical energy is released in a second step long after the first one but without displaying a second maximum in the reaction rate. The cellular structure is of a single cell whose size is characterized by the first exothermic step only, but the cell regularity and the wave sustenance are controlled by the total chemical energy release including the second elongated progressive reaction step (namely the two-step/one-cell detonation).

Hence, the notion of cellular structure integrates different physico-chemical features (induction and reaction length, chemical energy) of heat release laws, associated with the chemical composition, the detonation regime and initial conditions. The two-step/one-cell detonation type is more sensitive than the one-step/one-cell detonation to losses (expansion, wall losses, etc.). That brings the possibility to determine a priori the type of heat release law of a reactive mixture by putting its detonation into configurations leading to transient detonation behavior. Considering a self-sustained cellular detonation of a cell size λ and a hydrodynamic configuration of characteristic length L inducing transient detonation behavior, for the same ratio L/λ, the transient behavior remains nearly self-similar if detonation belongs to the one-step/one-cell detonation type. There is no self-similarity for the two-step detonation types. For instance, the critical tube diameter $d_c \sim 13\lambda$ criterion is valid only for the one-step/one-cell detonation type but not for the two-step/one-cell and the two-step/two-cell detonation types. For the two-step/one-cell detonation, $d_c/\lambda > 13$ and for the two-step/two-cell detonation, $d_c/\lambda_1 > 13$ and $d_c/\lambda_2 < 13$.

Finally, two main conclusions can be drawn from the current knowledge of detonation cellular structure:

- A detonation cellular structure is intimately linked to a heat release mechanism. This mechanism has to be known in its inherent complexity to establish to what detonation type a reactive system belongs for given initial conditions. Then a reduced comprehensive kinetic mechanism can be

adjusted to provide useful numerical simulations of multi-scaled detonation.
- The detonability of a reactive system depends closely on its heat release mechanism. Criteria for critical detonation sustenance or limits already established on the basis of the cell size must be re-examined for multi-step detonations, i.e., for nearly all reactive mixtures, thus opening a large field of research on detonation waves in gases.

Acknowledgements

The authors thank Boris Khasainov and Florent Virot for valuable discussions and help with organizing the manuscript.

References

1. Anderson, T.J., Dabora, E.K.: Measurements of normal detonation wave structure using Rayleigh imaging. In: 24th Symposium (International) on Combustion, pp. 1853–1860 (1992)
2. Auffret, Y., Desbordes, D., Presles, H.N.: Detonation structure of C_2H_4–O_2–N_2 mixtures at elevated initial temperature. Shock Waves **9**, 107–111 (1999)
3. Auffret, Y., Desbordes, D., Presles H.N.: Detonation structure and detonability of C_2H_2–O_2 mixtures at elevated initial temperature. Shock Waves **11**, 89–96 (2001)
4. Barthel, H.O.: Predicted spacings in hydrogen-oxygen-argon detonations. Phys. Fluids **17**(8), 1547–1553 (1974)
5. Benedick, W.B., Guirao, C.M., Knystautas, R., Lee, J.H.: Critical charge for the direct initiation of detonation in gaseous fuel-air mixtures. AIAA Prog. Astronaut. Aeronaut. **106**, 181–202 (1986)
6. Bone, W.A., Fraser, R.P.: A Photographic Investigation of Flame Movements in Carbonic Oxide-Oxygen Explosions. Philos. Trans. R. Soc. **A228**, 197–234 (1929)
7. Bozier, O., Sorin, R., Zitoun, R., Desbordes, D.: Detonation characteristics of H_2/natural gas with air mixtures at various equivalence ratio and molar fraction of hydrogen. In: 4th European Combustion Meeting (2009)
8. Branch, M.C., Sadequ, M.E., Alfarayedhi, A.A., Van Tiggelen, P.J.: Measurements of the structure of laminar, premixed flames of $CH_4/NO_2/O_2$ and $CH_2O/NO_2/O_2$ mixtures. Combust. Flame **83**, 228–239 (1991)
9. Campbell, C., Woodhead, D.W.: The ignition of gases by an explosion wave. Part I. Carbon monoxide and hydrogen mixtures. J. Chem. Soc. **130**, 1572–1578 (1927)
10. Chu, B.T.: Vibration of the gas column behind a strong detonation wave. In: Gas Dynamics Symposium on Aerothermochemistry, pp. 95–111. Northwestern University Press, Evanston (1956)
11. Ciccarelli, G.: Critical tube measurements at elevated initial mixture temperatures. In: CD: Proceedings of 18th International Colloquium on the Dynamics of Explosion and Reactive Systems (ICDERS), University of Washington, ISBN 0-9711740-0-8, Seattle (2001)

12. Ciccarelli, G., Boccio, J.L.: Detonation wave propagation through a single orifice plate in a circular tube. In: Proceedings of the 27th Symposium (International) on Combustion, pp. 2233–2239 (1998)
13. Ciccarelli, G., Ginsberg, T., Boccio, J.L.: Detonation cell size measurements and predictions in hydrogen-air-steam mixtures at elevated temperatures. Combust. Flame **99**, 212–220 (1994)
14. Ciccarelli, G., Boccio, J.L., Ginsberg, T., Tagawa, H.: The Influence of initial temperature on flame acceleration and deflagration to detonation transition. In: Proceedings of the 26th Symposium (International) on Combustion, pp. 2973–2979 (1996)
15. Ciccarelli, G., Ginsberg, T., Boccio, J.L.: The influence of initial temperature on the detonability characteristics of hydrogen-air-steam mixtures. Combust. Sci. Technol. **128**, 181–196 (1997)
16. Clavin, P., He, L.: Direct initiation of gaseous detonations. J. Phys. IV **5**, 431–440 (1995)
17. Crooker, A.J.: Phenomenological investigation of low mode marginal planar detonations. PhD thesis, University of Illinois, Urbana, Technical Report AAE 69-2 (1969)
18. Denisov, Yu.N., Troshin, Ya.K.: Pulsating and spinning detonation of gaseous mixtures in tubes. Dokl. Akad. Nauk. SSSR **125**, 110–113 (1959)
19. Desbordes, D.: Correlation between shock flame predetonation zone size and cell spacing in critically initiated spherical detonations. AIAA Prog. Astronaut. Aeronaut. **106**, 166–180 (1986)
20. Desbordes, D.: Transmission of overdriven plane detonations: critical diameter as a function of cell regularity and size. AIAA Prog. Astronaut. Aeronaut. **114**, 170–185 (1988)
21. Desbordes, D.: Aspects stationnaires et transitoires de la détonation dans les gaz: relation avec la structure "cellulaire" du front. PhD thesis, Thesis 498, University of Poitiers (1990)
22. Desbordes, D.: Critical initiation conditions for gaseous diverging spherical detonation. J. Phys. IV **5**, 155–162 (1995)
23. Desbordes, D., Lannoy, A.: Effects of a negative fuel concentration on critical diameter of diffraction of a detonation. AIAA Prog. Astronaut. Aeronaut. **133**, 187–201 (1991)
24. Desbordes, D., Vachon, M.: Critical diameter of diffraction for strong plane detonations. AIAA Prog. Astronaut. Aeronaut. **106**, 131–143 (1986)
25. Desbordes, D., Guerraud, C., Hamada, L., Presles, H.N.: Failure of the classical dynamic parameters relationships in highly regular cellular detonation systems. AIAA Prog. Astronaut. Aeronaut. **153**, 347–359 (1993)
26. Desbordes, D., Presles, H.N., Joubert, F., Douala, G.: Etude de la détonation de mélanges pauvres H_2–NO_2/N_2O_4. CRAS Méc. **332**, 993–999 (2004)
27. Desbordes, D., Joubert, F., Virot, F., Khasainov, B., Presles, H.-N.: The critical tube diameter in a two reaction-steps detonation: The H_2/NO_2 mixture case. Shock Waves **18**(4), 269–276 (2008)
28. Djebaïli Chaumeix, N., Abid, S., Paillard, C.E.: Shock tube study of the nitromethane decomposition and oxidation. In: 21st Symposium on Shock Waves vol. 1, pp. 121–126 (1997)
29. Dormal, M., Libouton, J.C., Van Tiggelen, P.J.: Etude experimentale des parametres a l'interieur d'une maille de detonation. Explosifs **36**, 76–94 (1983)

30. Duff, R.E.: Investigation of spinning detonation and detonation stability. Phys. Fluids **4**, 1427–1433 (1961)
31. Duff, R.E., Finger, M.: Stability of a spherical gas detonation. Phys. Fluids **8**, 764 (1965)
32. Edwards, D.H., Parry, D.J., Jones, A.T.: On the coupling between spinning detonation and oscillation behind the wave. Br. J. Appl. Phys. **17**, 1507–1510 (1966)
33. Edwards, D.H., Hooper, G., Job, E.M., Parry, D.J.: The behavior of the frontal and transverse shocks in gaseous detonation waves. Astronaut. Acta **15**, 323–333 (1970)
34. Edwards, D.H., Jones, A.T., Philipps, D.E.: The location of the Chapman-Jouguet surface in a multi-headed detonation wave. J. Phys. D. **9**, 1331–1342 (1976)
35. Edwards, D.H., Thomas, G.O., Nettleton, M.A.: Diffraction of a planar detonation in various fuel-oxygen mixtures at an area change. AIAA Prog. Astronaut. Aeronaut. **75**, 341–357 (1981)
36. Fay, J.A.: Mechanical theory of spinning detonations. J. Chem. Phys. **20**, 942–950 (1952)
37. Fickett, W., Davis, W.C.: Detonation. University of California Press, Berkeley (1979)
38. Gamezo, V.N., Desbordes, D., Oran, E.S.: Formation and evolution of two-dimensional cellular detonations. Combust. Flame **116**, 154–165 (1999)
39. Gamezo, V.N., Desbordes, D., Oran, E.S.: Two-dimensional reactive flow dynamics in cellular detonation waves. Shock Waves **9**, 11–17 (1999)
40. Gamezo, V.N., Vasil'ev, A.A., Khokhlov, A.M., Oran, E.S.: Fine cellular structures produced by marginal detonations. Proc. Combust. Inst. **28**(1), 611–617 (2000)
41. Gavrilenko, T.P., Prokhorov, E.S.: Overdriven gaseous detonation. AIAA Prog. Astronaut. Aeronaut. **87**, 244–250 (1983)
42. Gavrilenko, T.P., Nikolaev, Y.A., Topchian, M.R.: Supercompressed detonation waves. Combust. Explos. Shock Waves **15**(5), 659–692 (1980)
43. Gordeev, V.E.: Limiting velocity of overdriven detonation and the stability of shocks in the detonation wave spin. Dokl. Akad. Nauk. SSSR **226**, 619–662 (1976)
44. Guilly, V.: Etude de la diffraction de la détonation des mélanges C_2H_2/O_2 stoechiometriques dilues par l'argon. PhD thesis, University of Poitiers (2007)
45. Guilly, V., Khasainov, B., Presles, H.N., Desbordes, D.: Influence de la dilution par l'argon d'un mélange C_2H_2/O_2 sur les conditions critiques de diffraction de sa détonation. Méc. Ind. **6**, 269–273 (2005)
46. Guilly, V., Khasainov, B., Presles, H.N., Desbordes, D.: Simulation numérique des détonations à double structure cellulaire. CRAS **334**(10), 679–685 (2006)
47. Guilly, V., Khasainov, B., Presles, H.N., Desbordes, D.: Numerical study of detonation diffraction in argon diluted C_2H_2/O_2 mixtures. In: 3rd European Combustion Meeting, Chania, Greece (2007)
48. Guirguis, R., Oran, E.S., Kailasanath, K.: The effect of energy release on the regularity of detonation cells in liquid nitromethane. In: 21st Symposium (International) on Combustion, pp. 1659–1668 (1986)
49. Hall, A.R., Wolfhard, H.G.: Multiple reaction zones in low pressure flames with ethyl and methyl nitrate, methyl nitrite and nitromethane. In: 6th Symposium (International) on Combustion, pp. 190–199 (1957)

50. Hanana, M., Lefebvre, M.H., Van Tiggelen, P.: On rectangular and diagonal three-dimensional structures of detonation waves. In: Gaseous and Heterogeneous Detonation: Science to Applications, pp. 121–130. Enas Publishers, NC (1999)
51. Huang, Z.W., Van Tiggelen, P.J.: Experimental study of the fine structure in spin detonations. AIAA Prog. Astronaut. Aeronaut. **153**, 132–143 (1993)
52. Joubert, F., Desbordes, D., Presles, H.N.: Double cellular structure in the detonation of mixtures of compounds containing the NO_2 group. In: CD: Proceedings of the 19th International Colloquium on the Dynamics of Explosion and Reactive Systems (ICDERS), ISBN 4-9901744-1-0, Hakone (2003)
53. Joubert, F., Desbordes, D., Presles, H.N.: Structure cellulaire de la détonation des mélanges H_2–NO_2/N_2O_4. CRAS **331**, 365–372 (2003)
54. Joubert, F., Desbordes, D., Presles, H.N.: Cellular structure of the detonation of NO_2/N_2O_4-fuel mixtures. Comb. Flame **152**, 482–495 (2008)
55. Khasainov, B., Priault, C., Presles, H.N., Desbordes, D.: Influence d'un obstacle central sur la transmission d'une détonation d'un tube dans un grand volume. CRAS **329**, 679–685 (2001)
56. Khasainov, B., Presles, H.N., Desbordes, D., Demontis, P., Vidal, P.: Detonation diffraction from circular tubes to cones. Shock Waves **14**, 187–192 (2005)
57. Laberge, S., Knystautas, R., Lee, J.H.S.: Propagation and extinction of detonation waves in tubes bundles. AIAA Prog. Astronaut. Aeronaut. **153**, 381–396 (1993)
58. Lamoureux, N., Matignon, C., Sturtzer, M.O., Desbordes, D., Presles, H.N.: Interprétation de la double structure observée dans l'onde de détonation du nitrométhane gazeux. CRAS **329**, 687–692 (2001)
59. Lee, J.H., Ramamurthi, K.: On the concept of the critical size of a detonation kernel. Comb. Flame **27**, 331–340 (1976)
60. Lee, J.H., Soloukhin, R.I., Oppenheim, A.K.: Current views on gaseous detonation. Astronaut. Acta **14**, 565–584 (1969)
61. Libouton, J.C., Jacques, A., Van Tiggelen, P.: Cinétique, structure et entretien des ondes de détonation. Colloq. Int. Berthelot-Vieille-Mallard-Le Châtelier **2**, 437–442 (1981)
62. Luche, J., Desbordes, D., Presles, H.N.: Détonation des mélanges H_2–NO_2/N_2O_4–Ar. CRAS **334**, 323–327 (2006)
63. Makris, A., Papyrin, A., Kamel, M., Lee, J.H.S., Knystautas, R.: Mechanisms of detonation propagation in a porous medium. AIAA Prog. Aeronaut. **153**, 363–380 (1993)
64. Manson., N: Sur la structure de l'onde hélicoïdale dans les mélanges gazeux. CRAS **222**, 46–51 (1946)
65. Manzhalei, V.I.: On the detonation fine structure in gases. Fiz. Goreniya Vzryva **3**, 470–473 (1977)
66. Manzhalei, V.I., Subbotin, V.A.: Stability of an overcompressed detonation. Combust. Explos. Shock Waves **12**(6), 819–825 (1977)
67. Markov, V.V.: Numerical simulations of the formation of multifront structure of detonation wave. Dokl. Akad. Nauk. SSSR **258**, 314–317 (1981)
68. Matsui, H., Lee, J.H.: On the measure of the relative detonation hazards of gaseous fuel-oxygen and air mixtures. In: 17th Symposium (International) on Combustion, pp. 1269–1280 (1979)
69. Melius, C.F.: Thermochemistry and reaction mechanisms of nitromethane ignition. J. Phys. IV **5**, 535–548 (1995)

70. Meltzer, J., Shepherd, J.E., Akbar, R., Sabet, A.: Mach reflection of detonation waves. AIAA Prog. Astronaut. Aeronaut. **153**, 78–94 (1993)
71. Mitrofanov, V.V., Soloukhin, R.I.: On the instantaneous diffraction of detonation. Soviet Phys. Dokl. **159**(5), 1003–1006 (1964)
72. Moen, I.O., Sulmistras, A., Thomas, G.O., Bjerketvedt, D.J., Thibault, P.A.: Influence of cellular regularity on the behavior of gaseous detonations. AIAA Prog. Astronaut. Aeronaut. **106**, 220–243 (1986)
73. Oppenheim, A.K., Smolen, J.J., Kwak, D., Urtiew, P.A.: On the dynamics of shock intersections. In: 5th Symposium (International) on Detonation, ONR Department of the Navy, Washington, DC, pp. 119–136 (1970)
74. Oran, E.S., Boris, J.P., Young, T., Flanigan, M., Burks, T., Picone, M.: Numerical simulations of detonations in hydrogen-air and methane-air mixtures. In: 18th Symposium (International) on Combustion, pp. 1641–1649 (1981)
75. Oran, E.S., Young, T.R., Boris, J.P., Picone, J.M., Edwards, D.H.: A study of detonation structure: The formation of unreacted gas pockets. In: 19th Symposium (International) on Combustion, pp. 573–582 (1982)
76. Parker, W.G., Wolfhard, H.G.: Some characteristics of flames supported by NO and NO_2. In: 4th Symposium (International) on Combustion, pp. 420–428 (1953)
77. Pintgen, F., Eckett, C.A., Austin, J.M., Shepherd, J.E.: Direct observations of reaction zone structure in propagating detonations. Comb. Flame **133**, 211–229 (2003)
78. Presles, H.N., Desbordes, D., Bauer, P.: An optical method for the study of the detonation front structure in gaseous explosive mixtures. Comb. Flame **70**, 207–213 (1987)
79. Presles, H.N., Desbordes, D., Guirard, M.: Detonation in nitromethane and nitromethane-oxygen gaseous mixtures. In: Proceedings of the Zeldovich Memorial-International Conference on Combustion, pp. 382–385 (1994)
80. Presles, H.N., Desbordes, D., Guirard, M., Guerraud, C.: Nitromethane and nitromethane-oxygen mixtures: A new detonation structure. Shock Waves **6**, 111–114 (1996)
81. Presles, H.N., Virot, F., Desbordes, D., Khasainov, B.: Double cellular structure in the detonation of lean mixtures containing NO_2 group. In: 19th International Shock Wave Symposium, Moscow (2010)
82. Radulescu, M.I., Lee, J.H.S.: The failure mechanism of gaseous detonations: Experiments in porous wall tubes. Comb. Flame **131**, 29–46 (2002)
83. Schöffel, S.U.: Berechnung der dynamik zellularer detonations-strukturen ausgehend vom Zeldovich-Döring-v.Neumann-modell. Dipl.-Ing. Kaiserslautern, VDI-Verlag 7, 142, ISBN 3-18-144207-0 (1988)
84. Schöffel, S.U., Ebert, F.: Numerical analyses concerning the spatial dynamics of an initially plane gaseous ZDN detonation. AIAA Prog. Astronaut. Aeronaut. **114**, 3–31 (1988)
85. Schott, G.L.: Observation of the structure of spinning detonation. Phys. Fluids **8**, 850–865 (1965)
86. Schultz, E., Shepherd, J.E.: Detonation diffraction through a mixture gradient. Caltech, Galcit, EDL Report FM00-1 (2000)
87. Shchelkin, K.I., Troshin, Ya.K.: Gazodinamika Goreniya. Izdatel'stvo Akademii Nauk SSSR, Moscow (1963)
88. Shepherd, J.E.: Chemical kinetics of hydrogen-air-diluent detonations. AIAA Prog. Astronaut. Aeronaut. **106**, 263–293 (1986)

89. Shepherd, J.E.: Detonation: A look behind the front. In: CD: Proceedings of 19th International Colloquium on the Dynamics of Explosion and Reactive Systems (ICDERS), ISBN 4-9901744-1-0, Hakone (2003)
90. Shepherd, J.E.: http://www.galcit.caltech.edu/detn_db/html/db.html (2005)
91. Soloukhin, R.: Multi-headed Structure of Gaseous Detonation. G.I.F.M., Moscow (1963)
92. Stamps, D.W., Tieszen, S.R.: The influence of initial pressure and temperature on hydrogen-air-diluent detonations. Comb. Flame **83**(3), 353–364 (1991)
93. Steel, G.B., Urtiew, P.A., Oppenheim, A.K.: Experimental study of the wave structure of marginal detonation in a rectangular tube. University of California College of Engineering, Technical note 3-66 (1966)
94. Strehlow, R.A.: Gas phase detonations: Recent developments. Comb. Flame **12**, 81–101 (1968)
95. Strehlow, R.A.: The nature of transverse waves in detonations. Astronaut. Acta **14**, 539–548 (1969)
96. Strehlow, R.A., Biller, J.R.: On the strength of transverse waves in gaseous detonations. Comb. Flame **13**, 577–582 (1969)
97. Strehlow, R.A., Crooker, A.J.: The structure of marginal detonation waves. Acta Astronaut. **1**, 303–315 (1974)
98. Strehlow, R.A., Engel, C.D.: Transverse waves in detonation II: Structure and spacing in H_2–O_2, C_2H_2–O_2, C_2H_4–O_2, and CH_4–O_2 systems. AIAA J. **7**, 492–496 (1969)
99. Strehlow, R.A., Maurer, R.E., Rajan, S.: Transverse waves in detonation: I. Spacing in the hydrogen-oxygen system. AIAA J. **7**, 323–328 (1969)
100. Strehlow, R.A., Adamczyk, A.A., Stiles, R.J.: Transient studies on detonation waves. Astronaut. Acta **17**, 509–527 (1972)
101. Sturtzer, M.O.: Etude de la detonation de melanges gazeux de nitromethane et d'oxygene. PhD thesis, University of Poitiers (2006)
102. Sturtzer, M.O., Lamoureux, N., Matignon, C., Desbordes, D., Presles, H.N.: On the origin of the double cellular structure of detonation in gaseous nitromethane. Shock Waves **14**, 45–54 (2005)
103. Takai, R., Yoneda, K., Hikita, T.: Study of detonation wave structure. In: 15th Symposium (International) on Combustion, pp. 69–78 (1974)
104. Taki, S., Fujiwara, T.: Numerical analysis of two-dimensional nonsteady detonations. AIAA J. **16**, 73–77 (1978)
105. Trotsyuk, A.V., Ulyanitskii, Yu.V.: On the detonation wave characteristics initiated by concentrated energy source. Fiz. Goreniya Vzryva **6**, 76–82 (1983)
106. Tsuboi, N., Hayashi, K.A.: Numerical study on spinning detonations. In: 31st Symposium (International) on Combustion, pp. 2389–2396 (2007)
107. Urtiew, P.A.: Idealized two-dimensional detonation waves in gaseous mixtures. Acta Astronaut. **3**, 187–200 (1976)
108. Varatharajan, B., Williams, F.A.: Chemical-kinetic descriptions of high-temperature ignition and detonation of acetylene-oxygen-diluent systems. Comb. Flame **124**(4), 624–645 (2001)
109. Varatharajan, B., Williams, F.A.: Two-step chemical-kinetic descriptions for detonation studies. In: CD: Proceedings of 18th International Colloquium on the Dynamics of Explosion and Reactive Systems (ICDERS), University of Washington ISBN 0-9711740-0-8, Seattle (2001)
110. Vasil'ev, A.A., Grigoriev, V.V.: Critical conditions for gas detonation in sharply expanding channel. Fiz. Goreniya Vzryva **16**, 117–125 (1980)

111. Vasil'ev, A.A., Gavrilenko, T.P., Topchian, M.E.: On the Chapman-Jouguet surface in multi-headed detonations. Astronaut. Acta **17**, 499–502 (1972)
112. Virot, F.: Contribution a l'etude experimentale et numerique du regime helicoïdal de detonation dans les systemes H_2, CH_4, C_2H_6–O_2 dilues ou non par N_2 ou Ar. PhD thesis, ENSMA, Poitiers (2009)
113. Virot, F., Khasainov, B., Desbordes, D., Presles, H.-N.: Numerical spinning detonation structure. In: 3rd European Combustion Meeting, Chania, Greece (2007)
114. Virot, F., Khasainov, B., Desbordes, D., Presles, H.-N.: Spinning detonation: Experiments and simulations. In: CD: Proceedings of 21st International Colloquium on the Dynamics of Explosion and Reactive Systems (ICDERS), Poitiers (2007)
115. Virot, F., Khasainov, B., Desbordes, D., Presles, H.-N.: Numerical simulation of the influence of tube diameter on detonation regimes and structures in 2-step/2-cell mixtures. Combust. Explos. Shock Waves **45**(4), 435–441 (2009)
116. Virot, F., Khasainov, B., Desbordes, D., Presles, H.-N.: 2-cell detonation-losses effects on cellular structure and propagation in rich H_2–NO_2/N_2O_4 mixtures. Shock Waves **20**(6), 457–465 (2010)
117. Voitsekhovskii, B.V.: On Spinning Detonation (in Russian). Dokl. Akad. Nauk. SSSR **114**(4), 717–720 (1957)
118. Voitsekhovskii, B.V., Mitrofanov, V.V., Topchian, M.E.: Struktura fronta detonatsii v gazakh. Izd-vo Sibirsk, Odetl. Akad. Nauk. SSSR, Novosibirsk (1963)
119. Voitsekhovskii, B.V., Mitrofanov, V.V., Topchian, M.E.: Investigation of the structure of detonation waves in gases. In: 12th Symposium (International) on Combustion, pp. 829–837 (1969)
120. Washizu, T., Fujiwara, T.: A numerical study of spinning detonation. In: Proceedings of 14th International Colloquium on the Dynamics of Explosion and Reactive Systems (ICDERS), Coimbra, vol. 2, pp. D1–9 (1993)
121. Westbrook, C.K., Pitz, W.J., Urtiew, P.A.: Chemical kinetics of propane oxidation in gaseous detonation. AIAA Prog. Astronaut. Aeronaut. **94**, 151–174 (1984)
122. White, D.R.: Optical refractivity of high-temperature gases: III. The hydroxyl radical. Phys. Fluids **4**, 40–45 (1961)
123. White, D.R.: Turbulent structure of gaseous detonation. Phys. Fluids **4**, 465–480 (1961)
124. Zeldovich, Ya.B., Kogarko, S.M., Simonov, N.H.: An experimental investigation of spherical detonation of gases. Zh. Eksp. Teor. Fiz. **26**, 1744–1772 (1956)
125. Zhang, F., Grönig, H., Van de Ven, A.: DDT and detonation waves in dust-air mixtures. Shock Waves **11**, 53–71 (2001)

6

Condensed Matter Detonation: Theory and Practice

Craig M. Tarver

6.1 Introduction

Detonations of high density, high energy solid and liquid organic high explosives produce self-sustaining waves traveling at speeds approaching 10,000 $\mathrm{m\,s^{-1}}$ that reach approximately 40 GPa pressures and 4,000 K temperatures in nanoseconds. These pressure–density–temperature states and short time durations are unique to detonation and thus are extremely difficult to study experimentally and theoretically. However, a great deal of progress has been made in understanding the hydrodynamics and chemistry that occurs within the reaction zone of a condensed-phase detonation wave. These achievements have been reviewed in several books [14,19,20,45,69] and bibliographies [2,12]. This chapter discusses the progress made in recent years in two areas: the hydrodynamic theory of detonation and practical reactive flow modeling of detonation waves. The ultimate goal of condensed matter detonation research is to obtain an understanding of the underlying chemical and hydrodynamic phenomena equal to that of gas phase detonation. The current status of gas phase detonation theory and experimentation has been reviewed recently by Lee [40] and in several chapters of this book. The three-dimensional cellular structures of some gas detonations have been carefully measured and can be accurately modeled using detailed chemical reaction rate models. Since gaseous detonation waves produce only approximately 20 times the initial gas pressures, they can be studied in shock tubes at universities. The perfect gas law can be used to describe the unreacted and reaction product equations of state (EOS). The kinetics of the individual chemical reactions can be measured individually by shocking mixtures of the species involved in each reaction of the decomposition process. The temperatures, particles velocities, densities, and pressures within the reaction zones can be measured and calculated. The equilibrium Chapman–Jouguet (CJ) state and the "hydrodynamic thickness" of the reaction zone can be measured. Concentration limits for detonation of many gaseous mixtures are known.

F. Zhang (ed.), *Shock Wave Science and Technology Reference Library, Vol. 6*: Detonation Dynamics, DOI 10.1007/978-3-642-22967-1_6,

For condensed-phase detonation waves, the three-dimensional cellular structure has been frequently observed indirectly [53], but is too small to be completely quantified for pure liquid explosives and solid explosives. The one-dimensional average reaction zone structures of many condensed-phase explosives have been measured using embedded gauges [28], laser velocimetry techniques [59], and electrical conductivity probes [16] that average over the three-dimensional structure. These average states are modeled using phenomenological, multidimensional reactive flow models [83]. The temperatures at the CJ state of detonating transparent liquid and single crystal solid explosives have been measured by several groups [34] and can be calculated by chemical equilibrium computer codes, such as the CHEETAH code [3]. Many two-dimensional properties of condensed-phase detonation waves, such as confined and unconfined failure diameter, wavefront curvature, corner turning, divergence and convergence, and metal acceleration ability, have been experimentally measured and modeled [75]. Some three-dimensional experiments, for example, the Los Alamos National Laboratory Prism Failure Test [54], have been developed and modeled [23, 24]. Essentially, the average "mechanical" aspects of condensed-phase detonation (detonation velocity, von Neumann spike pressure, CJ pressure, reaction zone particle velocities, reaction product equation of state, etc.) are fairly well understood. However, many chemical aspects (detonation ignition mechanisms, chemical reaction rates following ignition, temperatures, species concentrations, etc.) are not well known even for pure explosives. Since most explosives are mixtures, bonded with plastics, and/or contain aluminum or other metal particles, their chemical aspects are even more difficult to study. This chapter discusses the current state of the hydrodynamic theory of detonation and practical reactive flow modeling of condensed-phase detonation and offers suggestions for future research.

6.2 Condensed-Phase Detonation Theory

One of the major developments in detonation theory over the last 30 years is the Nonequilibrium Zeldovich–von Neumann–Döring (NEZND) theory. It was developed to identify the nonequilibrium chemical processes that precede and follow exothermic chemical energy release within the reaction zones of self-sustaining detonation waves in gaseous, liquid, and solid explosives [64–68,70,80]. Prior to the development of the NEZND theory, the chemical energy released was treated as a heat of reaction in the conservation of energy equation in the Chapman–Jouguet (CJ) [8, 33], Zeldovich–von Neumann–Döring (ZND) [13, 89, 92], and curved detonation wavefront theories [91], and in hydrodynamic computer code reactive flow models [79]. NEZND theory has explained many experimentally observed detonation wave properties. These include: the induction time delay for the onset of chemical reaction, the rapid rates of the chain reactions that form the product molecules, the deexcitation rates of the initially highly vibrationally excited products, the feedback mechanism that allows the chemical energy to sustain the leading shock wavefront at the steady state detonation velocity, and the establishment of the complex three-dimensional Mach stem structure common to all detonation waves.

When the leading shock front of a detonation wave compresses an explosive molecule, the translational degrees of freedom are excited first. Dremin and co-workers [14] have speculated that this "transitional energy overshoot" can cause chemical reactions before the internal energy is distributed to the rotational and vibrational modes. Thus detonation reactions would occur in the picosecond time frame. However, it has long been known that translational energy alone cannot cause chemical decomposition reactions and that the vibrational modes must be excited before chemical reactions begin [64]. Therefore the thermal energy must be transported into the vibrational modes of the explosive molecule before endothermic bond breaking reactions can occur. The time required for internal energy equilibration behind strong shock waves in explosive molecules has been demonstrated to be on the order of tens of picoseconds both experimentally [30] and theoretically [31]. However, several experimental techniques have measured average unreacted von Neumann spike states lasting several nanoseconds in many detonating solid and liquid explosives [16, 28, 59]. What is the basic physical mechanism that causes the observed times for chemical decomposition? Direct chemical decomposition from an excited electronic state has been proposed [46], but excited electronic states relax to the ground state in picoseconds, depositing their energy in highly excited vibrational states that equilibrate in tens of picoseconds with neighboring vibrations through intramolecular vibrational relaxation (IVR) [90]. The most plausible explanation is that the vibrational mode (or modes) that becomes the transition state(s) for the initial endothermic unimolecular decomposition reaction(s) is in rapid equilibrium with its neighboring vibrational modes via IVR. Thus it cannot react as frequently as it would as an isolated transition state. Nanosecond induction times for the onset of the initial unimolecular endothermic reactions are predicted using high pressure, high temperature transition state theory, also known as Starvation Kinetics [17]. Following ignition, high temperature, high density chain reactions produce intermediate decomposition products, such as formaldehyde, nitrous oxide, hydrogen cyanide, and molecular ring fragments, that react further to produce the stable detonation reaction products H_2O, CO_2, N_2, CO, and solid carbon. The stable reaction products are initially created in highly vibrationally excited states that must relax to chemical and thermal equilibrium at the CJ state [65]. Since the chemical energy is released well behind the leading shock front of a detonation wave, a physical mechanism is required for this chemical energy to be able to reinforce the leading shock front and maintain its overall constant velocity. This mechanism is the amplification of pressure wavelets passing through the reaction zone during the process of deexcitation of initially highly vibrationally excited reaction product molecules to lower vibrational energy levels as thermal and chemical equilibrium is approached [61, 65]. The CJ state determines the energy delivery of the detonating explosive to its surroundings and thus must be accurately determined. Today's computers are not yet large or fast enough to include nonequilibrium reaction processes in two- and three-dimensional hydrodynamic calculations of condensed-phase detonation. Thus phenomenological ZND structure explosive reactive flow models have been developed in many computer codes.

Figure 6.1 illustrates the various processes that occur in the NEZND model of detonation in condensed explosives. At the head of every detonation wave is a complex three-dimensional Mach stem shock wavefront. Zeldovich and Raizer [93] defined shock wave width as the distance at which the viscosity and heat conduction become negligible. In gaseous explosives, the nonequilibrium processes that precede and follow chemical reaction can be observed [39]. Velocities, pressures, and temperatures are calculated using the perfect gas law. The high initial densities of liquids and solids make the measurements and calculations of the states attained behind strong shock waves more difficult, because the processes take place in tens of picoseconds and at pressures of tens of gigapascals. Understanding the distribution of the shock compression energy between the potential (cold compression) energy and the thermal energy of the unreacted liquid or solid explosive is essential for estimating chemical reaction rates within detonation waves.

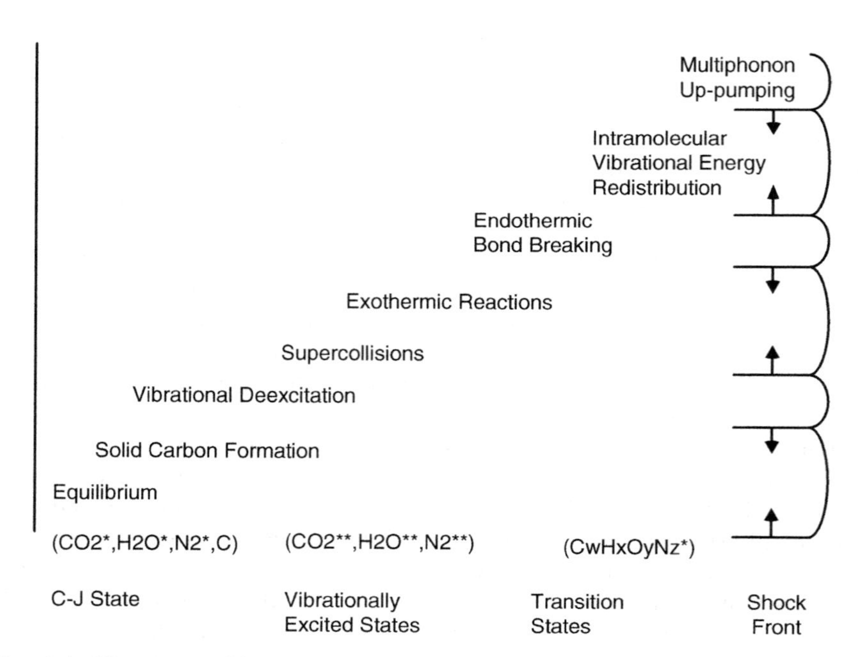

Fig. 6.1. The nonequilibrium Zeldovich–von Neumann–Döring (NEZND) theory of detonation [65]

Behind each shock in a detonation wave, the phonon modes are first excited, followed by multiphonon excitation of the lowest frequency vibrational (doorway) modes and then excitation of the higher frequency modes by multiphonon up-pumping and IVR. Internal energy equilibration has been studied experimentally in shocked liquid and solid explosives by several groups [29,60]. Hooper [31] summarized various studies of phonon up-pumping and IVR and concluded that equilibration of the vibrational degrees of freedom with the translational and rotational modes is complete in the tens of picosecond time frame. After the explosive molecules are vibrationally excited, endothermic bond-breaking reactions begin, followed by exothermic formation of stable products by rapid chain and recombination reactions. Many first principle molecular dynamic and reactive force field studies [32, 52, 55, 63, 94 and references therein] predict the onset of exothermic chemical reactions within picoseconds after the vibrational modes equilibrate.

Experimentally observed nanosecond induction times for initial endothermic bond breaking reactions [16, 28, 59] can be calculated using high pressure, high temperature transition state theory or Starvation Kinetics [17]. Experimentally, unimolecular reaction rates under low temperature ($< 1{,}000\,\mathrm{K}$) shock conditions obey the usual Arrhenius law:

$$K = A\mathrm{e}^{-E/RT}, \tag{6.1}$$

where K is the reaction rate constant, A is the frequency factor, E is the activation energy, R is the gas constant, and T is temperature. The frequency factor A need not be a constant and may contain pressure- and concentration-dependent terms. At extreme unreacted shock temperatures, pressures, and densities within detonation waves, reaction rates "fall off" [35] to slower rates than (6.1) predicts at high temperatures. Eyring [17] attributed this "fall-off" in unimolecular rates to the close proximity of vibrational states, which causes the high frequency mode that becomes the transition state for reaction to rapidly equilibrate with its neighboring modes by IVR. These modes form a "pool" of vibrational energy in which the energy required for decomposition is shared. Any large quantity of vibrational energy that a specific mode receives from an excitation process, such as electronic, is rapidly shared among the modes before reaction can occur. Conversely, sufficient vibrational energy from the entire pool of oscillators is statistically present in the transition state vibrational mode long enough to cause reaction. The unimolecular rate constant K for an isolated transition state in equilibrium with the associated translational and rotational degrees of freedom at temperature T is given by

$$K = (kT/h)\,\mathrm{e}^{-E/RT}, \tag{6.2}$$

where k, h, and R are Boltzmann's, Planck's, and the gas constants, respectively. For a bond with dissociation energy E in equilibrium with a reservoir of s vibrational modes, the probability p_s that this reservoir will have an energy E in s degrees of freedom is

$$p_s = \sum_{i=0}^{s-1} (E/RT)^i \mathrm{e}^{-E/RT}/i!. \tag{6.3}$$

The probability p_1 that a vibrational mode will have an energy E or more is then

$$p_1 = \mathrm{e}^{-E/\gamma}, \tag{6.4}$$

where γ is the average energy of a vibration in the reservoir. So the rate of decomposition of this transition state vibrational mode with energy E is approximately

$$K = (kT/h)\ p_s p_1. \tag{6.5}$$

Decomposition of this bond is clearly impossible until the energy in the $s + 1$ vibrational degrees of freedom exceeds E. When the total energy in the vibrational modes equals the activation energy E, $\gamma = E/s$ and the reaction rate constant K is

$$K = (kT/h)\mathrm{e}^{-s} \sum_{i=0}^{s-1} (E/RT)^i \mathrm{e}^{-E/RT}/i!, \tag{6.6}$$

where s is the number of neighboring vibrational modes interacting with the transition state. The main effect of rapid IVR among these $s + 1$ modes at high densities, temperatures, and pressures is to decrease the rate constant dependence on temperature.

Reaction rate constants have been calculated for detonating solids and liquids using (6.6) with realistic unreacted EOS. Reaction rate constants from (6.1) and (6.6) are compared to induction time results for liquid nitromethane and single crystal pentaerythritol tetranitrate (PETN) in Figs. 6.2 and 6.3, respectively. Despite uncertainties in the calculated shock temperatures from the two nitromethane EOS [10,43] in Fig. 6.2, it is clear that (6.6) with $s = 14$, the number of neighboring vibrational modes in nitromethane, predicts realistic reaction rates. This is also true for PETN in Fig. 6.3. Extrapolations to the highest unreacted shock temperatures (approximately 2,500 K) within the three-dimensional structures of nitromethane and PETN detonation waves show that (6.6) predicts nanosecond reaction times in agreement with experiments, while (6.1) predicts tens of picoseconds reaction times. Thus "Starvation Kinetics" or high pressure, high temperature transition state theory calculates realistic induction times for shock initiation and detonation of homogeneous liquid and heterogeneous solid explosives. Similar "fall-offs" in high temperature reaction rates have been measured for many gaseous unimolecular bond-breaking reactions in shock tube experiments [9].

After sufficient endothermic bond breaking has occurred, rapid exothermic chain reaction processes follow in which reaction product gases (CO_2, N_2, H_2O, CO, etc.) are formed in highly vibrationally excited states [64–68, 70, 81]. These excited products either undergo reactive collisions with unreacted explosive molecules or nonreactive collisions with other products

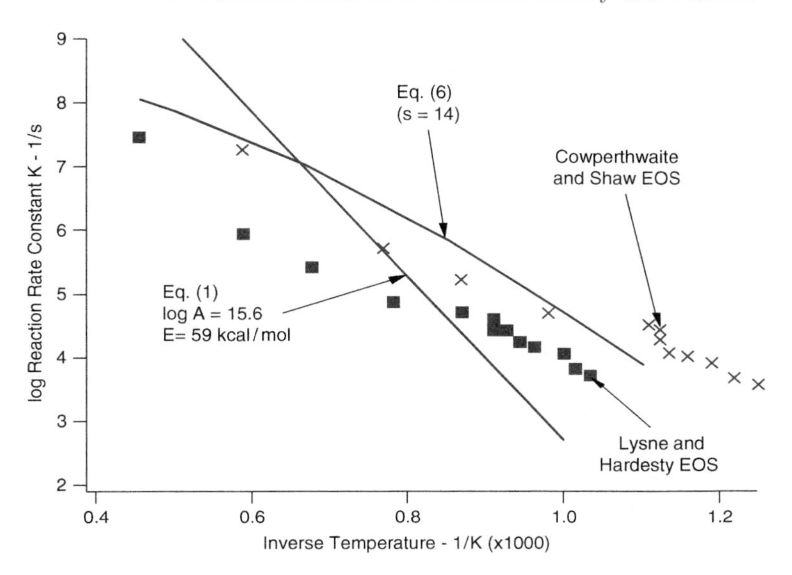

Fig. 6.2. Reaction rate constants for nitromethane as functions of shock temperature. *Lines*: predicted; *symbols*: experimental induction time data [10, 43, 70]

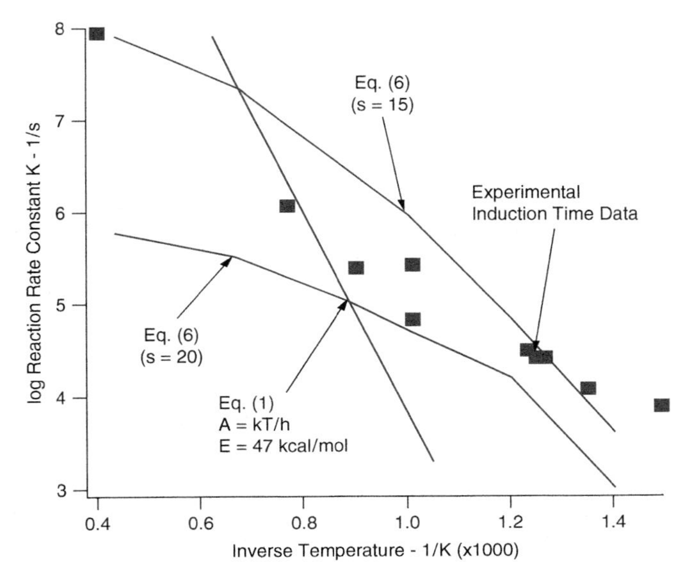

Fig. 6.3. Reaction rate constants for single crystal PETN as a function of shock temperature. *Lines*: predicted; *symbols*: experimental induction time data [70]

in which one or more quanta of vibrational energy are transferred. Some collisions are "supercollisions" [4] in which several quanta of vibrational energy are transferred. Since reaction rates increase rapidly with each quanta of vibrational energy available, reactive collisions dominate, and the main chemical energy release is extremely fast. Once the chain reactions are completed, the remainder of the reaction zone is dominated first by vibrational deexcitation of the gaseous molecules and then by solid carbon particle formation in underoxidized explosives.

This vibrational deexcitation process partially controls the length of the reaction zone and provides the chemical energy necessary for shock wave amplification during self-sustaining detonation [65]. As pressure wavelets pass through the subsonic reaction zone, they are amplified by discrete frequency vibrational deexcitation processes [61]. The opposite effect, shock wave damping by a nonequilibrium gas that lacks vibrational energy after expansion through a nozzle, is a well-known phenomenon [93]. These pressure wavelets then interact with the main shock front and replace the energy lost during compression, acceleration, and heating of the explosive molecules. The pressure wavelet amplification process provides the chemical energy required to develop the complex, three-dimensional Mach stem shock front structure shown in Fig. 6.1 [64]. This structure has been observed for gaseous, liquid, and solid explosives [40,53] and is currently being replicated for gaseous explosives in two- and three-dimensional hydrodynamic computer simulations using multiple reaction chemical kinetic schemes [87]. Simulations using simplified chemical kinetic schemes for condensed-phase detonation are also being developed [22].

Since most condensed-phase explosive formulations are under-oxidized, significant amounts of solid carbon particles form in the chemical reaction zones of most self-sustaining detonation waves. These particles can be diamond, graphite, or amorphous carbon depending on the temperatures and pressures attained in the reaction zone and have maximum diameters of about 10 nm independent of the amount of explosive detonated [88]. Since the solid carbon formation process is diffusion-controlled as carbon atoms attempt to form chains and particles in the presence of many gaseous molecules, this process requires more time than gaseous product formation and relaxation to thermal equilibrium. Thus the chemical energy release portion of a condensed-phase detonation wave exhibits two observable energy release rates: a fast reaction taking tens of nanoseconds in which the main gaseous products form and equilibrate followed by a slower reaction for the solid carbon particle formation requiring hundreds of nanoseconds [16,28,59]. These rates have been measured by several nanosecond time-resolved techniques: embedded particle velocity and pressure gauges, electrical conductivity probes, and laser interferometry [38]. Chemical and thermal equilibrium at the CJ state is closely approached as the nanometer size carbon particles form. A rarefaction wave in which the products expand and cool follows the detonation reaction zone. This expansion process does work on surrounding materials. The CJ state and subsequent expansion can be measured by metal acceleration experiments [82] and calculated using modern thermochemical equilibrium computer codes, such as the CHEETAH code [3].

Detonation reaction zones can be more complex for mixtures of explosive materials and for formulations containing metals that react with the explosive detonation product gases. Aluminum particles are added to organic explosives to provide later-time (microsecond to millisecond) energy release as the explosive gaseous products CO_2, CO, and H_2O penetrate the molten aluminum oxide outer layer and react with molten aluminum to form Al_2O_3 and various aluminum suboxides (AlO, Al_2O, AlO_2, etc.). Aluminum oxidation can liberate its large amount of thermal energy on very different time-scales [62]. Aluminum particles that do not react with the explosive detonation products may react with oxygen in air, when the cloud of detonation products turbulently mixes with air behind an expanding blast wave [21, 37]. Recently, Zhang et al. [95] presented evidence that ignition of fine aluminum particles may depend on surface kinetics rather than the classic molten aluminum diffusion model. Aluminum surface damage and breakup may then result in direct contact of oxide-free aluminum with oxygen, leading to shorter (microsecond) ignition delays and rapid surface reactions.

Even with today's large, fast multiprocessor computers, all of the aforementioned chemical processes cannot be included in practical one-, two-, and three-dimensional hydrodynamic code calculations of initiation and propagation of condensed-phase detonations in large explosive charges. Unlike gaseous detonations, the temperatures and species concentrations cannot yet be measured or calculated in the reaction zones of detonating solid and liquid explosives. Therefore, practical, phenomenological reactive flow models of detonation, such as the Ignition and Growth model [79], have been developed to calculate the main features of shock initiation and detonation reaction zones and subsequent metal acceleration during reaction product expansion. The practical application of the Ignition and Growth model to insensitive high explosives based on the triaminotrinitrobenzene (TATB) molecule is discussed in the next section.

6.3 Practical Modeling of Detonation: Ignition and Growth

All chemical reaction rates are mainly governed by the local temperature of the molecules that are about to react. However, since the temperature fields in detonating condensed-phase explosives have not been measured, phenomenological reactive flow models using rate laws based on properties that can be measured, such as pressure, particle velocity, and compression, are currently used to model condensed-phase shock initiation and detonation. All of these reactive flow models require as a minimum: two EOS, one for the unreacted explosive and one for its reaction products; a reaction rate law for the conversion of explosive to products; and a mixture rule to calculate partially reacted states in which both explosive and products are present. One of the most widely used models of this type is the Ignition and Growth reactive flow model [79], which uses two Jones–Wilkins–Lee (JWL) EOS, one for the unreacted explosive and one for the reaction products, in the temperature-dependent form

$$p = A\,\mathrm{e}^{-R_1 V} + B\,\mathrm{e}^{-R_2 V} + \omega C_v T/V, \tag{6.7}$$

where p is the pressure in Megabars, V is the relative volume, T is the temperature, ω is the Grüneisen coefficient, C_v is the average heat capacity, and A, B, R_1, and R_2 are constants. The unreacted explosive equation of state is fitted to the available shock Hugoniot data, and the reaction product equation of state is fitted to copper cylinder tests and other inert material acceleration data. At the high pressures involved in shock initiation and detonation of solid and liquid explosives, the pressures of the two phases must be equilibrated, because collisions between the hot gases and the explosive molecules at hundreds of kilobars pressure occur on subnanosecond time-scales based on the sound velocities and close proximity of the components. Obviously, the use of additive pressures assumes a Dalton's law gaseous mixture, which cannot be valid at tens of gigapascals. Various assumptions have been made about the temperatures in the explosive mixture, because heat transfer from the hot products to the cooler explosive is slower than pressure equilibration. In the Ignition and Growth model, the temperatures of the unreacted explosive and its reaction products are equilibrated. Temperature equilibration is used, because heat transfer becomes increasingly efficient as the reacting "hot spots" grow and consume more explosive particles as the high pressures and temperatures associated with detonation are approached. Fine enough zoning must be used in all reactive flow calculations so that the results have converged to solutions that do not change with finer zoning. Generally, this requires a resolution of at least 20 zones in the detonation reaction zone. The insensitive solid explosive LX-17 (92.5% TATB and 7.5% Kel-F binder) has an experimentally measured reaction zone length of approximately 3 mm [58,78], so using 10 zones per mm spreads the reaction over 30 zones and generally yields converged calculations.

The Ignition and Growth reaction rate equation is given by

$$\begin{array}{ccc} \mathrm{d}F/\mathrm{d}t = I(1-F)^b(\rho/\rho_\mathrm{o}-1-a)^x & + G_1(1-F)^c F^d p^y & + G_2(1-F)^e F^g p^z \\ 0 < F < F_{ig\max} & 0 < F < F_{G1\max} & F_{G2\min} < F < 1, \end{array} \tag{6.8}$$

where F is the fraction reacted (actually the fraction of available energy released), t is time in µs, ρ is the current density in g cm^{-3}, ρ_o is the initial density, p is pressure in Mbars, and I, G_1, G_2, a, b, c, d, e, g, x, y, z, $F_{ig\max}$, $F_{G1\max}$, and $F_{G2\min}$ are constants. This three-term reaction rate law represents the three stages of reaction observed using embedded pressure and particle velocity gauges during shock initiation and detonation of solid explosives [79]. The first stage of reaction is the formation and ignition of "hot spots" caused by various possible mechanisms (void collapse, friction, shear, etc.) as the initial shock or compression wave interacts with the unreacted explosive. The fraction of solid explosive heated to high temperatures in "hot spots" during shock compression is approximately equal to the original void volume [44]. For shock initiation modeling, the second term in (6.8) then describes the relatively slow process of the inward (represented by the $(1-F)^c$ term) and/or outward (represented by F^d term) pressure-dependent growth rate of the isolated "hot spots" in deflagration-type processes. The third term in (6.8) represents the rapid completion rate of reaction as the "hot spots" coalesce at high pressures and temperatures, rapidly heating the remaining explosive and causing a rapid transition to detonation.

For detonation modeling, the first term also reacts a quantity of explosive less than or equal to the void volume after the explosive is compressed to the unreacted von Neumann spike state. The second term in (6.8) models the fast decomposition of the solid into stable reaction product gases (CO_2, H_2O, N_2, CO, etc.). The third term describes the relatively slow diffusion-limited formation of solid carbon (amorphous, diamond, or graphite) as the chemical and thermodynamic equilibrium CJ state is approached. Experimentally, only nanometer size solid carbon particles are recovered after detonation of extremely large explosive charges [36]. This implies that the growth of the carbon particles becomes very slow at pressures and temperatures less than those of the CJ state. These reaction zone stages have been observed experimentally using embedded gauges, laser interferometry, and electrical conductivity probes [25, 57, 58].

The Ignition and Growth reactive flow model has been applied to several one-, two-, and three-dimensional hydrodynamic codes and compared to a great deal of experimental data. For shock initiation, it has successfully reproduced data of embedded gauge, run distance to detonation, short pulse duration, multiple shock, reflected shock, ramp wave compression, gap test, and divergent flow experiments on many high explosives at various initial porosities and temperatures [84–86]. For detonation, the model has successfully calculated embedded gauge, laser interferometric metal acceleration, failure diameter, corner turning, converging, diverging, and overdriven experiments [26]. Examples of one-, two- and three-dimensional applications are shown for the TATB-based explosives LX-17 (92.5% TATB and 7.5% Kel-F binder) and PBX 9502 (95% TATB and 5% Kel-F). Table 6.1 contains the equation of state and reaction rate parameters used for detonating LX-17.

Two types of one-dimensional nanosecond time-resolved experimental records and Ignition and Growth calculations are shown in Figs. 6.4 and 6.5. Figure 6.4 shows the measured and calculated interface velocity histories for detonating LX-17 impacting various salt crystals [83]. The von Neumann spike state, a relatively fast reaction, a slower reaction, and finally the initial expansion of the products are clearly evident in Fig. 6.4. Figure 6.5 illustrates the

Table 6.1. Ignition and growth model parameters for detonating LX-17

25°C LX-17	$\rho_o = 1.905\,\mathrm{g\,cm^{-3}}$		
Unreacted JWL	Product JWL	Reaction rates	
$A = 77{,}810$ GPa	$A = 1{,}481.05$ GPa	$I = 4.0 \times 10^6\ \mu s^{-1}$	
$B = -5.031$ GPa	$B = 63.79$ GPa	$a = 0.22$	
$R_1 = 11.3$	$R_1 = 6.2$	$b = 0.667$	
$R_2 = 1.13$	$R_2 = 2.2$	$x = 7.0$	$F_{igmax} = 0.02$
$\omega = 0.8938$	$\omega = 0.5$	$G_1 =$ 0.0045 $\mathrm{GPa^{-3}\ \mu s^{-1}}$	
$C_v = 2.487 \times 10^{-3}\ \mathrm{GPa\,K^{-1}}$	$C_v = 1.0 \times 10^{-3}\ \mathrm{GPa\,K^{-1}}$	$c = 0.667$	
$T_o = 298$ K	$E_o =$ 6.9 GPa-cm^3/cm^3-g	$d = 1.0$	
Shear modulus = 3.54 GPa		$y = 3.0$	$F_{G1max} = 0.8$
Yield strength = 0.2 GPa		$G_2 =$ 0.3 $\mathrm{GPa^{-1}\ \mu s^{-1}}$	
		$e = 0.667$	
		$g = 0.667$	
		$z = 1.0$	$F_{G2min} = 0.8$

measured and calculated free surface velocities of a 0.267-mm-thick tantalum disc driven by 19.871 mm of detonating LX-17 [83]. The momentum associated with the LX-17 reaction zone and early product expansion are accurately measured and calculated. These and other one-dimensional experiments were used to calibrate the LX-17 and PBX 9502 Ignition and Growth reaction rate parameters, which are then tested against two- and three-dimensional experimental data from several laboratories.

The main two-dimensional detonation experiment at LLNL is the cylinder expansion performance test [75]. Figure 6.6 (left) shows the radial velocity histories for a 2.54 cm radius LX-17 charge confined by 0.272 cm of copper. To calculate the exact momentum imparted to the cylinder wall, the Ignition and Growth model and the Steinberg–Guinan or Johnson–Cook metal model are used. Over a range of LX-17 to copper thickness ratios, the copper wall spalls and the wall acceleration profile is not as regular. Figure 6.6 (right) shows the experimental and calculated radial wall velocities for the case of copper spall calculated using the Steinberg–Guinan model which includes spall [75].

Another example of a unique two-dimensional TATB detonation wave structure is shown in Fig. 6.7 in which EDC35 (95% TATB and 5% Kel-F, the same composition as PBX 9502) is sandwiched between brass (left) and beryllium (right) [75]. Brass, like most metals, has a lower shock velocity than the detonation velocity of EDC35, so the brass shock front lags behind the detonation wave. Beryllium has a higher shock velocity than the EDC35

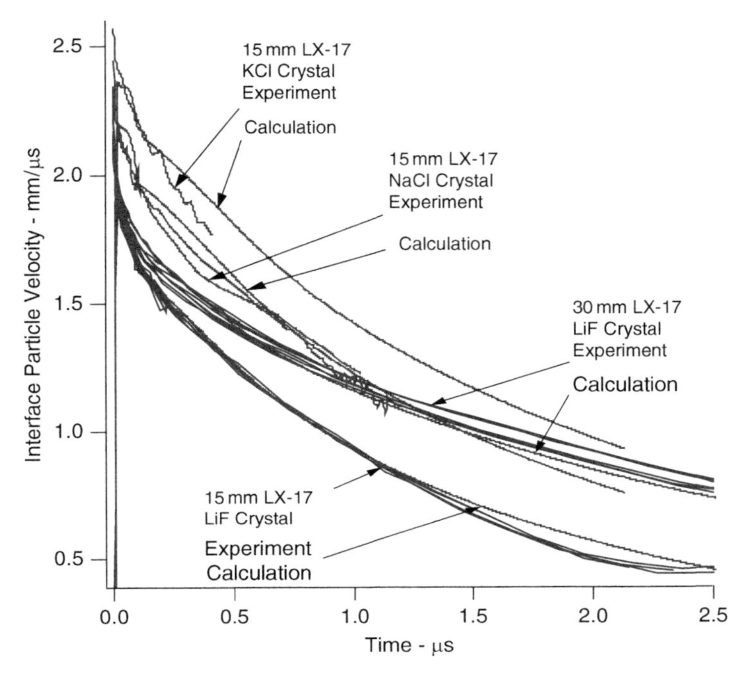

Fig. 6.4. Interface particle histories for detonating LX-17 and various salt crystals. The differences in von Neumann peaks are due to impedance differences between crystals [83]

detonation velocity, and thus its shock wave pulls the detonation wave along at higher velocity than its CJ value. The curved shapes of the EDC35 detonation waves and the arrival times of these waves at both edges after various propagation lengths are accurately calculated by the Ignition and Growth model parameters for PBX 9502 [75].

The detonation wavefront curvature is often measured as the detonation wave reaches the surface of an unconfined or confined cylindrical explosive charge. This wave curvature and the associated decrease in detonation velocity for a finite diameter charge is the result of the interaction of the chemical reaction rates with the radial rarefaction wave that propagates from the outer explosive boundary and reduces the pressure, temperature, and reaction rates in the reaction zone [75]. Since TATB has very high activation energies for its decomposition reactions, its plastic bonded explosives (PBXs) like LX-17 and PBX 9502 fail to detonate at velocities below 97% of the CJ detonation velocity [6]. This is similar to homogeneous liquid explosives, which also have high activation energies for decomposition and no "hot spot" sites [77]. The measured and calculated detonation wavefront curvatures for LX-17 confined by copper, PMMA, and tantalum (calculation only) are shown in Fig. 6.8 [75].

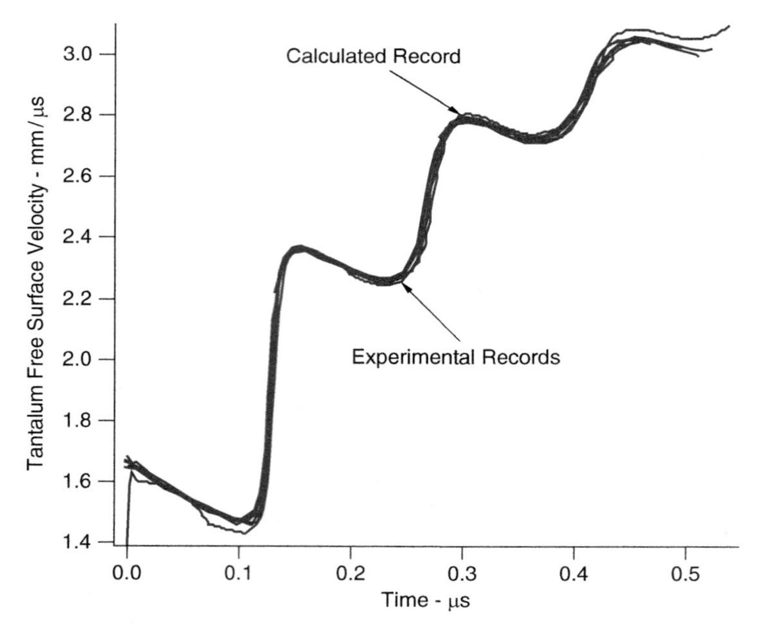

Fig. 6.5. Free surface velocity of a 0.267-mm-thick Ta disc driven by 19.871 mm of LX-17 [83]

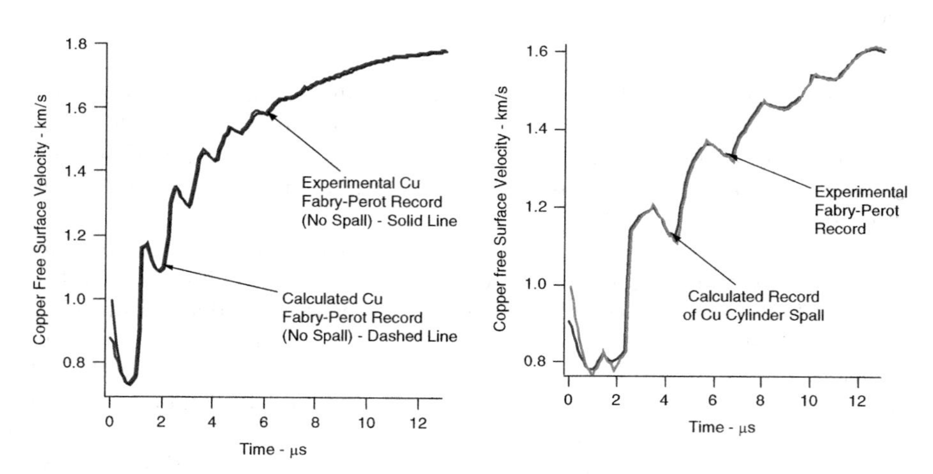

Fig. 6.6. Experimental and calculated LX-17 copper cylinder test radial free surface velocities without spall (*left*) and with spall (*right*) [75]

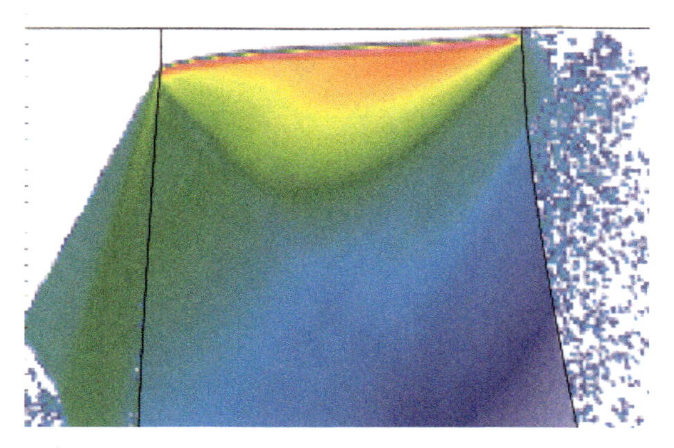

Fig. 6.7. LX-17 detonation wave propagating upward in brass (*left*) and beryllium (*right*) [75]

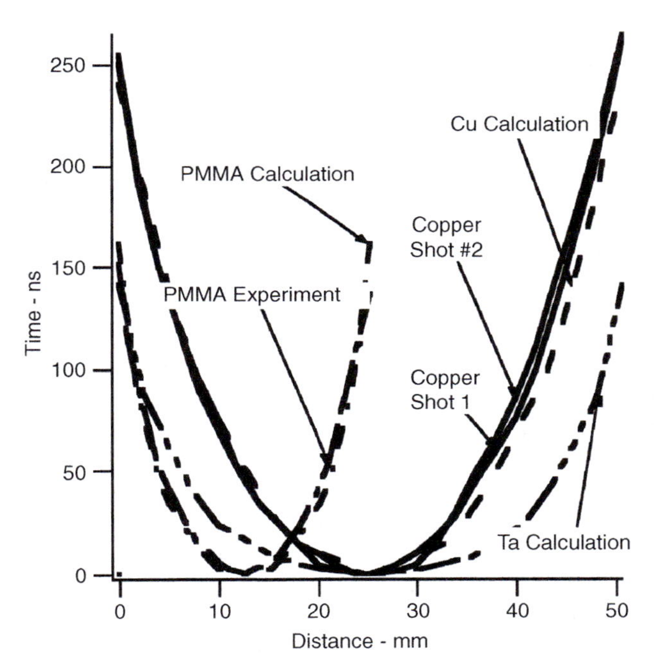

Fig. 6.8. LX-17 detonation wavefront curvature for copper, PMMA, and tantalum cylinders [75]

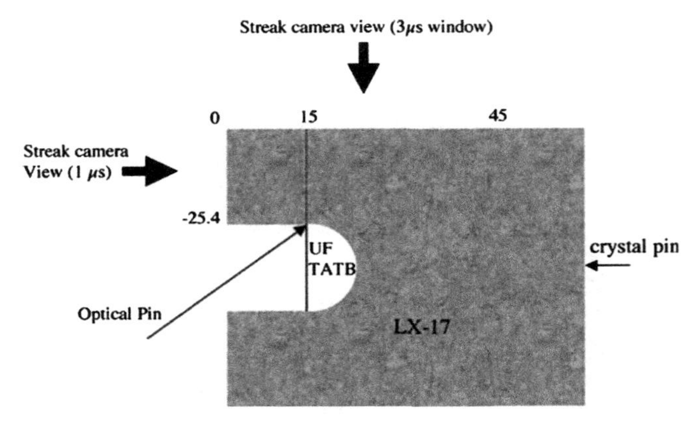

Fig. 6.9. Experimental geometry for a 25.4 mm wide LX-17 "Hockey Puck" [71]

Most other solid explosives have lower activation energies and exhibit larger detonation velocity decreases before detonation failure [6, 41, 70].

Since TATB-based detonation waves are very sensitive to rarefaction waves, they have strong interactions with changes in geometry and exhibit regions of zero or partial reaction in spherical divergence [15] or corner turning [18, 44] experiments. To ensure correct prediction of wave propagation and impulse delivery when modeling complex geometries that cannot be tested, reactive flow models must closely simulate all available two-dimensional experimental data. One well-instrumented corner turning experiment is the "Hockey Puck" experiment shown in Fig. 6.9 [71]. A 19.05-mm-radius hemispherical booster of Ultrafine (UF) TATB, which is fine particle TATB pressed to $1.8\,\mathrm{g\,cm^{-3}}$, is initiated at the center of the hemisphere. A spherically divergent detonation wave is established which initiates the LX-17 charge. The LX-17 detonates spherically outward, but also attempts to turn the corner and detonate the LX-17 in the region from 0 to 15 mm in Fig. 6.9. As it turns the corner, the LX-17 detonation wave leaves a region of unreacted or partially reacted explosive near the corner. This "dead zone" region is very similar to those observed using X-rays [44] and proton radiography [18]. More importantly, the "Hockey Puck" experiments measured the arrival times of LX-17 detonations along their perimeters. The experimental and calculated arrival times for the LX-17 detonation wave in a 12.7 mm wide "Hockey Puck" experiment are shown in Fig. 6.10. Excellent agreement was obtained using previously determined LX-17 Ignition and Growth parameters. "Hockey Pucks" have been fired and accurately modeled using different widths of LX-17 and PBX 9502 acceptor charges, and LX-07 (90% HMX and 10% Viton binder) as the booster explosive [74].

Besides turning 90° corners, detonation waves are propagated around 90 and 180° arcs of various inner and outer radii with unconfined or confined boundaries. Unconfined LX-17 90° arc tests fired at LLNL were modeled by

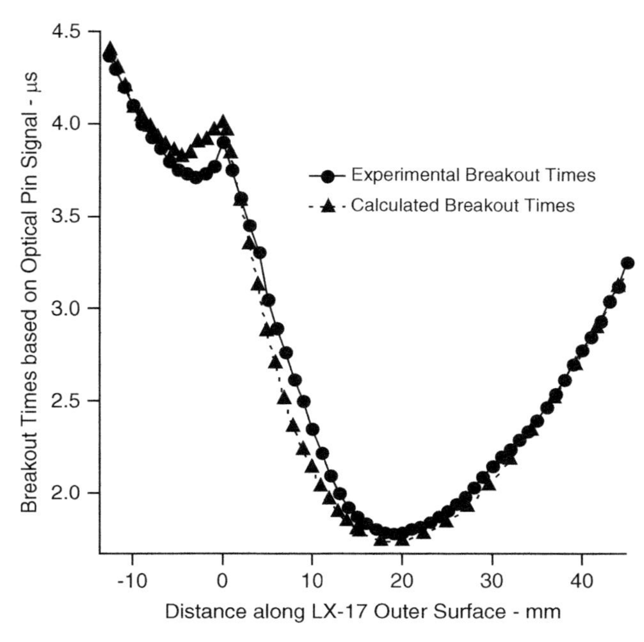

Fig. 6.10. Experimental and calculated breakout times along the LX-17 perimeter for a 12.7 mm wide "Hockey Puck" [71]

Tarver and Chidester [73]. One arc had an inner radius of 8.89 cm and an outer radius of 11.43 cm, while another arc had a 6.35 cm inner radius and a 10.16 cm outer radius. Figure 6.11 shows the experimental and calculated wave velocities at various angles along the outer edge of the 11.43 cm arc. The average calculated phase velocity from 0 to 90° is 8.781 km s^{-1}, while the measured value is 8.667 km s^{-1} for 0–85°. The inner surface pins measured a constant velocity of 7.289 km s^{-1} from 0 to 84°, while the calculated inner surface velocity equaled 7.241 km s^{-1}. For the smaller inner radius arc, the measured inner edge velocity was 7.03 km s^{-1}, while the calculated value was 7.042 km s^{-1}. The measured outer surface phase velocity was 10.07 km s^{-1}, and the calculated value was 9.75 km s^{-1}. Both the experimental and calculated initial breakouts of the LX-17 detonation waves occurred about 3 mm from the inner surface.

A set of confined TATB arc turning experiments was published by Lubyatinsky et al. [42] and modeled by Tarver and Chidester [73]. These experiments used 180° TATB explosive arcs with outer radii of 6 cm and inner radii of 3, 4, and 5 cm confined on both edges by 1 cm of steel or PMMA. Time of arrival pins were placed every 15° along both explosive edges. The edge wave velocities measured for the 180° arcs were slightly less than those calculated by the LX-17 model. Figure 6.12 contains the experimental and calculated arrival time differences for the three LX-17 thicknesses (10, 20, and 30 mm) with

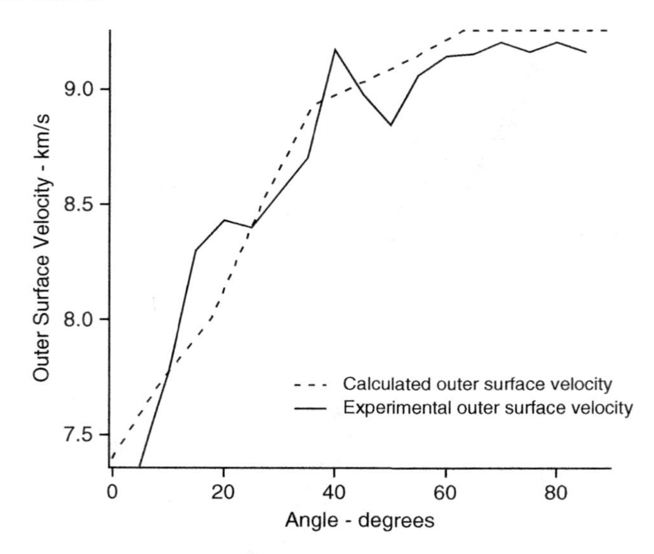

Fig. 6.11. Experimental and calculated outer surface velocities versus angle for the 11.43 cm arc [73]

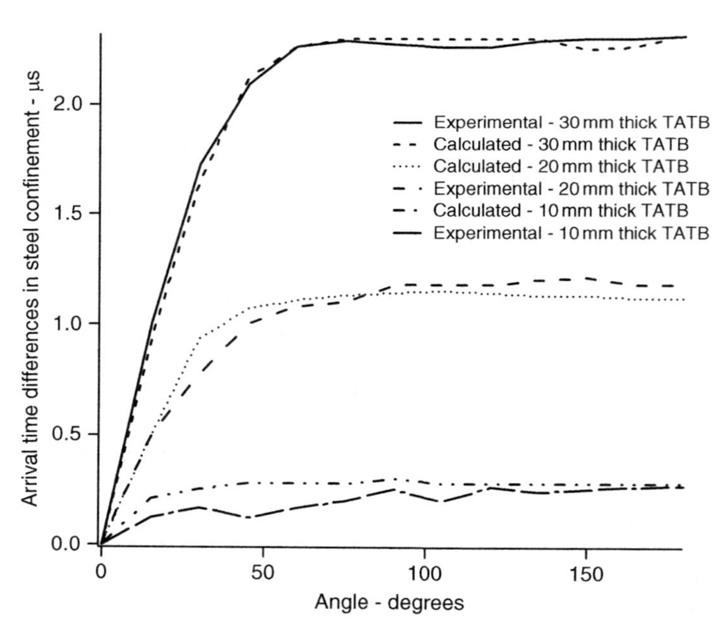

Fig. 6.12. Arrival time differences in steel confinement versus angle for three TATB arcs [42, 73]

steel confinement. The calculated and experimental arrival time differences as functions of angle of propagation agree closely for the three steel confined charges. Another two-dimensional detonation wave propagation comparison of the LX-17/PBX 9502 Ignition and Growth model [73] with experiments is for the cones of various areas reported by Salyer and Hill [56]. They reported the edge detonation velocities as functions of inverse radius of PBX 9502 cones with included angles of 10, 20, 30, 40, 80, and 90°. As steady PBX 9502 waves entered the converging cones, they became overdriven to higher velocities and shock pressures and temperatures than the CJ and von Neumann spike values [26, 59]. At 40, 80 and 90°, the converging detonation waves propagated through the entire cones at greater than CJ velocities. At 10, 20 and 30°, the convergence effects were overcome by rarefaction waves, which reduced the reaction rates until the reactions separated from the shock fronts. Then the detonation waves rapidly failed. To simulate these effects, the 2D conical tests were simulated using 20 zones per mm and the PBX 9502 parameters. Figure 6.13 shows the experimental and calculated edge velocities as functions of inverse cone radius for the 40, 80, and 90° cones, along with the unconfined detonation velocity-inverse radius curve [6]. The calculated detonation waves also detonated to the tips of the cones. The 80 and 90° cones remained overdriven at edge velocities exceeding $10\,\mathrm{km\,s^{-1}}$. Figure 6.14 shows the 10, 20, and 30° comparisons. These calculated detonation waves failed to detonate

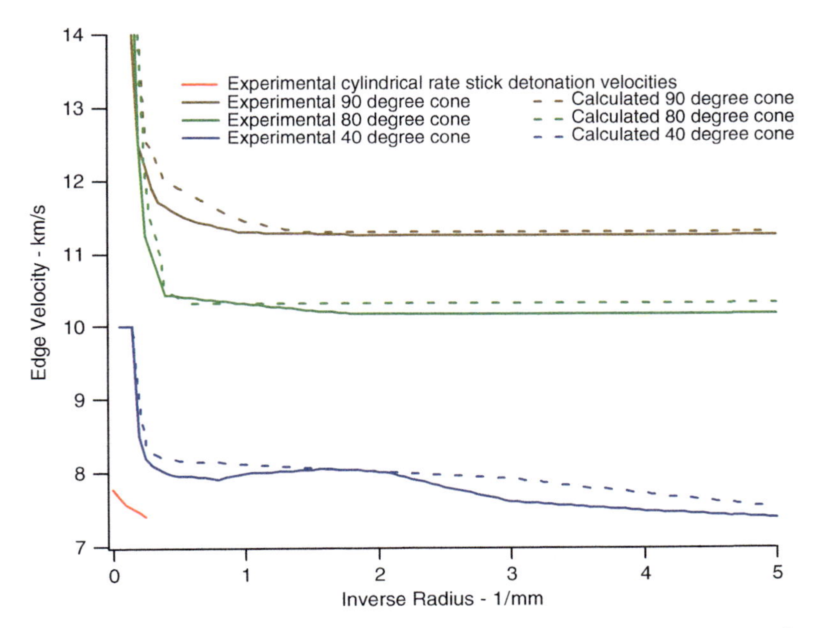

Fig. 6.13. Experimental and calculated edge velocities for the 40, 80 and 90° cones [56, 73]

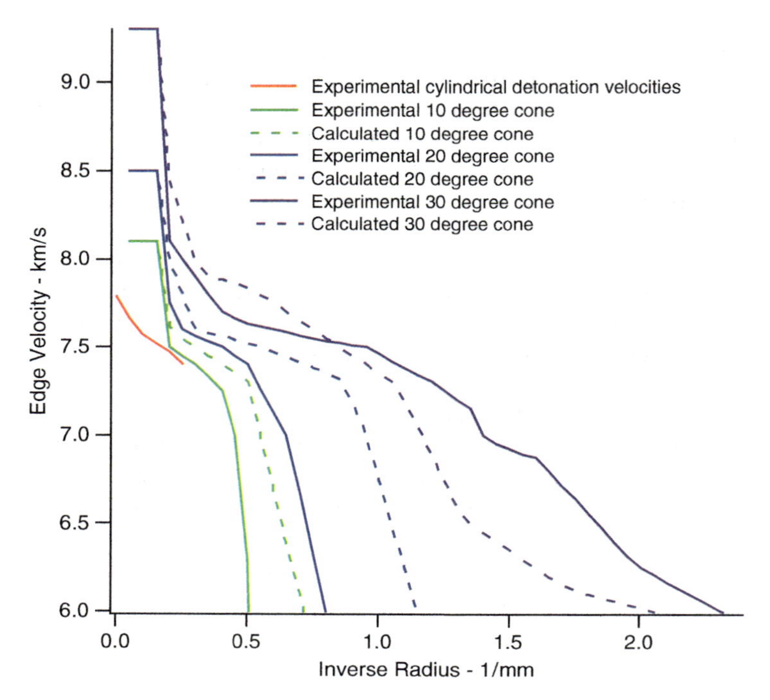

Fig. 6.14. Experimental and calculated edges velocities for the 10, 20, and 30° cones [56, 73]

to the tips of the cones. The calculated failures occurred at a slightly larger radius than the 30° cone test and at slightly smaller radii than the 10 and 20° cone tests. The overall agreement is excellent for these detonation waves that are initially highly overdriven and then fail extremely rapidly.

Another fascinating detonation phenomenon for which new quantitative two-dimensional experimental data were recently obtained is "shock desensitization." This phenomenon was previously known as "dead pressing" and has been observed in several explosives, including very sensitive primary explosives [5]. In the first quantitative study of shock desensitization, Campbell and Travis [7] impacted large PBX 9404 (94% HMX, 3% nitrocellulose, and 3% chloroethylphosphate) and Composition B (65% RDX/35% TNT) charges with weak shocks on one edge and initiated detonation waves on another edge. They then measured the interactions of these detonation waves with weak shocks of various strengths. For a certain shock pressure regime (1–2.4 GPa) in both PBX 9404 and Composition B, the detonation waves propagated a few millimeters into the precompressed explosive and then abruptly failed. The measured time duration before failure was close to the experimentally measured shock initiation time for that specific shock pressure. For shock pressures below 1 GPa, which did not precompress the unreacted explosives to their maximum densities, the detonation waves wavered slightly but continued to detonate through the precompressed explosive that still contained some hot spot reaction sites. At shock pressures greater than 2.4 GPa, the detonation waves encountered shock-compressed explosives with growing hot spots so they continued to detonate through these compressed, reacting explosives.

Recently, five new experiments were designed to precompress LX-17 in the 1–2 GPa pressure range by diverging shocks propagating through steel shadow plates, while the LX-17 detonation waves are propagating around two corners [27]. The LX-17 detonation waves then arrive in weakly shocked regions of precompressed LX-17. If the diverging shock pressures are high enough and are applied long enough, the preshocked LX-17 regions are shock desensitized and the detonation waves fail. If the shocks are not strong enough or too strong, the detonation waves continue to detonate. The Ignition and Growth model including shock desensitization correctly predicted that the detonation waves would turn both corners of the steel discs and that desensitization would occur in all five experiments in the LX-17 between the steal and aluminum plates.

Figure 6.15 is a cross section of one of the double corner turning and shock desensitization experiments. Under the stainless steel shadow plate, a small charge of the PETN-based explosive LX-16 initiates detonation of a UF TATB hemispherical booster. The UF TATB is pressed to 1.80 g cm^{-3} or 93% theoretical maximum density. The main LX-17 charge surrounds the UF TATB booster. On the top and bottom are the 6 mm thick aluminum wit-

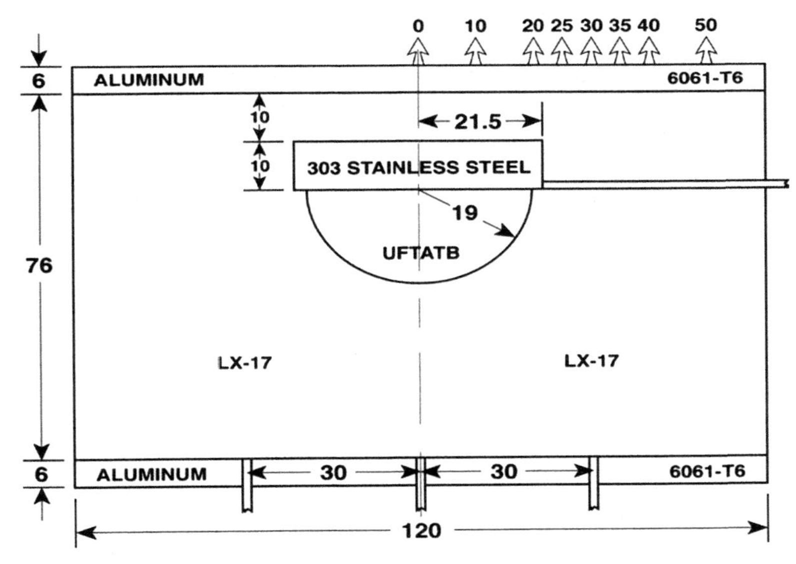

Fig. 6.15. Cross section of a double corner turning and shock desensitization test (in mm) [27]

ness plates. Upon firing, the small LX-16 explosive charge initiates the hemispherical UF TATB booster, which in turn initiates a LX-17 hemispherical detonation. The LX-17 detonation propagates outward until it reaches the aluminum plates. The bottom aluminum plate contains three time-of-arrival pins to confirm that a diverging LX-17 detonation was initiated. The top aluminum plate is instrumented with eight photonic Doppler velocimetry (PDV) probes to measure the free surface velocity at eight radii. X-ray radiographs and framing camera images are taken at various times. The LX-17 detonation propagates around the two corners of the steel shadow plates and into a thin LX-17 region between the steel shadow plate and the top aluminum plate. These LX-17 regions are compressed to 1–2 GPa by diverging shock waves that had propagated through the steel shadow plates. These weak shocks desensitize the LX-17, resulting in failures of LX-17 detonation waves when they reach the LX-17 above the steel discs.

To model shock desensitization in doubly shocked LX-17, a second compression constant was added to the first term of the reaction rate law in (6.2) by Tarver et al. [80]. This forced the reaction rate to be zero when the LX-17 was shocked within a range of compressions. This assumption worked well for LX-17 shock desensitization due to long duration reflected shocks in the experiments described above. However, it has been shown that shock desensitization in other solid explosives is very time dependent [7]. The failure of detonation in the precompressed explosive requires about the same amount of time as shock initiation at that shock pressure. For TATB PBXs, it has been shown using proton radiography that dead zones can exist for relatively long times after detonation waves turn corners [18]. Figure 6.16 shows a proton radiographic image of a diverging detonation wave in PBX 9502 and the attached dead zone [18]. The regular Ignition and Growth model creates dead zones and propagates around corners accurately, but allows the partially reacted dead zone regions to slowly react at very late times [71]. To model these longer lasting dead zones, a time-dependent desensitization rate law was added to the ignition reaction rate term in (6.8) by DeOliveira et al. [11]. The desensitization rate S is defined as

$$S = A_p(1-\phi)(\phi+\varepsilon), \tag{6.9}$$

where A is a constant, p is the shock pressure, ε is a small constant, and ϕ varies from zero in a pristine explosive to one in a fully desensitized explosive. The density threshold a in (6.8) is redefined to be a linear function of ϕ:

$$a(\phi) = a_0(1-\phi) + a_1(\phi), \tag{6.10}$$

where a_0 and a_1 are constants. The relative density threshold for ignition of the pristine explosive becomes $1 + a_0$, and, for the fully desensitized explosive, the relative density for ignition becomes $1 + a_1$. Additionally, the second reaction rate term in (6.8) is modified so that it turns on only when F exceeds a minimum $F_{G1\text{min}}$, which is a linear function of ϕ:

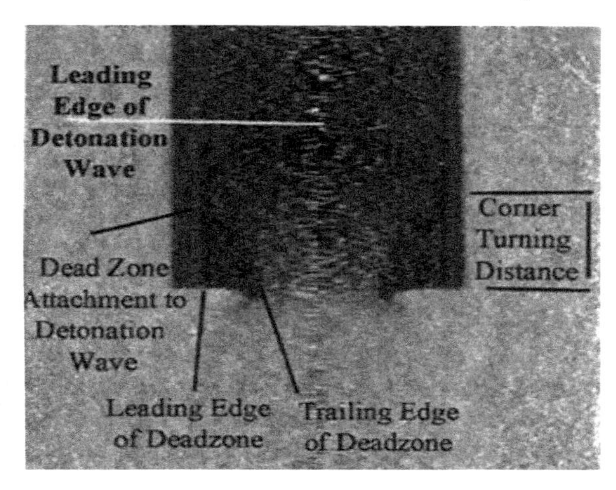

Fig. 6.16. PBX 9502 diverging detonation wave and attached dead zone. The detonation wave propagates upward [75]

$$F_{G1\min}(\phi) = F_c\phi, \tag{6.11}$$

where F_c is a constant related to the initial porosity. This modification provides a competition between desensitization and reaction growth and thus determines an extinction mechanism. Four new parameters are required: A, ε, a_1, and F_c. No time-resolved experiments like those of Campbell and Travis [7] have yet been done on TATB PBX's, but DeOliveria et al. [11] found that values of $A = 1,000$, $\varepsilon = 0.001$, $a_1 = 0.50$, and $F_c = 0.01$ produce reasonable dead zones for the Hockey Puck corner turning experiments. The estimated desensitization times were 1.29 µs at 1 GPa shock and 0.26 µs at 5 GPa.

Using these desensitization values and the regular LX-17 Ignition and Growth parameters [72], the measured and calculated PDV axial free surface velocity histories for one of the Hockey Puck experiments are compared in Fig. 6.17. The calculated jump-off times for all 8 PDV probes agree well with experiment. At radii greater than the steel plate radii, LX-17 detonation waves accelerate the aluminum plates to velocities above 2,000 m s^{-1}. At radii less than those of the steel plates, i.e., in the zone between the steel and aluminum plate, the jump-off velocities are lower and increase when later shocks arrive. These low jump-off velocities are clear evidence of shock desensitization.

Perhaps the ultimate test of a detonation reactive flow model is a quantitative three-dimensional experiment in which the detonation wave propagates or fails to propagate. The prism test [54] developed by Ramsay at LANL for insensitive high explosives is the only such experiment. The unconfined version of this test is shown in Fig. 6.18. It consists of a line wave generator to initiate the top surface of a PBX 9501 (95% HMX, 2.5% BDNPA/F, and 2.5% Estane) charge. The detonating PBX 9501 sends a detonation wave into

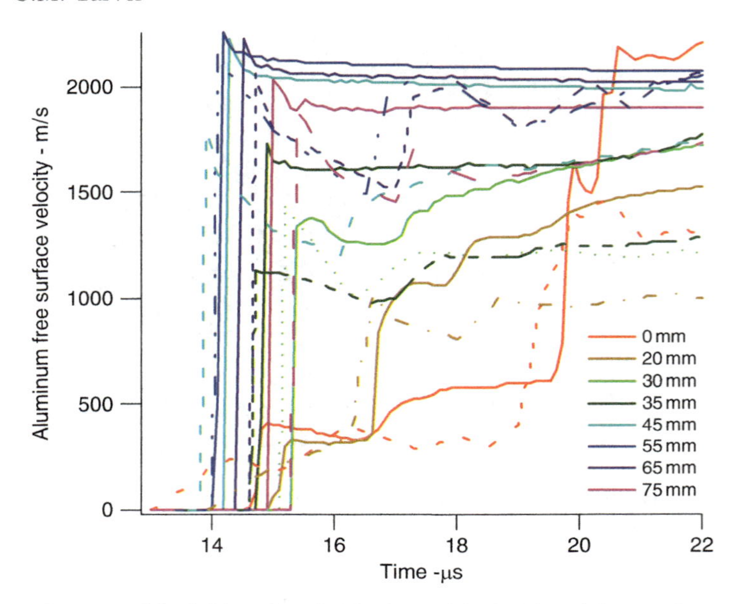

Fig. 6.17. Measured (*solid lines*) and calculated (*dashed lines*) free surface velocities for a LX-17 shock desensitization test [27, 72]

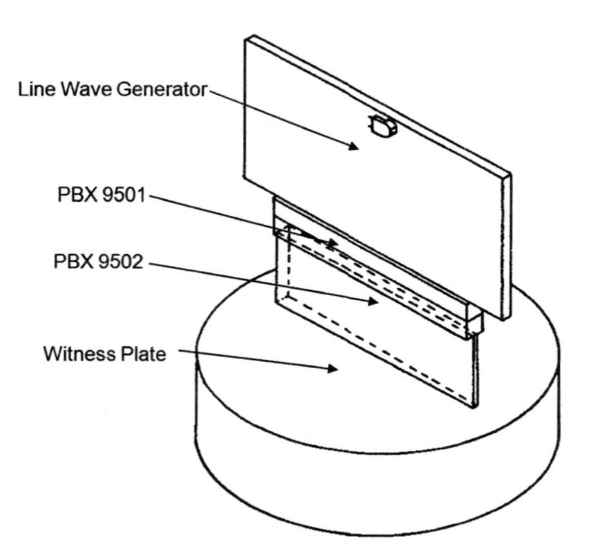

Fig. 6.18. The unconfined LANL prism test geometry [54]

PBX 9502, which is initially overdriven. The PBX 9502 then detonates downward in a trapezoid of decreasing thickness. At a thickness approximately equal to one half of the cylindrical failure diameter, the PBX 9502 detonation wave fails to propagate [54]. The remaining PBX 9502 detonation wave propagates downward to an aluminum witness plate, creating a crater that clearly shows the boundary between failed and propagated detonation. Various confinement materials (water, PMMA, aluminum, copper, water, and lead) were used on the PBX 9502 boundary to change its confined failure thickness. The measured experimental failure thickness was inversely proportional to the impedance of the confinement material. Ramsay [54] studied three initial temperatures (−55, 25, and 75°C), and Asay and McAfee [1] fired several 250°C PBX 9502 prisms. This large experimental data set is an excellent test for three-dimensional reactive flow modeling.

Three-dimensional Ignition and Growth model parameters were tested on previously discussed two-dimensional experiments and on ambient temperature prism tests by Garcia and Tarver [23]. Recently, Garcia and Tarver [24] extended the prism test modeling to all four initial temperatures and several confinement materials using finer three-dimensional zoning. Table 6.2 lists the values of B and G_1 used in (6.7) and (6.8), respectively, for the four temperatures. The other parameters are those in Table 6.1. Table 6.3 lists the experimental and calculated failure thicknesses for confined 25°C PBX 9502.

The experimental and calculated PBX 9502 unconfined failure thicknesses at the four initial temperatures are compared in Fig. 6.19. Considering that the

Table 6.2. Values of B and G_1 used in (6.7) and (6.8) for four temperatures of PBX 9502

Temperature (°C)	G_1(Mbar^{-3} μs^{-1})	B(Mbar)
−54	3,750	−0.004488
25	4,613	−0.05031
75	7,200	−0.05376
250	8,000	−0.06580

Table 6.3. Measured and calculated prism failure thicknesses of 25°C PBX 9502

Material	Measured (mm)	Calculated (mm)
Unconfined	3.5	4.0
Al (0.075 mm)	2.6	3.8
Al (0.5 mm)	1.9	1.3
Al (1.0 mm)	1.6	<1
Steel (0.5 mm)	–	<1
Copper (0.5 mm)	–	<1
PMMA (1.0 mm)	3.5	2.0
PMMA (3.0 mm)	3.5	2.0

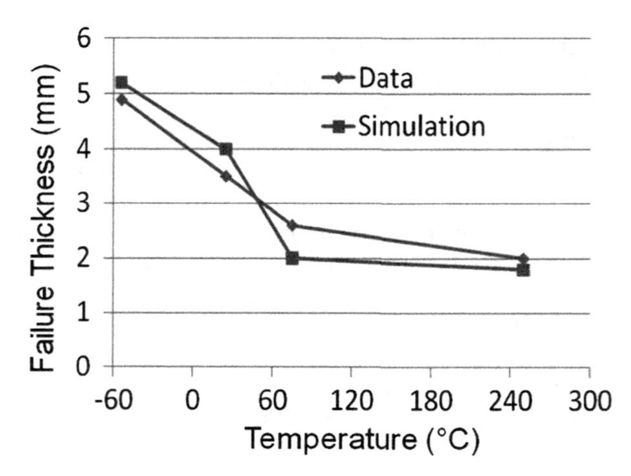

Fig. 6.19. Experimental versus calculated PBX 9502 failure thickness for four initial temperatures [1, 24, 54]

−54, 75, and 250°C model parameters are based on only unconfined cylinder failure diameter data [75], the agreement with this three-dimensional experiment is encouraging.

Quantitative three-dimensional reactive flow modeling of condensed-phase detonation waves is therefore possible given the size and speed of existing and future computers. The current calculations were done without automatic mesh refinement (AMR) techniques that make even larger scale, longer time duration calculations possible. It also appears that a set of reactive flow model parameters for a particular explosive developed using one- and two-dimensional experimental data can be used for calculating three-dimensional problems. Currently, only the TATB-based explosives LX-17 and PBX 9502 have been studied with all of the aforementioned experimental techniques and thus have the most fully developed parameter sets. HMX, RDX, TNT, and Composition B have been studied to a lesser extent [38, 70, 78, 82]. Average two-dimensional reaction zone properties of detonation waves in liquid explosives, such as nitromethane, can also be calculated using the Ignition and Growth model [76]. Reactive flow modeling of very "nonideal" solid explosives, such as ammonium nitrate–fuel oil (ANFO) and aluminized explosives, is more difficult due to their extremely long reaction zone lengths and buildup distances to full CJ detonation, but it is possible [41].

6.4 Concluding Remarks

While a great deal has been learned in recent years about the extreme chemistry occurring in a detonation reaction zone, much more research is required to fully understand the nonequilibrium processes, the reaction pathways, and the approach to equilibrium CJ mixtures occurring within a detonation wave. A long-term goal is to understand condensed-phase explosive detonation toward the same level as our understanding of the gas phase detonation. A tightly coupled experimental and theoretical approach is required to produce such an understanding. Experimental efforts are underway to measure the rates of vibrational excitation by phonon up-pumping, IVR, and initial decomposition reactions. Molecular dynamics and reactive force field reaction pathway modeling is rapidly becoming more sophisticated, and larger scale systems can now be studied using parallel computers and AMR techniques. Improved potentials are being developed to better describe partially reacted and equilibrium thermochemical states of product mixtures.

Since chemical reaction rates and equilibrium concentrations are mainly controlled by the local temperature of a small volume of molecules, the most urgent need in explosives research is for time-resolved experimental measurements of temperature in all regions of reacting explosives: impact- and shock-induced hot spots, deflagration waves, reactive flows behind shock fronts, and detonation waves. Knowing the unreacted explosive temperature as a function of shock pressure will complete its EOS description and allow more accurate predictions of the induction time delay for the onset of bond breaking behind each individual shock front of a three-dimensional detonation wave. Accurate temperature measurements will enable molecular dynamics and reactive force field simulations to be done at the exact pressure, density, and temperature conditions attained in various regions of a detonation wave. Temperature measurements in the vicinity of the CJ plane and in the subsequent reaction product expansion flow will eliminate the last remaining (and most important) unknown in thermochemical equilibrium CJ predictions. Improved potentials can be developed to predict the distribution of internal and potential energies under all of the conditions attained in the flows produced by detonation waves. Then the impulse delivered to an adjoining material can be more accurately modeled.

Since not all of the scenarios involving detonation waves can be tested experimentally, hydrodynamic computer models have to be improved to predict the safety and performance properties of the reactive flows produced by shock initiating and detonating explosives. Assuming that the necessary activation energy and temperature data will become available, the next generation of hydrodynamic computer code reactive flow models for simulating ideal explosive detonation waves in one-, two-, and three-dimensions will need to be based on temperature-dependent Arrhenius rate laws, replacing current compression- and pressure-dependent rate laws. Mesoscale models are being formulated in which individual particles of a solid explosive plus their binders and voids are resolved, shocked heated, and either react or fail to react using Arrhenius kinetics. Modeling reactive flows of individual particles is impractical for large-scale simulations even with today's parallel supercomputers. So, models like the continuum Statistical Hot Spot reactive flow model [47, 49–51] in the ALE3D hydrodynamic computer code [48] at LLNL are being developed. The ALE3D code enables the complete coupling of heat transfer, chemical reactions, hydrodynamics, and chemical species equilibrium [48]. In the Statistical Hot Spot model, a realistic number of hot spots of various sizes, shapes, and temperatures based on the original void volume, particle size distribution, and temperature of the solid explosive are assumed to be created as an initiating shock front compresses the explosive particles. The hot spots then either react and grow into the surrounding explosive or fail to react and die out based on multistep Arrhenius kinetics rates. The Statistical Hot Spot reactive flow model simulates shock desensitization without an additional desensitization rate law [50]. The coalescence of growing hot spots at high pressures and temperatures, the creation of additional surface area available to the reacting sites as the pressure and temperature rise, the rapid transition to detonation, and the formation of the three-dimensional cellular structure of self-sustaining detonation are four of the most challenging current problems under investigation in hydrodynamic reactive flow modeling efforts on homogeneous and heterogeneous condensed-phase explosives.

Acknowledgement

It is impossible to individually thank all of the great experimentalists and theoreticians who developed the current understanding of condensed-phase detonation over the past 40 years. The author thanks them all collectively for their hard word and enthusiasm. This work was performed under the auspices of the U.S. Department of Energy by the Lawrence Livermore National Laboratory under Contract DE-AC52-07NA27344.

References

1. Asay, B.W., McAfee, J.M.: Temperature effects on failure thickness and the deflagration-to-detonation transition in PBX 9502 and TATB. In: Tenth International Detonation Symposium, pp. 485–489. Office of Naval Research, Boston (1993). ONR 33395-12
2. Asay, B.W., Kennedy, J.E., Ramsay, J.B., Schelling, F.J., Takala, B. E.: Indices of the Proceedings for the 13th International Symposium on Detonation 1951–2006, Los Alamos National Laboratory Report LA-UR-07-4550 (2007)
3. Bastea, S., Glaesemann, K.R., Fried, L.E.: Equation of state for high explosives detonation products with polar and ionic species. In: Thirteenth International Detonation Symposium, pp. 1137–1143. Office of Naval Research, Norfolk (2006). ONR 351-07-01
4. Bernshtein, V., Oref, I.: Energy transfer rate coefficients from trajectory calculations and contributions of supercollisions to reaction rate coefficients. J. Phys. Chem. **100**, 9738–9744 (1996)
5. Bowden, F.P., Yoffe, A.D.: Ignition and Growth of Explosion in Liquids and Solids. Cambridge University Press, Cambridge (1985)

6. Campbell, A.W., Engelke, R.: The diameter effect in high-density heterogeneous explosives. In: Sixth Symposium (International) on Detonation, pp. 642–652. Office of Naval Research, Coronado (1976). ACR-221
7. Campbell, A.W., Travis, J.R.: The shock desensitization of PBX-9404 and composition B-3. In: Eighth Symposium (International) on Detonation, pp. 1057–1068. Naval Surface Weapons Center NSWC MP 86–194, Albuquerque (1985)
8. Chapman, D.L.: Detonation states. Phil. Mag. **213**, Series 5, 47–90 (1899)
9. Cobos, C.J., Croce, A.E., Luther, K., Troe, J.: Shock tube study of the thermal decomposition of CF_3 and CF_2 radicals. J. Phys. Chem. A **114**, 4755–4761 (2010)
10. Cowperthwaite, M., Shaw, R.: $C_v(T)$ equation of state for liquids. J. Chem. Phys. **53**, 555–560 (1970)
11. DeOliveira, G., Kapila, A.K., Schwendeman, D.W., Bdzil, J.B., Henshaw, W.D., Tarver, C.M.: Detonation diffraction, dead zones and the ignition and growth model. In: Thirteenth International Detonation Symposium, pp. 13–23. Office of Naval Research, Norfolk (2006). ONR 351-07-01
12. Dobratz, B.: Detonation and combustion of explosives: A selected bibliography. In: Eleventh International Detonation Symposium, pp. 1101–1110. Office of Naval Research, Snowmass (1998). ONR 33300-5
13. Döring, W.: Theory of detonation waves. Ann. Physik **43**, 421–436 (1943)
14. Dremin, A.N.: Toward Detonation Theory. Springer, New York (1999)
15. Druce, R.L., Roeske, F., Souers, P.C., Tarver, C.M., Chow, C.T.S., Lee, R.S., McGuire, E.M., Overturf, G.E., Vitello, P.A.: Propagation of axially symmetric detonation waves. In: Twelfth International Detonation Symposium, pp. 675–683. Office of Naval Research, San Diego (2002). ONR 333-05-2
16. Ershov, A.P., Satonkina, N.P., Ivanov, G.M.: Reaction zones and conductive zones in dense explosives. In: Thirteenth International Detonation Symposium, pp. 79–88. Office of Naval Research, Norfolk (2006). ONR 351-07-01
17. Eyring, H.: Starvation kinetics. Science **199**, 740–743 (1978)
18. Ferm, E.N., Morris, C.L., Quintana, J.P., Pazuchanic, P., Stacy, H., Zumbro, J. D., Hogan, G., King, N.: Proton radiography examination of unburned regions in PBX 9502 corner turning experiments. In: Shock Compression of Condensed Matter – 2001, AIP Conference Proceedings, vol. 620, pp. 966–969. Atlanta (2002)
19. Fickett, W.: Introduction to Detonation Theory. University of California Press, Berkeley (1984)
20. Fickett, W., Davis, W.C.: Detonation Theory and Experiment. Dover Publications, Mineola (2000)
21. Frost, D.L., Goroshin, S., Ripley, R., Zhang, F.: Interaction of a blast wave with a metalized explosive fireball. In: Fourteenth International Detonation Symposium, pp. 696–705. Office of Naval Research, Coeur d'Alene (2011). ONR-351-10-185
22. Gamezo, V.N., Desbordes, D., Oran, E.S.: Two-dimensional reactive flow dynamics in cellular detonation waves. Shock Waves **9**, 11–17 (1999)
23. Garcia, M.L., Tarver, C.M.: Three-dimensional ignition and growth reactive flow modeling of prism failure tests on PBX 9502. In: Thirteenth International Detonation Symposium, pp. 63–70. Office of Naval Research, Norfolk (2006). ONR 351-07-01

24. Garcia, M.L., Tarver, C.M.: Three-dimensional ignition and growth reactive flow modelling of confined and hot prism tests. In: Fourteenth International Detonation Symposium, pp. 1229–1236. Office of Naval Research, Coeur d'Alene (2011). ONR-351-10-185
25. Gorshkov, M.M., Grebenkin, K.F., Slobodenyukov, V.M., Tkachev, O.V., Zaikin, V.T., Zherebtsov, A.L.: Shock-induced electro conductivity in the insensitive high explosive TATB. In: Thirteenth International Detonation Symposium, pp. 435–444. Office of Naval Research, Norfolk (2006). ONR 351-07-01
26. Green, L.G., Tarver, C.M., Erskine, D.J.: Reaction zone structure in supracompressed detonating explosives. In: Ninth Symposium (International) on Detonation, pp. 670–682. Office of the Chief of Naval Research, Portland (1989). OCNR 113291-7
27. Hart, M.R.: Jack rabbit investigation of TATB IHE detonation chemical kinetics. In: Fourteenth International Detonation Symposium, pp. 282–291. Office of Naval Research, Coeur d'Alene (2011). ONR-351-10-185
28. Hayes, B., Tarver, C.M.: Interpolation of detonation parameters from experimental particle velocity records. In: Seventh Symposium (International) on Detonation, pp. 1029–1039. Naval Surface Weapons Center NSWC MP 82-334, Annapolis (1981)
29. Holmes, W., Francis, R.S., Fayer, M.D.: Crack propagation induced heating in crystalline energetic materials. J. Chem. Phys. **110**, 3576–3583 (1999)
30. Hong, X., Chen, S., Dlott, D.D.: Ultrafast mode-specific intermolecular vibrational energy transfer to liquid nitromethane. J. Phys. Chem. **99**, 9102–9109 (1995)
31. Hooper, J.: Vibrational energy transfer in shocked molecular crystals. J. Phys. Chem. **132**, 014507 (2010)
32. Jaramillo, E., Sewell, T. D., Strachan, A.: Atomic-level view of inelastic deformation in a shock loaded molecular crystal. Phys. Rev. B **76**, 064112 (2007)
33. Jouguet, E.: Theory of detonation. Pure Appl. Math. **70**(1), Series 6, 347–388 (1904)
34. Kato, Y., Mori, N., Sakai, H., Tanaka, K., Sakurai, T., Hikita, T.: Detonation temperature of nitromethane and some solid high explosives. In: Eighth Symposium (International) on Detonation, pp. 558–566. Naval Surface Weapons Center NSWC MP 86-194, Albuquerque (1985)
35. Kiefer, J.H., Kumaram, S.S., Sundaram, S.: Vibrational relaxation, dissociation, and dissociation incubation times in norbornene. J. Chem. Phys. **99**, 3531–3541 (1993)
36. Kozyrev, N.V., Larionov, B.V., Sakovich, G.V.: Influence of HMX particle size on the synthesis of nanodiamonds in detonation waves. Combust. Explos. Shock Waves **44**, 193–197 (2008)
37. Kuhl, A.L., Reichenbach, H., Bell, J.B., Beckner, V.E.: Reactive blast waves from composite charges. In: Fourteenth International Detonation Symposium, pp. 806–815. Office of Naval Research, Coeur d'Alene (2011). ONR-351-10-185
38. Kury, J.W., Breithaupt, R.D., Tarver, C.M.: Detonation waves in trinitrotoluene. Shock Waves **9**, 227–237 (1999)
39. Lee, J.H.S.: The universal role of turbulence in the propagation of strong shocks and detonation waves. In: Horie, Y., Davidson, L., Thadhani, N.N. (eds.) High-Pressure Shock Compression of Solids VI, pp. 121–148. Springer, New York (2003)

40. Lee, J.H.S.: The Detonation Phenomenon. Cambridge University Press, Cambridge (2008)
41. Leiper, G.A. Cooper, J.: Reaction rates and the charge diameter effect in heterogeneous explosives. In: Ninth Symposium (International) on Detonation, pp. 197–208. Office of the Chief of Naval Research, Portland (1989). OCNR 113291-7
42. Lubyatinsky, S.N., Batalov, S.V., Garmashev, A.Y., Israelyan, V.G., Kostitsyn, O.V., Loboiko, B.G., Pashentsev, V.A., Sibilev, V.A., Smirnov, E.B., Filin, V.P.: Detonation propagation in 180° ribs of an insensitive high explosive. In: Shock Compression of Condensed Matter-2003, AIP Conference Proceedings, vol. 706, pp. 859–862. Portland (2003)
43. Lysne, P.C., Hardesty, D.R.: Fundamental equation of state of liquid nitromethane to 100 kbar. J. Chem. Phys. **59**, 6512–6523 (1976)
44. Mader, C.L.: Two-dimensional homogeneous and heterogeneous detonation wave propagation. In: Sixth Symposium (International) on Detonation, pp. 405–413. Office of Naval Research, Coronado (1976). ACR-221
45. Manaa, M.R. (ed.): Chemistry at Extreme Conditions. Elsevier, Amsterdam (2005)
46. Manaa, M.R., Fried, L.E.: Intersystem crossings in model energetic materials. J. Phys. Chem. A **103**, 9349–9354 (1999)
47. Nichols III, A.L.: Statistical hot spot model for explosive detonation. In: Shock Compression of Condensed Matter-2005, AIP Conference Proceedings, vol. 845, pp. 465–470. Baltimore (2005)
48. Nichols III, A.L. (ed.): User's Manual for ALE3D, an Arbitrary Lagrange/Eulerian 2D and 3D Code System. Lawrence Livermore National Laboratory Report LLNL-SM-404490 Rev. 1 (2009)
49. Nichols III, A.L.: Comparison of the growth of pore and shear band driven detonations. In: Fourteenth International Detonation Symposium, pp. 1549–1559. Office of Naval Research, Coeur d'Alene (2011). ONR-351-10-185
50. Nichols III, A.L., Tarver, C.M.: A statistical hot spot model for shock initiation and detonation of solid high explosives. In: Twelfth International Detonation Symposium, pp. 489–496. Office of Naval Research, San Diego (2002). ONR 333-05-2
51. Nichols III, A.L., Tarver, C.M., McGuire, E.M.: ALE3D statistical hot spot model results for LX-17. In: Shock Compression of Condensed Matter-2003, AIP Conference Proceedings, vol. 706, pp. 397–400. Portland (2003)
52. Nomura, K., Kalia, R.K., Nakano, A., Vashishta, P., van Duin, A.C.T., Goddard III, W.A.: Dynamic transition in the structure of an energetic crystal during chemical reactions at shock front prior to detonation. Phys. Rev. Lett. **99**, 148303 (2007)
53. Plaksin, I., Coffey, C.S., Mendes, R., Riberio, J., Campos, J., Direito, J.: Formation of CRZ 3D structure at SDT and at Shear Initiation of PBX: Effects of Front Irradiation and Reaction Localization in HMX Crystals. In: Thirteenth International Detonation Symposium, pp. 319–330. Office of Naval Research, Norfolk (2006). ONR 351-07-01
54. Ramsay, J.B.: Effect of confinement on failure in 95 TATB/5 Kel-F. In: Eighth Symposium (International) on Detonation, pp. 372–379. Naval Surface Weapons Center NSWC MP 86-194, Albuquerque (1985)
55. Reed, E.J., Manaa, M.R., Fried, L.E., Glaesemann, K.R., Tarver, C.M., Joannopoulos, J.D.: Ab-initio discovery of ultrafast detonation, metallization

and Chapman–Jouguet states in nitromethane and hydrozoic acid. In: Fourteenth International Detonation Symposium, pp. 224–232. Office of Naval Research, Coeur d'Alene (2011). ONR-351-10-185
56. Salyer, T.L., Hill, L.G.: The dynamics of detonation failure in conical PBX 9502 charges. In: Thirteenth International Detonation Symposium, pp. 24–34. Office of Naval Research, Norfolk (2006). ONR 351-07-01
57. Seitz, W.L., Stacy, H.L., Engelke, R., Tang, P.K., Wackerle, J.: Detonation reaction-zone structure of PBX 9502. In: Ninth Symposium (International) on Detonation, pp. 657–669. Office of the Chief of Naval Research, Portland (1989). OCNR 113291-7
58. Sheffield, S.A., Bloomquist, D.D., Tarver, C.M.: Subnanosecond measurements of detonation fronts in solid high explosives. J. Chem. Phys. **80**, 3831–3844 (1984)
59. Sheffield, S.A., Engelke, R., Alcon, R.R., Gustavsen, R.L., Robbins, D.L., Stahl, D.B., Stacy, H.L., Whitehead, M.C.: Particle velocity measurements of the reaction zone of nitromethane. In: Twelfth International Detonation Symposium, pp. 159–166. Office of Naval Research, San Diego (2002). ONR 333-05-2
60. Shigato, S., Pang, Y., Fang, Y., Dlott, D.D.: Vibrational relaxation of normal and deuterated nitromethane. J. Phys. Chem. A **112**, 232–241 (2008)
61. Srinivasan, J., Vincenti, W.G.: Criteria for acoustic instability in a gas with ambient vibrational and radiative nonequilibrium. Phys. Fluids **18**, 1670–1677 (1975)
62. Stiel, L.I., Baker, E.L., Capellos, C.: Characteristic melt times and onset of reaction for aluminized explosives. In: Fourteenth International Detonation Symposium, pp. 1035–1042. Office of Naval Research, Coeur d'Alene (2011). ONR-351-10-185
63. Strachan, A., Holian, B.L.: Energy exchange between mesoparticles and their internal degrees of freedom. Phys. Rev. Lett. **94**, 014301 (2005)
64. Tarver, C.M.: Comb. Flame **46**, 111–134, 135–156, 157–176 (1982)
65. Tarver, C.M.: Multiple roles of highly vibrationally excited molecules in the reaction zones of detonation waves. J. Phys. Chem. A **101**, 4845–4851 (1997)
66. Tarver, C.M.: Chemical reaction and equilibration mechanisms in detonation waves. In: Shock Compression of Condensed Matter-1997, AIP Conference Proceedings 429, pp. 301–304. Amherst (1998)
67. Tarver, C.M.: Next generation experiments and models for shock initiation and detonation of solid explosives. In: Shock Compression of Condensed Matter-1999. AIP Conference Proceedings, vol. 505, pp. 873–877. Snowbird (2000)
68. Tarver, C.M.: What is a shock wave to an explosive molecule? In: Shock Compression of Condensed Matter-2001. AIP Conference Proceedings, vol. 620, pp. 42–49. Atlanta (2002)
69. Tarver, C.M.: What is a shock wave to an explosive molecule? In: Horie, Y., Davidson, L., Thadhani, N.N. (eds.) High-Pressure Shock Compression of Solids VI, pp. 323–340. Springer, New York (2003)
70. Tarver, C.M.: Detonation reaction zones in condensed explosives. In: Shock Compression of Condensed Matter-2005, AIP Conference Proceedings, vol. 845, pp. 1026–1029. Baltimore (2005)
71. Tarver, C.M.: Ignition and growth modeling of LX-17 hockey puck experiments. Propellants Explos. Pyrotech. **30**, 109–117 (2005)

72. Tarver, C.M.: Corner turning and shock desensitization experiments plus numerical modeling of detonation waves in the triaminotrinirobenzene bases explosive LX-17. J. Phys. Chem. A **114**, 2727–2736 (2010)
73. Tarver, C.M., Chidester, S.K.: Ignition and growth modeling of detonating TATB cones and arcs. In: Shock Compression of Condensed Matter – 2007, AIP Conference Proceedings, vol. 955, pp. 429–432. Waikoloa (2007)
74. Tarver, C.M., Chidester, S.K.: Modeling LX-17 detonation growth and decay using the ignition and growth model. In: Shock Compression of Condensed Matter-2009, AIP Conference Proceedings, vol. 1195, pp. 249–254. Nashville (2009)
75. Tarver, C.M., McGuire, E.M.: Reactive flow modeling of the interaction of TATB detonation waves with inert materials. In: Twelfth International Detonation Symposium, pp. 641–649. Office of Naval Research, San Diego (2002). ONR 333-05-2
76. Tarver, C.M., Urtiew, P.A.: Theory and modeling of liquid explosive detonation. J. Ener. Mat. **28**, 299–317 (2010)
77. Tarver, C.M., Shaw, R., Cowperthwaite, M.: Detonation failure diameter studies of four liquid explosives. J. Chem. Phys. **64**, 2665–2673 (1976)
78. Tarver, C.M., Parker, N.L., Palmer, H.G., Hayes, B., Erickson, L.M.: Reactive flow modeling of recent embedded gauge and metal acceleration experiments on detonating PBX-9404 and LX-17. J. Ener. Mat. **1**, 213–250 (1983)
79. Tarver, C.M., Hallquist, J.O., Erickson, L.M.: Modeling short pulse duration shock initiation of solid explosives. In: Eighth Symposium (International) on Detonation, pp. 951–961. Naval Surface Weapons Center NSWC MP 86-194, Albuquerque (1985)
80. Tarver, C.M., Fried, L.E., Ruggiero, A.J., Calef, D.F.: Energy transfer in solid explosives. In: Tenth International Detonation Symposium, pp. 3–10. Office of Naval Research, Boston (1993). ONR 33395-12
81. Tarver, C.M., Cook, T.M., Urtiew, P.A., Tao, W.C.: Multiple shock initiation of LX-17. In: Tenth International Detonation Symposium, pp. 696–703. Office of Naval Research, Boston (1993). ONR 33395-12
82. Tarver, C.M., Tao, W.C., Lee, C.G.: Sideways plate push test for detonating solid explosives. Propellants Explos. Pyrotech. **21**, 238–246 (1996)
83. Tarver, C.M., Kury, J.W., Breithaupt, R.D.: Detonation waves in triaminotrinitrobenzene. J. Appl. Phys. **82**, 3771–3782 (1997)
84. Urtiew, P.A., Tarver, C.M.: Shock initiation of energetic materials at different initial temperatures. Combust. Explos. Shock Waves **41**, 766–776 (2005)
85. Urtiew, P.A., Vandersall, K.S., Tarver, C.M., Garcia, F., Forbes, J.W.: Shock initiation experiments and modeling of composition B and C-4. In: Thirteenth International Detonation Symposium, pp. 929–939. Office of Naval Research, Norfolk (2006). ONR 351-07-01
86. Vandersall, K.S., Tarver, C.M., Garcia, F., Chidester, S.K.: On the low pressure shock initiation of octahydro-1,3,5,7-tetranitro-1,3,5,7-tetrazocine based plastic bonded explosives. J. Appl. Phys. **107**, 094906 (2010)
87. Vasil'ev, A.A., Trotsyuk, A.V.: Experimental and numerical simulation of an expanding multifront detonation wave. Combust. Explos. Shock Waves **39**, 80–90 (2003)
88. Viecelli, J.A., Glosli, J.N.: Carbon cluster coagulation and fragmentation kinetics in shocked hydrocarbons. J. Chem. Phys. **117**, 11352–11358 (2002)

89. Von Neumann, J.: Hydrodynamic theory of detonation. Office of Science Research and Development Report No. 549 (1942)
90. Weston, R.E. Jr., Flynn, G.W.: Relaxation of Molecules with Chemically Significant Amounts of Vibrational Energy: The dawn of the Quantum State Resolved Era. Ann. Rev. Phys. Chem. **43**, 559–589 (1993)
91. Wood, W.W., Kirkwood, J.G.: Present status of detonation theory. J. Chem. Phys. **29**, 957–958 (1958)
92. Zeldovich, Y.B.: Detonation. J. Exper. Theor. Phys. (USSR) **10**, 542–561 (1940)
93. Zel'dovich, Y.B., Raizer, Y.P.: Physics of Shock Waves and High-Temperature Hydrodynamic Phenomena. Academic, New York (1966)
94. Zhang, F., Alavi, S., Hu, A., Woo, T.K.: First principles molecular simulation of energetic materials at high pressures. In: Horie, Y. (ed.) Shock Wave Science and Technology Reference Library, vol. 3, pp. 65–107. Springer, Berlin (2009)
95. Zhang, F., Gerrard, K., Ripley, R.C.: Reaction mechanism of aluminum particle-air detonation. J. Prop. Power **25**, 845–858 (2009)

7

Theory of Detonation Shock Dynamics

John B. Bdzil and D. Scott Stewart

7.1 Introduction

To accurately predict the propagation of detonation through an explosive, one needs to model the physics that occurs on the chemical reaction-zone length scale $\eta_{\rm rz}$. In sharp contrast to the structured shocks observed for gaseous detonation, those for heterogeneous solid explosives are broadly curved on the $\eta_{\rm rz}$ scale due to the interaction with boundaries. The speed of the detonation is influenced by the curvature of the shock κ, with reductions of speed of 40% in strongly divergent flows. To ensure that explosives used in devices are both safe from accidental initiation of detonation and provide performance very near that of ideal detonation, a class of insensitive explosives is often used. Satisfying these dual goals requires that the ratio of the detonation reaction-zone length relative to the device dimension, L, must be of the order of $\eta_{\rm rz}/L \sim 10^{-2} - 10^{-3}$; that is, the ratio must be large enough to provide safe handling characteristics, yet small enough to give good detonation performance. Curvature effects have a strong influence on detonation in these systems, and highly resolved multidimensional simulations are "expensive." A body of theory and supporting experiments, called Detonation Shock Dynamics (DSD) [8,31], has been developed that treats these curvature effects [10]. The DSD front theory derives a function $D_n(\kappa)$ for the relationship between the normal detonation shock speed D_n and the total shock curvature κ based on a weakly divergent, quasi-one-dimensional (1D) model of the detonation reaction zone. This relationship can also be determined directly from experiments. The regions of strongest flow divergence are found near the explosives' boundaries. The $D_n(\kappa)$ relationship, now commonly referred to as a "D-kappa" relation, is an intrinsic, coordinate-free front propagation law that can be used to compute the location of the shock front, given some information about its initial location. The initial geometrical configuration that describes the location of the explosive in relation to adjacent and confining inerts must be supplied (in addition to boundary conditions (BCs) at

F. Zhang (ed.), *Shock Wave Science and Technology Reference Library, Vol. 6*: Detonation Dynamics, DOI 10.1007/978-3-642-22967-1_7,

the confinement edges). The boundary conditions model how the detonation shock described by DSD interacts with confinement edges through a narrow boundary layer a few η_{rz} thick, where the flow is both reactive and fully multidimensional.

7.2 Overview of DSD Theory

DSD is the acronym and common name given to the body of multidimensional detonation theory and experiments that can describe the dynamics of detonation that have shocks with small curvature when measured on the length scale of the reaction-zone η_{rz}. It is sufficient to consider two-dimensional flows for most considerations. The model equations that have been used to develop most of the theory are the two-dimensional reactive Euler equations transformed to shock-attached, intrinsic coordinates [10]. A diagram showing a two-dimensional, curved detonation reaction-zone flow is displayed in Fig. 7.1, inset (a).

7.2.1 Detonation Dynamics

The reactive Euler equations [15]

$$\dot{\rho} + \rho\nabla\cdot\mathbf{u} = 0, \quad \rho\dot{\mathbf{u}} + \nabla P = 0, \quad \dot{E} + P\dot{v} = 0, \quad \dot{\lambda} = R(P, T, \lambda), \qquad (7.1)$$

form the basis of detonation modeling, where ρ is density, $v = 1/\rho$, P is pressure, E is the internal energy, λ is the degree of reaction and satisfies $0 \le \lambda \le 1$, R is the reaction rate, $\mathbf{u}$ is the laboratory-frame particle velocity vector, and $\dot{(\ )} = \frac{\partial}{\partial t} + \mathbf{u}\cdot\nabla$ is the material or Lagrangian time derivative. At shock discontinuities, these equations are appended with the Rankine–Hugoniot (RH) relations

$$[\rho U_n] = 0, \quad [P + \rho U_n^2] = 0, \quad [U_t] = 0, \quad [E + Pv + U_n^2/2] = 0, \quad [\lambda] = 0, \qquad (7.2)$$

where $\mathbf{U} = \mathbf{u} - \mathbf{n}D_n$, is the particle velocity in the shock-attached frame, with $U_n = \mathbf{n}\cdot\mathbf{U}$, $U_t = \mathbf{t}\cdot\mathbf{U}$, where $\mathbf{n}$ and $\mathbf{t}$ are the local shock normal and tangent vectors, respectively, D_n is the normal-shock velocity, and $[\ \ ]$ denotes the change across the shock. Given an equation of state (EOS), $E(P, v, \lambda)$, for the explosive and a prescribed ambient state, an elementary one-dimensional steady analysis, assuming λ as a parameter and based on (7.2) for a detonation traveling at speed D_n, defines the steady one-dimensional λ-dependent profile through the detonation reaction zone ($0 \le \lambda \le 1$) (see Sect. 7.3 and also Chap. 2 for more details). These results show that the minimum allowed detonation speed is D_{CJ} and that the pressure in the reaction zone drops from its value at the shock, referred to as the Zeldovich–von Neumann–Döring

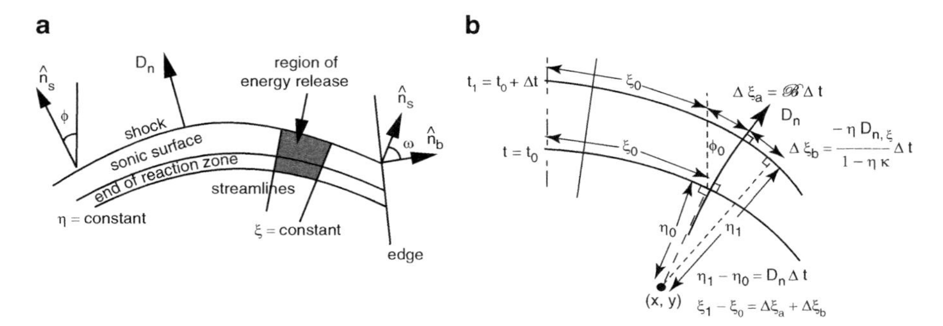

Fig. 7.1. The speed of the detonation is influenced by two factors, the divergence of the flow and the location of the sonic surface. Two diagrams showing the intrinsic coordinate net we use, including the definition of the boundary-edge angle ω and shock-normal angle ϕ. (**a**) Snapshot of the intrinsic coordinate net (the orthogonal ξ = constant lines and η = constant detonation shock parallel curves) for a diverging detonation. (**b**) Evolution of the shock-attached intrinsic coordinates over the time interval Δt. The shock-attached observer moves with the shock along the curved detonation ray, which remains locally normal to the shock. The transformation of the time derivative in going from the laboratory Cartesian to the shock-attached intrinsic coordinates is indicated

(ZND) state where $\lambda = 0$, to its value at the end of the reaction zone, referred to as the Chapman–Jouguet (CJ) state where $\lambda = 1$ [15]. We discuss this issue at length in Sect. 7.3.

7.2.1.1 Shock-Attached Intrinsic Coordinate Representation

Figure 7.1 shows shock-attached orthogonal coordinates that are straight line normal distances to the shock and curves parallel to the shock, all moving with the shock normal speed D_n. The two-dimensional (2D) coordinate transformation between laboratory-frame coordinates (x, y, t) and intrinsic coordinates (η, ξ, t) is given by

$$\mathbf{x} = \mathbf{x}_s(\xi, t) - \eta \mathbf{n}(\xi, t). \tag{7.3}$$

In these coordinates D_n is a function of ξ and t, $D_n(\xi, t)$. The shock surface in the laboratory frame is represented generally as a zero level curve $\psi(x, y, t) = 0$ that constrains the laboratory coordinate position vectors to $\mathbf{x} = \mathbf{x}_s$. Then the normal to the shock surface $\mathbf{n}$ that appears in (7.3) is chosen to be positive in the direction of the unreacted explosive and is computed from the formula $\mathbf{n} = \nabla\psi/|\nabla\psi|$. The shock surface can be represented by any monotonic surface parameterization, but in 2D the best choice is the arc length ξ that measures the physical length along the coordinate surface. The unit vector tangent to it is $\mathbf{t}$. The normal coordinate η measures distance normal to the shock and is positive in the direction of increasing depth into the reaction-zone (i.e., the reaction zone structure lies in a region where $\eta > 0$). The shock moves into the region where $\eta < 0$.

Simple consequences of the change of coordinates follow. For example, the laboratory-frame particle velocity that appears in the governing equations in the shock-attached frame is $\mathbf{u} = u_\eta \mathbf{n} + u_\xi \mathbf{t}$. The time derivative in the laboratory Cartesian frame transforms to

$$\left(\frac{\partial}{\partial t}\right)_{x,y} = \left(\frac{\partial}{\partial t}\right)_{\eta,\xi} + D_n \frac{\partial}{\partial \eta} + \left(\mathscr{B} - \frac{\eta D_{n,\xi}}{1-\eta\kappa}\right)\frac{\partial}{\partial \xi}, \tag{7.4}$$

in the shock-attached intrinsic frame (see inset (b) of Fig. 7.1), while the relevant laboratory-frame vector operators transform as

$$\mathbf{u}\cdot\boldsymbol{\nabla} = -u_\eta \frac{\partial}{\partial \eta} + \frac{u_\xi}{1-\eta\kappa}\frac{\partial}{\partial \xi}, \tag{7.5}$$

$$\boldsymbol{\nabla}\cdot\mathbf{u} = -u_{\eta,\eta} + \frac{u_\eta\kappa + u_{\xi,\xi}}{1-\eta\kappa}. \tag{7.6}$$

The comma notation $(\ )_{,x}$ denotes partial derivative with respect to x, and $\kappa = \phi_{,\xi}$ is the curvature shown in Fig. 7.1 and is the transverse derivative of the change of the angle of the normal with a fixed direction (shown as ϕ in Fig. 7.1). Thus it follows that the laboratory-frame material derivative, $\dot{(\ )} = \frac{\partial}{\partial t} + \mathbf{u}\cdot\nabla$, becomes in the shock-attached intrinsic frame

$$\dot{(\ )} = \mathscr{L}(\ ) + (D_n - u_\eta)\frac{\partial}{\partial \eta}, \tag{7.7}$$

and where

$$\mathscr{L}() = \frac{D}{Dt} + \frac{(u_\xi - \eta D_{n,\xi})}{(1-\eta\kappa)}\left(\frac{\partial}{\partial \xi}\right)_{t,\eta}, \quad \frac{D}{Dt} = \left(\frac{\partial}{\partial t}\right)_{\eta,\xi} + \mathscr{B}\left(\frac{\partial}{\partial \xi}\right)_{t,\eta}. \tag{7.8}$$

The operator $\mathscr{L}(\ )$ is the portion of the material time derivative in these coordinates not associated with the η-direction, $\frac{D}{Dt}$ is the time rate of change as measured by an observer moving with the shock in the shock normal direction (intrinsic, time derivative at the shock), and $\mathscr{B}$ is the rate of change of arc length along the shock as measured by the observer moving with the shock and along the path of a detonation ray. A complete development of the 2D coordinates and their 3D extension can be found in [39] and [23], respectively.

Using the transformation (7.3), the first three equations of (7.1) are rewritten as

$$\mathscr{L}(\rho) - (U_n\rho)_{,\eta} = -\rho\mathscr{G} \equiv -\frac{\rho(\kappa u_\eta + u_{\xi,\xi})}{1-\eta\kappa}, \tag{7.9}$$

$$\mathscr{L}(u_\eta) - U_n U_{n,\eta} - \frac{P_{,\eta}}{\rho} = -u_\xi\mathscr{H} \equiv -\frac{u_\xi(D_{n,\xi} - \kappa u_\xi)}{1-\eta\kappa}, \tag{7.10}$$

$$\mathscr{L}(u_\xi) - U_n u_{\xi,\eta} = \mathscr{J} \equiv -\frac{P_{,\xi}}{\rho(1-\eta\kappa)} + u_\eta\mathscr{H}, \tag{7.11}$$

$$\mathscr{L}(E) - U_n E_{,\eta} - \frac{P}{\rho^2}\left[\mathscr{L}(\rho) - U_n\rho_{,\eta}\right] = 0, \tag{7.12}$$

where $u_\eta = U_n + D_n$, $u_\xi = U_t$. Herein we limit consideration to simple, one-step, state-dependent reaction rate laws of the form $R = k(1-\lambda)^m\mathscr{F}(P, T, D_n)$.

7.2.1.2 The Master Equation

The "Master Equation" is an auxiliary equation that is easily derived from the above equations by starting from the energy equation (7.12) and using the other equations to eliminate the density and pressure gradients in favor of the velocity gradient. Thus it can be used in the analysis in place of one of the Euler equations. Its value lies in that it reveals the interaction of front curvature, time dependence, and heat-release rate that controls multidimensional detonation and highlights the role played by the transonic character of the flow. Since an unsupported detonation is transonic, the Master Equation is used extensively. It can be expressed as

$$\left(U_n^2 - c^2\right)U_{n,\eta} = c^2\left(\sigma R - \mathscr{G}\right) + U_n\left[\mathscr{L}(u_\eta) + u_\xi\mathscr{H}\right] - \mathscr{L}(P)/\rho. \tag{7.13}$$

Specialized to the shock (i.e., $\eta = 0$ and $u_\xi = 0$) (7.13) becomes the shock-change equation [15]. In these coordinates $(U_n^2 - c^2)$ is the shock-based sonic parameter, $c^2 = v^2(P + E_{,v})/E_{,P}$ is the sound speed squared, $\rho c^2\sigma = -(E_{,\lambda}/E_{,P})$, with σ the thermicity coefficient for the reaction, $\kappa/(1-\eta\kappa)$ is the component of divergence associated with the flow through a 1D nozzle (where A is the local nozzling streamtube area and $\kappa/(1-\eta\kappa) \equiv \mathrm{d}\ln(A)/\mathrm{d}\eta$), and $u_{\xi,\xi}$ measures the effect on flow divergence due to streamline curvature.

7.2.2 Weak-Curvature Limit

The weak shock-curvature limit defined by $\kappa = O(\epsilon^2)$, where $\epsilon^2 \equiv$ (reaction-zone length)/(shock radius of curvature), is the basis of most theoretical analysis of multidimensional detonation. This limit envisions that $O(1)$ changes in the field variables occur over large $O(\epsilon^{-1})$ distances in the ξ-direction, and the flow velocity in the ξ-direction is no greater than $O(\epsilon^2)$. Under these assumptions, the DSD-related time derivatives are no faster than $O(\epsilon)$, the flow is "nozzle"-like and principally in the η-direction (see Fig. 7.1), and where 2D effects enter the flow only parametrically, via $\kappa(\xi, t)$. This limit allows for time dependence on the $O(\epsilon)$ scale, provided no $O(1)$ velocities are generated in the ξ-direction. Then various streamtubes communicate with one another only as neighbors through their mutual connection to the multidimensional detonation shock.

The $O(\epsilon)$-scale time dependence appears prominently near the sonic transition, the point where $(U_n^2 - c^2)$ approaches zero in the detonation reaction zone. Near the sonic transition, the perturbation to the standard ZND reaction-zone solution, which is $O(\epsilon^2)$ most places in the reaction zone, is promoted to $O(\epsilon)$. Then one has $U_n \sim (U_n)_{\text{ZND}} + O(\epsilon)$, $(U_n^2 - c^2) \sim O(\epsilon)$, which then requires that the time variation in the operator $\mathscr{L}()$ be measured on a slow $O(\epsilon^{-1})$ scale, to be compatible with the overall $O(\epsilon^2)$ perturbation coming from the shock curvature, κ.

In the limit that the curvature of the shock, κ, acts as an $O(\epsilon^2)$ perturbation to a basic ZND reaction-zone flow, then these scalings define the standard DSD limit and are compatible with taking

$$\mathscr{G} \to \kappa u_\eta = O(\epsilon^2) \text{ nozzling flow,} \tag{7.14}$$

$$\mathscr{L}(u_\eta) \to (\partial u_\eta / \partial t)_{\eta,\xi} = O(\epsilon^2), \tag{7.15}$$

$$\mathscr{L}(P) \to (\partial P / \partial t)_{\eta,\xi} = O(\epsilon^2), \text{ time dependence,} \tag{7.16}$$

$$u_\xi \mathscr{H} = O(\epsilon^4), \tag{7.17}$$

$$u_\xi = O(\epsilon^2). \text{ transverse flow,} \tag{7.18}$$

In this limit, Euler equations (7.9)–(7.12), and Master Equation (7.13), reduce to

$$(\dot{\rho}) - \rho U_\eta = -\rho u \kappa \ , \tag{7.19}$$

$$\rho(\dot{u}) - P_\eta = 0 \ , \tag{7.20}$$

$$(\dot{E}) - \frac{P}{\rho^2}(\dot{\rho}) = 0 \ , \tag{7.21}$$

$$(\dot{\lambda}) = R \ , \tag{7.22}$$

where now we identify the normal component of the laboratory-frame particle velocity as $u_\eta \equiv u$, the shock attached-frame normal velocity as $U = u - D_n$, and define $(\dot{\ }) \equiv (\partial/\partial t)_{\eta,\xi} - U(\partial/\partial\eta)_{t,\xi}$. The "Master Equation" becomes

$$\frac{\partial P}{\partial t} - \rho U \frac{\partial u}{\partial t} = \rho\left(c^2 - U^2\right)\frac{\partial U}{\partial \eta} + \rho c^2\left(\sigma R - \kappa u\right) \ . \tag{7.23}$$

Henceforth, when t, η, ξ, etc. are used as subscripts, they denote partial derivatives.

The truncated equations (7.19)–(7.22) can be written in a conservative form by using the normal flux variables, M, Π, and H and are defined by

$$M = -\rho U, \quad \Pi = P + \rho U^2, \quad H = E + \frac{P}{\rho} + \frac{1}{2}U^2, \tag{7.24}$$

in which case (7.19)–(7.22), recast, become

$$\rho_t + M_\eta + \rho u \kappa = 0 \ , \tag{7.25}$$

$$M_t + \Pi_\eta - \rho D_{nt} + \kappa M u = 0 \ , \tag{7.26}$$

$$H_t - U H_\eta + U D_{nt} - \frac{P_t}{\rho} = 0 \ , \tag{7.27}$$

$$\lambda_t - U \lambda_\eta = R \ . \tag{7.28}$$

The flux variables are exactly the same as those defined in the Rankine–Hugoniot algebra (7.2), used for discontinuity analysis for fixed values of M, Π, H, and λ. Their defining relations in terms of the primitive variables are the same algebraic equations and can be inverted to obtain the primitive states ρ, P, and U (or u) in terms of M, Π, H, and λ.

In the standard DSD limit, the multidimensional reaction-zone equations map to the Euler equations for two-dimensional, cylindrically symmetric flow. The one exception is that the shock curvature, κ, is an independent parameter and *not* simply constrained to be $\kappa = 1/r$, with r the radial coordinate for cylindrically symmetric flow. One of the principal results associated with the solution of the standard DSD-limit equations is the establishment of a detonation front propagation law. The simplest example of such a law is the $D_n(\kappa)$-law, whereby the normal speed of the detonation front is related simply to the local curvature of the detonation shock front. We describe in some detail how this relation can be calculated from the truncated DSD equations in Sect. 7.4.2.

7.2.3 Boundary Conditions

The slow variations in the ξ-direction and the small transverse flow velocities, $u_\xi = O(\epsilon^2)$, which are a part of the DSD limit, can break down at explosive boundaries. Near the boundary interface, the flow can have $O(1)$ transverse velocities and be time dependent on the $O(1)$ reaction-zone scale [7]. However, when viewed on the slow-time DSD scale, this region can appear to be steady. Then the DSD and boundary regions are coupled as follows: (1) the DSD region drives the boundary with information on the shock slope and D_n, and in turn, (2) the boundary region uses this data to return a (possibly) modified shock slope, etc. [5]. Using the boundary angle ω, defined in Fig. 7.1, DSD supplies ω_{in} to the boundary flow, which then returns ω_{out} to the DSD region (see [1], Fig. 8). The boundary is itself characterized by two angles that depend on the explosive/inert pair being considered: a critical angle ω_s and a confinement angle ω_c. We examine these more fully two-dimensional flow regions in later sections of this chapter.

7.3 Basic 1D ZND Model of Detonation

The basic model for a detonation in an explosive is based on continuum mixture theory for a molecularly premixed set of reactants that decompose to products. The specific internal energy in the products is less than that of the reactants, and the difference accounts for the chemical energy that is converted to pressure and kinetic energy in the products which serve as the working fluid of the detonation. Mixing rules and weighting of the thermodynamic properties of the constituents that make up both reactants and products are used to determine a closure model for the specific internal energy of the form $E(P, v, \lambda)$ and a reaction-rate law of the form $R(P, v, \lambda)$. Some models for the reaction-rate allow for dependence of the rate on the strength of the shock wave that processes the reactants, and hence, in that case, R is assumed to take the form $R(P, v, D_n, \lambda)$.

7.3.1 Ideal EOS Model and Reaction Rate Laws

Consider a simple reaction in which reactants convert to products $r \to p$. For the same pressure and specific volume the reactants will have more specific internal energy than the products; the difference is the chemical energy that is released during decomposition. For a simple ideal EOS the reactant and product $E(P, v)$ forms are given by

$$E_r(P_r, v_r) = \frac{P_r v_r}{\gamma - 1}, \quad E_p(P_p, v_p) = \frac{P_p v_p}{\gamma - 1} - Q,$$

and the difference in the energies at the same pressure and volume, Q is the heat of combustion or detonation. However, the functional form of the products and reactants EOS are generally different. The specific energies and specific volumes (energy/mass and volume/mass) sum according to the mass fractions of each constituent. If one makes very simple closure assumptions, such as $v = v_r = v_p$, $P = P_r = P_p$, combined with a mass-weighted sum of the reactants and products, one obtains

$$E = \frac{Pv}{\gamma - 1} - Q\lambda, \tag{7.29}$$

which is of the required $E(P, v, \lambda)$ form. From the general equations for the sound speed squared c^2 and the thermicity coefficient σ, given in the previous section, their formulas for the ideal EOS model are

$$c^2 = \gamma P v, \quad \sigma = (\gamma - 1)\frac{Q}{c^2}. \tag{7.30}$$

The ideal EOS model is important for understanding the presentation of DSD theory because of its simplicity of form. But the procedures described in this section can be extended to reactants and products that have more complex EOS forms [37]. Common fitting forms used for the reactant and product EOS include variations of the Mie–Grüneisen (MG) EOS. One matches the experimentally obtained shock wave data for the reactants to determine the principal shock Hugoniot of the reactants used as the reference curve for the MG EOS. Data from detonation product expansion experiments are used to determine the isentrope for the products as the reference curve for the MG EOS [13]. Common forms include a shock velocity versus particle velocity, "U_s versus u," based MG form for the reactants and the "JWL" MG form for the products. Both are well-tested empirical EOS forms.

Depletion forms for the reaction-rate law $R(P, v, D_n, \lambda)$ are often assumed, based on arguments from collisional reaction-rate theory (in the case of gases), with Arrhenius or purely empirical, posited, pressure or temperature-dependent laws, such as

$$R = k(1-\lambda)^m \mathrm{e}^{-E^\ddagger/\bar{R}T}, \bar{R}T = Pv \quad \text{an Arrhenius rate with } m\text{-order depletion}$$

$$R = k(1-\lambda)^m P^n \quad \text{a pressure-dependent rate with } m\text{-order depletion}$$

$$R = k(1-\lambda)^m \mathscr{R}(D_n) \quad \text{a shock-state dependent rate with } m\text{-order depletion} \tag{7.31}$$

Here, $E^\ddagger$ is an activation energy, $\bar{R}$ is the ideal gas constant, n is a pressure exponent, m is a depletion exponent, k is a rate pre-factor, and $\mathscr{R}(D_n)$ must be regarded as a material property whose value is advected with the flow, similar to the advection used in an explosive strength model. More complex forms of the reaction-rate law can be assumed or considered, motivated by the need to match predictions of the model with calibration experiments.

7.3.1.1 Rankine–Hugoniot Algebra

The normal mass, momentum and energy fluxes are defined in terms of the primitive variables by formulas (7.24), and for known values of those fluxes the formulas can be inverted. The formulas and their inverses define the Rankine–Hugoniot algebra. We illustrate this algebra for the ideal EOS and the case of motionless, fresh upstream reactants. The first two equations of (7.24) imply that

$$U = -Mv, \quad P = \Pi - M^2 v, \tag{7.32}$$

and substitution of these results into (7.24c) leads to a quadratic equation for v,

$$v^2 - \frac{2\gamma}{\gamma+1}\frac{\Pi}{M^2}v + \frac{2(\gamma-1)}{(\gamma+1)}\frac{(H+Q\lambda)}{M^2} = 0. \tag{7.33}$$

The shock branch solution is

$$v = \frac{\gamma}{\gamma+1}\frac{\Pi}{M^2}(1-\delta), \quad \text{where} \quad \delta = \sqrt{1 - \frac{2(\gamma^2-1)}{\gamma^2}\frac{M^2}{\Pi^2}(H+\lambda Q)}. \tag{7.34}$$

Once v is determined, U and P are computed from (7.32). If we introduce the normal Mach number (squared) in the shock-attached frame $\mathsf{M}^2 = U^2/c^2$ then simple algebra shows that δ^2 can also be re-written compactly as

$$\delta^2 = \left(\frac{1 - \mathsf{M}^2}{1 + \gamma \mathsf{M}^2}\right)^2 . \tag{7.35}$$

Importantly, this shows that the argument of the square root that defines δ is positive.

For the case when the upstream state is ambient (denoted by a 0 subscript) and motionless ($u_0 = 0$) then the flux constants and the reaction progress variable evaluated at the shock are

$$M = \rho_0 D_n, \quad \Pi = P_0 + \rho_0 D_n^2, \quad H = \frac{c_0^2}{\gamma - 1} + \frac{1}{2} D_n^2, \quad \lambda = 0. \tag{7.36}$$

The strong-shock approximation is expressed as $1/\mathsf{M}_0^2 = c_0^2/D_n^2 \ll 1$. The ratio of the ambient pressure (i.e., one atmosphere) to the shock pressure (on the order of 100 kbar $= 10^5$ atm) is on the order of 10^{-5}–10^{-6}, and so it is an excellent approximation for condensed-phase explosives to set P_0 and c_0^2 equal to zero. We use the strong-shock approximation in the remainder of this chapter.

7.3.1.2 The Shock and CJ States

The solution for v allows us to easily compute the normal-shock states which we list here for convenience. At the shock where $\lambda = 0$ one finds

$$v_s = \frac{\gamma - 1}{\gamma + 1} v_0, \quad U_s = -D_n \frac{\gamma - 1}{\gamma + 1}, \quad P_s = \frac{2\rho_0}{\gamma + 1} D_n^2, \quad \lambda_s = 0.$$

The Chapman–Jouguet (CJ) detonation states correspond to complete reaction at the sonic point, where $\mathsf{M}^2 = 1$. In this case, we set $\delta = 0$ at $\lambda = 1$. If we use the values of the flux using the flux constant values defined by (7.36), the zero of the argument of the square root that defines δ becomes a quadratic for D_n^2 with solution

$$D_{\mathrm{CJ}} = \sqrt{q}, \quad \text{where} \quad q = 2\,Q(\gamma^2 - 1). \tag{7.37}$$

The CJ states are found by setting $\delta = 0$ with $D_n = D_{\mathrm{CJ}}$, whereby

$$v_{\mathrm{CJ}} = \frac{\gamma}{\gamma + 1} v_0, \quad U_{\mathrm{CJ}} = -\frac{\gamma}{\gamma + 1} D_{\mathrm{CJ}}, \quad P_{\mathrm{CJ}} = \frac{\rho_0 D_{\mathrm{CJ}}^2}{\gamma + 1}, \quad \lambda_{\mathrm{CJ}} = 1. \tag{7.38}$$

7.3.2 ZND Spatial Structure

We illustrate the calculation of the spatial structure for the ZND detonation wave for the ideal EOS model. For a steady plane traveling detonation, the statement that the material rate of reaction is equal to the reaction rate leads to the following differential equation and initial condition for λ:

$$-U\frac{\mathrm{d}\lambda}{\mathrm{d}\eta} = R, \quad \text{with} \quad \lambda = 0 \quad \text{at} \quad \eta = 0. \tag{7.39}$$

The rate law $R(P, v, D_n, \lambda)$ must be specified, and U is found in terms of λ as the solution to the same Rankine–Hugniot algebra, with the same flux constants (7.36). The resulting formulas for v, U, and P that derive from (7.34), (7.32a, b) are

$$\begin{aligned} v &= \frac{v_0}{\gamma+1}\left(\gamma - \sqrt{1-\lambda}\right), \quad U = -\frac{D_{\mathrm{CJ}}}{\gamma+1}\left(\gamma - \sqrt{1-\lambda}\right), \\ P &= \frac{\rho_0 D_{\mathrm{CJ}}^2}{\gamma+1}(1+\sqrt{1-\lambda}). \end{aligned} \tag{7.40}$$

The spatial distribution of reactants is then computed via the integration of the ODE (7.39) as

$$\eta = -\int_0^\lambda \frac{U(\bar{\lambda})}{R(P(\bar{\lambda}), v(\bar{\lambda}), D_n, \bar{\lambda})}\mathrm{d}\bar{\lambda}. \tag{7.41}$$

For illustration consider an extremely simple rate law that is not state sensitive but does have a fractional reaction order on the rate of depletion

$$R = k(1-\lambda)^m. \tag{7.42}$$

For the values $\gamma = 3$, $Q = 4\,\mathrm{MJ}$, $\rho_0 = 2\,\mathrm{gm/cc}$, corresponding to a value of $D_{\mathrm{CJ}} = 8\,\mathrm{mm}\,\mu\mathrm{sec}^{-1}$, we use $k = 2.514718626\,(\mu\mathrm{sec})^{-1}$ and $m = 1/2$ to set the half reaction-zone length, $\lambda = 1/2$, at $\eta = 1\,\mathrm{mm}$, and thus the spatial scale. Figure 7.2a–f shows the resulting ZND spatial structure.

7.4 Detonation Front Propagation Laws

7.4.1 Characteristic Analysis and DSD Theory

The asymptotic approximations that lead to a propagation law for a free-running detonation at the heart of DSD theory are based on the following assumptions: (1) The detonation is broadly curved so that the radius of curvature of the detonation front is large compared with the reaction-zone length, i.e., $\kappa = O(\epsilon^2)$, $\epsilon^2 \ll 1$. (2) Shape changes to the detonation's reaction-zone profile and multidimensional shock shape occur on a timescale that is long when compared with the passage time of a particle through the reaction zone, i.e., the evolutionary timescale in the heart of the reaction zone is $O(\epsilon^{-2})$. (3) The detonation reaction zone is isolated from the flow that follows the reaction-zone region, i.e., for a right-going detonation, there exists a right-going, forward characteristic (limiting characteristic) which propagates in the reaction zone that causally isolates the reaction zone from the flow following the reaction zone [10]. This limiting characteristic locally propagates near

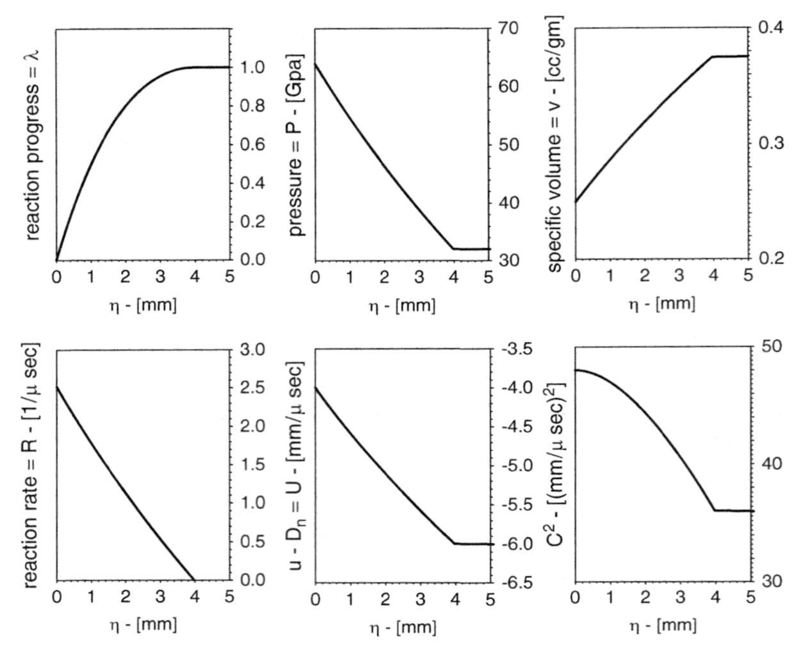

Fig. 7.2. ZND structure for a condensed-phase explosive; $Q = 4\,\mathrm{MJ}$, $\gamma = 3$ ideal test case, with $m = 1/2$ and $\lambda = 1/2$ at $\eta = 1\,\mathrm{mm}$ reaction-zone length scale

speed $u+c \approx D_n$, where D_n is the local shock speed, so that $(u-D_n)/c \approx -1$. That is, the relevant flow Mach number squared is near one; i.e., at each instant, there is an embedded, evolving sonic locus behind the detonation shock.

Displayed in Fig. 7.3, inset (a) are snapshots of the particle velocity of such an evolving, cylindrically symmetric detonation at times $t = 0$ and $t = 100$. For the same simulation, Fig. 7.3, inset (b) shows a wave diagram (for $0 \le t \le 18$) which displays the shock (short-dashed, right-most red line) and a fan consisting of nine forward characteristics. A shock-facing, forward characteristic (right-most, long-dashed black line in the fan) runs into the detonation shock at $t = 14$, and the left-most forward characteristic in the fan (left-most, heavy-solid black line in the fan) runs into the end of the computational domain at $t = 16$. The characteristics are computed by solving $\mathrm{d}x/\mathrm{d}t = u+c$ and represent the track followed by acoustic disturbances as they propagate through the reaction zone. The characteristics are all seeded very near one another and in the vicinity of the location of the sonic point in the $t = 0$ particle-velocity profile. The total length of the computational domain is twice that of the $t = 0$ reaction-zone length. Only disturbances propagating along the forward characteristics that reach the shock can influence the shock state and hence the detonation speed. Even though the characteristics in the

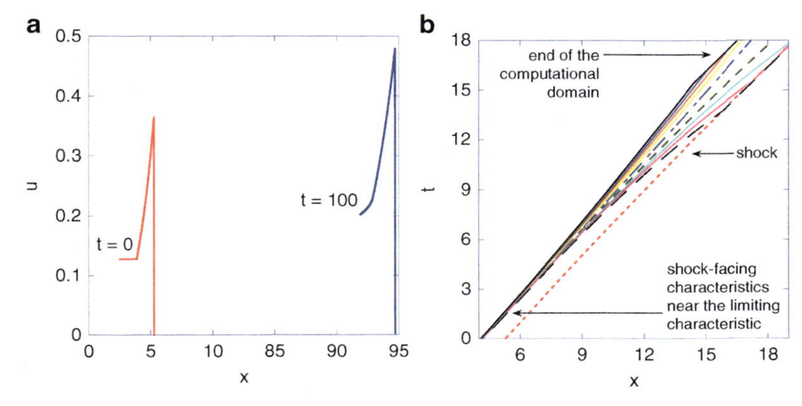

Fig. 7.3. A detonation propagating cylindrically outwards. Shown in inset (**a**) is a quasi-steady, initial detonation reaction-zone profile ($t = 0$) and the profile after detonation propagation to $t = 100$. Inset (**b**) displays the tracks of various waves over the interval $0 \leq t \leq 18$. The shock is shown as the right-most *short-dashed red line*. A fan formed of nine forward characteristics is also shown. These characteristics emanate at $t = 0$ for the immediate vicinity of the point in the reaction zone where the flow in the shock-attached frame is nearly sonic $(u - D_n)/c \approx -1$. The right-most four of these characteristics eventually run into the shock; the right-most of these, shown as *long-dashed black line*, intersects the shock at $t = 14$. The left-most five of these characteristics eventually run past the end of the reaction zone and to the end of the computational domain; the left-most of these, shown as a *heavy-solid black line*, runs into the end of the computational domain at $t = 16$. There exists a limiting characteristic (not shown and difficult to determine) that separates those characteristics that eventually hit the shock from those that run into the end of the computational domain. Only information traveling along the characteristics to the right of the limiting characteristic influence the solution at the shock. This limiting characteristic provides the isolation of a free-running detonation from its following flow. All of this information comes from a very localized, near-sonic region in the $t = 0$ flow

fan emanate at $t = 0$ from nearly the same, nearly sonic, location (see Fig. 7.3, inset (b)), it is clear that a continuous family of forward characteristics emanating from near the $t = 0$ sonic point would fill the entire reaction zone. A limiting characteristic (not shown and difficult to initialize) separates those that intersect the lead detonation shock from those that propagate toward the end of the computational domain, and hence away from the shock. The limiting characteristic for this problem is located between the medium-dashed dark-green line and the chain-dashed blue line of inset (b).

The data needed to propagate the detonation emanates from within the reaction zone in the absence of any faster, overtaking shocks from the rear. It is this locality of data, in the domain of the in-reaction-zone flow between the shock and a bounding characteristic (limiting characteristic), which allows the construction of a theory, like DSD, by purely considering the reaction zone.

An analysis based on the standard DSD limit is tractable since the flow in the interior of the explosive (away from the edges of the explosive) is quasi-one-dimensional. The terms that describe the shock motion enter as either leading order steady terms or perturbations to a steady one-dimensional structure. The perturbed equations of motion can be solved either with formal perturbation techniques or with successive iterative, numerical, integration techniques.

Next we illustrate the basic ideas for the ideal EOS and one exothermic reaction, which is the standard model. These ideas can be extended to real equations of state and more reactions. The ideal EOS is an important case to illustrate the underlying theory and methods of calculation, since the simplicity of the ideal EOS allows for explicit solution of the Rankine–Hugoniot relations which can be reduced to a single quadratic equation in the specific volume (say). In many cases an evolution equation for the DSD can be derived analytically.

7.4.1.1 Consideration of the Characteristics

Recall that the truncated DSD equations are isomorphic to the cylindrical equations to $O(\kappa)$ and are given in primitive form by (7.19)–(7.22) and in conservative form by (7.25)–(7.28). Written in primitive form, the governing equations for a general EOS $E(P, v, \lambda)$, truncated at $O(\kappa^2)$, are given by (7.19)–(7.22), and establish the basis for our discussion

$$\frac{\partial \rho}{\partial t} - U\frac{\partial \rho}{\partial \eta} - \rho\left(\frac{\partial u}{\partial \eta} - \kappa u\right) = 0, \tag{7.43}$$

$$\frac{\partial u}{\partial t} - U\frac{\partial u}{\partial \eta} - \frac{1}{\rho}\frac{\partial P}{\partial \eta} = 0, \tag{7.44}$$

$$\frac{\partial E}{\partial t} - U\frac{\partial E}{\partial \eta} - \frac{P}{\rho^2}\left(\frac{\partial \rho}{\partial t} - U\frac{\partial \rho}{\partial \eta}\right) = 0, \tag{7.45}$$

$$\frac{\partial \lambda}{\partial t} - U\frac{\partial \lambda}{\partial \eta} = R. \tag{7.46}$$

Equation (7.45) can be rewritten using the definition of c^2 given earlier as

$$\frac{\partial P}{\partial t} - U\frac{\partial P}{\partial \eta} - \rho c^2\left(\frac{\partial u}{\partial \eta} - \kappa u\right) = \rho c^2 \sigma R\ . \tag{7.47}$$

Following on the discussion of characteristics, it is very instructive to write these equations in characteristic form (which can be obtained by simple linear combinations of (7.43), (7.44), (7.47) and (7.46)) as

$$\frac{\partial P}{\partial t} - (U-c)\frac{\partial P}{\partial \eta} - \rho c\left(\frac{\partial u}{\partial t} - (U-c)\frac{\partial u}{\partial \eta}\right) = \rho c^2(\sigma R - \kappa\, u), \tag{7.48}$$

$$\frac{\partial P}{\partial t} - (U+c)\frac{\partial P}{\partial \eta} + \rho c\left(\frac{\partial u}{\partial t} - (U+c)\frac{\partial u}{\partial \eta}\right) = \rho c^2(\sigma R - \kappa u), \tag{7.49}$$

$$\frac{\partial P}{\partial t} - U\frac{\partial P}{\partial \eta} - c^2\left(\frac{\partial \rho}{\partial t} - U\frac{\partial \rho}{\partial \eta}\right) = \rho c^2 \sigma R, \tag{7.50}$$

$$\frac{\partial \lambda}{\partial t} - U\frac{\partial \lambda}{\partial \eta} = R. \tag{7.51}$$

Recast as ordinary differential equations along defined space–time (s.t.) curves, on the characteristic curves, these equations are [15, 36].

$$C_- : \ \frac{\mathrm{d}P}{\mathrm{d}t_-} - \rho c\frac{\mathrm{d}u}{\mathrm{d}t_-} = \rho c^2(\sigma R - \kappa u) \,\text{on}\, (\eta, t) \text{ s.t. } \frac{\mathrm{d}\eta}{\mathrm{d}t_-} = -(U-c), \tag{7.52}$$

$$C_+ : \ \frac{\mathrm{d}P}{\mathrm{d}t_+} + \rho c\frac{\mathrm{d}u}{\mathrm{d}t_+} = \rho c^2(\sigma R - \kappa u) \,\text{on}\, (\eta, t) \text{ s.t. } \frac{\mathrm{d}\eta}{\mathrm{d}t_+} = -(U+c), \tag{7.53}$$

$$C_0 : \ \frac{\mathrm{d}P}{\mathrm{d}t_0} - c^2\frac{\mathrm{d}\rho}{\mathrm{d}t_0} = \rho c^2 \sigma R \ \text{ on } (\eta, t) \text{ s.t. } \frac{\mathrm{d}\eta}{\mathrm{d}t_0} = -U, \tag{7.54}$$

$$C_0 : \ \frac{\mathrm{d}\lambda}{\mathrm{d}t_0} = R \ \text{ on } (\eta, t) \text{ s.t. } \frac{\mathrm{d}\eta}{\mathrm{d}t_0} = -U. \tag{7.55}$$

The characteristic form of the governing equations shows that in the shock-attached frame, disturbances in the smooth regions of the flow evolve and propagate on a family of waves that travel at speeds

$$\frac{\mathrm{d}\eta}{\mathrm{d}t} = -(U-c), \quad \frac{\mathrm{d}\eta}{\mathrm{d}t} = -(U+c), \quad \text{and} \quad \frac{\mathrm{d}\eta}{\mathrm{d}t} = -U.$$

This is shown schematically in Fig. 7.4. The $C_\pm$ characteristics are the acoustic characteristics, and C_0 is the material or entropic characteristic. With the lead shock wave fixed in the shock-attached frame, then the forward C_+ characteristic speed $\mathrm{d}\eta/\mathrm{d}t = -(U+c)$ is negative at the shock and consequently the shock-frame Mach number $\mathsf{M} \equiv |-U/c| < 1$ and the flow just behind the shock is *subsonic*. But far in the reaction zone, there is a point where the flow is sonic (denoted by a * subscript) where $\mathrm{d}\eta_*/\mathrm{d}t = 0$ at that point on a forward C_+ characteristic, hence $-U_* = c_*$ and $\mathsf{M}_* \equiv |-U_*/c_*| = 1$ and the flow is both characteristic and, at that special point, sonic. Therefore, on that characteristic and at that point

$$C_+ : \ \frac{\mathrm{d}P_*}{\mathrm{d}t} + \rho_* c_* \frac{\mathrm{d}u_*}{\mathrm{d}t} = \rho_* c_*^2(\sigma R - \kappa u)_* \ \text{ on } (\eta, t) \text{ s.t. } \frac{\mathrm{d}\eta_*}{\mathrm{d}t} = -(U+c)_*. \tag{7.56}$$

The locally slow time evolution along any near-sonic forward-going characteristics argues for the introduction of a slow-time variable, $\hat{t} = \delta t$, where $0 < \delta \ll 1$, into (7.56)

$$C_+ : \frac{\mathrm{d}P_*}{\mathrm{d}\hat{t}} + \rho_* c_* \frac{\mathrm{d}u_*}{\mathrm{d}\hat{t}} = \frac{\rho_* c_*^2(\sigma R - \kappa\, u)_*}{\delta} = O(1) \text{ on } (\eta, t) \text{ s.t.}$$
$$\frac{\mathrm{d}\eta_*}{\mathrm{d}\hat{t}} = \frac{-(U+c)_*}{\delta} = O(1). \tag{7.57}$$

Taken together, these conditions constrain the solution such that when $-(U + c)_* = O(\delta)$, i.e., when near the instantaneous sonic locus, then $(\sigma R - \kappa u)_* = O(\delta)$. Enforcing (7.57) in addition to the shock conditions will constrain the propagation speed of the detonation to be compatible with the shock curvature, κ, being imposed and the free-running properties of self-sustaining detonation.

The wave diagram shown in Fig. 7.5 is for the problem whose solution is displayed in Fig. 7.3. Shown in the inset to Fig. 7.5 are the sonic locus (short-dashed curve) and two characteristics immediately to the right (chain-dashed curve) and immediately to the left (medium-dashed curve) of the limiting characteristic (not shown). At an earlier time, these characteristics and the limiting characteristic are in a supersonic flow. As the characteristics evolve in time, the right (chain-dashed curve) characteristics cross into the region of subsonic flow and is seen intersecting the shock. The left (medium-dashed curve) characteristic remains in a region of supersonic flow and moves relentlessly toward the end of the reaction zone and the end of the computational

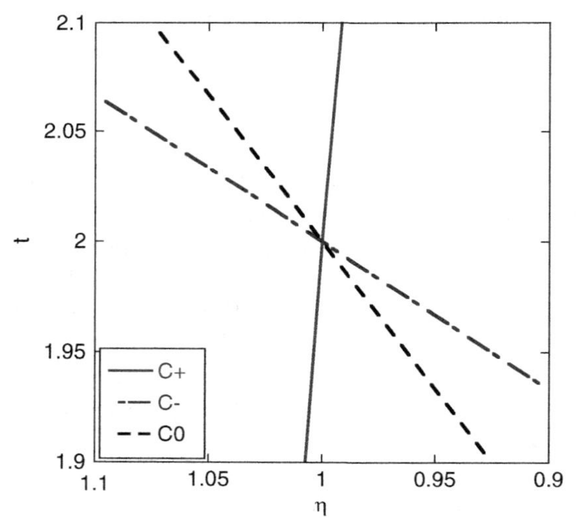

Fig. 7.4. A schematic representation of the C_+, C_- and C_0 characteristics at a point near the end of a detonation reaction zone. Data propagating (t increasing) along all three characteristics is needed to define the solution at the point of the characteristic intersection ($\eta = 1$). The near-sonic flow along the C_+ characteristic (a nearly *vertical line*) implies the evolution time along the C_+ characteristic is much longer than it is along either the C_- or C_0 characteristics

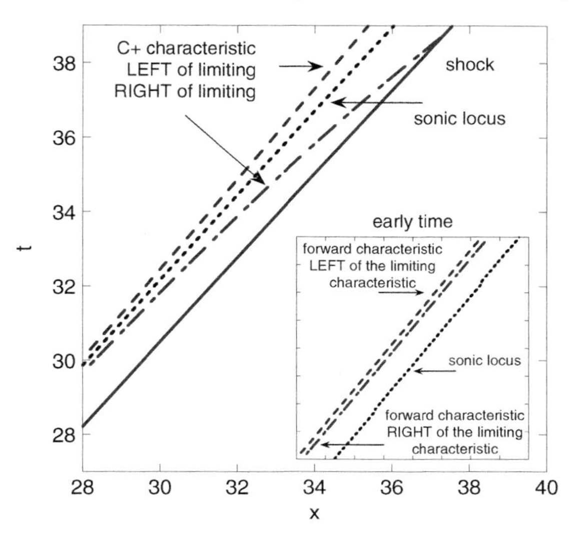

Fig. 7.5. Wave diagram for the problem of Fig. 7.3 showing the shock (right-most *heavy-solid line*), the sonic locus (*short-dashed line*), and two C_+ forward characteristics that at $t = 0$ is tightly clustered around the limiting characteristic (not shown). At the late time shown in the main figure, the right characteristic (*chain-dashed curve*) is propagating in a subsonic flow region, while the left characteristic (*medium-dashed curve*) is propagating in a supersonic flow region. The inset shows the two characteristics and sonic locus at an earlier time. Then both characteristics are following the limiting characteristic, all of which are in a region of supersonic flow

domain. The limiting characteristic is the effective rear boundary for the part of the reaction-zone flow that can influence the shock. The compatibility conditions of (7.56) and (7.57) should be enforced along this curve.

The above discussion was based on the results from a single, although very high resolution, shock-fitted, numerical simulation [16]. Details of the limiting characteristic path are not available from any numerical simulation, even the very high resolution results we have presented. Here we provide quantitative support for the constraining conditions that we argued for qualitatively in the above and for the role of the limiting characteristic. We restrict our investigation of the characteristic equations and of the limiting characteristic by making special choices for the EOS, etc. We follow Courant and Friedrichs [12] and Thompson [36], and introduce the Riemann variable

$$\mathscr{I}(P,S) = \int_{P_0}^{P} \frac{\mathrm{d}P}{\rho c}, \tag{7.58}$$

where S is the entropy, associated with a material particle, and the functional integration on P is performed at fixed S. To make rapid progress with the analysis and closely parallel the developments in [12, 36], we restrict our attention to the polytropic EOS we used earlier,

$$E = \frac{P}{(\gamma - 1)\rho} - Q\lambda, \tag{7.59}$$

where γ is the constant polytropic exponent, $Q = (2(\gamma^2 - 1))^{-1}$ is the scaled heat of detonation, and λ, with $0 \leq \lambda \leq 1$, is the degree of reaction. The expression for Q assumes that the ideal detonation speed, D_{CJ}, is scaled to one, $D_{\mathrm{CJ}} = 1$. We integrate the energy equation (7.50), along the C_0 characteristic, to get

$$P/\rho^{\gamma} = \exp\left(\gamma \int \sigma R \mathrm{d}t_0\right) \equiv \exp\left(\gamma \int \sigma \mathrm{d}\lambda\right), \tag{7.60}$$

which if we assume an ideal gas for the thermal response of the fluid can be written as

$$\rho = P^{1/\gamma}\left(\frac{\rho_r}{P_r^{1/\gamma}}\right)\exp\left(\frac{-(S - S_r)}{\gamma C_v}\right), \tag{7.61}$$

$$S = \gamma C_v \int \sigma R \mathrm{d}t_0 \equiv \gamma C_v \int \sigma \mathrm{d}\lambda, \tag{7.62}$$

where C_v is the heat capacity and the subscript r denotes a reference state (the shock state can serve as the reference state). Thus, we have

$$\rho c = \gamma^{1/2} P^{(\gamma+1)/(2\gamma)}\left(\frac{\rho_r}{P_r^{1/\gamma}}\right)^{1/2} \exp\left(\frac{-(S - S_r)}{2\gamma C_v}\right), \tag{7.63}$$

and since S is an independent variable, integration in (7.58) yields the following expression for the Riemann variable, $\mathscr{I}(P, S)$:

$$\mathscr{I}(P, S) = \frac{2c}{\gamma - 1}. \tag{7.64}$$

Thus, the evolution equations along the C_- and C_+ characteristics can be rewritten as

$$C_- : \ \frac{\mathrm{d}}{\mathrm{d}t_-}\left(\frac{2c}{\gamma - 1} - U\right) = c(\sigma R - \kappa u) + \frac{\mathrm{d}D_n}{\mathrm{d}t_-}, \tag{7.65}$$

$$C_+ : \ \frac{\mathrm{d}}{\mathrm{d}t_+}\left(\frac{2c}{\gamma - 1} + U\right) = c(\sigma R - \kappa u) - \frac{\mathrm{d}D_n}{\mathrm{d}t_+}. \tag{7.66}$$

Even in the absence of the source terms on the right-hand sides, these equations in general remain coupled. As our earlier discussion of characteristics suggested, one cannot simply identify the appropriate initial state for the limiting characteristic, and then integrate to both find the limiting characteristic and the value $2c/(\gamma - 1) - U$ and $2c/(\gamma - 1) + U$ along the characteristics.

However, for the special EOS value, $\gamma = 3$ (a value often used to describe condensed-phase explosives), a portion of the coupling is broken. Then, except for the coupling provided by the source terms, the evolution along the C_- and C_+ is decoupled, yielding

$$C_- : \frac{\partial(c-U)}{\partial t} + (c-U)\frac{\partial(c-U)}{\partial \eta} = c(\sigma R - \kappa u) + \frac{\mathrm{d}D_n}{\mathrm{d}t_-}, \tag{7.67}$$

$$C_+ : \frac{\partial(c+U)}{\partial t} - (c+U)\frac{\partial(c+U)}{\partial \eta} = c(\sigma R - \kappa u) - \frac{\mathrm{d}D_n}{\mathrm{d}t_+}. \tag{7.68}$$

Then irrespective of how fast or slow the evolutionary timescales are, the solution along a shock-facing, forward-going, characteristic can be found once the source term is determined.

To demonstrate the properties of the solution to (7.68), here we select a form for the thermicity consistent with the problem described in Fig. 7.3

$$c(\sigma R - \kappa u) - \frac{\mathrm{d}D_n}{\mathrm{d}t_+} = (\eta_e(t) - \eta)A, \tag{7.69}$$

where A is a constant and $\eta_e(t)$, although not specified, satisfies $\mathrm{d}(\eta_e(t))/\mathrm{d}t \geq 0$; that is, the thermicity locus evolves toward the end of the reaction zone. Equation (7.68) is then reduced to

$$f_{\tilde{t}} + f = \eta_e(\tilde{t}/B), \tag{7.70}$$

with $c+U = (f(\tilde{t})-\eta)B$, and with $B^2 = A$, a constant, and $\tilde{t} = Bt$. If initially the flow is sonic at the point of zero thermicity, then we find as the solution

$$f(\tilde{t}) = \eta_e(\tilde{t}/B) - \exp(-\tilde{t}) \int_0^{\tilde{t}} \exp(\tau)\frac{\mathrm{d}\eta_e(\tau/B)}{\mathrm{d}\tau}\mathrm{d}\tau. \tag{7.71}$$

A sonic flow, $\eta = f(\tilde{t})$, at the point of zero thermicity, $\eta = \eta_e(\tilde{t}/B)$, is clearly not a compatible condition. Given that $\mathrm{d}(\eta_e(\tilde{t}/B))/\mathrm{d}\tilde{t} > 0$, it follows that $f(\tilde{t}) < \eta_e(\tilde{t}/B)$, and thus the sonic locus lags behind the point in the flow where the thermicity is zero, approaching the zero thermicity locus only in the limit of $\tilde{t} \to \infty$. Turning our attention to the equation for the characteristics,

$$\frac{\mathrm{d}\eta}{\mathrm{d}\tilde{t}} = (\eta - f(\tilde{t})), \tag{7.72}$$

it follows that a sonic flow, $(\eta - f(\tilde{t})) = 0$, is not compatible with a shock-facing, forward characteristic, except at an isolated point. For $(\eta - f(\tilde{t})) = 0$ would imply $\eta =$ constant along such a forward characteristic, the general solution for any characteristic path is

$$\eta_c(\tilde{t}) = (\eta_c(0) - \eta_e(0))\exp(\tilde{t}) + f(\tilde{t}) - \int_0^{\tilde{t}} \sinh(\tilde{t} - \tau)\frac{\mathrm{d}\eta_e(\tau/B)}{\mathrm{d}\tau}\mathrm{d}\tau. \tag{7.73}$$

Now, there is a special characteristic, called the limiting characteristic, that approaches $f(\tilde{t})$ as $\tilde{t} \to \infty$, that is, $\eta_c(\tilde{t}) \to f(\tilde{t})$. This characteristic, selected by setting $\eta_c(0)$, has the property such that only those characteristics ahead of it (nearer the shock) can influence the evolution of the detonation front. Selecting

$$\eta_c(0) = \eta_e(0) + \lim_{\tilde{t}\to\infty} \exp(-\tilde{t}) \int_0^{\tilde{t}} \sinh(\tilde{t} - \tau) \frac{\mathrm{d}\eta_e(\tau/B)}{\mathrm{d}\tau} \mathrm{d}\tau, \tag{7.74}$$

ensures that the limiting characteristic merges into the sonic characteristic at long times ($t \to \infty$). The locus of sonic flow is forward of the zero-thermicity locus at finite time, given $\mathrm{d}\eta_e(\tilde{t}/B)/\mathrm{d}\tilde{t} > 0$. Since as $\tilde{t} \to \infty$ the two loci become coincident as does the limiting characteristic; it follows that with increasing time

$$\frac{\mathrm{d}\eta_c}{\mathrm{d}\tilde{t}} = (\eta_c - f(\tilde{t})) = (\eta_c(0) - \eta_e(0)) \exp(\tilde{t}) - \int_0^{\tilde{t}} \sinh(\tilde{t} - \tau) \frac{\mathrm{d}\eta_e(\tau/B)}{\mathrm{d}\tau} \mathrm{d}\tau > 0, \tag{7.75}$$

which corresponds to a supersonic flow along the limiting characteristic. For hydrodynamically stable, diverging detonations the limiting characteristic is generally found between the zero thermicity and sonic loci.

As an example, we consider the special case $A = 1$ and

$$\eta_e(t) = 2.25 - \frac{1.125}{1 + 0.5t}, \tag{7.76}$$

which provides an analog for a cylindrically expanding detonation, and as was considered in Fig. 7.3, provides a prototypical example of the evolution of curved detonation waves. The t versus η wave diagram for this example is displayed in Fig. 7.6. As discussed above, the zero-thermicity locus (medium-dashed curve) is furthest from the shock (to the left), while the limiting characteristic (short-dashed curve) sits between the zero-thermicity locus and the sonic locus (chain-dashed curve). As the detonation flattens out with increasing time, all three loci become coincident, as one finds them in a one-dimensional, steady-state detonation. For finite time, these three curves are distinct, with the evolution of the solution along the limiting characteristic being the slowest. As characteristics peel away from the limiting characteristic and head towards the shock, the evolution along them speeds up as the shock-facing characteristics cross over and break ahead of the sonic locus.

The limiting characteristic is unique and the most important of the shock-facing characteristics. It is relatively close to the zero-thermicity locus. The evolution of the flow along it is slow and nearly of zero thermicity, and it does provide the back boundary for information relevant to the evolution of the detonation shock. Thus, it does represent the rear boundary for the relevant part of the reaction-zone flow evolution. Unfortunately, except for simple examples as we considered here, it is difficult to precisely define the limiting characteristic a priori. On the other hand, the points of intersection of the sonic locus and the many other shock-facing characteristics also see a flow that is slowly evolving (vertical slope), as the various characteristics cross the instantaneous sonic locus. Here we propose using an asymptotic analysis, which uses the notion that the thermicity locus, sonic locus, and limiting characteristic are instantaneously found in the same neighborhood to build a rear-boundary for time-dependent DSD analysis. In the next section, we describe how a PDE can be constructed which provides an effective rear-boundary condition for self-sustaining, freely propagating detonation.

7.4.1.2 Multi-Scale Transients and DSD Theory

As we have already alluded to, implicit in DSD theory is the notion of slow-time dynamics, with the evolutionary timescale being $O(\epsilon^{-1})$ or longer. That is, equilibration of perturbations to the main part of the detonation reaction zone, herein called the main-reaction layer (MRL), occurs on the $O(1)$ reaction-rate scale (particle passage time through the reaction zone). Thus, on the longer timescales envisioned by DSD theory, the MRL should appear

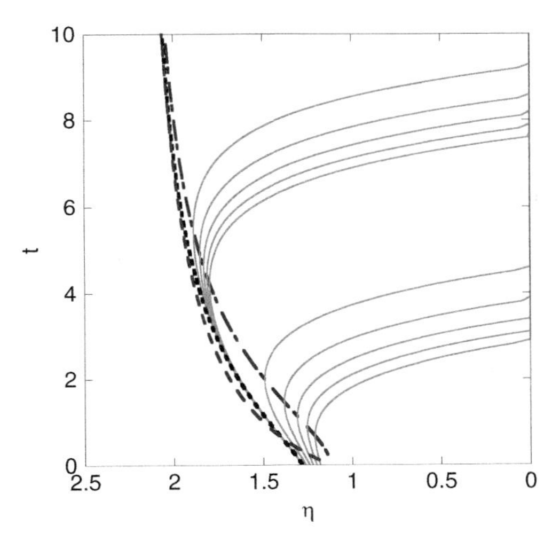

Fig. 7.6. Wave diagram for analog of the reaction-zone structure shown in Fig. 7.3, with $\eta_e(t)$ given by (7.69) and (7.76). The *medium-dashed curve* is the zero-thermicity locus, the *chain-dashed curve* is the sonic locus, and the *short-dashed curve* is the limiting characteristic, which is defined in the text. Ten shock-facing characteristics that begin ahead of the limiting characteristics are shown as *thin-solid gray curves*. Only a narrow band of characteristics beginning near the origin of the limiting characteristic influences the detonation flow for any significant amount of time

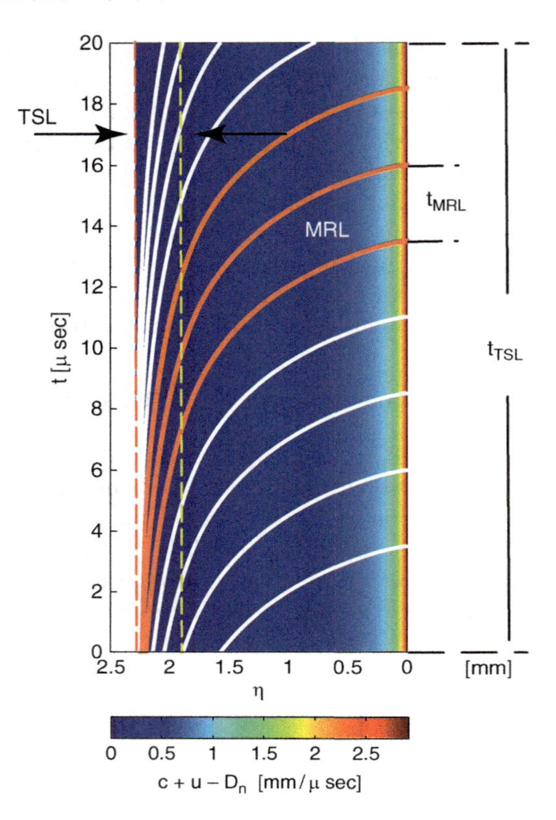

Fig. 7.7. A diagram showing the shock-facing characteristics in a one-dimensional, shock-attached frame and their intersection with the lead shock and the TSL to MRL transition (Adapted from [21] for the reaction-zone structure shown in Fig. 7.3 and where $\gamma = 3$)

to be steady, actually, quasi-steady. By examining various asymptotic limiting forms of the characteristic form of the Euler equations, we delineate what the various flow sub-layers are in the evolving detonation reaction zone, once the detonation described in the above example becomes nearly steady, and the zero-thermicity locus, sonic locus, and limiting characteristic are all nearly coincident.

Klein [21] identified three relevant timescales for the problem: (1) the $t \sim O(1)$ chemical reaction time, associated with the particle and backward-going characteristic waves as we have discussed; (2) the $\tau = \epsilon^2 t \sim O(1)$ MRL time, associated with standard DSD limit scaling quasi-steadiness; and (3) the $\bar{t} = \epsilon t \sim O(1)$ transonic layer (TSL) time, associated with trapping of the forward characteristic rays near the end of the reaction zone. This trapping is shown in Fig. 7.7 and can be understood with the aid of the equations we developed in the previous section, when their solution is carried to very long times, where the sonic locus and limiting characteristic are nearly coincident. Then the PDE that governs the TSL and that provides the rear boundary condition can be developed in the limit of a near ZND detonation perturbed by an $O(\epsilon^2)$ curvature, $\kappa = O(\epsilon^2)$. We focus attention on (7.67) and 7.68), and on the region of nearly sonic flow, $(c+u-D_n) = O(\epsilon)$, slow-time evolution along C_+ characteristics, $\bar{t} = \epsilon t$, $O(1)$ time evolution along C_- characteristics, and an $O(\epsilon^2)$ thermicity term $c(\sigma R - \kappa u)/\epsilon^2 = O(1)$. All perturbations are taken off of a base ZND, Chapman–Jouguet detonation state, with $D_n = 1 + O(\epsilon^2)$. With these approximations, $\frac{\mathrm{d}D_n}{\mathrm{d}t_+} = O(\epsilon^3)$ and $\frac{\mathrm{d}D_n}{\mathrm{d}t_-} = O(\epsilon^2)$ and we get at the first, nontrivial, leading order

$$c + u - D_n = \epsilon(c_1 + u_1) + \cdots, \tag{7.77}$$

$$c - u + D_n = \frac{2\gamma}{\gamma+1} + \epsilon(c_1 - u_1) + \cdots = \text{constant}, \tag{7.78}$$

from (7.67) and 7.68), respectively, and thus $c_1 = u_1$, which results in the leading-order, TSL evolution equation

$$\frac{\partial u_1}{\partial \bar{t}} - 2u_1\frac{\partial u_1}{\partial \eta} = \frac{\gamma}{2(\gamma+1)^2}\frac{1}{\epsilon^2}\left(\frac{(\gamma+1)^2}{2\gamma^2}R - \kappa\right), \tag{7.79}$$

where $\sigma = (\gamma+1)/(2\gamma^2)$.

When the flow in the MRL is stable and adjusts on the $t \sim O(1)$ time to become quasi-steady, the fastest time in the problem is the TSL time $\bar{t} = \epsilon t \sim O(1)$. If the flow in the TSL reaches a steady state, the solution in the MRL will evolve on the $\tau = \epsilon^2 t \sim O(1)$ time, and so will be quasi-steady. If the flow in the TSL does not approach steadiness but continues to evolve on the TSL time $\bar{t} = \epsilon t \sim O(1)$, the MRL will not be quasi-steady. For situations where either the detonation is highly unstable (e.g., for a strongly, state-sensitive Arrhenius rate) or for the case of a converging flow, where unsteadiness is maintained by forward propagation of acoustic energy from behind the TSL [21], the TSL will not become steady and is not consistent with quasi-steady evolution in the MRL. Here we assume none of those situations occur, and that the flow in the MRL is slowly varying (at least at leading order).

As witnessed by the shock-facing characteristics displayed in Fig. 7.7, we see a continuous flow of information from the TSL-region, where the evolution is slow, into the MRL-region, where the evolution occurs on the $O(1)$ timescale. This diagram points out the importance of the TSL, the region which embeds the limiting characteristic, and how the near-sonic and near-end of reaction-zone flow provides the closure for the slow-time dynamics of detonation shock evolution.

The above discussion establishes the basic concepts for self-sustaining, multidimensional, curved detonation-front propagation. The limiting characteristic provides the required forward-characteristic, C_+, condition that must be applied in the vicinity of the sonic locus and thermicity locus. When combined with information on the solution that comes along the particle, C_0, and backward characteristic, C_-, equations, this generates the PDE (7.79), which provides the rear boundary condition for the reaction-zone flow. In the next section, we examine the dynamics of the basic MRL, quasi-steady, detonation dynamics.

7.4.2 DSD: Quasi-Steady Limit Using the Master Equation Formulation

Here we explicitly neglect the time dependence in the truncated equations (7.25)–(7.28), specifically neglecting terms related to derivatives of D_n and partial-time derivatives of the flow variables. The theory of thin, broadly curved front evolution that governs the quasi-steady, quasi-one-dimensional reaction-zone structure can be solved by many different techniques, asymptotic and numerical. Here we describe one of the simplest formulations that can be distilled to the analysis of solution trajectories in a u, λ plane for the reaction-zone structure between the shock and the sonic point.

Neglecting all explicit time variation in the Master Equation (7.23), we have

$$\left(U^2 - c^2\right) u_\eta = c^2(\sigma R - \kappa u), \tag{7.80}$$

and in (7.25)–(7.28), we have

$$M_\eta = -\rho u \kappa, \tag{7.81}$$

$$\Pi_\eta = -M u \kappa, \tag{7.82}$$

$$H_\eta = 0, \tag{7.83}$$

$$-U\lambda_\eta = R. \tag{7.84}$$

A first, and powerful, consequence of the quasi-steady limit is that the energy flux, H, is a constant and equal to the shock value

$$H = E_0 + \frac{1}{2} D_n^2. \tag{7.85}$$

[Note: We assume that $\rho_0 = 2$, $u_0 = 0$, $P_0 = 0$, and $\lambda_0 = 0$ in the state ahead of the shock.] Assuming the polytropic EOS form (7.59), and using the scalings introduced in the characteristic analysis section, we have the heat of detonation, $Q = 1/(2(\gamma^2 - 1))$, the thermicity, $\sigma = 1/(2(\gamma + 1)c^2)$, and the ideal Chapman–Jouguet detonation velocity is one ($D_{\mathrm{CJ}} = 1$), then (7.83) immediately returns an expression for the sound speed

$$c^2 = \left(\frac{\gamma - 1}{2}\right)\left(D_n^2 - U^2 + \frac{\lambda}{(\gamma^2 - 1)}\right). \tag{7.86}$$

If the reaction-rate law, R, contains no explicit dependence on either P or ρ independently, but only on c^2, where $c^2 = \gamma P/\rho$, then the Master Equation is closed with the addition of only the reaction-rate equation

$$\left(U^2 - c^2\right) u_\eta = \frac{R}{2(\gamma+1)} - \kappa u c^2, \tag{7.87}$$

$$-U\lambda_\eta = R(c^2, \lambda). \tag{7.88}$$

If the reaction-rate law were to contain state sensitivity on P, ρ, and/or T (temperature) then either the mass or momentum flux equations (7.81) and (7.82), would have needed to be included in the system of ordinary-differential equations (ODEs). This is not the situation we consider here, although we return to it later. (Note: The reaction rate can be generalized to include a dependence of R on the shock state, through D_n, without it affecting this analysis.)

With all the above assumptions being made, the solution of (7.87) and (7.88) can most conveniently be carried forward in the u versus λ phase-plane space. Eliminating η between these two equations and substituting $(u - D_n)$ for U yield the single equation, u versus λ plane version for the problem

$$\left((D-u)^2 - c^2\right) u_\lambda = (D-u)\left(\frac{1}{2(\gamma+1)} - \frac{\tilde{\kappa} u c^2}{\tilde{R}}\right), \tag{7.89}$$

whose properties and solution we now consider in detail. In the above, the tilded variables are dimensionless, scaled according to

$$\tilde{R} = R/k, \quad \tilde{\kappa} = \kappa D_{\mathrm{CJ}}/k, \quad D = D_n/D_{\mathrm{CJ}}, \tag{7.90}$$

where k is the reaction-rate constant and D_{CJ} is the Chapman–Jouguet detonation speed.

We note the prominence of the sonic parameter, $c^2(\mathsf{M}^2 - 1) = \left((D-u)^2 - c^2\right)$, with M the flow Mach number in the shock-attached frame, and the thermicity parameter,

$$\Phi = \left(1/(2(\gamma+1)) - \tilde{\kappa} u c^2/\tilde{R}\right) \tag{7.91}$$

in (7.89). We recall the significant role played by both of these parameters in the discussion of the time-dependent, characteristic equations (7.68) and the rear boundary PDE (7.79). In the quasi-steady analysis considered here, these important entities enter a little differently, through their definition of a critical point for the ODE. This can best be appreciated by recasting (7.89) as

$$\frac{\mathrm{d}u}{\mathrm{d}\lambda} = \frac{(D-u)\Phi}{c^2(\mathsf{M}^2 - 1)}, \tag{7.92}$$

where a critical point occurs when $\Phi = 0$ and $\mathsf{M}^2 - 1 = 0$ simultaneously at a point in the u versus λ phase plane.

For this demonstration, we select a reaction-rate law form that allows for simple algebra. The reaction rate we use for the remainder of our analysis is

$$R = k\mathscr{R}(D)c^2(1-\lambda)^{1/2} = k\tilde{R}, \tag{7.93}$$

where $\mathscr{R}(D)$ describes the shock-state-dependent component of the reaction rate.

7.4.2.1 Detonation-Eigenvalue Problem

Equation (7.92) is solved subject to the shock conditions (7.2), being satisfied at $\lambda = 0$. The solution to this boundary value problem defines how the problem parameters such as the heat of detonation, Q, the EOS parameter γ, the reaction-rate parameters and the detonation-front curvature, κ, work to set the detonation speed, D. The solution curves, which take u from its prescribed value at $\lambda = 0$ to its value at the end of the reaction zone at $\lambda = 1$, are called integral curves. A good overview of the theory of integral curves and the integration of ODEs can be found in Birkhoff and Rota [11].

We can get a qualitative estimate of what said integral curves look like by following the changes in sign that $\mathrm{d}u/\mathrm{d}\lambda$ experiences as we move through the u versus λ phase plane. In turn, these sign changes are set by the zeros of the thermicity, Φ, and the sonic parameter, $c^2(\mathsf{M}^2 - 1)$. The loci of the zeros of these two functions are called the separatrices of the phase plane. The places where these separatrices intersect correspond to points of indeterminacy for the ODE, since $\mathrm{d}u/\mathrm{d}\lambda = 0/0$. These points, where the separatrices intersect, are the phase-plane critical points. The behavior of the solution in the neighborhood of these critical points is not immediately obvious, and it requires that mathematical analysis [11] be performed in their vicinity. This analysis defines a reduced ODE whose eigenvalues determine the solution near these points. For the special rate law form we have selected, the thermicity zero curve (thermicity locus) and sonic zero curve (sonic locus) are given by the solution of one linear and one quadratic algebraic equation

$$u = \frac{\sqrt{1-\lambda}}{2(\gamma+1)} \frac{1}{\bar{\kappa}}, \tag{7.94}$$

$$u = D \mp \sqrt{\left(\frac{\gamma-1}{\gamma+1}\right)\left(D^2 + \frac{\lambda}{(\gamma^2-1)}\right)}, \tag{7.95}$$

and where $\bar{\kappa} = \tilde{\kappa}/\tilde{\mathscr{R}}(\tilde{D})$. [Note: The scaling of $\tilde{\kappa}$ by $\tilde{\mathscr{R}}(\tilde{D})$ is the sole appearance of $\tilde{\mathscr{R}}(\tilde{D})$ in the analysis. Thus, this popular form of reaction-rate state sensitivity [19] leads to analysis that is isomorphic with the analysis for a state-independent reaction rate.]

Displayed in Fig. 7.8 is the phase plane for (7.92) for a value of D that fully solves the equation. The thermicity locus is shown as the medium-dashed curve and the sonic locus (negative sign in (7.95)) as the chain-dashed curve. Also shown are four trial, unsuccessful integral curves, drawn as solid curves that begin at $\lambda = 0$ and integrate in the direction of increasing λ. The thermicity, Φ, is endothermic above $\Phi = 0$ and the flow is subsonic above the sonic locus $c^2(\mathsf{M}^2 - 1) = 0$. Thus, the trial integral curve that begins at $u = 0.65$, $\lambda = 0$ (upper solid curve) moves upwards in a region that is subsonic and endothermic. Not shown is the zero of the factor, $(D - u)$, in (7.92), which

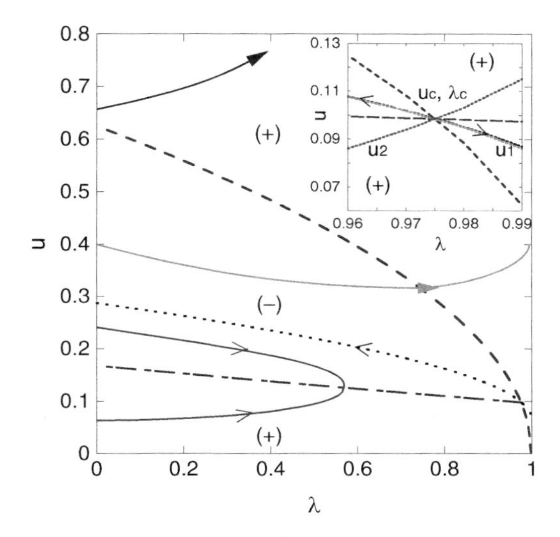

Fig. 7.8. The u versus λ phase plane for (7.92) with the reaction rate of (7.93). The *medium-dashed curve* is the thermicity locus, $\Phi = 0$, and the *chain-dashed line* is the sonic locus, $c^2(\mathsf{M}^2 - 1) = 0$. The *solid curves* are "trial" integral curves for (7.92). Integration starts from the shock ($\lambda = 0$) and proceeds as the arrows indicate. All the "trial" curves are deficient in some way. The location of the critical point is shown in the expanded view in the inset. The curves $u1$ and $u2$ shown in the inset represent the exact local solution in the neighborhood of the critical point. The *short-dashed curve* is the successful integrated solution

ultimately limits the increase of u. This solution is not of interest for a number of reasons. Among these reasons are that it does not satisfy the shock conditions, and if the curve is able to reach $\lambda = 1$, the flow there is subsonic and therefore unstable to perturbations from a subsonic, post-reaction zone, following flow. This does not represent a self-sustaining detonation. The middle solid trial curve is problematic for some of the same reasons as those for the upper solid trial curve. The two lower trial curves starting at $u = 0.24$, $\lambda = 0$ and at $u = 0.07$, $\lambda = 0$, both terminate at $\lambda < 1$ on the sonic locus, and so cannot reach the end of the reaction zone.

An acceptable solution (and in fact the solution) would be the short-dashed integral curve. It corresponds to a mostly exothermic flow and passes through the critical point, ending at a stable, supersonic flow at $\lambda = 1$. Whether a solution can pass through the critical point, rather than be attracted to it, as to a sink, requires that we analyze the character of the solution in the neighborhood of the critical point and construct a local solution there. For the remainder of the discussion on this problem, we introduce $\zeta = \sqrt{1 - \lambda}$ in place of λ as the independent variable.

The simultaneous solution of (7.94) and (7.95) yields the possibilities for the coordinates of the critical point

$$u_c = \frac{D \mp \sqrt{D^2 - \frac{2(1+4\bar{\kappa}^2)}{(\gamma+1)}\left(D^2 - \frac{1}{(\gamma+1)}\right)}}{(1+4\bar{\kappa}^2)}, \tag{7.96}$$

$$\zeta_c = \sqrt{1-\lambda_c} = 2(\gamma+1)u_c\bar{\kappa}, \tag{7.97}$$

where we select the minus sign, since it corresponds to the intersection of the thermicity locus with the lower branch of the sonic locus shown in Fig. 7.8. Expanding u and λ about the critical point values

$$u = u_c + \delta u, \tag{7.98}$$

$$\zeta = \zeta_c + \delta\zeta, \tag{7.99}$$

and substituting the result into (7.92) and retaining only linear terms in δu and $\delta\zeta$ yield the critical point ODE

$$\frac{\mathrm{d}(\delta u)}{\mathrm{d}(\delta\zeta)} = \frac{(\delta\zeta)c_1 + (\delta u)d_1}{(\delta\zeta)a_1 + (\delta u)b_1}, \tag{7.100}$$

and where

$$a_1 = -\zeta_c/(D-u_c), \qquad b_1 = (\gamma+1)^2, \tag{7.101}$$

$$c_1 = 1\ , \qquad d_1 = -2(\gamma+1)\bar{\kappa}. \tag{7.102}$$

Two eigenvalues can be found for (7.100)

$$\Lambda_{1,2} = \frac{(a_1+d_1) \pm \sqrt{(a_1+d_1)^2 - 4(a_1d_1 - b_1c_1)}}{2}, \tag{7.103}$$

where Λ_1 corresponds to the positive root, and where

$$(a_1d_1 - b_1c_1) = -\frac{(\gamma+1)^2}{(D-u_c)}\left(D - (1+4\bar{\kappa}^2)u_c\right) < 0. \tag{7.104}$$

Since we have two real roots of opposite sign, the critical point is of saddle-type. The two orthogonal solutions are

$$\delta u_{1,2} = \frac{1}{b_1}\left(\Lambda_{1,2} - a_1\right)(\delta\zeta). \tag{7.105}$$

The nature of the critical point and its saddle character, along with the relevant local solution, u_1, is displayed in the inset of Fig. 7.8. The second solution, u_2, is orthogonal to u_1 as it passes through the critical point. The solution u_2 is not of interest, since it takes the solution into an endothermic, subsonic region, as does the trial, middle integral curve. Having u_1 and u_2 is important to being able to solve (7.92) in the entire domain, $0 \le \lambda \le 1$. A numerical solution of the ODE with a packaged solver can easily send the integral curves off into unfruitful directions. For this problem, the only two solution paths through the critical point are the two linear solutions we have presented. The solution u_2 is not of interest for reasons already discussed.

The phase plane presented here is relatively simple. Small changes to the reaction rate, adding an Arrhenius, local temperature dependence, for example, can introduce more critical points into the u versus λ phase plane [15]. Then a successful integration from $\lambda = 0$ to $\lambda = 1$ would need to tread its way through two possible critical points.

Having only the local, near critical point solution does not completely solve the problem. The integral curve must satisfy both the shock condition, $u(\lambda = 0) = 2D/(\gamma + 1)$, and extend to the end of the reaction zone. For the example shown here, we used the default Maple numerical ODE solver. The solution was initialized near the critical point with the above described local solution for a trial value of D. The analytical solution was used in the vicinity of the critical point to establish a first point on the integral curve, then the numerical solver was used to integrate the trial solution curve to the shock $u(\lambda = 0)$. The trial solution value at the shock was subtracted from the value defined by the shock condition, $u_s = 2D/(\gamma + 1)$, to define a merit function (residual), $MF = u(\lambda = 0) - 2D/(\gamma+1)$. With D as a parameter and using (for example) Newton's method to solve $MF = 0$, the global, eigenvalue solution, $D(\bar{\kappa}, \gamma, Q, \text{etc.})$, is obtained with just a few iterations. The successful integral curve, displayed in Fig. 7.8 as the short-dashed curve, was obtained via such a process. This process is known as solving the detonation-eigenvalue problem, using a shooting method. Once the detonation-eigenvalue problem is solved, u versus λ, M versus λ, and Π versus λ can be solved for with two integrations: (1) an integration from near the shock-side of the critical point to the shock and (2) a second integration from the end of reaction-zone side of the critical point to $\lambda = 1$.

The principal result obtained by solving the detonation-eigenvalue problem is the D versus $\bar{\kappa}$ propagation law. For the EOS (7.59) and reaction-rate law (7.93) considered here, we get the extended propagation law D versus $\kappa/\mathscr{R}(D)$, once $\mathscr{R}(D)$ is exposed as part of the definition of $\bar{\kappa} = \kappa/\mathscr{R}(D)$. The shooting problem results for D versus κ, for $\mathscr{R} = D^n$, are displayed in Fig. 7.9. As the reaction rate, state dependence is increased (as n is increased), steady-state solutions are only possible for a limited range of κ. The portion of the D versus κ curve, where κ decreases as D decreases, is found to be unstable for both DSD propagation (diffusion-style evolution, because of a negative diffusion coefficient) and from time-dependent, direct numerical simulation (DNS) results.

Although one can compute a D versus κ law as we have done here given the EOS and reaction-rate models, D versus κ laws can also be determined independently, directly by experimental means with detonation-rate stick experiments. In these experiments, detonation is initiated at one end (typically the bottom) of a long, circular cylinder of explosive, and the axial progress of detonation is monitored. Once the detonation establishes a steady-state phase speed, this speed and the shape of the detonation front as it breaks

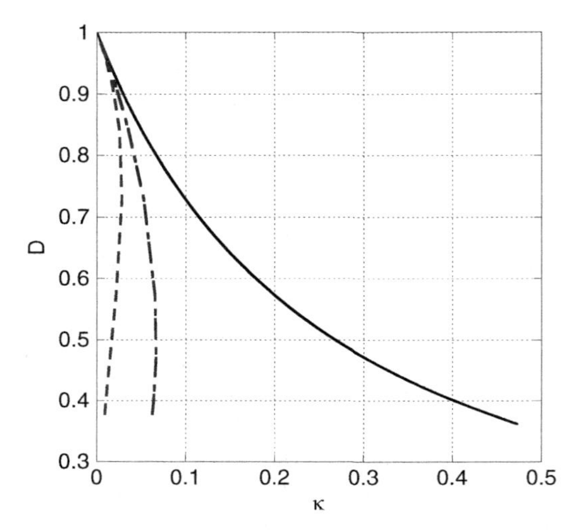

Fig. 7.9. The D versus κ law resulting from the solution of the detonation-eigenvalue problem for the EOS of (7.59), and the rate law of (7.93). The plotted results are for the shock-state function, $\mathscr{R}(D) = D^n$, with $n = 0, 2, 4$. The *solid curve* is for $n = 0$, the *chain-dashed curve* is for $n = 2$ and the *medium-dashed curve* is for $n = 4$. When $n = 0$, κ increases monotonically with decreasing D. For both $n = 2$ and $n = 4$, a maximum value of κ exists, after which κ decreases for a further decrease in D

out (typically at the top) of the stick are measured. The assumed steady, experimentally determined shock front locus and phase velocity can be used to determine D versus radius and κ versus radius variation across the detonation front. By repeating the experiment for a variety of different radii cylinders, sufficient data can be collected to experimentally map out a D versus κ curve [4]. In the final section of this chapter, we show how given a D versus κ function and an appropriate set of DSD boundary-condition angles (see the next major section), the rate stick and other detonation-propagation problems can be modeled using DSD.

7.4.2.2 D versus κ for More Complex EOS and Rate Laws

In the previous section, we showed how a D versus κ relation could be developed for a model that assumes a polytropic EOS and a reaction rate that contained a special dependence on local, thermodynamic state. With these two assumptions, it was sufficient to integrate only the Master Equation and the rate law in order to solve the detonation-eigenvalue problem. For more complex EOS models, integration of the energy equation (7.83) to the Bernoulli law does not return an equation for the sound speed. Also, for rate law forms that depend on local temperature and/or pressure such as Arrhenius rates, local thermodynamic variables other than the sound speed are needed. Either one of these two conditions requires that we know other thermodynamic state variables, such as density or pressure. This requires that the conservation law for either the mass flux or the momentum flux be added to the Master Equation and rate law and solved together as a system. This not only complicates the solution procedure, since more equations are involved, but can significantly complicate the phase plane to a three-dimensional phase space, u versus P versus λ (see Chap. 5 in [15]).

Suppose the basic structure of the u versus λ phase plane is unchanged and that it continues to control the solution of the detonation-eigenvalue problem. As we said, the energy equation still integrates to a Bernoulli law and so gives an algebraic relation among $(D-u)$, λ, P and ρ, say, $P(\rho,(D-u)^2,\lambda)$. Thus, one additional variable, ρ, must be accounted for. We assume that the thermicity locus, $\Phi=0$, and sonic locus, $c^2(\mathsf{M}^2-1)=0$, play the same role as before, and have only one crossing point, and thus one critical point. Then we can treat ρ at the critical point as an additional unknown parameter, ρ_c. Given that the mass-flux equation is apparently regular at the critical point, no special procedures are required to integrate it. Since both u and ρ must satisfy the shock conditions ($u(\lambda=0)=u_s$ and $\rho(\lambda=0)=\rho_s$), then there is an additional condition on the solution with which to determine the additional parameter, ρ_c. Iterating on D and ρ_c, and solving the shooting problem with the expanded number of equations, will ultimately return a D versus κ relation. As before, once the detonation-eigenvalue problem is solved, then the entire flow field can be solved for, from the shock, through the critical point, and to the end of the reaction zone. This shooting procedure is stable and robust, provided that the basic, one critical point (of saddle-type) phase plane in u versus λ remains intact.

There are a number of DSD solver packages available to solve the detonation-eigenvalue problem. The DSDTools software bundle [Stewart, et al., private communication] includes a robust $D_n-\kappa$ solver package that is well tested.

7.4.2.3 Vanishing of the Critical Point

Examination of the critical-point equations (7.94) and (7.95) reveals that once u_c drops below zero, the critical point is lost. From (7.96), we find that this occurs at

$$D_f^2=\frac{1}{2(\gamma+1)}. \tag{7.106}$$

For $u<0$ and $\kappa>0$, the thermicity function, Φ, no longer has a zero. With $u<0$ and the flow reversed, $u\kappa$ acts as would a convergent nozzle section. The more negative u becomes, the stronger is the exothermic effect of $u\kappa$, which seems unphysical. For the polytropic EOS with $\gamma=3$ and the rate law model of (7.93), the loss of the critical point occurs at $\kappa\approx 0.4875$ and with $D=0.3536\ldots$. Although the values of $\kappa>0.4875$ and $D<0.3536$ are clearly well outside of the range where our derived small κ theory can be expected to apply, we examine this issue here in the interest of completeness. We note that the existence of this property of the critical point is tied most strongly to the structure of the sonic locus, and is not strongly influenced by the rate law form used. With the critical point lost, the steady-state analysis provides little guidance. Next, we examine the time-dependent solutions of (7.25)–(7.28) for guidance.

7.4.3 Time-Dependent Solutions

Motivated by the loss of any clear statement concerning eigenvalue detonations for $u < 0$ coming from the steady analysis, and seeking to understand the accuracy of the quasi-steady, DSD approximation, in this section we examine the initial-value, time-dependent solutions of (7.25)–(7.28). We solve these equations using η and t as the independent variables. To obtain solutions of sufficiently high accuracy to address the subtleties that are of concern, we use a highly accurate, shock-fitted, numerical Euler solver, developed by Henrick et al.[16]. Their scheme solves the flow in a shock-attached frame (the shock is at $\eta = 0$). The shock state is fitted into the flow, consistent with the shock conditions, by solving a shock-change equation (similar to the Master Equation) to update the shock state. The solutions are fifth-order accurate in space and time. Kasimov and Stewart [20] first proposed this approach and also developed a software that solves the equations in the shock-attached frame. Taylor et al. [35] recently used this technique to study the nonlinear stability of the quasi-steady solutions and give results similar to those shown in Fig. 7.12. The advantage of simulations that use shock-attached schemes over shock-capturing schemes is that they can directly return a very accurate result for D. Shock-capturing schemes are only first-order accurate and diffuse the shock, so extensive data processing is required to accurately measure the local shock speed.

For illustration, we solve the Euler equations displayed in (7.25)–(7.28) for two limiting forms: (1) for the $\kappa =$ constant DSD limit and (2) for the $\kappa = 1/r$ limit, where $r = r_0 + \int D dt$ is the cylinder symmetry, r-variable.

To serve as a crosscheck, the D and reaction-zone profiles developed with the shooting analysis we have described in the previous section were compared with the solutions obtained with the shock-fitted, time-dependent code. The solutions obtained with the quasi-steady, shooting code were used as initial data for the κ-fixed version of the time-dependent, shock-fitted code. Figure 7.10 shows a comparison of the detonation speed for these two solutions. The model considered is that described in the section on the quasi-steady analysis, with $\kappa = 0.2$ and $n = 0$ being used in the rate-function factor, $\mathscr{R}(D) = D^n$. Fifty grid points spanned the reaction zone in the shock-fitted solution. The solutions are validated by the fact that their difference occurs in the sixth-significant figure.

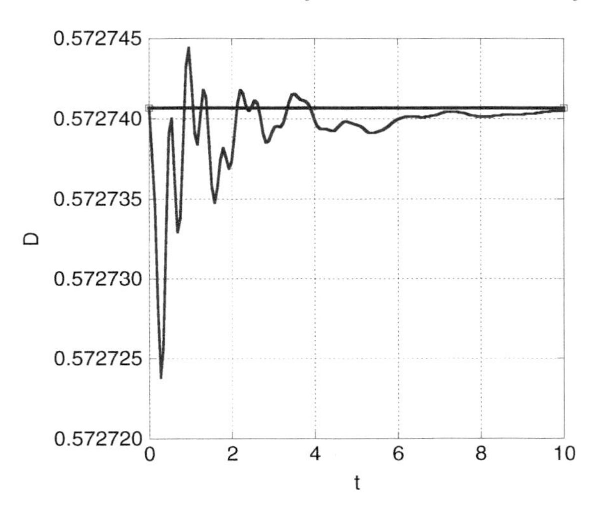

Fig. 7.10. The shooting code provided the input to the shock-fitted code. The shock curvature was fixed and equal to $\kappa = 0.2$ and the rate law factor, $\mathscr{R}(D) = D^n$, exponent was $n = 0$. The *horizontal line* is the D obtained with the shooting code. The oscillatory *solid curve* is the $D(t)$ obtained with the shock-fitted code. The differences are in the sixth-significant figure

7.4.3.1 The Vanishing Critical Point

The normal velocity, D, curvature, κ, relation, defined through the solution of the steady-state Master Equation (7.92), is largely controlled by a saddle-type critical point in the u versus λ phase plane, whose coordinates are given by (7.96) and (7.97)

$$u_c = \frac{D - \sqrt{D^2 - \frac{2(1+4\bar{\kappa}^2)}{(\gamma+1)}\left(D^2 - \frac{1}{2(\gamma+1)}\right)}}{1 + 4\bar{\kappa}^2}, \tag{7.107}$$

$$\zeta_c = \sqrt{1 - \lambda_c} = 2(\gamma + 1)u_c\bar{\kappa}. \tag{7.108}$$

This critical point collapses to $u_c = 0$, $\lambda_c = 1$ when $D^2 = 1/(2(\gamma + 1))$, at which point

$$\Lambda_{1,2} = -(\gamma + 1)\bar{\kappa} \pm (\gamma + 1)\sqrt{1 + \bar{\kappa}^2}, \tag{7.109}$$

thus the critical point continues to be of saddle-type, given that the roots are real and of opposite sign. No critical point exists for $D^2 < 1/(2(\gamma+1))$, since the thermicity locus is then lost.

One then asks, does a steady-state (quasi-steady) solution exist beyond that point? If so, is there a continuation of a D versus κ relationship beyond that point and what condition sets it? Some of the possibilities that have been discussed are: (1) the solution ends on the sonic locus, with $u < 0$ and $\lambda = 1$, at a positive value of the thermicity, $\Phi > 0$, (2) the solution ends at $u = 0$,

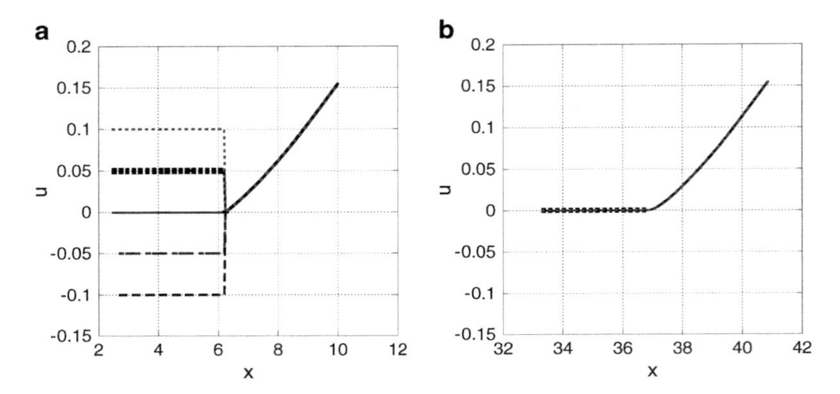

Fig. 7.11. The four initial data states (inset (**a**)) obtained with the shooting code, and the four essentially identical final states (inset (**b**)) obtained with the shock-fitted, direct numerical simulation, show the stability of $u(\lambda = 1) = 0$ to perturbations. This case, with $\bar{\kappa} = 0.6$ and $n = 0$, is an example of a case for which there is no critical point in the u versus λ phase plane

$\lambda = 1$, at a zero thermicity state and a subsonic flow, and (3) no steady-state eigenvalue detonation exists. To test whether one of these or as yet some other possibility exists, we seeded a series of time-dependent solutions with initial conditions corresponding to variations on a theme for the first two situations described above. Importantly, since the steady-state analysis does not provide an unambiguous choice, we must look elsewhere. That elsewhere is the time-dependent solution.

The summary conclusion from the time-dependent study is that for a fixed $\bar{\kappa} > 0.4875$, the solution equilibrates to $u = 0$ at $\lambda = 1$, with $D^2 < 1/(2(\gamma+1))$ and D set by the value of $\bar{\kappa}$. The flow is subsonic at $\lambda = 1$, and is followed by a non-reactive, thermally neutral, $\Phi = 0$, following constant-state region. A composite plot for the case $\bar{\kappa} = 0.6$ and $n = 0$, showing four sets of initial data in inset (a) and the resulting final states in inset (b), are displayed in Fig. 7.11. The final state has $D = 0.30852$ and $u(\lambda = 1) = 0$. Solving the shooting problem, using $\bar{\kappa} = 0.6$, $n = 0$ and $u(\lambda = 1) = 0$, returns essentially the same value of D, at $D = 0.30851$.

A rough physical argument, for the stability of $u(\lambda = 1) = 0$, can be made. The focus is on the sign of the thermicity in the flow downstream of the reaction zone, which was part of the region modeled in the simulations. When $u > 0$ in the flow downstream, $\Phi < 0$, and thus the flow acts endothermically. This endothermicity then acts on the $u > 0$ flow, until it is eroded to the neutral state, $\Phi = 0$. Correspondingly, when $u < 0$ in the flow downstream, $\Phi > 0$, thus the flow acts exothermically. This exothermicity then acts to build up (enhance) the flow downstream, by increasing u until the neutral state, $\Phi = 0$, is achieved. In any event, we find that the reaction zone reaches a quasi-steady state with $u(\lambda = 1) = 0$, with a thermally neutral, subsonic flow following the reaction zone.

7.4.3.2 Time-Dependent Versus Quasi-Steady Evolution

The quasi-steady, D versus κ solutions we have developed in the previous sections are intended to be used as propagation laws when modeling detonation propagation using DSD. They were derived from the reduced DSD theory equations (7.25)–(7.28) in the limit of leading-order, quasi-steady DSD approximation. The limiting DSD partial-differential equations (7.25)–(7.28) contain κ as a prescribed parameter. These equations become identical to the Euler equations for cylindrically symmetric flow when κ is replaced by $1/r$, $\kappa = 1/r$, where r is the spatial coordinate for cylindrically symmetric flow. In this section, we solve the cylindrically symmetric Euler equations, using the shock-fitted code described earlier. The $\kappa =$ constant, DSD quasi-steady, shooting code solutions are used as the initial data. We then plot D versus κ for both the locally constant κ quasi-steady solution and the cylindrically symmetric, time-dependent solutions, for which we take

$$\kappa(t) = \frac{1}{r} = \frac{1}{r_0 + \int_0^t D\mathrm{d}t}. \tag{7.110}$$

Such a comparison is a bit asymmetric, since $\kappa =$ constant at the shock value in the entire reaction zone for the quasi-steady solutions, while $\kappa = (r_0 + \int_0^t D\mathrm{d}t)^{-1}$ increases with increasing distance in the reaction zone, for the time-dependent solutions of the cylindrically symmetric Euler equations. At large radius (small curvature) this discrepancy is small, but for larger curvatures the differences are significant. By way of example, for the reaction-rate case $n = 0$ and $\kappa_0 = 0.2$, the reaction-zone length is ≈ 2 (see Fig. 7.3), and so the radius of curvature of the shock is only a little larger than two reaction-zone lengths. Thus, with $\kappa = 0.2$ at the shock for both cases, the quasi-steady solution has κ constant throughout the reaction zone, while the time-dependent solution has κ increasing to $\kappa = 0.33$ at the end of the reaction zone. This 50% increase in κ through the reaction zone for the time-dependent solution contributes significantly to the lower values of D observed early in the time-dependent solutions.

Noting these differences, we compare the quasi-steady and time-dependent solutions for two cases of the reaction rate, shock-state sensitivity, $\mathscr{R}(D) = D^n$. These results are displayed in Fig. 7.12, for $n = 0$ (a) and for $n = 2$ (b). The attracting property of the quasi-steady solution is clear. For $D > 0.85$, the time-dependent solution closely tracks the quasi-steady solution. Given the larger curvatures in the time-dependent solutions as noted above, the agreement shown between the two is good. This comparison provides some measure of validation for the consistency of the DSD approximation for modeling slowly evolving, self-sustaining, detonation flows.

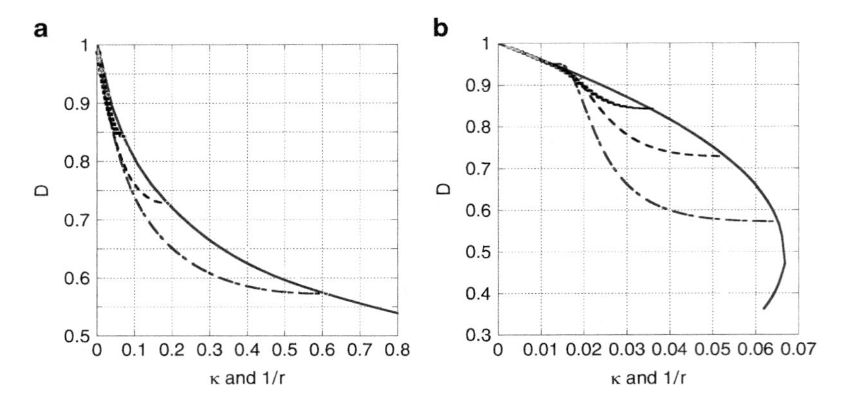

Fig. 7.12. Shock-fitted, time-dependent solutions of the cylindrically-symmetric Euler equations are compared with the quasi-steady, $\kappa =$ constant DSD approximation solutions obtained with the shooting method D versus κ_0 solver (*solid curve*). $D(t)$ versus $\kappa_0^{-1} + \int_0^t D(t)\mathrm{d}t$ is plotted for the time-dependent solutions (various *dashed curves*). The shooting solutions are used as initial conditions for the time-dependent solutions. Two cases of the reaction-rate factor, $\mathscr{R}(D) = D^n$, are studied, $n = 0$ (**a**) and $n = 2$ (**b**). The time-dependent solutions follow the quasi-steady track closely for $D > 0.85$

7.4.4 Flux Formulation: Numerics and Asymptotics

Here we discuss calculation of DSD propagation laws using the flux formulation. In the flux formulation the dependent variables are normal fluxes of mass, momentum, and energy M, Π, H, and λ. The analysis is carried out in the shock-attached coordinates for which the fluxes are natural variables. The eigenvalue character of the formulation is maintained in the quasi-one-dimensional, quasi-steady limit. Since a conversion between the flux variable and the primitive variables is *always* required, the Rankine–Hugoniot algebra is used to go back and forth between the two formulations. The main advantage of the flux formulation is that for quasi-one dimensional, quasi-steady flows the fluxes are nearly constant throughout the flow and nearly equal to their values determined at the lead shock. Hence powerful and fairly simple iterative methods can be used to compute the solutions. The flux formulation can be used to carry out a numerical solution of the truncated DSD formulation, or to generate asymptotic expansions in the limit as $\kappa \to 0$, and here we provide details for both solution procedures.

Briefly, the flux formulation rewrites the governing equations for the fluxes (7.25)–(7.28) as spatial ODES in η by placing derivatives of the fluxes on the left-hand side of the equation and small time-dependent and curvature terms on the right-hand side. The equations are integrated with respect to η, and the shock boundary conditions are applied at $\eta = 0$ to obtain, in general, a set of integro-differential equations. To illustrate the main points, we restrict ourselves here and retain only the curvature effect and neglect other higher order time and transverse effects, while specializing to the ideal EOS in the strong-shock approximation. The reader is referred to [26] for extensions of the methods described here to multiple reactions, more complex time dependence, and nonideal EOS.

Integration of (7.25)–(7.27) gives the following formal expressions for M, Π, H,

$$M = \rho_0 D_n + \kappa I_1, \quad \Pi = \rho_0 D_n^2 + \kappa I_2, \quad H = \frac{1}{2} D_n^2, \tag{7.111}$$

$$I_1 = -\int_0^\eta \rho u \, \mathrm{d}\eta, \quad I_2 = -\int_0^\eta M u \, \mathrm{d}\eta, \tag{7.112}$$

where we remind the reader of the definition of the flux variables, M, Π, and H defined earlier

$$M = -\rho U, \quad \Pi = P + \rho U^2, \quad H = E + Pv + \frac{1}{2} U^2. \tag{7.113}$$

The reaction progress distribution is given by solving the rate equation (7.28), subject to $\lambda = 0$ at $\eta = 0$. The shock conditions have been applied and are satisfied exactly. The state variables $v = 1/\rho, u$, and P are related to the fluxes and λ by the inversion of the Rankine–Hugoniot algebra, which we laid out in Sects. 7.3.1.1–7.3.2.

The formulation for $\kappa \neq 0$ follows the development in Sect. 7.3 for the planar detonation. Neglecting any explicit time dependence, then the limiting characteristic and sonic locus coalesce and the compatibility and speed relation in (7.57) (whose state is denoted with a * subscript) are recognized as the sonic locus and thermicity locus conditions,

$$(D_n - u)_* = c_* \text{ and } (\gamma - 1) Q R_* = \kappa \, c_*^2 \, u_* \quad \text{at} \quad \lambda = \lambda_* \text{ and } \eta = \eta_*. \tag{7.114}$$

The boundary conditions at the sonic point depend on all the state variables $v_*, (D_n - u)_*, P_*$ and λ_*, and occur at a position η_* in the structure. The dependence of $v, (D_n - u), P$ on λ and the flux variable is known through the Rankine–Hugoniot algebra. The location of the sonic point is computed from the rate integral

$$\eta_* = \int_0^{\lambda_*} \frac{(D_n - u(\bar{\lambda}))}{R(P(\bar{\lambda}, v(\bar{\lambda})), \bar{\lambda})} \, \mathrm{d}\bar{\lambda}, \tag{7.115}$$

which is an additional condition that must be imposed.

Given a fixed value of D_n and some trial value or guess for κ (say), one can determine the λ dependence on the state variables, $v, (D_n - u), P$ in terms of λ, D_n and κ, and in turn a trial, spatial distribution obtained by integration of the rate equation. If D_n is given, the unknowns are κ, λ_*, and η_*. Then the sonic and thermicity conditions (7.114) and the η_*, λ_* relation given by (7.115) are the three conditions needed to close the system. Note that small values of κ require specifically that the reaction rate at the sonic point be correspondingly small, either by virtue of complete reaction or by a very small reaction rate (in the case of state sensitivity of the reaction rate). The normal detonation speed D_n does not necessarily need to be close to its CJ value. One can then use either iterative numerical methods (treating κ as a finite value), or derive asymptotic formulas for the truncated DSD equations in the limit of small curvature. We describe a simple, easy-to-use numerical treatment next.

7.4.4.1 Flux Formulation: Numerical Approach

Here we give a brief description of the procedures for a numerical approach that is fairly robust. One needs an integration package for a system of stiff ODEs, and a nonlinear root solver package.

1. For a given EOS and rate law, compute the CJstates, as they provide the seed for $D_n = D_{\mathrm{CJ}}$ with $\kappa = 0$.
2. Main loop: Give $D_n \neq D_{\mathrm{CJ}}$ and guesses (seeds) for κ, v_*, and P_*. Since the CJ states are known, those can be used for initial guesses.
3. Use the thermicity condition (7.114b) to compute an estimate for λ_*. This can be done since $c(P_*, v_*) = (D_n - u)_*$ can be used to estimate the particle velocity u_*. Compute M_* and Π_*. Use the fact that H is constant to generate a separate estimate for $(D_n - u)_*$.
4. Compute a * state sonic residual based on the previous trial step, $R_1 = c_* - (D_n - u)_*$.
5. Now that a trial, sonic * state is known, integrate the flux ODEs, for M and Π
$$\frac{\partial M}{\partial \eta} = -\kappa \rho u, \frac{\partial \Pi}{\partial \eta} = -\kappa M u, \frac{\partial \lambda}{\partial \eta} = \frac{R}{(D_n - u)} \tag{7.116}$$
from the trial sonic (*) point to the shock, where $\lambda = 0$.
6. Compute two flux residuals at the shock, $R_2 = M(\lambda = 0) - \rho_0 D_n$ and $R_3 = \Pi(\lambda = 0) - \rho_0 D_n^2$
7. Iterate on values of κ, v_* and P_* until the three residuals R_1, R_2, and R_3 go to zero. The distance from the sonic point to the shock determines η_*. With the * state determined, the spatial structure of the detonation reaction zone is determined as well as the D_n, κ pair. These can then be displayed.
8. Increment D_n and go to step 2.

One can use the spatial coordinate or use the rate equation to carry out the integration in λ instead of η. If one uses the reaction progress variable as an independent variable there is an apparent singularity that can be removed by a simple coordinate transformation as well. For example, we further change to a space-like independent variable by introducing a new variable y in order to remove the singularity that appears on the RHS at complete reaction, as $\lambda \to 1$. For the reaction-rate form considered, the reaction-rate equation is

$$\frac{\mathrm{d}\lambda}{\mathrm{d}\eta} = k\frac{(1-\lambda)^m}{(D_n - u)}. \tag{7.117}$$

Introduce

$$\frac{\mathrm{d}\lambda}{\mathrm{d}y} = (1-\lambda)^m, \quad \text{with solution} \quad y = (1-(1-\lambda)^{1-m})/(1-m) \quad \text{for} \quad m < 1, \tag{7.118}$$

in which case the equation for M (and similarly for Π) can be recast as

$$\frac{\partial M}{\partial y} = -\kappa\rho u\left(\frac{D_n - u}{k}\right), \tag{7.119}$$

which has no singular behavior.

The main advantage of this method is that one always starts from the sonic point and integrates toward the shock. Since the sonic point has a singular behavior (associated with its saddle character) and the shock is a regular point, this method eliminates the problems associated with integration into a singular fixed point. The method can be made quite general.

7.4.4.2 Flux Formulation: Asymptotic Approach

Here we give a brief description of the procedures that can be used to generate asymptotic approximations and explicit formulas with information about the behavior of the explosive's response as a function of EOS and rate law parameters. As before, we illustrate the general ideas for the ideal EOS in the steady-state analysis.

Asymptotic expansions can be generated in various ways. One way is to use the method of matched asymptotic expansions (MAE) and propose an asymptotic expansion and then develop a consistent approximation based on the proposed forms. If the expansions become nonuniform in the domain of the independent variable, then one has to introduce additional expansion in the region of non-uniformity, which almost always requires a rescaling of coordinates. Then the solutions must be matched across the disparate regions to develop a uniform description. Another way is to use the method of successive approximations (MSA) whereby a first approximation is assumed (a seed) and corrections to the solution are generated by an explicit series of recursive evaluations, in which case an asymptotic expansion is generated more or less automatically. The appearance of certain terms in the generated expansions in each recursion depends on the seed; the art of the method is to pick a seed that generates the correct expansion with the fewest recursions (i.e., steps). If a valid asymptotic description for the problem exists, MAE and MSA must give the same results. When the flux formulation is used, the MSA is a natural one to develop asymptotic expansions, and is almost identical to the numerical procedures treated in the last section.

To illustrate, we use the MSA for D_n close to D_{CJ} and only treat curvature effects and again neglect the possibility of explicit time dependence. We use a 0 superscript to denote the seed (or zeroth-iterate) from (7.40), with $D_n = D_{\rm CJ}$

$$v^{(0)} = v_0 \left(\frac{\gamma - \ell}{\gamma + 1}\right), \quad u^{(0)} = D_{\rm CJ} \left(\frac{1 + \ell}{\gamma + 1}\right), \quad P^{(0)} = \rho_0 D_{\rm CJ}^2 \left(\frac{1 + \ell}{\gamma + 1}\right), \tag{7.120}$$

where $\ell = \sqrt{1 - \lambda}$. The spatial distribution of reactant progress is given by the integral

$$\eta = \int_0^{\lambda^{(0)}} \frac{D_{\rm CJ} - u^{(0)}}{R^{(0)}} \,\mathrm{d}\bar{\lambda} \tag{7.121}$$

and defines $\lambda^{(0)}(\eta)$, which in turn is used in the above appearance of ℓ in formulas (7.120). It follows that $(D_n - u)^{(0)} = D_{\rm CJ}(\gamma - \ell)/(\gamma + 1)$ and $\rho^{(0)} = \rho_0(\gamma + 1)/(\gamma - \ell)$.

The first iteration assumes a general value of D_n and uses the approximation of the (0)th iteration to approximate the corrections to the fluxes

$$M^{(1)} = \rho_0 D_n + \kappa^{(1)} I_1^{(0)}, \quad \Pi^{(1)} = \rho_0 D_n^2 + \kappa^{(1)} I_2^{(0)}, \quad H^{(1)} = \frac{1}{2} D_n^2, \tag{7.122}$$

$$I_1^{(0)} = -\int_0^n \rho^{(0)} u^{(0)} \,\mathrm{d}\eta, \quad I_2^{(0)} = -\int_0^n M^{(0)} u^{(0)} \,\mathrm{d}\eta. \tag{7.123}$$

It is convenient to use (7.121) to convert the integral over η to one over $\lambda^{(0)}(\eta)$,

$$d\eta = \frac{D_{\rm CJ}(\gamma - \ell)}{(\gamma + 1) R^{(0)}} \mathrm{d}\bar{\lambda}. \tag{7.124}$$

Using $M^{(0)} = \rho_0 D_{\rm CJ}$ and using the (0)th approximation lead to expressions for the integrals

$$I_1^{(0)} = -\rho_0 D_{\rm CJ}^2 \int_0^{\lambda^{(0)}} \frac{1 + \ell}{\gamma + 1} \frac{\mathrm{d}\bar{\lambda}}{R^{(0)}}, \quad I_2^{(0)} = -\rho_0 D_{\rm CJ}^3 \int_0^{\lambda^{(0)}} \frac{(1 + \ell)(\gamma - \ell)}{(\gamma + 1)^2} \frac{\mathrm{d}\bar{\lambda}}{R^{(0)}}. \tag{7.125}$$

At this point we have an approximation to the flux functions through the reaction zone. Next we turn to the computation of the conditions at the sonic point.

In the first iteration, the speed relation equation reduces to the quasi-steady condition

$$0 = u_*^{(1)} + c_*^{(1)} - D_{n*}^{(1)}, \tag{7.126}$$

which is equivalent to the vanishing of the discriminant, i.e., $\delta = 0$ in the formulas for v_* in (7.34), so that

$$\gamma^2 (\Pi_*^{(1)})^2 = 2(\gamma^2 - 1)(M_*^{(1)})^2 \left(H_*^{(1)} + \lambda_*^{(1)} Q\right). \tag{7.127}$$

It follows directly from the Rankine-Hugoniot algebra that

$$v_*^{(1)} = \frac{\gamma}{\gamma+1}\frac{\Pi_*^{(1)}}{(M_*^{(1)})^2}, \quad (D_n - u)_*^{(1)} = \frac{\gamma}{\gamma+1}\frac{\Pi_*^{(1)}}{M_*^{(1)}}, \quad P_*^{(1)} = \frac{1}{\gamma+1}\Pi_*^{(1)}. \tag{7.128}$$

So now given an approximation for the updated $M_*^{(1)}, \Pi_*^{(1)}, H_*^{(1)}$, and $\lambda_*^{(1)}$, we can insert them into (7.127) and generate the constraint from the sonic condition,

$$\gamma^2(\Pi_*^{(1)})^2 = 2(\gamma^2-1)(M_*^{(1)})^2\left(H_*^{(1)} + \lambda_*^{(1)}Q\right)$$

and expand to get a constraint relation on values at the sonic point. Write the updated fluxes as

$$M_*^{(1)} = \rho_0 D + M_{*1}^{(1)}, \quad \Pi_*^{(1)} = \rho_0 D^2 + \Pi_{*1}^{(1)}, \quad H_{*1}^{(1)} = \frac{c_0^2}{\gamma-1} + \frac{1}{2}D^2, \tag{7.129}$$

with

$$M_{*1}^{(1)} = \kappa^{(1)}(I_1^{(0)})_*, \quad \Pi_{*1}^{(1)} = \kappa^{(1)}(I_2^{(0)})_*, \quad H_{*1}^{(1)} = 0, \tag{7.130}$$

and represent the updated value of the reaction progress variable at the sonic point as

$$\lambda_*^{(1)} = 1 - (1-\lambda_*^{(1)}) \quad \text{or} \quad \lambda_*^{(1)} = 1 + \lambda_{*1}^{(1)}, \tag{7.131}$$

with $\lambda_{*1}^{(1)} = \lambda_*^{(1)} - 1 \leq 0$ to represent a value of $\lambda_*^{(1)}$ close to 1. If we expand (7.127) and use $D_{\mathrm{CJ}}^2 = 2(\gamma^2-1)Q$, we obtain

$$\begin{aligned}\rho_0^2\left[D^2(D^2 - D_{\mathrm{CJ}}^2)\right] &= -2\rho_0 D^2\gamma^2\Pi_{*1}^{(1)} + 2\rho_0 D M_{*1}^{(1)}\left[(\gamma^2-1)D^2 + D_{\mathrm{CJ}}^2\right] \\ &\quad + \rho_0^2 D^2 D_{\mathrm{CJ}}^2\lambda_{*1}^{(1)} + o(P_{*1}^{(1)}) + o(M_{*1}^{(1)}).\end{aligned} \tag{7.132}$$

What remains is to compute approximations to the perturbations: $M_{*1}^{(1)}, \Pi_{*1}^{(1)}, H_{*1}^{(1)}$ and $\lambda_{*1}^{(1)}$. Recall the definitions $M_{*1}^{(1)} = \kappa^{(1)}(I_1^{(0)})_*, \Pi_{*1}^{(1)} = \kappa^{(1)}(I_2^{(0)})_*$, $H_{*1}^{(1)} = 0$. Thus we need to compute values of the two integrals at the updated sonic point, $(I_1^{(0)})_*$ and $(I_2^{(0)})_*$, which are defined as

$$\begin{aligned}(I_1^{(0)})_* &= -\rho_0 D^2\int_0^{\lambda_*^{(1)}} \frac{1+\ell}{\gamma+1}\frac{\mathrm{d}\bar{\lambda}}{R^{(0)}}, \\ (I_2^{(0)})_* &= -\rho_0 D^3\int_0^{\lambda_*^{(1)}} \frac{(1+\ell)(\gamma-\ell)}{(\gamma+1)^2}\frac{\mathrm{d}\bar{\lambda}}{R^{(0)}}.\end{aligned} \tag{7.133}$$

This requires an estimate for $\lambda_*^{(1)}$ which can be obtained from the thermicity condition, which has not yet been invoked.

We will consider a simple depletion rate form that represents cases that are important examples of interest, such as the pressure-dependent rate, or an Arrhenius-like rate. Let

$$R = k(1-\lambda)^m, \tag{7.134}$$

then k can be a function of P, ρ or even shock state-dependent as a function of D_n. The updated "thermicity condition"

$$\kappa^{(1)} (c_*^{2\,(1)})\, u_*^{(1)} = (\gamma - 1)\, Q\, R_*^{(1)}, \tag{7.135}$$

with $\lambda_*^{(1)} = 1 + \lambda_{*1}^{(1)}$, leads to the approximation

$$\lambda_{*1}^{(1)} = -(\kappa^{(1)})^{1/m} (z_*)^{1/m}, \quad \text{where} \quad z_* = \left[\frac{2\gamma^2}{(\gamma+1)^2} \frac{D_{\mathrm{CJ}}}{k_*^{(1)}} \right]. \tag{7.136}$$

The integrals are then computed by splitting the domain over λ from 0 to 1 and then from 1 to $1 + \lambda_{*1}^{(1)}$, which can be done provided the first integral is finite. That is the case for this special form of the rate for $0 \leq m < 1$. Thus for $(I_1^{(0)})_*$, we write

$$(I_1^{(0)})_* = -\rho_0 D_{\mathrm{CJ}}^2 \mathscr{S}_1 - \rho_0 D_{\mathrm{CJ}}^2 \int_1^{1-(\kappa^{(1)} z_*)^{1/m}} \left(\frac{1+\ell}{\gamma+1} \right) \frac{\mathrm{d}\bar{\lambda}}{R^{(0)}}, \tag{7.137}$$

where we have defined the integrals

$$\mathscr{S}_1 = \int_0^1 \left(\frac{\gamma+\ell}{\gamma+1} \right) \frac{\mathrm{d}\bar{\lambda}}{R^{(0)}}, \quad \mathscr{S}_2 = \int_0^1 \left(\frac{(\gamma+\ell)(1+\ell)}{(\gamma+1)^2} \right) \frac{\mathrm{d}\bar{\lambda}}{R^{(0)}}. \tag{7.138}$$

We use the leading order approximations at the sonic point to write

$$\begin{aligned} \rho_0 D_{\mathrm{CJ}}^2 \int_1^{1-(\kappa^{(1)} z_*)^{1/m}} \frac{1+\ell}{\gamma+1} \frac{\mathrm{d}\bar{\lambda}}{R^{(0)}} &= \frac{\rho_0 D_{\mathrm{CJ}}^2}{(\gamma+1) k_*^{(0)}} \int_1^{1-(\kappa^{(1)} z_*)^{1/m}} \frac{\mathrm{d}\bar{\lambda}}{(1-\lambda)^m} \\ &= -\frac{\rho_0 D_{\mathrm{CJ}}^2}{(\gamma+1) k_*^{(0)}} \frac{1}{1-m} (\kappa^{(1)} z_*)^{(1-m)/m} \end{aligned} \tag{7.139}$$

so that

$$(I_1^{(0)})_* = -\rho_0 D_{\mathrm{CJ}}^2 \int_0^1 \left(\frac{\gamma+\ell}{\gamma+1} \right) \frac{\mathrm{d}\bar{\lambda}}{r^{(0)}} + \frac{\rho_0 D_{\mathrm{CJ}}^2}{(\gamma+1) k_*^{(0)}} \frac{1}{1-m} (\kappa^{(1)} z_*)^{(1-m)/m}.$$

Likewise

$$(I_2^{(0)})_* = -\rho_0 D_{\mathrm{CJ}}^3 \mathscr{S}_2 + \rho_0 D_{\mathrm{CJ}}^3 \left[\frac{\gamma + c_0^2/D_{\mathrm{CJ}}^2}{(\gamma+1)^2 k_*^{(0)}} \right] \frac{1}{1-m} (\kappa^{(1)} z_*)^{(1-m)/m}. \tag{7.140}$$

Then the updated fluxes and the reaction progress variable at the sonic point are given by

$$M_{*1}^{(1)} = -\kappa^{(1)} \rho_0 D_{\mathrm{CJ}}^2 \mathscr{S}_1 + \kappa^{(1)} \frac{\rho_0 D_{\mathrm{CJ}}^2}{(\gamma+1) k_*^{(0)}} \frac{1}{1-m} (\kappa^{(1)} z_*)^{(1-m)/m} + \cdots$$

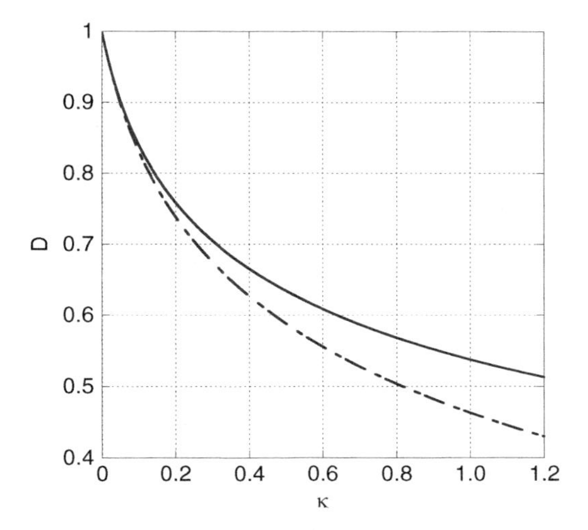

Fig. 7.13. The D-kappa relation generated by the numerical approach described in the previous section and by the asymptotic formula (7.142). The numerical flux approach solution (the *upper curve*) lies above the asymptotic flux approach solution displayed as *chain-dashed*. The explosive modeling parameters are the same ones used in previous sections, with $\gamma = 3$, $\rho_0 = 2$, $Q = 1/(2(\gamma^2 - 1)) = 0.0625$, $k = 1$, and $m = 1/2$

$$\Pi_{*1}^{(1)} = -\kappa^{(1)}\rho_0 D_{\mathrm{CJ}}^3 \mathscr{S}_2 + \kappa^{(1)} \frac{\gamma\rho_0 D^3}{(\gamma+1)^2 k_*^{(0)}} \frac{1}{1-m} (\kappa^{(1)} z_*)^{(1-m)/m} + \cdots$$

$$H_{*1}^{(1)} = O(\kappa^2), \quad \lambda_{*1}^{(1)} = -(\kappa^{(1)})^{1/m} (z_*)^{1/m}, \tag{7.141}$$

and after some algebra obtains a D-kappa relation:

$$D^2 - D_{\mathrm{CJ}}^2 = 2\gamma^2 D_{\mathrm{CJ}}^3 (\mathscr{S}_2 - \mathscr{S}_1)\kappa^{(1)} + \frac{m}{(1-m)} D_{\mathrm{CJ}}^2 (\kappa^{(1)} z_*)^{1/m} + \cdots \tag{7.142}$$

Figure 7.13 shows a comparison of the D-kappa relation generated by the numerical approach described in the previous section, and that from the asymptotic formula.

In the following sections, we develop the understanding of the detonation/inert material boundary interaction problem, which we need to formulate the boundary conditions required to propagate detonation as described by DSD.

7.5 Interaction of Detonation with Explosive/Inert Material Boundaries

The oblique interaction of a detonation wave with an explosive/inert material boundary can accelerate the inert material to a high velocity. In the process, the detonation reaction zone is altered as reflected waves generated by the interaction enter the reaction zone. The degree to which the detonation reaction zone is affected depends on the angle of attack between the normal to the incident detonation shock and the normal to the unshocked explosive/inert interface. A leading-order description of this interaction can be developed by constructing shock-polar diagrams for the explosive and inert material and then using them to match the pressure and streamline deflection angle (or particle velocity normal to the interface) at their common interface [27,30,36].

Shock-polar analysis is based on steady two-dimensional simple wave solutions that are admitted in the neighborhood of a common interaction point along a shared material interface of the explosive and inert confinement. The analysis is local and hence scale-free at the point where the incident detonation shock intersects the explosive/inert interface. It uses the appropriate simple wave solutions to construct a local solution to the flow which connects the detonation shock to the induced shock in the inert material and the effect the resulting interaction has on the flow state in the explosive. Along their common interface the pressure and streamline angle must match. These two-dimensional, steady wave solutions include the incident shock, a reflected shock, a Prandtl-Meyer (PM) fan, a constant state, and a possible Mach-reflection shock.

Although such a local analysis does not bring in all the details of the reaction-zone flow and the post-shock inert material flow, it does provide an excellent estimate away from the edge of the global, far-field structure of the interaction. It thus provides information for angle boundary conditions that can be used in conjunction with the DSD propagation law model. The pressure and the sound speed, and hence the acoustic signaling speed, are highest at or nearest the detonation shock and then are much lower toward the end of the reaction zone. Thus, the fastest transverse propagating waves in the reaction zone are those propagating along the shock. A local shock-polar analysis, that only requires knowledge of the unreacted explosive EOS, defines which of such waves are admissible and the type of interaction, depending on the nature of the confinement.

When the high-pressure detonation impacts a relatively dense, high shock-impedance inert material, a reflected shock is returned into the explosive. When detonation impacts a relatively low-density, low shock-impedance inert material, then a reflected rarefaction, affixed to the incident-detonation shock, is reflected into the explosive. Figure 7.14 depicts the types of interactions that can be expected for the case when the angle between the normal to the incident detonation shock and the normal to the undeflected interface, ω, is small.

For ω small, the disturbances generated by the interface interaction do not influence the incident shock. This is a consequence of the supersonic flow downstream of the incident shock (when ω is small) as measured from the perspective of an observer riding with the incident shock/interface intersection point. As ω is increased, a value is reached where disturbances can propagate

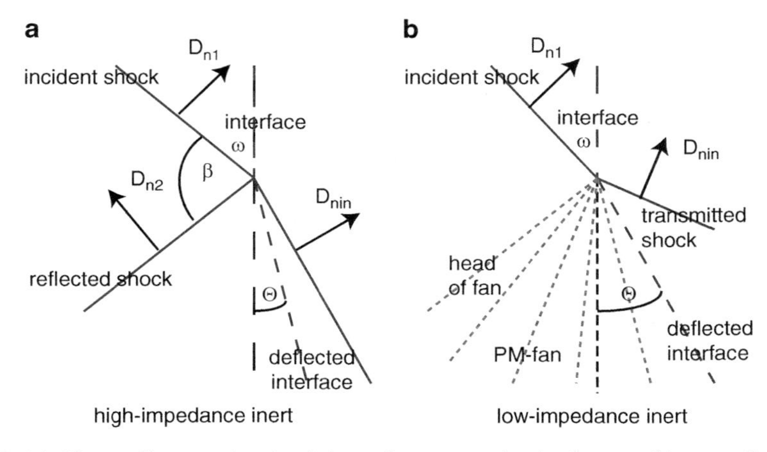

Fig. 7.14. Depending on the shock impedance, two basic classes of interactions can be expected to occur when a detonation drives into an adjacent inert material: a reflected shock for a high-impedance inert material and a reflected rarefaction (PM-fan) for a low-impedance inert material. Depicted are the cases for small ω. The interaction causes the interface to be deflected by the angle, Θ

out from the interface and thereby modify the shape of the incident shock. Whether disturbances propagate away from the interface is a function of the angle, ω, and the explosive's unreacted EOS. As the shock impedance of the adjacent inert material is reduced, one obtains only the flow type depicted in Fig. 7.14, inset (b). In this circumstance, the incident shock remains undisturbed until, in the incident shock/interface intersection-point frame, the flow transitions to subsonic. Once the incident shock angle, ω, becomes greater than the value corresponding to the sonic transition angle, ω_s, disturbances from the interface (for all inert material impedance classes) will be able to influence the detonation shock and cause a change to the incident shock which is felt globally.

In the following sections, we first detail the shock-polar calculations. We then compare those results with the results for detonating explosive/inert material interaction calculations performed using DNS. Finally, we present the DSD boundary condition description and how the boundary condition parameters can be developed from a shock-polar analysis.

7.6 Detonation Shock-Polar Analysis

The detonation shock-polar analysis is performed in the reference frame of the moving incident shock/material interface intersection point. The class of simple waves that are admitted into the analysis are shock and rarefaction waves. Given an incident ZND detonation shock, making an angle ω with respect to the interface, we look for the collection of simple wave solutions that can change pressure and the streamline angle in the flow downstream of the lead ZND detonation shock. We make a number of assumptions: (1) the state upstream of the incident detonation is quiescent and at zero pressure, (2) the incident detonation is propagating at the Chapman–Jouguet (CJ) speed, (3) the ZND model of detonation describes the detonation (the detonation is running in fully fresh explosive), and (4) we perform only a local analysis, centered about the incident detonation shock/interface intersection point.

7.6.1 Reflected Shock Simple Wave Solution

The reflected shock solution to the detonation/inert material interaction problem is appropriate when the detonation interacts with a relatively high shock-impedance inert material. Figure 7.15 shows the configuration and definition labels when an incident detonation generates a transmitted shock in the inert and a simple reflected shock in the explosive. The oblique shock conditions are used to describe both the lead incident and reflected shocks in the explosive and the transmitted shock wave into the inert, and are given by

$$[\rho\,(\mathbf{n}\cdot\mathbf{u}-D_n)]=0, \tag{7.143}$$

$$\left[P+\rho\,(\mathbf{n}\cdot\mathbf{u}-D_n)^2\right]=0, \tag{7.144}$$

$$[\mathbf{t}\cdot\mathbf{u}]=0, \tag{7.145}$$

$$\left[E+\frac{P}{\rho}+\frac{1}{2}\,(\mathbf{n}\cdot\mathbf{u}-D_n)^2\right]=0, \tag{7.146}$$

where $\mathbf{n}$ is the outward pointing normal to the shock, $\mathbf{t}$ is the tangent to the shock (right hand rule convention), and $\mathbf{u}$ is the laboratory-based particle velocity, respectively.

Given a pre-incident shock quiescent state and using the strong-shock approximation, the oblique shock conditions become

$$\frac{\rho_0}{\rho_1}=\frac{D_{n1}-\mathbf{n}_1\cdot\mathbf{u}_1}{D_{n1}}, \tag{7.147}$$

$$P_1=\rho_0 D_{n1}\mathbf{n}_1\cdot\mathbf{u}_1, \tag{7.148}$$

$$\mathbf{t}_1\cdot\mathbf{u}_1=0, \tag{7.149}$$

$$E_0-E_1+\frac{1}{2}P_1\left(\frac{1}{\rho_0}-\frac{1}{\rho_1}\right)=0. \tag{7.150}$$

For the configuration in Fig. 7.15, we have

$$D_{n1}=D_0\sin(\omega), \tag{7.151}$$

$$\mathbf{n}_1=\mathbf{i}\cos(\omega)+\mathbf{j}\sin(\omega), \tag{7.152}$$

$$\mathbf{t}_1=-\mathbf{i}\sin(\omega)+\mathbf{j}\cos(\omega). \tag{7.153}$$

The incident shock state, $(\rho_1, \mathbf{u}_1, P_1)$, can be written as a function of D_{n1}. Since the collision of the incident shock with the interface causes the interface to deflect, a secondary wave (either a shock or a rarefaction) is generally needed in the explosive to maintain the streamlines along the interface in contact.

For sufficiently high shock-impedance inert materials, a secondary shock is required to turn the streamlines in the explosive. We assume that the angle between the incident and reflected shocks is β (see Fig. 7.15). The shock conditions (7.143)–(7.146) across the reflected shock yield

$$\frac{\rho_1}{\rho_2} = \frac{(\mathbf{n}_2 \cdot \mathbf{u}_2 - D_{n2})}{(\mathbf{n}_1 \cdot \mathbf{n}_2(\mathbf{n}_1 \cdot \mathbf{u}_1) - D_{n2})}, \tag{7.154}$$

$$P_2 = P_1 + \rho_1 \left(1 - \frac{\rho_1}{\rho_2}\right) (\mathbf{n}_1 \cdot \mathbf{n}_2(\mathbf{n}_1 \cdot \mathbf{u}_1) - D_{n2})^2, \tag{7.155}$$

$$\mathbf{t}_2 \cdot \mathbf{u}_2 = \mathbf{t}_2 \cdot \mathbf{n}_1(\mathbf{n}_1 \cdot \mathbf{u}_1), \tag{7.156}$$

$$E_1 - E_2 + \frac{1}{2}(P_1 + P_2)\left(\frac{1}{\rho_1} - \frac{1}{\rho_2}\right) = 0, \tag{7.157}$$

where

$$D_{n2} = D_0 \sin(\omega + \beta), \tag{7.158}$$

$$\mathbf{n}_2 = \mathbf{i}\cos(\omega + \beta) + \mathbf{j}\sin(\omega + \beta), \tag{7.159}$$

$$\mathbf{t}_2 = -\mathbf{i}\sin(\omega + \beta) + \mathbf{j}\cos(\omega + \beta). \tag{7.160}$$

From the conditions for the incident and reflected shocks, we develop the relation between the pressure, P_2, and the streamline turning angle, Θ_2.

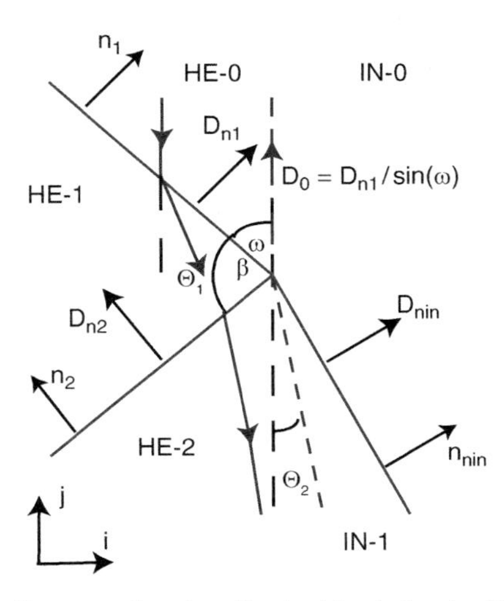

Fig. 7.15. Schematic diagram showing the incident shock, D_{n1}, the reflected explosive shock, D_{n2}, and the streamline deflection at the explosive/inert material interface

Due to the nature of the point analysis (the reaction rate can be considered to be frozen), we need to specify only the EOS of the reactants (unreacted explosive). For demonstration, we use the same, simple polytropic form for our EOS description that we have been using

$$E = \frac{1}{\gamma - 1}\frac{P}{\rho}. \tag{7.161}$$

From the Hugoniot equation (7.157), it immediately follows that

$$1 - \frac{\rho_1}{\rho_2} = \frac{2}{\gamma - 1}\left(P_2 - P_1\right) / \left[\left(\frac{\gamma + 1}{\gamma - 1}\right) P_2 + P_1\right]. \tag{7.162}$$

Substituting (7.162) into (7.155) returns an equation for P_2 as a function of the specified phase velocity and inter-shock angle, β,

$$P_2 = P_1 + \frac{2\rho_1}{\gamma + 1}\left[\left(\mathbf{n}_1 \cdot \mathbf{n}_2(\mathbf{n}_1 \cdot \mathbf{u}_1) - D_{n2}\right)^2 - \gamma\frac{P_1}{\rho_1}\right]. \tag{7.163}$$

Substituting (7.163) back into (7.162) and then the result into (7.154) yields an expression for the particle velocity in the post-reflected shock state

$$\mathbf{n}_2 \cdot \mathbf{u}_2 - D_{n2} = \frac{2\gamma}{\gamma + 1}\frac{P_1}{\rho_1}\frac{1}{\left(\mathbf{n}_1 \cdot \mathbf{n}_2(\mathbf{n}_1 \cdot \mathbf{u}_1) - D_{n2}\right)} \tag{7.164}$$

$$+ \left(\frac{\gamma - 1}{\gamma + 1}\right)\left(\mathbf{n}_1 \cdot \mathbf{n}_2(\mathbf{n}_1 \cdot \mathbf{u}_1) - D_{n2}\right), \tag{7.165}$$

and where

$$\mathbf{t}_2 \cdot \mathbf{u}_2 = \mathbf{n}_1 \cdot \mathbf{t}_2(\mathbf{n}_1 \cdot \mathbf{u}_1). \tag{7.166}$$

The tangent of the streamline turning angle behind the reflected shock is then given as

$$\tan(\Theta_2) = \frac{\cos(\omega + \beta)(\mathbf{n}_2 \cdot \mathbf{u}_2) - \sin(\omega + \beta)(\mathbf{t}_2 \cdot \mathbf{u}_2)}{D_0 - \sin(\omega + \beta)(\mathbf{n}_2 \cdot \mathbf{u}_2) - \cos(\omega + \beta)(\mathbf{t}_2 \cdot \mathbf{u}_2)}. \tag{7.167}$$

Using the results for the incident shock state, which follow immediately from (7.147)–(7.150),

$$\left(\mathbf{n}_1 \cdot \mathbf{n}_2(\mathbf{n}_1 \cdot \mathbf{u}_1) - D_{n2}\right) = \left(\frac{2}{\gamma + 1}\cos(\beta) - \frac{\sin(\omega + \beta)}{\sin(\omega)}\right) D_{\mathrm{CJ}}, \tag{7.168}$$

$$\mathbf{n}_1 \cdot \mathbf{u}_1 = \frac{2D_{\mathrm{CJ}}}{\gamma + 1}, \tag{7.169}$$

$$\gamma\frac{P_1}{\rho_1} = \frac{2\gamma(\gamma - 1)}{\gamma + 1}D_{\mathrm{CJ}}^2, \tag{7.170}$$

yields explicit expressions for P_2 and Θ_2 in terms of ω, β, γ, and the CJ detonation velocity, D_{CJ}.

In mapping out the shock-polar function, P_2 versus Θ_2, β is spanned over the range

$$\beta_{\min} \leq \beta \leq \beta_{\max}, \tag{7.171}$$

with

$$\beta_{\min} = \arctan(\mathrm{B}_+), \tag{7.172}$$

$$\beta_{\max} = \arctan(\mathrm{B}_-) + \frac{\pi}{2}\left(1 + \mathrm{sign}\left(\cot^2(\omega) - \frac{2\gamma(\gamma-1)}{(\gamma+1)^2}\right)\right), \tag{7.173}$$

$$\mathrm{B}_\pm = \frac{-\frac{\gamma-1}{\gamma+1}\cot(\omega) \pm \sqrt{\frac{2\gamma(\gamma-1)}{(\gamma+1)^2}\left(\cot^2(\omega) - \frac{\gamma-1}{\gamma+1}\right)}}{\left(\cot^2(\omega) - \frac{2\gamma(\gamma-1)}{(\gamma+1)^2}\right)}. \tag{7.174}$$

7.6.2 Reflected Rarefaction Simple Wave Solution

When the interaction of the detonation is with a low-impedance inert material, a rarefaction is reflected into the explosive behind the lead detonation shock. The flow analysis is best performed using the polar coordinate frame shown in Fig. 7.16, specialized to a steady-state flow, in the reference frame of the incident shock/material interface intersection point. The Euler equations given earlier (7.1) provide a description of the isentropic rarefaction wave interaction. For the local analysis being considered here and where (1) the incident shock is straight, (2) the reaction rate can be considered to be frozen, and

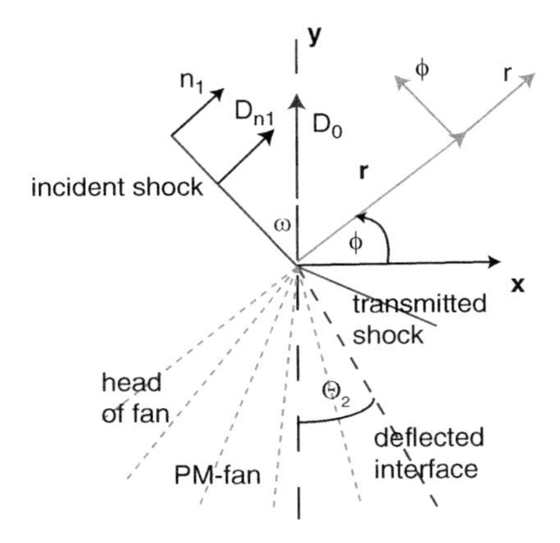

Fig. 7.16. A schematic diagram showing the incident detonation shock and the shock/interface intersection coordinate frame we use, expressed in both local Cartesian and local polar coordinates. D_0 is the phase velocity of the coordinate system origin in the laboratory reference frame. The polar unit vectors are $\mathbf{r} = \mathbf{i}\cos(\phi) + \mathbf{j}\sin(\phi)$ and $\boldsymbol{\phi} = -\mathbf{i}\sin(\phi) + \mathbf{j}\cos(\phi)$

(3) time dependence is suppressed, the two-dimensional steady-state Euler equations are

$$c^2 \mathbf{\nabla} \cdot \mathbf{U} - \mathbf{U} \cdot \mathbf{\nabla} \left(\frac{1}{2} \mathbf{U} \cdot \mathbf{U} \right) = 0, \tag{7.175}$$

$$\mathbf{\nabla} \times \mathbf{U} = 0, \tag{7.176}$$

$$E + \frac{P}{\rho} + \frac{1}{2} \mathbf{U} \cdot \mathbf{U} = \frac{1}{2} D_0^2. \tag{7.177}$$

In the above, c^2 is the frozen sound speed, $\mathbf{U}$ is the particle velocity in the moving frame, $\mathbf{U} = \mathbf{u} - \mathbf{j} D_0$, D_0 is the phase velocity of the detonation along the undeflected explosive/inert material interface, and (7.177) is the Bernoulli law. Expressing (7.175) and (7.176) in polar coordinates and enforcing the simple wave nature of the solution (the flow is independent of the radial coordinate, r), we have

$$(c^2 - U_\phi^2) \left(\frac{\partial U_\phi}{\partial \phi} + U_r \right) = 0, \tag{7.178}$$

$$\frac{\partial U_r}{\partial \phi} - U_\phi = 0, \tag{7.179}$$

where U_ϕ and U_r are the components of particle velocity in the moving frame and expressed in polar coordinates.

Equation (7.178) can be satisfied in two ways: either with (1) $c^2 - U_\phi^2 = 0$ or (2) $\partial U_\phi / \partial \phi + U_r = 0$. It is then readily shown that the solution to case (2) corresponds to a constant state, since

$$\frac{\partial^2 U_\phi}{\partial \phi^2} + U_\phi = 0, \tag{7.180}$$

from which it follows that

$$U_\phi = A \cos(\phi + K_2), \tag{7.181}$$

$$U_r = A \sin(\phi + K_2), \tag{7.182}$$

and therefore, $U_x = A \sin(K_2)$ and $U_y = A \cos(K_2)$. The case (1) condition, $(c^2 - U_\phi^2) = 0$, leads to the PM-fan rarefaction solution. As in the previous section, if we assume a polytropic EOS form (7.161), the Bernoulli law (7.177), then yields

$$c^2 = \left(\frac{\gamma - 1}{2} \right) (D_0^2 - U_r^2 - U_\phi^2), \tag{7.183}$$

and therefore

$$U_\phi^2 = \left(\frac{\gamma - 1}{\gamma + 1} \right) (D_0^2 - U_r^2), \tag{7.184}$$

and thus

$$\frac{\partial U_r}{\partial \phi} = \pm \sqrt{\frac{\gamma - 1}{\gamma + 1}} \sqrt{D_0^2 - U_r^2}, \tag{7.185}$$

which yields the PM-fan solution

$$U_r = -D_0 \cos\left(\pm\sqrt{\frac{\gamma-1}{\gamma+1}}\phi + K_1\right), \tag{7.186}$$

$$U_\phi = \pm D_0\sqrt{\frac{\gamma-1}{\gamma+1}} \sin\left(\pm\sqrt{\frac{\gamma-1}{\gamma+1}}\phi + K_1\right), \tag{7.187}$$

and where the streamline deflection angle in the PM-fan is given by

$$\tan(\Theta) = \frac{\sin(\phi)U_\phi - \cos(\phi)U_r}{\cos(\phi)U_\phi + \sin(\phi)U_r}. \tag{7.188}$$

Since the flow in the PM-fan is isentropic, it also follows that

$$\frac{P_{\rm PM}}{P_1} = \left(\frac{U_\phi^2}{c_1^2}\right)^{\frac{\gamma}{\gamma-1}}. \tag{7.189}$$

Given that the sonic parameter in the PM-fan is

$$c^2 - U_\phi^2 - U_r^2 = -U_r^2 \leq 0, \tag{7.190}$$

it follows that the flow is supersonic in the PM-fan, with $U_r = 0$ at the beginning (head) of the fan for the case of a sonic state behind the incident detonation shock. The magnitude of U_r must increase and that of U_ϕ must decrease as the flow expands through the fan. If we consider that the sin( ) and cos( ) functions in (7.186) and (7.187) assume their principal values, and that $|U_r|$ increases and $|U_\phi|$ decreases through the fan, then we are led to take the $(-)$ square root, $-\sqrt{(\gamma-1)/(\gamma+1)}\phi + K_1$, in the argument for the trigonometric functions. Further, matching the sound speed and the streamline deflection angle at the interface between the incident shock state and the beginning (head) of the PM-fan yields the matching conditions

$$\tan(\Theta_M) = \frac{\left(\sin(\omega)\frac{2D_{\rm CJ}}{\gamma+1} - D_0\right)U_r - \cos(\omega)\frac{2D_{\rm CJ}}{\gamma+1}U_\phi}{\cos(\omega)\frac{2D_{\rm CJ}}{\gamma+1}U_r + \left(\sin(\omega)\frac{2D_{\rm CJ}}{\gamma+1} - D_0\right)U_\phi}, \tag{7.191}$$

with

$$U_\phi = -D_0\sqrt{\frac{\gamma-1}{\gamma+1}} \sin\left(-\sqrt{\frac{\gamma-1}{\gamma+1}}\phi_M + K_1\right) = -c_1, \tag{7.192}$$

$$U_r = -D_0 \cos\left(-\sqrt{\frac{\gamma-1}{\gamma+1}}\phi_M + K_1\right) = D_0\sqrt{1 - \frac{(\gamma+1)c_1^2}{(\gamma-1)D_0^2}}, \tag{7.193}$$

and

$$D_0 = D_{\rm CJ}/\sin(\omega). \tag{7.194}$$

In solving for ϕ_M and K_1, we take

$$\phi_M = \pi + \arctan\left(\tan(\phi_M)\right), \tag{7.195}$$

and

$$K_1 = \sqrt{\frac{\gamma-1}{\gamma+1}}\phi_M + \arcsin\left(\sqrt{\frac{\gamma+1}{\gamma-1}}\frac{c_1}{D_0}\right). \tag{7.196}$$

In mapping out the PM-fan solution, we span ϕ over the range

$$\phi_m \le \phi \le K_1 / \sqrt{\frac{\gamma-1}{\gamma+1}}. \tag{7.197}$$

7.7 Shock Polars for the ZND Model of Detonating Explosive

In this section we consider a series of cases where the angle of incidence of the explosive shock relative to the inert boundary varies from a parallel oblique angle down to an angle that corresponds to a sonic incidence angle. To illustrate the basic methods of analysis required to either specify or constrain the DSD angle boundary condition, we use specific choices for both the explosive and confinement in order to illustrate the nature of how the flow at the confinement edge changes. We use a polytropic explosive reactant EOS modeled by (7.161) and (7.198) with $\gamma = 4.75$, and choices of copper and Lexan for inerts that represent typical heavy and light (or high- and low-impedance) confining materials. The detonation shock incidence angle ω, goes from a head-on incidence angle, $\omega = 0$ (the detonation shock is parallel to the interface), to a sonic incidence angle ($\omega = \omega_s = 51.07°$).

We identify interaction regimes and sub-class by specific values of the angles based on our material choices used for this illustration. However, the limiting values of the regimes change with other choices that depend on the application. There are three basic interaction regimes that need to be considered.

7.7.1 Regime I: Supersonic Interaction

When $0 \le \omega < 24.25°$, Regime I, the interaction of the detonation's shock produces a postincident shock supersonic flow state, which contains either a reflected shock or PM-fan. These supersonic interactions leave the incident detonation shock undisturbed and produce only a perturbation to the incident detonation reaction-zone flow which is very localized near the interface (the perturbation to the reaction zone extends only a fraction of a reaction-zone length laterally from the interface).

7.7.2 Regime II: Interaction of the Detonation's Reaction Zone with the Adjacent Inert Material

When $24.25^\circ \leq \omega < 51.07^\circ$, Regime II, the interaction of the detonation's reaction zone with the adjacent inert material layer, can be divided into three classes.

1. *Regime II Class-1* comprises those low shock-impedance inert materials (such as Lexan), where the interaction with the detonation is via a reflected PM-fan. With this group of inert materials, the intersection with the detonation produces a supersonic interaction, similar to what we have described for Regime I.
2. *Regime II Class-2* comprises those moderate shock-impedance inert materials (such as high-density organics), for which the interactions consist of a reflected shock, with a postreflected shock supersonic flow. That flow type is shown in Fig. 7.22 inset (a).
3. *Regime II Class-3* includes those higher shock-impedance inert materials (such as copper). Depending on the angle of the incident detonation shock, ω, there are two possible scenarios (see Fig. 7.22 inset (a) and (b)): (1) a reflected shock in the explosive, with a supersonic following flow, for smaller ω's and less-high shock impedance inert materials, and (2) a subsonic Mach-stem involving flow with inert materials whose shock impedance is higher. For Class-3, higher shock impedance inert materials such as copper, a regular reflected shock in the explosive (supersonic flow) can be expected for attack angles such as $\omega = 30^\circ$, and a subsonic Mach-reflection flow for larger attack angles such as $\omega = 40^\circ$. As the shock impedance increases, such as would be the case in going from copper to tantalum, it can be expected that Mach-reflection will extend down to $\omega = 30^\circ$. With further increases of the shock impedance, one approaches the rigid wall condition, $\Theta_2 = 0$. Then Mach-style reflection is found over the entire range, $24.25^\circ \leq \omega < 51.07^\circ$.

7.7.3 Regime III: Subsonic Transition to Sonic Flow in the HE Reaction Zone at the Edge of Confinement

Regime III has relatively high angles of detonation shock incidences, but the flow in the detonation shock reaction zone is subsonic with a transition to sonic at the confinement's edge. Specifically for the cases illustrated here, when $51.07^\circ \leq \omega \leq 90^\circ$, all the material classes we have considered yield a sonic/subsonic flow in the reaction zone, for which the angle of the incident detonation cannot be prescribed. For such flows, the lead shock is similar to the subsonic flow behind the stem of a Mach-stem-involved reflection. Then the detonation shock is parameterized by the shock angle, ω, with the value of ω at the detonating explosive/inert material interface given by the solution of

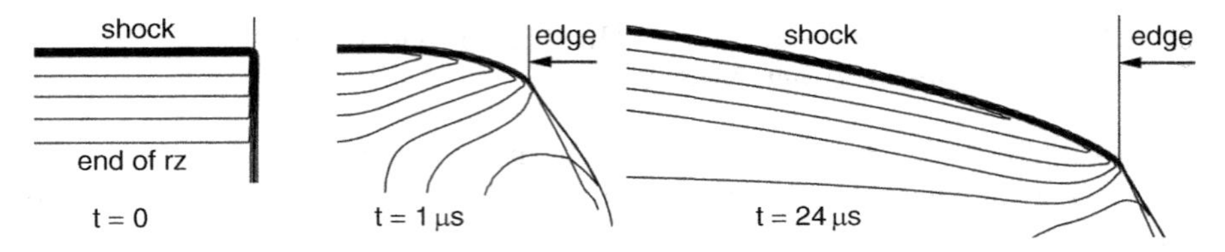

Fig. 7.17. A snapshot from a direct numerical simulation (DNS) of the two-dimensional detonation that develops when a flat detonation front, with a shock incidence angle of $\omega = 90°$, interacts with an adjacent, low-impedance material layer. The explosive model is that described above, and the detonation is initially moving upwards. Displayed are snapshots of the detonation reaction-zone pressure isobars at three times. A PM-fan can be seen to develop at the interface, where the shock angle has evolved to $\omega = \omega_s$

the Mach-stem polar interaction we will describe. The flow is perturbed by the interface interaction over the entire lateral extent of the reaction zone. Once a quasi-steady detonation develops, the local detonation speed is parameterized by ω. Figure 7.17 shows the evolution of a curved detonation reaction zone for Lexan confinement.

7.7.4 Regular Shock Reflection

Regular shock reflection uses the results presented in Sect. 7.6.1. Given the functions we have developed for the pressure and streamline turning states behind the incident detonation, here we map out the P_2 versus Θ_2 function for a detonating explosive, over a range of incident detonation-shock angles, ω, for a representative condensed-phase explosive. The fresh explosive is described by

$$\rho_0 = 2\,\mathrm{gm/cc}, \quad \gamma = 4.75, \quad D_{\mathrm{CJ}} = 8\,\mathrm{mm/\mu sec}. \tag{7.198}$$

Figure 7.18 displays the pressure and streamline turning angle behind the reflected shock and the PM-fan for four values of the incident detonation angle, ω. For ω in the range, $5° \leq \omega < 24.25°$, the incident shock remains undisturbed at its prescribed input value, $P = 44.5\,\mathrm{GPa}$, and the state behind the incident shock is supersonic. The reflected shock states are parameterized by β, while the rarefaction, PM-fan states are a function of ϕ. The states with larger values of $\beta \sim 120°$ correspond to the left side of the reflected shock polars (medium-dashed curve), while $\beta \sim 30°$ correspond to the right side of the reflected shock polar (chain-dashed curve). The closed, reverse teardrop shaped loops represent the shock polar for the reflected shock, P_2 versus Θ_2. The flow behind the reflected shock is subsonic on the upper portion for pressures above the vertical tangency, and supersonic on the lower portion. The curve emanating from the right side of the reverse teardrop and moving down and to the right is the PM-fan and is shown as a short-dashed curve. The pressure state at the bottom of the reverse teardrop corresponds to the pressure

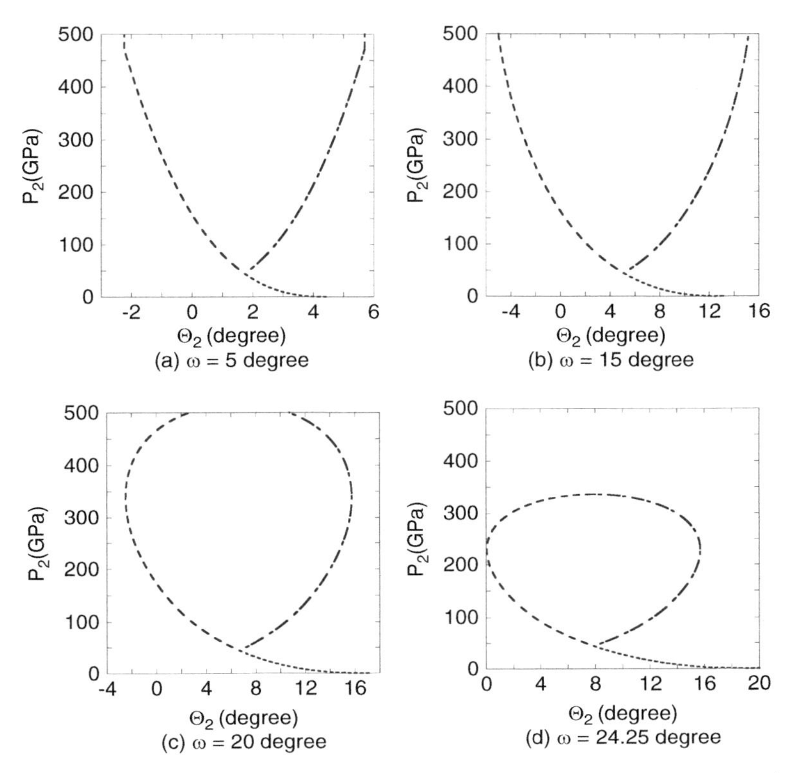

Fig. 7.18. The shock polars for our model, resolved reaction-zone detonation (7.161) and (7.198). The ZND state for our CJ detonation has $P_{\text{ZND}} = 44.5\,\text{GPa}$. The incident shock is not disturbed by the interaction. Pressures above 500 GPa are not displayed

behind the incident shock. One finds that as ω is increased, the pressure at a rigid wall (the place where $\Theta_2 = 0$) increases until it reaches a maximum for $\omega \approx 24.25°$.

Matching the explosive flow to that in the adjacent inert material shows that higher-density inert materials typically have a steeper shock polar (restricted to smaller angles, Θ). Lower-density materials have a shallower shock polar, which encompasses a larger range of angles, Θ. Figure 7.19 shows shock polars for copper and Lexan, in addition to the explosive polars, for the case $\omega = 20°$. A Mie–Grüneisen EOS is used for both copper and Lexan, where

$$E(P, \rho) = E_r(\rho) + (P - P_r(\rho))/(\rho\Gamma). \tag{7.199}$$

The functions $E_r(\rho)$ and $P_r(\rho)$ are the reference states based on the principal shock Hugoniot, where

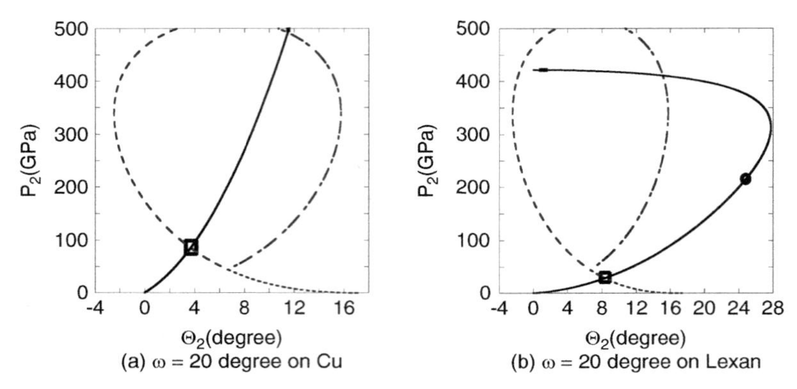

Fig. 7.19. The shock polars for our model explosive (7.161) and (7.198), with $\omega = 20°$. The solution points for the case of copper and Lexan inert materials are shown in inset (**a**) and (**b**), respectively. The higher density and sound speed in the copper lead to a reflected shock state, while the lower density and sound speed in the Lexan produce a PM-fan solution state. The solution state for both copper and Lexan are supersonic. The *black circle* on the Lexan polar denotes the point of sonic flow, with the region below the circle being supersonic. Additionally, there can be a higher pressure and larger streamline deflection solution, for both copper and Lexan confinement. These correspond to subsonic explosive reflected shock states. It takes a special set of circumstances (a very highly convergent flow) to lead to these higher-pressure solutions

$$D_{n\text{in}} = C_{0\text{in}} + S_{\text{in}}(\mathbf{n}_{\text{in}} \cdot \mathbf{u}_{\text{in}}), \tag{7.200}$$

and Γ is the Grüneisen gamma. The incident-shock relations (7.147)–(7.149), and (7.200) for the Hugoniot lead to the following expressions for an inert material shock polar:

$$P_{\text{in}} = \rho_{0\text{in}} D_{n\text{in}} \left(D_{n\text{in}} - C_{0\text{in}}\right)/S_{\text{in}}, \tag{7.201}$$

$$\tan(\Theta_{\text{in}}) = \frac{\cos(\omega_{\text{in}})\left(D_{n\text{in}} - C_{0\text{in}}\right)/S_{\text{in}}}{D_0 - \sin(\omega_{\text{in}})\left(D_{n\text{in}} - C_{0\text{in}}\right)/S_{\text{in}}}, \tag{7.202}$$

and where the standard values are used for the material parameters [24].

Figure 7.19 shows that the intersection point (solution point) for the higher density and higher sound speed copper has a higher pressure and smaller streamline turning angle than the solution point for Lexan, which has a lower pressure and larger streamline turning angle solution. The solution for copper involves a reflected shock in the explosive with a large angle β, while that for the Lexan involves the PM-fan branch. The reflected wave states, which all correspond to supersonic states in the explosive, cover the full range of such states that are possible for this range of ω, $0 \leq \omega < 24.25°$.

The value of ω that corresponds to a sonic flow state behind the incident shock is $\omega_s = 51.07°$ for a value of $\gamma = 4.75$. For values of $\omega \geq \omega_s$, disturbances arising from the interaction of the detonation with the adjacent inert material can be expected to propagate into the reaction zone and disturb the incident shock. From Fig. 7.18, we see that the shock polar corresponding to a reflected shock, for the case of $\omega = 24.25°$, is just tangent to the $\Theta_2 = 0$ line. For values of $\omega > 24.25°$, a gap opens in available polar states for small values of Θ_2. In the next section, we argue that in some situations a Mach-stem shock is needed to fill this gap as ω increases beyond $\omega \geq 24.25°$.

7.7.5 Mach-Stem Solution

As ω increases beyond $\omega = 24.25°$ and before it reaches the sonic angle, $\omega_s = 51.07°$, an increasing gap opens up in the available streamline turning angles, Θ_2. The transformation of the reflected shock polar as a function of increasing ω is shown in Fig. 7.20. Since we are seeking to match the flow

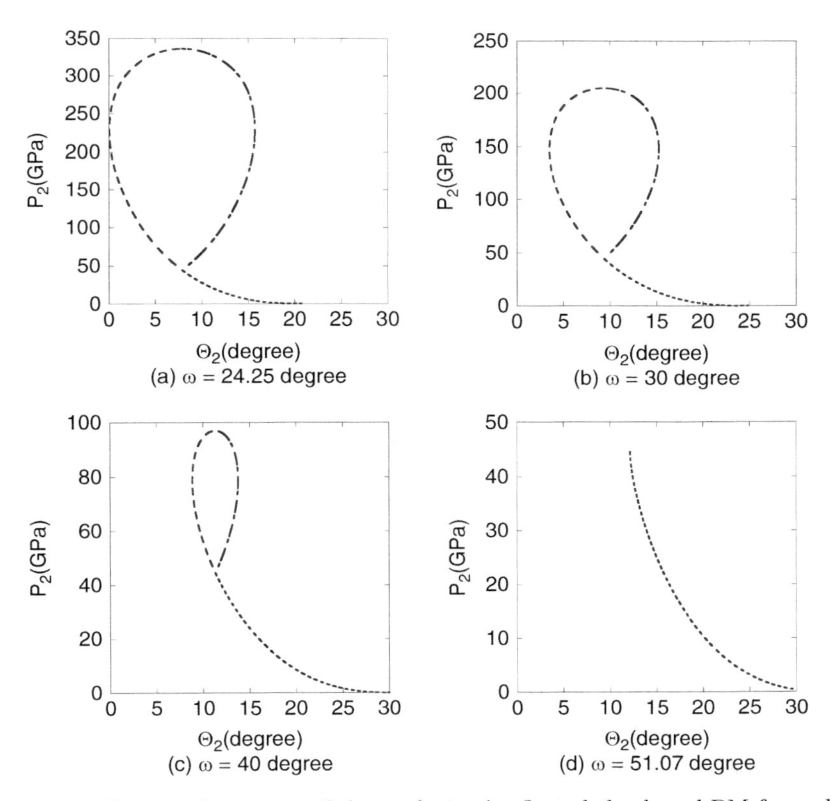

Fig. 7.20. The transformation of the explosives' reflected shock and PM-fan polars as ω is increased from $\omega = 24.25°$ to $\omega = \omega_s = 51.07°$. Although the pressure state of the incident ZND detonation (exhibited at the tip of the reverse teardrop shape) remains unchanged, the accessible pressures via the reflected shock are diminished and the available turning angles move further to the right

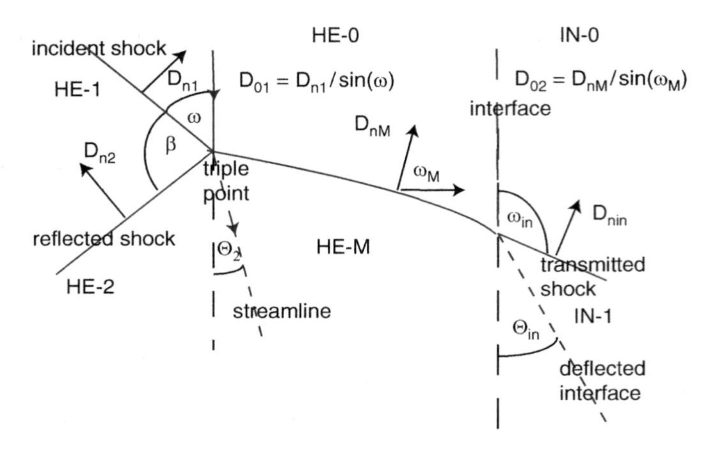

Fig. 7.21. The Mach-stem configuration proposed as the "filler" to span the gap between incident/reflected explosive shock-polar structure that is disconnected from the interface and a shock/interface polar structure. Here we assume that $\mathbf{D}_{01} = \mathbf{D}_{02}$, with the direction of these vectors being in the positive y-direction (upwards). The flow downstream of the Mach-stem is subsonic in the moving frame. Our diagram of the stem region is a schematic representation. The details of the Mach-stem's shape and evolution can only be determined via a complete solution of the flow field

for the explosive, P_2 and Θ_2, with the corresponding pressure and streamline turning angle for the adjacent inert material, P_{in} and Θ_{in}, an intervening wave structure may be needed to span between $\Theta_2 = 0$ and the minimum value of Θ_2 on the reflected shock polar. The classical filler used to span the gap is a Mach-stem structure. This is accomplished by inserting a curved shock, with a subsonic downstream state, between the structure consisting of the incident and the reflected explosive shocks, and the material interface. A schematic representation of such a structure is displayed in Fig. 7.21. In most respects, the Mach-structure is a single wave that looks like the subsonic reaction zone wave structure of a standard, resolved reaction-zone detonation.

The phase velocity of the Mach-reflection shock along the explosive/inert interface, D_{02}, does not need to be equal to the phase velocity of the triple point, D_{01}. Additionally, the Mach-stem can be curved. Based mostly on our numerical simulation results [9] and some experimental results [18], the transverse velocity of the triple point is observed to be small (velocity in the negative x-direction is small).

Here we assume that the transverse velocity is zero, and that D_{01} moves only in the positive y-direction. Further, if the Mach-structure is to retain some coherence over time (be nearly quasi-steady), the upward speed of the triple point and that of the stem/interface point should be roughly equal. Without any specific information to the contrary, here we adopt the simplest scenario and take

$$\mathbf{D}_{01} = \mathbf{D}_{02} = D_{02}\mathbf{j}. \tag{7.203}$$

With the assumption of quasi-steadiness, it then follows that the normal velocity of any point along the Mach-stem lead shock is given by

$$D_{nM} = D_{02} \sin(\omega_M), \tag{7.204}$$

where ω_M is the angle between the normal to the un-deflected interface and the local normal to the Mach-stem shock. We span ω_M over the range, $\omega_s \leq \omega_M \leq 90°$.

The Mach-stem shock state is given by (7.147)–(7.150), which with the polytropic EOS (7.161) yields

$$P_M = \frac{2\rho_{01} D_{nM}^2}{\gamma + 1}, \tag{7.205}$$

$$\tan(\Theta_M) = \frac{\cos(\omega_M)\sin(\omega_M)}{\frac{\gamma+1}{2} - \sin^2(\omega_M)}. \tag{7.206}$$

The Mach-stem polar is constructed by plotting P_M and Θ_M, as parameterized by ω_M.

Figure 7.22 displays the explosive shock polars, including the Mach-stem region, and the shock polar for copper for $\omega = 30, 40,$ and $51.07°$, where $D_{02} = D_{\mathrm{CJ}}/\sin(\omega)$. The Mach-stem polar begins at the $\Theta_2 = 0$ axis (where it is normal to the axis) and then slopes downward, ending interior to the reflected shock reverse teardrop, having there an infinite slope. Also shown is the polar for $\omega = 51.07°$ and $D_{02} = D_{\mathrm{CJ}}$, which corresponds to the ultimate steady state that develops, following the collapse of the initial Mach-stem due to the sonic/subsonic flow and the ultimate collapse of the incident wave.
Figure 7.22, inset (a), shows that for $\omega = 30°$, the regular reflection case persists, even though the reflected polar has moved to the right of the $\Theta_2 = 0$ axis. This solution corresponds to a supersonic explosive flow and a supersonic copper flow. Figure 7.22, inset (b), shows that for $\omega = 40°$, there are two solution points: (1) the intersection of the supersonic copper polar with the subsonic explosive Mach-stem, and (2) the subsonic–subsonic intersections of the explosive Mach-stem polar and the reflected shock polar. These two intersections, and the intervening section of the Mach-stem polar, define the Mach interaction. Figure 7.22, inset (c), shows the case for the incident shock having the sonic attack angle, $\omega = \omega_s = 51.07°$. At this point, the detonation reflected shock polar collapses, and the single Mach shock structure and PM-fan are all that remain. Since the flow described in Fig. 7.22, inset (c), corresponds to a sonic flow state behind the incident shock, the incident shock itself collapses, and one loses the ability to prescribe the phase velocity as $D_{02} = D_{\mathrm{CJ}}/\sin(\omega_s)$. The likely replacement for the flow is what is shown in Fig. 7.22, inset (d), where the phase velocity drops to $D_{02} = D_{01} = D_{\mathrm{CJ}}$. This would certainly be the state that develops for an explosive charge of infinite extent at long times, where the phase velocity would be no greater than D_{CJ}.

The solutions for the explosive/Lexan interaction problem are shown in Fig. 7.23 for $\omega = 40°$ with $D_{01} = D_{\mathrm{CJ}}/\sin(\omega)$ and for $\omega = 51.07°$ with

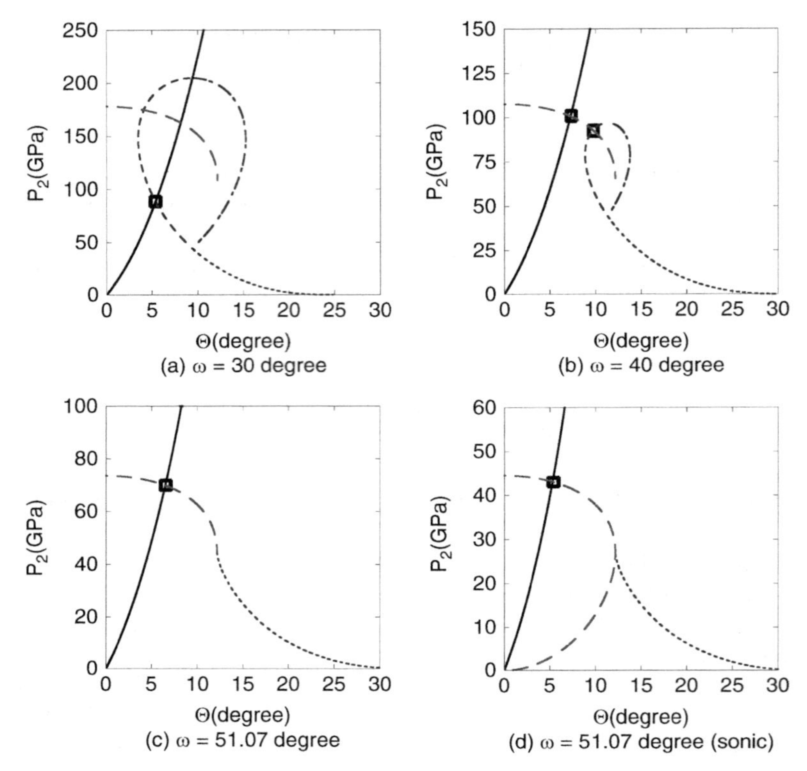

Fig. 7.22. Examples of the explosive and copper polars for $\omega = 30, 40,$ and $51.07°$, the region where the explosive reflected shock polar disconnects from the $\Theta_2 = 0$ axis. Regular shock reflection is observed for $\omega = 30°$, while a Mach-reflection is needed for $\omega = 40°$. The *two squares* on inset (**b**) correspond to the ends of the Mach-reflection stem, at which points the flow is subsonic. Insets (**c**) and (**d**) show two scenarios for the $\omega = 51.07°$ case. Inset (**c**) shows the higher-pressure states that would develop if the incident shock were maintained. For this sonic/subsonic incident flow shock, the shock will collapse to the state shown in inset (**d**), a self-propagating detonation

$D_{01} = D_{\mathrm{CJ}}$. Figure 7.23, inset (a), shows a supersonic–supersonic intersection that involves the explosive's PM-fan, whereas Fig. 7.23, inset (b), shows a higher-pressure solution that would apply in the case of thin layer of Lexan backed up on its right by a high-density third material layer [2].

For ω greater than the sonic angle, $\omega > \omega_s = 51.07°$, the incident shock cannot be sustained. Given the subsonic nature of the flow downstream of the incident shock and the tendency of the reaction-zone flow to focus disturbances toward the shock, the entirety of the reaction-zone flow will eventually sense the change of pressure generated at the explosive/inert interface. For all such

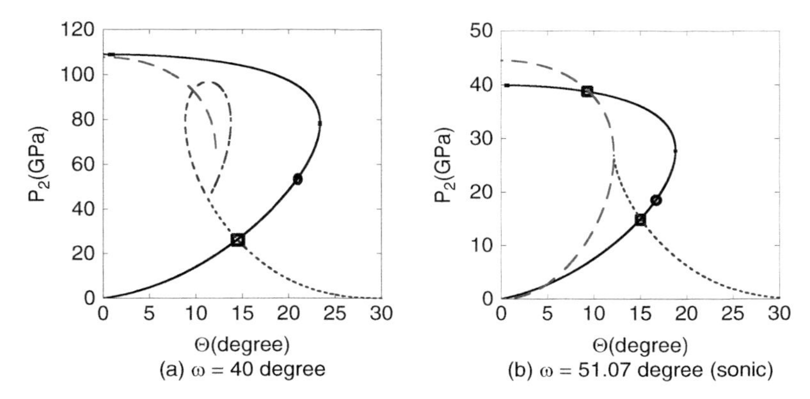

Fig. 7.23. Examples of the explosive and Lexan shock polars for $\omega = 40^\circ$ and $\omega = 51.07^\circ$. Unlike that for copper, no Mach-stem region is required, since the lower shock impedance Lexan yields a polar with a larger streamline deflection, which permits a solution on the PM-fan. There is also a second, higher pressure and smaller streamline deflection angle solution that is possible when $\omega = 40^\circ$. Although this solution state is generally harder to achieve, it can be found in cases where the Lexan confinement is thin and immediately backed up to its left with a high-density material, like copper. The lower pressure solution is more typical and found for thicker layers of Lexan confinement

cases, the detonation reaction zone will be perturbed by the presence of the edge of the explosive.

7.7.6 Direct Numerical Simulation of Mach-Style Reflection

The shock-polar analysis gives a reasonably clear picture of the detonation reaction zone/inert material interaction problem for $0 \le \omega < 24.25^\circ$ and $51.07^\circ \le \omega \le 90^\circ$. In the possible Mach-reflection range, $24.25^\circ \le \omega < 51.07^\circ$, and for higher-impedance inert materials, additional assumptions are required to proceed. Chief among these is that a local analysis can be performed at two separate points (the two ends of the Mach shock) and that these points move with the same speed and in the same direction. That is, we have assumed $|\mathbf{D}_{01}| \approx |\mathbf{D}_{02}|$ and $\mathbf{D}_{01}$ points nearly in the direction of the interface. As a strong test of the assumptions made in our analysis we present DNS results for two different attack angles, ω, for the case of a rigid confining wall, a case for which there is no streamline deflection at the wall. The question is how broadly felt is the Mach-reflection over the lateral extent of the reaction zone.

Consider the interaction of an originally steady-state, one-dimensional ZND detonation, supported at the CJ state, which passes over and interacts with a rigid wedge. The one-dimensional, ZND reaction-zone length is 4 mm. The detonation travels from right to left in these simulations. The one-dimensional detonation is originally at $x = 220$ mm and first meets the wedge

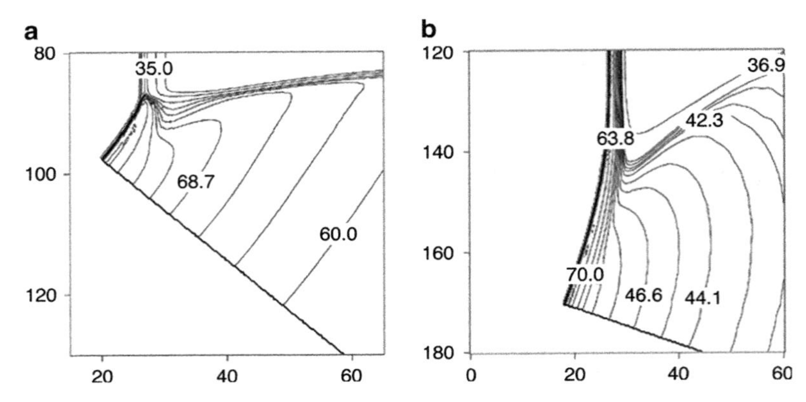

Fig. 7.24. Two snapshots showing the results from direct numerical simulations (DNS) of the two-dimensional detonations that develop when an initially flat detonation front meets two cases of a rigid wedge: inset (**a**) shows the 40° wedge angle case and inset (**b**) shows the 20° wedge angle case. The explosive model is that described previously. The pressure contours (numbers are in GPa) are shown for detonations that have run a horizontal distance of approximately 40 reaction-zone lengths over the wedge. Only a small transverse velocity is found for the triple point that develops in the Mach-stem configuration that is shown in inset (**a**), for the case $\omega = 50°$. The maximum stem width is a couple of reaction-zone lengths. A broadly curved front, with no evidence of any triple point, is shown in inset (**b**), for the case $\omega = 70°$. The influence of the wedge on the detonation extends laterally further for the case of $\omega = 70°$ shown in inset (**b**)

at $x = 190\,\mathrm{mm}$. The calculations assume a two-dimensional slab-geometry configuration. Two cases are considered: (1) a wedge angle of 40°, which corresponds to $\omega = 50°$, and (2) a wedge angle of 20°, which corresponds to $\omega = 70°$. The incident detonation angle of $\omega = 50°$ is in the Mach-reflection interaction regime, while the angle $\omega = 70°$ is squarely in the subsonic interaction regime. The $\omega = 50°$ case can be expected to show the fastest growth of a Mach-stem, since the flow relative to the incident detonation is only weakly supersonic.

Figure 7.24 shows the locations of the detonation fronts at $t = 24\,\mu\mathrm{s}$ for the cases of a wedge angle of 40° and 20°, in inset (a) and inset (b), respectively. This corresponds to a horizontal travel of approximately 40 reaction-zone lengths past the points of the wedges. During this distance of travel, the Mach-stem has grown laterally by a distance of about two reaction-zone lengths. This stem growth is slow and not incompatible with our assumption of an approximately zero transverse velocity of the Mach configuration triple point. Thus, the lateral reach of the Mach-reflection is limited, and most of the incident detonation shock remains undisturbed. Since the pressure behind the Mach-stem is identical to that behind the reflected shock emanating from the incident detonation, reflected shock, and Mach-stem triple point, the pressure induced into the inert material by the Mach configuration will be similar to what the reflected shock alone would have generated (see Fig. 7.24 inset (a)). It is worthwhile to note that as we transition from a rigid confiner to a high-impedance compliant boundary, the strength (pressure) and range of available incident shock angle for Mach-reflection both decrease (see Fig. 7.22). Thus, for most real high-impedance material confiners in the Mach-reflection range, $24.25° \leq \omega < 51.07°$, the extent to which the Mach wave propagates into the detonation reaction zone will be small.

For the $\omega = 70°$ case, shown in Fig. 7.24 and inset (b), we find a very different behavior. As expected, there is no Mach-reflection for this case. What is observed is a broadly curved incident detonation front. The detonation does show an increased pressure near the rigid wall of the wedge. The lateral extent to which the detonation reaction zone is perturbed by the interaction now reaches further laterally: about six reaction-zone lengths. Thus, when the angle of the incident detonation front, ω, is in the subsonic flow range, one sees a higher pressure detonation near the wedge due to flow convergence, but the front remains smooth. This converging flow solution case is similar in the smoothness of the detonation front and extended lateral reach of the interface perturbation to what is observed for the diverging flow shown in Fig. 7.17.

7.8 DSD Boundary Conditions

In the last few sections, we showed how a detonation that approaches the interface between the explosive and an inert material at an oblique angle interacts with and accelerates the inert material. A ZND-type model, with a fully resolved reaction zone, was used as the explosive description. Local analyses centered about the incident detonation and material interface intersection point were used to describe which wave families were needed to describe the explosive/inert interactions, and in particular those in the explosive. The purpose was to describe how one formulates physically-based boundary conditions for describing how a DSD front interacts with a material interface. In this section we give a summary of the boundary conditions that a DSD wave must satisfy to properly describe the interaction of detonation with an inert material boundary. We then use DNS results to compare with those stated boundary conditions.

7.8.1 Summary of Oblique Detonation Interaction Study

In our study, we considered the oblique attack of a distributed reaction zone detonation on the explosive/inert material interface. The angle of attack, ω, was varied over the range, $0 \leq \omega \leq 90°$. That interaction problem breaks down into three distinct regions: (1) $0 \leq \omega < 24.25°$, where all the waves that resolved the interaction were supersonic, (2) $24.25° \leq \omega < 51.07°$, where all waves, except for the subsonic Mach interaction, were supersonic, and (3) $51.07° \leq \omega \leq 90°$, where the majority of waves were subsonic and sonic.

The values of the transition angles are determined by the explosive EOS for which we use (7.161) and by the parameter values, particularly the value of γ, which was $\gamma = 4.75$. This led to $\omega_s = 51.07°$ as the sonic transition angle, with a smaller ω indicating a supersonic flow and a larger ω a subsonic flow. Here we summarize the results by the range of ω.

7.8.1.1 Regime I: Supersonic Interaction; $0 \leq \omega < 24.25°$

In Regime I, the interaction of the detonation with both high and low shock-impedance inert materials produces a reflected wave with a following supersonic flow. High-impedance inert materials produce a reflected shock, while low-impedance inert materials produce a reflected PM-fan (see Fig. 7.19). The incident shock is not perturbed by the interaction, and only a very small region of the reaction zone, immediately adjacent to the interface, is perturbed. A diagram of the flow for Regime I is displayed in Fig. 7.25 and the features of the flow are described in the caption.

7.8.1.2 Regime II: Supersonic Mach Interaction; $24.25° \leq \omega < 51.07°$

For Regime II, the interaction of the detonation with low-impedance inert materials remains the same as described above for Regime I. For high-impedance inert materials, the interaction can take one of two forms. For

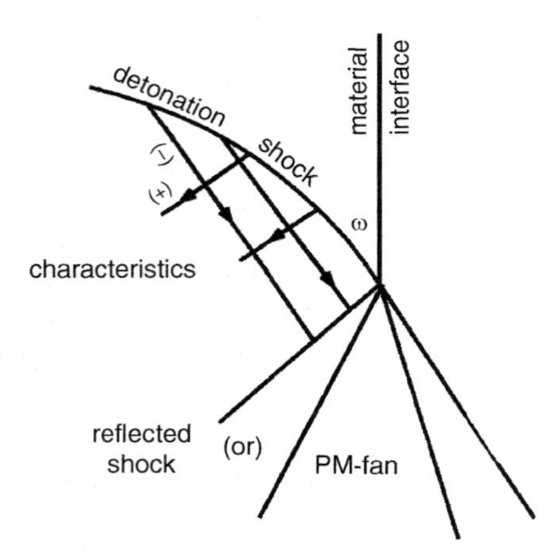

Fig. 7.25. The reaction-zone flow configuration for $0 \leq \omega < 24.25°$. All acoustic information, on both characteristic families, flows away from the vicinity of the shock. Either a reflected shock or PM-fan can be used to adjust the pressure, P, and streamline deflection, Θ, to match the detonating explosive flow with the adjacent inert material flow. The incident shock remains undisturbed, and only the reaction-zone flow very near the interface is disturbed

very high impedance inert materials and for a rigid, immovable interface, a Mach-reflection shock (a regular subsonic shock) must be interposed between the material interface and the incident/reflected shock structure in the explosive. For the case of a rigid interface, a Mach wave is needed over the entire range of ω being considered in Regime II. Our DNS results, for the case of a rigid interface (see Fig. 7.24) and for $\omega = 50°$ shown in inset (a), show a very slow growing Mach-stem region. As ω is decreased from $\omega = 50°$, the flow downstream of the incident shock becomes more supersonic and the Mach stem growth becomes even slower. In all these cases, the pressure behind the Mach-stem shock is comparable to what is generated by the incident/reflected shock structure (a result of pressure continuity).

Moving from rigid confinement, $\Theta_2 = 0$, to a more typical high-impedance inert confiner material such as copper, the requirement that there be a Mach-stem region is pushed off to larger values of ω; Mach-reflection is required for $\omega = 40°$ but not for $\omega = 30°$. Thus for Regime II, we find a Mach-stem shock is not required over the entire domain, $24.25° \leq \omega < 51.07°$, and as the shock-impedance of the inert material is reduced, is only required at the high-end of the angle limits for Regime II. The growth of the Mach-stem region laterally into the reaction zone is slower than for the case of rigid confinement. The extent of the Mach-stem growth is less than a couple of ZND reaction-zone lengths as the detonation travels 40 reaction-zone lengths in the horizontal direction. The flow beyond (further into the explosive) the triple point maintains the undisturbed incident shock/reflected shock structure observed in Regime I (both streamline flow and pressure). Thus, the perturbation to the reaction-zone flow for the Mach-stem required flows, as in Regime I, is restricted to the near interface region.

7.8.1.3 Regime III: Sonic/Subsonic confinement; $51.07° \leq \omega \leq 90°$

For Regime III, the prescribed incident shock is lost and replaced by a shock whose shape is determined by the subsonic, reaction-zone flow. Roughly speaking, it is like having the lateral reach of the Mach-stem be infinite at late time. This replacement flow quickly develops a quasi-steady and subsonic character and, therefore, the flow that develops is smooth with no sharp wave front intersections. Once the flow is in this configuration, the shock-polar analysis shown in Fig. 7.22, inset (d), and Fig. 7.23, inset (b), can be used to determine ω_c, the critical angle for the specified detonating explosive/inert material pair, and the pressure and streamline turning angle at the explosive/inert material boundary. A diagram of the flow for Regime III is displayed in Fig. 7.26, and the features of the flow are described in the caption.

7.8.2 Statement of DSD Boundary Conditions

The shock-polar and DNS results provide information that can be used to formulate boundary conditions that the incident, interior detonation-shock

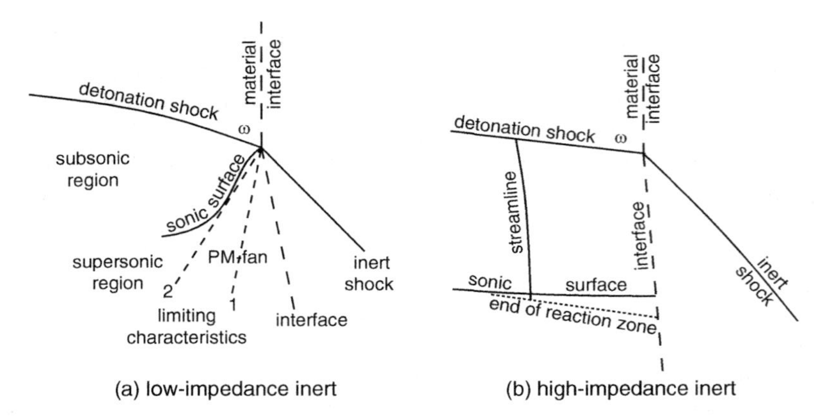

Fig. 7.26. The reaction-zone flow configuration for $51.07° \le \omega \le 90°$. Inset (**a**) shows the reaction-zone flow that develops for the case of a low-impedance adjacent inert material, while inset (**b**) shows the reaction-zone flow that develops for the case of a high-impedance adjacent inert material. For the low-impedance case, inset (**a**), the pressure-lowering PM-fan can only intrude into the reaction zone up to the sonic locus, and where the sonic locus intersects the detonation shock and serves to insulate the bulk of the reaction zone from any inert-material specific information. The details of the material properties of the inert do not influence the bulk of the reaction zone. For the high-impedance case, inset (**b**), the pressure-raising reflected shock penetrates into the reaction zone and is quickly degraded to a laterally propagating compression wave by the subsonic character of the flow. In both cases, a quasi-steady, mostly subsonic flow is setup in the reaction zone, for which the boundary conditions for the reaction zone are given by the shock-polar analysis described

angle must make at the explosive/inert interface. Given that the detonation shock in the interior of the explosive propagates according to a DSD $D_n(\kappa)$ relation, [1], the boundary conditions for the DSD wave can be specified as follows:

1. First measure the size of the detonation attack angle, ω, by linearly extrapolating the DSD-front shape to the interface.
2. Given the extrapolated ω, determine the sonic character of the flow by comparing ω with the sonic value, ω_s.
3. When the extrapolated angle, ω, corresponds to a supersonic flow, $\omega < \omega_s$, then the extrapolated value of ω is set as the DSD boundary condition angle. (In developing this prescription, a possible Mach-reflection is not considered. This is based on two observations. First, the lateral intrusion of the Mach-stem into the reaction zone is limited, and thus only a region of a few reaction-zone widths is affected by the Mach-stem, with the remainder possessing the incident shock as the lead wave. Second, the pressures developed behind the Mach-stem are roughly the same as those behind the reflected shock, which is part of the Mach triple-point structure. Thus,

neglecting the Mach-stem does not affect the pressure delivered to the inert material by the detonation.)

4. When the extrapolated angle, ω, falls in the range, $\omega_s \leq \omega \leq 90°$, we perform a test to determine if the adjacent inert material is either a high-impedance or low-impedance material. When the shock-polar analysis indicates that ω_s is the confinement angle (low-impedance inert material case), then we set $\omega = \omega_s$ as the boundary condition (see Fig. 7.23, inset (b)). When the shock-polar analysis indicates that ω_c is the confinement angle, where $\omega_c \geq \omega_s$, then we set $\omega = \omega_c$ as the boundary condition (see Fig. 7.22, inset (d)).

There are other inert material types that can lead to other modes for the detonation/inert material interaction problem, modes that we have not considered here. As an example, there are materials whose ambient pressure sound speed is higher than the D_{CJ} of the explosive. Those situations were studied via a theoretical development in a paper by Sharpe and Bdzil [28], and more recently via DNS by Short et al. [29]. We do not consider those cases here.

7.8.3 Comparison of DSD Boundary Conditions with DNS Predictions

Much of the earlier calibration work for the DSD model focused on a class of so-called insensitive explosives, such as the triaminotrinitrobenzene (TATB) pressed, high-density, plastic-bonded explosive PBX 9502. Work at Los Alamos National Laboratory and The University of Illinois was carried out to systematically compare the confinement angle, ω_c, determined from shock-polar analysis, with the results from high-resolution DNS and experimental studies. We briefly describe how the DNS results compare with the DSD boundary condition prescription given above.

Craig Tarver's Ignition and Growth reactive flow model [34] was used to model PBX 9502. A Mie–Grüneisen EOS, with constant Grüneisen-gamma, was used to model the various inert materials considered as confinement material. The problem geometry was a layered, two-dimensional slab configuration consisting of a slab of PBX 9502 adjacent from the left to a slab of inert material, with their common interface being the y-axis containing the vertical plane. In these simulations, the detonation propagates upwards (in the positive y-direction) and eventually reaches a constant propagation speed in the frame of the detonation shock and the material interface intersection point. Once the flow becomes steady in this frame, the detonation shock edge angle, ω_c, is measured. The measured value determined with the DNS is then compared with the results obtained from the shock-polar analysis we have described above. In this study, the thickness of the inert material layer can be considered as being infinite.

The simulations were performed using a CENO shock-capturing algorithm, which uses a formally fifth-order interior flow solver for the Euler equations

Table 7.1. A comparison of ω_c determined using DNS with those obtained from shock-polar analysis for the explosive PBX 9502 and various inert material confiners

Inert material	DNS ω_c radians	Shock polar ω_c radians
S5370	0.76 ± 0.02	0.776
Sylgard	0.76 ± 0.02	0.776
PMMA	0.80 ± 0.01	0.776
Teflon	0.86 ± 0.01	0.776
Ti-6Al-4V	1.34 ± 0.01	1.345
SS304	1.41 ± 0.01	1.416
Tantalum	1.44 ± 0.01	1.445
Composite	1.41 ± 0.01	2.601

to update the solution. The material interface was treated using a ghost-fluid algorithm. This numerical, high-speed, Euler solver is described in Fedkiw et al. [14]. The multi-aterial Euler solver was embedded in the James Quirk Amrita environment [25]. Amrita provides an adaptive mesh refinement (AMR) capability as well as a modular environment with which to orchestrate the resolution and parameter study for our problems.

The results of this study were reported in the 13th Detonation Symposium by Aslam and Bdzil [3] and are reproduced in Table 7.1. The inert materials considered in this study were: (1) S5370 (a polymeric material), Sylgard (a silicone-base polymer), PMMA (polymethyl methacrylate), Teflon, Ti-6Al-4V (a titanium, aluminum, and vanadium alloy), SS304 (stainless steel), tantalum, and a composite (a moderate density, high sound speed metal-loaded polymer). The inert material EOS parameters are given in [3]. In most cases, the agreement of shock-polar analysis with the DNS results is very good. The exception is with the composite material, which has a high sound speed. The shock transmitted into the composite by the detonating explosive is able to move slightly ahead of the detonation shock. More details on the composite case can be found by examining the shock-polar analysis given by Aslam and Bdzil [3] and also in the discussion of subsonic confiner flow given by Sharpe and Bdzil [28].

A DNS study, using the polytropic EOS for the explosive described earlier (7.161), was undertaken to examine the effect that the inert material layer thickness has on detonation confinement. The problem geometry was similar to that described above, except that the inert confiner thickness was varied, and a rigid-interface boundary condition was applied on the inert material layer (i.e., the inert material was "sandwiched" between the explosive and a rigid wall). That study is described in [2]. The confinement layer was also modeled with the polytropic EOS (7.161), using parameter values that simulated the shock response of Lexan.

As described earlier, the solution state that is obtained for Regime III, where $51.07° \leq \omega \leq 90°$, has two possible shock polar defined solution states.

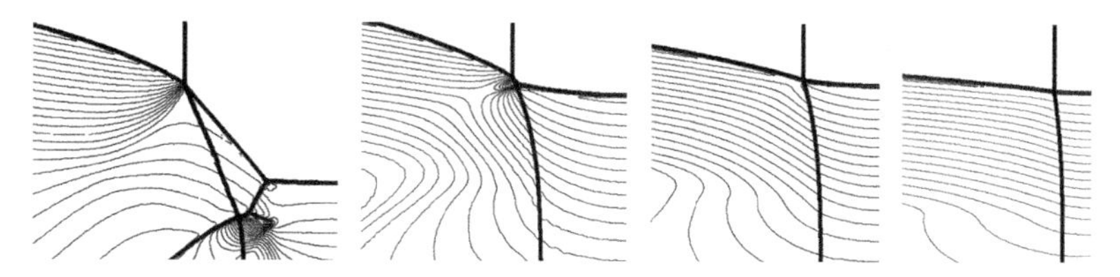

Fig. 7.27. Pressure plots of the region around the explosive/inert material interface, showing the pressure contours as a function of the confiner thickness. From *left* to *right*, the confiner thickness is 4, 3, 2 and 1 mm. The explosives one-dimensional, ZND reaction-zone length is 4 mm

For a sufficiently thick polymeric material layer, the solution state that is obtained corresponds to the lower-pressure shock-polar solution displayed in Fig. 7.23, inset (b). The critical confinement layer thickness to get this solution requires that the confinement layer be at least as thick as the explosives reaction-zone length. Those simulations showed that as the thickness of the confinement layer is diminished, a transition to the smaller streamline deflection angle and higher pressure solution state (and larger ω_c value) occurs. This is the higher-pressure shock-polar solution shown in Fig. 7.23, inset (b). These results are displayed in Fig. 7.27. The lead shock shape is shown as a function of the polymer layer thickness.

7.9 Examples of DSD Front Propagation: Application of the $D_n(\kappa)$ Relation with Boundary Conditions

Level-set based algorithms can be used to propagate DSD fronts because they lead to robust code that can handle the complex topological changes that detonation fronts experience when they propagate in complex-shaped explosive charges. Since the level-set method embeds the two-dimensional detonation front (curve) in a higher, three-dimensional representation, $\psi(x, y, t)$, collisions and bifurcations of the detonation front are handled automatically. The level-set function, $\psi(x, y, t)$, can be envisioned as being the z-coordinate in a three-dimensional Cartesian space. The zero contour of the level-set function, $\psi(x, y, t) = 0$, then corresponds to the location of the detonation front, and can be extracted from ψ by the zero-contour of the level-set function, $\psi(x, y, t) = 0$.

In this section we briefly describe a full, level-set method for a two-dimensional (2D) DSD solver, DSD2D, and show some representative test problems. The algorithm for the interior solver has been demonstrated to be second-order accurate in space and time. For problems dominated by boundary conditions, the accuracy drops to first order. A detailed description of the solver algorithm and comparison of computed results with exact solutions can be found in [6] and [17]. Other techniques that use higher-order methods and hybrid level-set methods [40] can also be used with success.

7.9.1 DSD Level-Set Method

The full, level-set DSD2D algorithm solves the modified level-set equation to propagate the detonation front

$$\frac{\partial \psi}{\partial t} + D_n(\kappa)|\nabla \psi| = \frac{\psi}{\epsilon\sqrt{\delta^2 + \psi^2}}(1 - |\nabla \psi|), \tag{7.207}$$

or

$$\frac{\partial \psi}{\partial t} + \left(D_n(\kappa) + \frac{\psi}{\epsilon\sqrt{\delta^2 + \psi^2}}\right)|\nabla \psi| = \frac{\psi}{\epsilon\sqrt{\delta^2 + \psi^2}}, \tag{7.208}$$

where here $\psi(x, y, t)$ is the level-set function, D_n is the speed of the detonation in the shock-normal direction and ϵ and δ^2 are small parameters used as part of the numerical implementation. Hence, the variables $\psi(x, y, t)$, ϵ, and δ are specific to this section. The variables, x, y, are the Cartesian coordinates. The detonation-front curvature, κ, is given by the following expression for two-dimensional, slab geometry:

$$\kappa = \nabla \cdot \mathbf{n}_s = \frac{\psi_{xx}\psi_y^2 - 2\psi_{xy}\psi_x\psi_y + \psi_{yy}\psi_x^2}{\left(\psi_x^2 + \psi_y^2\right)^{3/2}}. \tag{7.209}$$

The operator on the left-hand side of (7.207) is the standard level-set operator, while that on the right-hand side of (7.207) has been called the re-distancing operator. Roughly speaking, the operator on the left-hand side of (7.207) causes the level-set function to drop, which moves the zero contour of the level-set function through space (see Fig. 7.28). Away from the zero contour of the level-set function, the operator on the right-hand side of (7.207) acts to drive the gradient of the level-set function to one. It accomplishes this

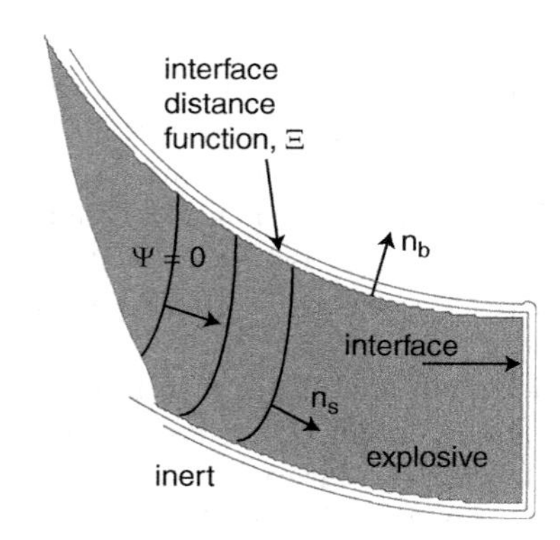

Fig. 7.28. A 2D schematic representation of an explosive piece, showing the contours of the interface distance function, Ξ, the local normal to the explosive interface, $\mathbf{n}_b$, the detonation level-set function, ψ, and the local normal to the detonation level-set function, $\mathbf{n}_s$

by moving data from the vicinity of the zero contour of the level-set function both upstream and downstream of the zero contour of the level-set function as shown by (7.208). This equation is solved numerically as described in [6] and [1].

The solution of (7.207), (7.208) is initialized as a distance function, with $|\nabla\psi| = 1$ initially. Although the re-distancing term in (7.207) tends to maintain this condition when ϵ is sufficiently small, this increases the stiffness of the equation which then substantially reduces the step size for the time integration. Also, near explosive boundaries it can be challenging to maintain $|\nabla\psi| = 1$ with the re-distancing operator. To avoid these issues, the DSD2D solver makes no assumption about the magnitude of $|\nabla\psi|$ and uses re-distancing only to keep accuracy robbing plateaus from developing in the level-set function.

The solution to (7.207), (7.208) is subjected to the DSD boundary condition; this requires our knowing the value of the DSD sonic angle, ω_s, for the explosive. We have described the DSD boundary conditions and the role played by ω_s in an earlier section. Our description of the boundary of the explosive is given by a second, stationary level-set function, $\Xi(x, y)$, whose zero contour corresponds to the initial boundary of the undetonated explosive. The stationary $\Xi(x, y)$ function is defined such that its contours correspond to the normal distance to the interface, as displayed in Fig. 7.28. The gradient of this interface distance function, evaluated at the interface, yields the outward-pointing normal to the undeformed explosive interface. Applying the DSD boundary condition then involves computing a condition on ω, the included angle between the unit normal to the level-set function,

$$\mathbf{n}_s = \frac{\nabla\psi}{|\nabla\psi|}, \tag{7.210}$$

and the unit normal to the boundary level-set function,

$$\mathbf{n}_b = \frac{\nabla\Xi}{|\nabla\Xi|}, \tag{7.211}$$

with $\mathbf{n}_s \cdot \mathbf{n}_b = \cos(\omega)$ and where all quantities are evaluated at the interface.

The boundary condition is applied in two steps, as we have also described earlier in this chapter. More details can be found in previous sections. First, ω is extrapolated from the interior of the explosive to the explosive–inert boundary, under the condition $\mathbf{n}_b \cdot \nabla\omega = 0$. When the extrapolated ω satisfies the inequality, $\mathbf{n}_b \cdot \mathbf{n}_s = \cos(\omega) \leq \cos(\omega_s)$ then ω is reset to $\omega = \omega_c$ where ω_c is the characteristic angle for the explosive–inert pair. When $\mathbf{n}_b \cdot \mathbf{n}_s = \cos(\omega) > \cos(\omega_s)$ then the extrapolated value of ω is maintained at the boundary.

7.9.2 DSD Solution Examples

Here we give some examples of DSD detonation front propagation solutions for two different explosive geometries. We use a simple linear form for the

DSD propagation law

$$D_n = 1 - 0.1\kappa, \tag{7.212}$$

and take ω_s and ω_c to be equal,

$$\omega_s = \omega_c = \pi/4. \tag{7.213}$$

For both examples, the DSD front is taken to be initially a flat wave.

The explosive geometry must be provided in the format of a distance function, $\Xi(x, y)$, to the explosive boundary, where the explosive boundary is the $\Xi(x, y) = 0$ contour. The value of the distance function at any point not on the explosive boundary is given by the normal distance from $\Xi(x, y) = 0$ to the point. Near sharp corners of the boundary, the distance function is either given by a simple fan, with $\mathbf{n}_b$ centered at the outside of the sharp corner and rotating about it, or by a shock, where $\mathbf{n}_b$ is centered at the inside of the corner and jumps discontinuously in transitioning the corner.

The example of an explosive rate stick is displayed in Fig. 7.29. The detonation-front contours, shown in inset (b) of Fig. 7.29, which are equally spaced in time, show that the detonation front, which is initially flat, quickly establishes a front contour of fixed shape. The transient phase of the problem is short, and the detonation propagates steadily at a detonation phase speed,

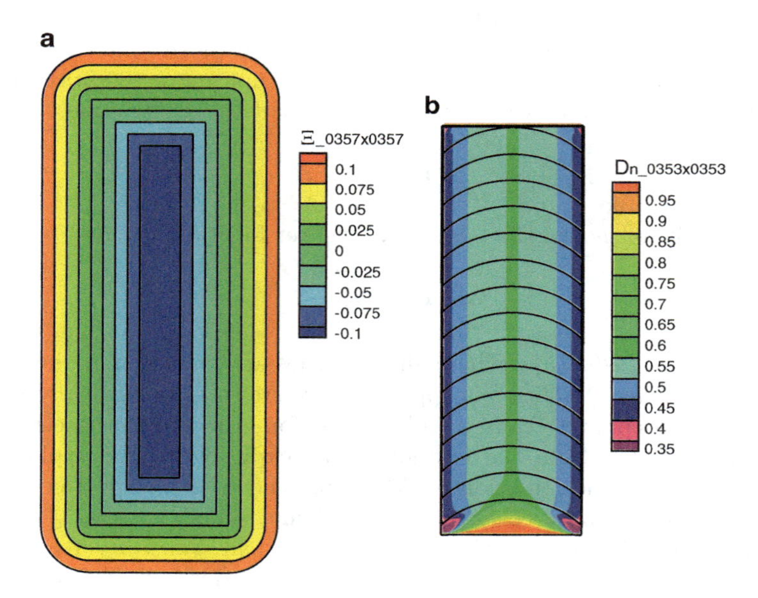

Fig. 7.29. DSD calculation of detonation in an explosive rate stick. The stationary, distance level-set function, $\Xi(x, y)$, is shown in inset (**a**). In inset (**b**) the detonation front contours, at equal-spaced times, are shown overlayed over the values of D_n at the time of detonation passage. Once a curved detonation front is established, the detonation propagates at a constant phase velocity, D_0, with $D_0 < D_{\mathrm{CJ}} = 1$

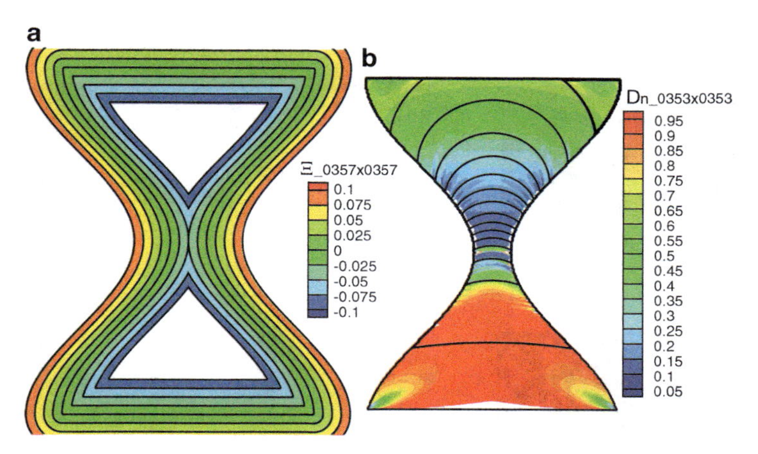

Fig. 7.30. DSD calculation of detonation in an "hourglass"-shaped explosive piece. The stationary, distance level-set function, $\Xi(x, y)$, is shown in inset (**a**). In inset (**b**) the detonation front contours, at equal-spaced times, are shown overlayed over the instantaneous value of D_n at the time of passage of the detonation front. The detonation front senses the boundaries only weakly in the lower-section of the "hourglass" explosive piece, and propagates at nearly $D_{\rm CJ} = 1$, as witnessed by the relatively flat, widely spaced contours. Once the detonation enters the upper section of the "hourglass," it strongly senses the boundaries, and slows dramatically, as witnessed by the very closely spaced contours

D_0, which is below $D_{\rm CJ}$, with $D_0 < D_{\rm CJ} = 1$. The DSD angle boundary condition locks into $\omega = \omega_c$ at a very early time and maintains this constant value.

The example of an "hourglass" shaped explosive is displayed in Fig. 7.30. The detonation-front contours, shown in inset (b) of Fig. 7.30, are equally spaced in time. The detonation, which is initially flat when at the bottom of the "hourglass," is only weakly disturbed by the boundaries in the lower section of the "hourglass," and runs at nearly $D_{\rm CJ}$. As the detonation leaves the narrow section of the "hourglass" and then expands into the upper section, the detonation strongly senses the boundaries and is dramatically slowed. In this example, the extrapolated value of ω is mostly maintained for the lower part of the "hourglass." Once the detonation moves through the narrow section and proceeds to the upper section of the "hourglass," the DSD boundary condition locks into $\omega = \omega_c$. This example provides a demonstration of the strong and varying influence that DSD boundary conditions can have on detonation propagation.

7.9.3 Application of DSD to Detonation Diffraction Past a Disk

In this section we examine how DSD can be used in a representative explosive system application of detonation diffraction over an embedded inert, dense disk, also referred to as a "detonation passover" experiment. The detonation

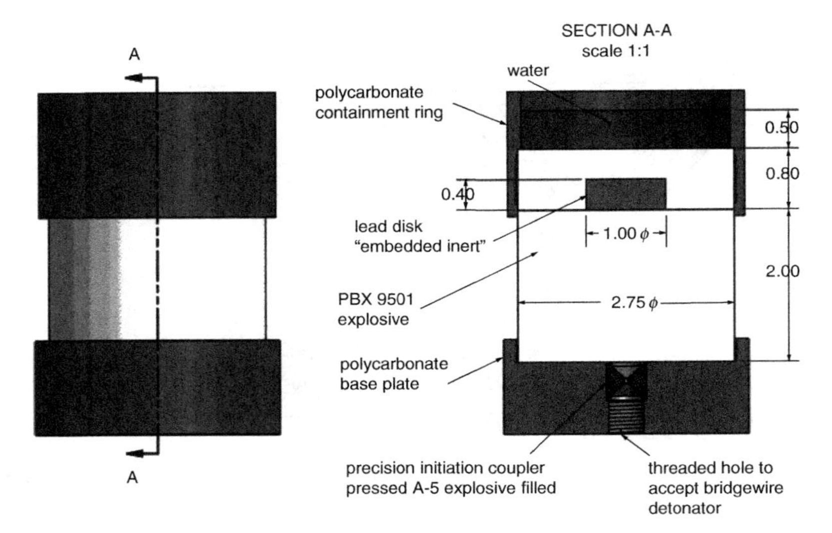

Fig. 7.31. Assembly sketch for DSD "passover" validation experiment

dynamics obtained from the diffraction experiment described here was simulated with both a multimaterial DNS and with DSD, which were found to be in close agreement. A full account with complete references is found in [22].

The experimental setup is shown in Fig. 7.31. A dense lead (gray) disk was embedded along the central axis of a right circular cylinder HMX-based, PBX 9501 explosive charge. Lead has a high shock-impedance and well-characterized shock Hugoniot properties. The PBX 9501 charge is initiated at the bottom. As the detonation shock front propagates away from the detonator, for the first 50 mm it does so as a simply connected hemispherical surface with convex, positive curvature. When it encounters the lead disk within the top piece of the charge, the wave is split into two pieces. The shock speed in the inert lead is much lower than the detonation velocity in the PBX 9501, and a diffraction event occurs as the detonation sweeps about the disk and encompasses it. Light from a single slit aperture plate is captured and recorded by a fast framing camera. This record indicates the time of arrival (TOA) of the detonation shock. Four passover experiments, labeled as "PASS-x," were conducted with identical hardware by D. Lambert and are shown in Fig. 7.32.

In order to model the experiment, one must specify the geometry and choose constitutive models for both the explosive and the inert confinements, i.e., the lead disk and the air and plastic that surround the boundaries of the main charge. The reader is referred to [22] for the full details of the constitutive description used for the modeling. Briefly, the explosive EOS was chosen to be a binary mixture EOS for conversion of reactants (fresh explosive) to products, and the product mass fraction λ was used as the reaction progress variable. The functional forms of both the reactant and

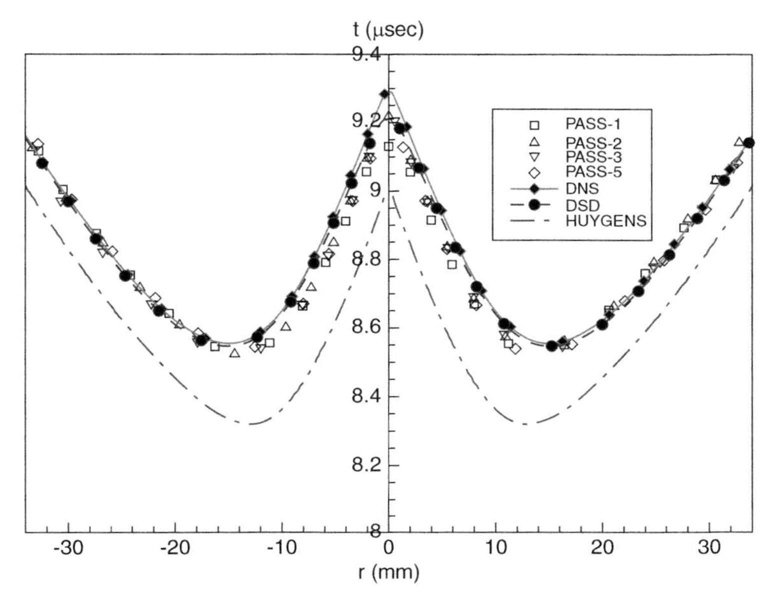

Fig. 7.32. TOA comparison of experiments, DNS, DSD simulations and the ideal Huygens construction, at the top of the charge assembly shown in Fig. 7.31 (from [22])

product EOS were based on the Mie–Grüniesen EOS. They were calibrated to fit the experimental unreacted explosive shock Hugoniot and detonation Hugoniot, respectively. The methodology used to calibrate the EOS is reported in [37]. The general functional form for the $E(P, v, \lambda)$ EOS can be reduced to $E = f(v, \lambda) + g(v, \lambda)P$ which is linear in the pressure and nonlinear in v and λ. In addition, a reaction-rate law was used with depletion and pressure-dependent rate coefficients of the form

$$R(P, v, \lambda) = k(1 - \lambda)^m \left(\frac{P}{P_{\mathrm{CJ}}}\right)^n . \tag{7.214}$$

Shock to detonation run-up distance data ("Pop"-plot) and detonation shock speed and curvature data were used to calibrate the parameters of the rate law. One-dimensional, reverse impact simulations were carried out using the specified EOS and rate law form to adjust their parameters to match the published experimental data for PBX 9501; the pressure exponent n and rate constant k were adjusted to match the shock initiation data. In addition, some negative curvature data for PBX 9501, inferred from overdriven experimental data obtained by L. Hull of Los Alamos National Laboratory [18], were used to set the depletion exponent m, see Fig. 7.33. Complete specification of the explosive's EOS form $E(P, v, \lambda)$ and the rate law allows one to compute the $D_n(\kappa)$ relation with the methods explained previously in this chapter. Figure 7.33 shows the $D_n(\kappa)$ relation computed for this model of PBX 9501.

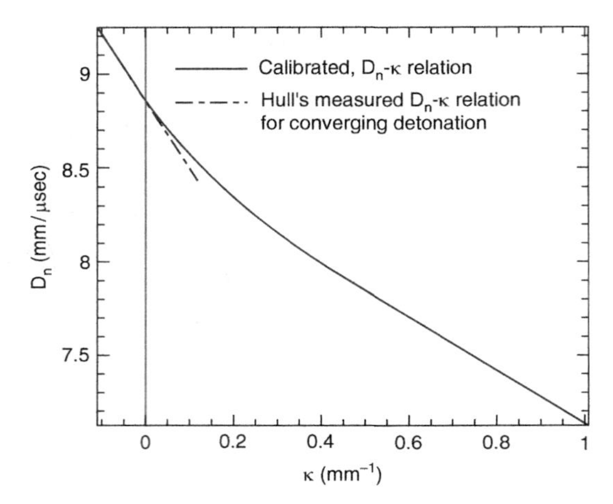

Fig. 7.33. The $D_n - \kappa$ relation for PBX 9501 calculated from the wide-ranging EOS and rate law model

With the specified EOS and rate law forms, it is also possible to carry out multimaterial DNS for comparison. The identification of the EOS for the lead, and the reactant EOS for the explosive also enable the computation of both the confinement angle ω and its sonic value ω_c as a function of the normal-shock velocity at the interface of the lead and the air. These angles are then used for the DSD boundary conditions in the DSD simulation. For the simulation of the passover experiment, shock-polar analysis for the DSD confinement boundary conditions was used to derive the angles at the PBX 9501/lead interface (the interior angle between the shock and interface normals) as $\omega_s = 35°$ (sonic) and $\omega_c = 66°$ (subsonic).

The shock TOA results obtained in the passover experiments was compared to the results of two different types of simulations: (1) a DSD simulation and (2) a DNS simulation. The DSD simulation used the $D_n - \kappa$ detonation motion rule shown in Fig. 7.33, subject to inert angle confinement boundary conditions. The initial shock in the DSD simulation was a hemisphere of radius 5 mm centered at the bottom of the charge. Using the methods described in [33], a multimaterial numerical simulation (DNS), with outflow lateral boundary conditions, was carried out with the wide-ranging EOS and rate law for the explosive and the Mie–Grüneisen EOS for the lead. The DNS simulation also used an initial condition of a hemispherical hot spot of radius $r = 5$ mm, centered at the bottom. The corresponding DSD simulation is shown in Fig. 7.34. The DSD shock contours are overlaid on the shock pressure. Those pressures correspond to the shock state when the detonation passes over the point (x,y). Figure 7.35 shows a comparison of the DSD and DNS simulation to show that they give consistent results (i.e., the shocks overlay). Figure 7.32 shows a composite of shock TOA records for the passover experiments that simultaneously show results for experiments, DNS, DSD simulations, and the ideal Huygens

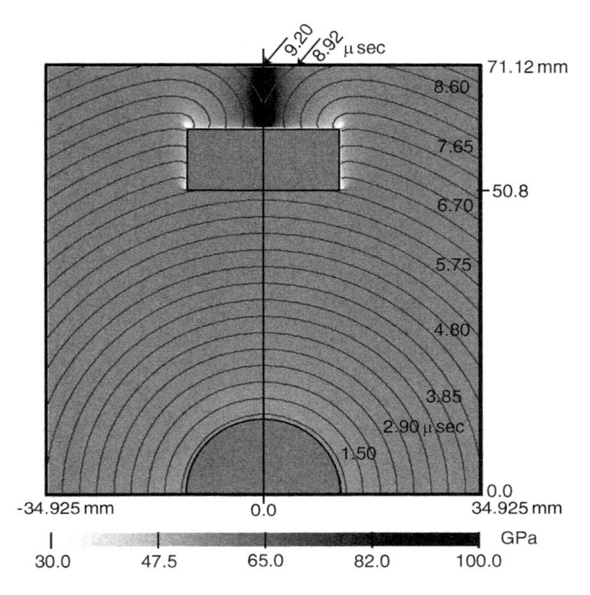

Fig. 7.34. DSD simulation of the axisymmetric passover experiment. The *gray-scale* shows the shock pressure in GPa when a shock passes a point (x, y) in the explosive. Contours show the shock front at the various times indicated

construction, where the shock is simply propagated at $D_{\mathrm{CJ}} = 8.86\,\mathrm{mm}\,\mu\mathrm{sec}^{-1}$ for PBX 9501. The Huygens construction assumes that the detonation propagates at a constant normal-shock speed and has no curvature correction; thus it does not account for the slowing of the wave. The level of agreement, both qualitative and quantitative, between experiment, DSD and DNS is excellent. This indicates that one can use the wide ranging EOS/rate law and the corresponding DSD description effectively to model real explosives and predict complex dynamic behaviors.

7.10 Concluding Remarks

In this chapter, we have presented a self-contained theoretical development of the DSD detonation-front propagation model. We have shown how a quasi-one-dimensional detonation description captures multidimensional detonation effects, and thus leads us to study curved, cylindrically symmetric detonation waves. Via characteristic analysis, we showed the existence of a limiting characteristic for self-sustaining detonation. This characteristic, which is located near the end of reaction zone, demarks a narrow region near the detonation shock, which is solely responsible for describing the propagation of the detonation. We then showed how the quasi-steady limit of this quasi-one-dimensional model yields a detonation-propagation law, where the nor-

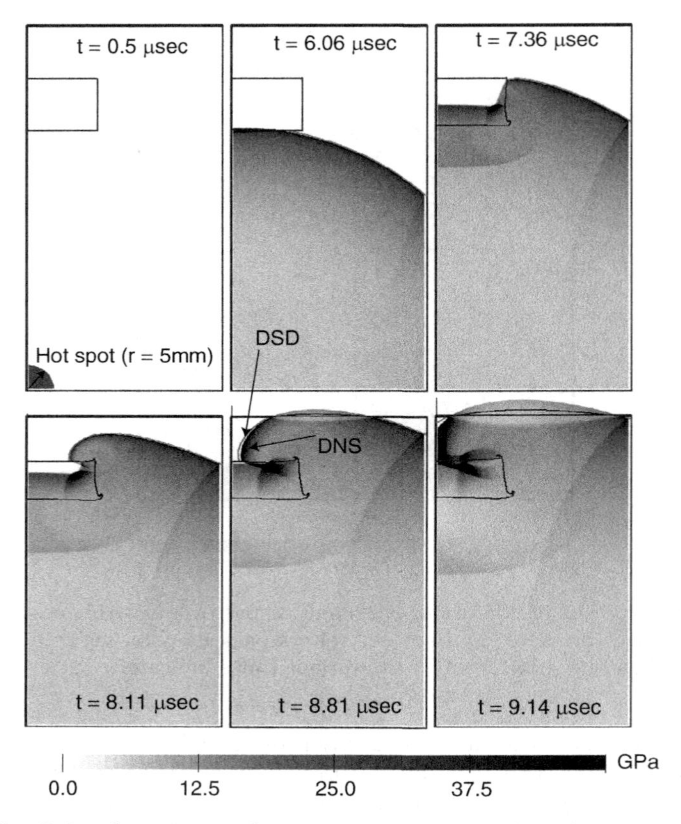

Fig. 7.35. Superimposed comparison of DSD and DNS simulation

mal detonation speed, D_n, is a function of the curvature of the detonation front, κ. With time-dependent analysis, we showed that this $D_n(\kappa)$ quasi-steady law describes time-dependent detonation. This detonation-propagation law applies in the interior of the explosive charge.

We also presented the theory, simulation, and arguments that were instrumental in developing the DSD boundary condition model. What this work shows is that simple shock-polar analysis can be used to obtain the two DSD boundary condition angles, ω_s and ω_c. This conclusion is supported for the most part by high-resolution DNS studies. Using this body of theoretical and numerical work, we have established what boundary conditions need to be applied at the explosive/inert material interface, when the explosive is modeled with first-order, $D_n(\kappa)$, DSD theory.

Finally, we have demonstrated how a DSD propagation law and DSD boundary conditions are used. The level-set function-based, DSD-front propagation code, DSD2D, implements the DSD model and allows detonations to be modeled in complex explosive charges. Versions of DSD2D and DSD3D that can treat multi-component explosive and inert material problems are available. This DSD software automatically reads the explosive geometries from parent hydrodynamic codes and returns the detonation-lighting times to the hydrodynamic codes for the application of the programmed-burn detonation model.

Acknowledgements

Acknowledgement to AFRL/RW, AFOSR/NM, ARDEC, ARL, ONR, DTRA, DOE/LANL, DOE/LLNL for funding support. We thank Tariq Aslam for his help with the shock-fitted calculations reported in Sect. 7.4.

References

1. Aslam, T.D., Bdzil, J.B., Stewart, D.S.: Level set methods applied to modeling detonation shock dynamics. J. Comput. Phys. **126**, 390–409 (1996)
2. Aslam, T.D., Bdzil, J.B.: Numerical and theoretical investigations on detonation-inert confinement interaction. In: Twelfth (International) Detonation Symposium, pp. 483–488. Office of Naval Research, Arlington (2002). ONR 333-05-2
3. Aslam, T.D., Bdzil, J.B.: Numerical and theoretical investigations on detonation confinement sandwich tests. In: Thirteenth (International) Detonation Symposium, pp. 761–769. Office of Naval Research, Arlington (2006). ONR 351-07-01
4. Aslam, T.D., Bdzil, J.B., Hill, L.G.: Extensions to DSD theory: Analysis of PBX 9502 rate stick data. In: Eleventh (International) Detonation Symposium, pp. 21–29. Office of Naval Research, Arlington (1998). ONR 33300-5
5. Bdzil, J.B.: Steady-state two-dimensional detonation. J. Fluid Mech. **108**, 195–226 (1981)
6. Bdzil, J.B., Aida, T., Henninger, R.J., Walter, J.W.: Test problems for DSD3D, Los Alamos Natl. Lab. Rep. LA-14336 (2006)
7. Bdzil, J.B., Stewart, D.S.: Two-dimensional time-dependent detonation. J. Fluid Mech. **171**, 1–26 (1986)
8. Bdzil, J.B., Stewart, D.S.: Modeling two-dimensional detonations with detonation shock dynamics. Phys. Fluids A. **1**, 1261–1267 (1989)
9. Bdzil, J.B., Aslam, T.D., Stewart, D.S.: Curved detonation fronts in solid explosives: Collisions and boundary interactions. In: Hornung, H.G., Shepherd, J.E., Sturtevant, B. (eds.) Twentieth Symposium on Shock Waves, pp. 97–106. Springer, Berlin (1995)
10. Bdzil, J.B., Stewart, D.S.: The dynamics of detonation in explosive systems. In: Annual Review of Fluid Mechanics, vol. 39, pp. 263–292. Annual Reviews, Palo Alto (2007)
11. Birkhoff, G., Rota, G.-C.: Ordinary Differential Equations. Wiley, New York (1969)
12. Courant, R., Friedrichs, K.O.: Supersonic Flow and Shock Waves. Interscience Publishers, New York (1948)

13. Davis, W.C.: Shock waves; Rarefaction waves; Equations of state. In: Zukas, J.A., Walters, W.P. (eds.) Explosive Effects and Applications, pp. 97–106. Springer, Berlin (2003)
14. Fedkiw, R.P., Aslam, T.D., Merriman, B., Osher, S.: A non-oscillatory Eulerian approach to interfaces in multimaterial flows (the ghost fluid method). J. Comput. Phys. **152**, 457–492 (1999)
15. Fickett, W., Davis, W.C.: Detonation: Theory and Experiment. Dover, New York (1979)
16. Henrick, A.K., Aslam, T.D., Powers, J.M.: Simulations of pulsating one-dimensional detonations with true fifth order accuracy. J. Comput. Phys. **213**, 311–329 (2006)
17. Hernandez, A.M., Bdzil, J.B., Stewart, D.S.: A MPI parallel level set algorithm for propagating front curvature dependent detonation shock fronts in complex geometries. Combust. Theory Mod. (submitted) (2012)
18. Hull, L.M.: Mach reflection of spherical detonation waves. In: Tenth (International) Detonation Symposium, pp. 11–18. Office of Naval Research, Arlington (1994). ONR 33395-12
19. James, H.R., Lambourn, B.D.: On the systematics of particle histories in the shock-to-detonation transition regime. J. Appl. Phys. **100**, 084906, (2006)
20. Kasimov, A.R., Stewart, D.S.: On the dynamics of self-sustained one-dimensional detonations: A numerical study in the shock-attached frame. Phys. Fluids **16**(10), 3566–3578 (2004)
21. Klein, R. On the dynamics of weakly-curved detonation. In: Fife, P.C., Linan, A, Williams, F.A. (eds.) Dynamical Issues in Combustion Theory. IMA Volumes in Mathematics and Its Applications, vol. 35, pp. 127–166. Springer, New York (1991)
22. Lambert, D.L., Stewart, D.S., Yoo, S. and Wescott, B.L.: Experimental validation of detonation shock dynamics in condensed explosives. J. Fluid Mech. **546**, 227–253 (2006)
23. Matalon, M., Cui, C., Bechtold, J.K.: Hydrodynamics theory of premixed flames: Effects of stoichiometry, varialble transport coefficients and arbitrary reaction orders. J. Fluid Mech. **487**, 179–210 (2003)
24. McQueen, R.G., Marsh, S.P., Taylor, J.W., Fritz, J.N., Carter, W.J.: The equation of state of solids from shock wave studies. In: Kinslow, R. (ed.) High-Velocity Impact Phenomena, pp. 293–417, 521–568. Academic, New York (1970)
25. Quirk, J.J.: A computational facility for CFD modeling. In: Twenty-Ninth Computational Fluid Dynamics, VKI lecture series, Chap. 5. von Karman Institute, Rhode-St-Genese (1998)
26. Saenz, J., Taylor, B.D., Stewart, D.S.: Asymptotic calculations of the dynamics of self-sustained detonations in condensed phase explosives. J. Fluid Mech. (submitted) (2012)
27. Shapiro, A.: The Dynamics and Thermodynamics of Compressible Fluid Flow. Ronald, New York (1953)
28. Sharpe, G.J., Bdzil, J.B.: Interaction of inert confiners with explosives. J. Eng. Math. **54**(3), 273–298 (2006)
29. Short, M., Quirk, J.J., Kiyanda, C.B., Jackson, S.I., Briggs, M.E., Shinas, M.A.: Simulation of detonation of ammonium nitrate fuel oil mixture confined by aluminum: Edge angles for DSD. In: Fourteenth (International) Detonation Symposium, pp. 769–778. Office of Naval Research, Arlington (2010). ONR 351-10-185

30. Sternberg, H.M., Piacesi, D.: Interaction of oblique detonation waves with iron. Phys. Fluids **9**(7), 1307–1315 (1966)
31. Stewart, D.S., Bdzil, J.B.: A lecture on detonation shock dynamics. In: Buckmaster, J.D., Takeno, T. (eds.) Mathematical Modeling in Combustion Science. Lecture Notes in Physics, vol. 299, pp. 17–30. Springer, Berlin (1988)
32. Stewart, D.S., Bdzil, J.B.: The shock dynamics of stable multidimensional detonation. Combust. Flame. **72**, 311–323 (1988)
33. Stewart, D.S., Yoo, S., Wescott, B.L.: High-order numerical simulation and modeling of the interaction of energetic and inert materials. Combust. Theory Mod. **11**(2), 305–332 (2007)
34. Tarver, C.M., McGuire, E.M.: Reactive flow modeling of the interaction of TATB detonation waves with inert materials. In: Twelfth (International) Detonation Symposium, pp. 641–649. Office of Naval Research, Arlington (2002). ONR 333-05-2
35. Taylor, B.D., Kasimov, A.R., Stewart, D.S.: Shock-fitted simulations of radially symmetric detonations. J. Fluid Mech. (submitted) (2012)
36. Thompson, P.: Compressible-Fluid Dynamics. McGraw-Hill, New York (1972)
37. Wescott, B.L., Stewart, D.S., Davis, W.C.: Equation of state and reaction rate for condensed-phase explosives. J. Appl. Phys. **98**, 053514 (2005)
38. Wood, W.W., Kirkwood, J.G.: Diameter effect in condensed explosives: The relation between velocity and radius of curvature of the detonation wave. J. Chem. Phys. **22**(11), 1920–1924 (1954)
39. Yao J., Stewart, D.S.: On the dynamics of multi-dimensional detonation. J. Fluid Mech. **309**, 225–275 (1996)
40. Yoo, S., Stewart, D.S.: A Hybrid level-set method for modeling detonation and combustion problems in complex geometries. Combust. Theory Mod. **9**(2), 219–254 (2005)

Index

Author Biographies for Volume "Detonation Dynamics"

Authors of Chapter 1

Sorin Bastea

Lawrence Livermore National Laboratory, Physical and Life Sciences Directorate, L-350, 7000 East Ave., Livermore, CA 94550 USA
sbastea@llnl.gov

Sorin Bastea is a staff scientist in the Physical and Life Sciences Directorate at the Lawrence Livermore National Laboratory. He received a Ph.D. in Physics from Rutgers University in 1997 and has been at Lawrence Livermore National Laboratory since 1998, following postdoctoral work at Michigan State University. His main research interests are the statistical mechanics, thermodynamics, transport and kinetics of multi-component fluids, suspensions, multi-phase systems, etc., and applications to high temperature and pressure phenomena. He was a contributor to the CHEQ thermochemical code from 1998 to 2005 and has been leading the Cheetah thermochemical code project since 2006.

Laurence E. Fried

Lawrence Livermore National Laboratory, Physical and Life Sciences Directorate, L-282, 7000 East Ave., Livermore, CA 94550 USA
lfried@llnl.gov

Dr. Laurence Fried is the leader of the Chemistry at Extreme Conditions Group at Lawrence Livermore National Laboratory. Dr. Fried received a Ph.D. in Theoretical Chemistry from Cornell University in 1988. He was a National Science Foundation postdoctoral Fellow at the University of Rochester. He has been a member of the staff at Lawrence Livermore National Laboratory since 1992. Dr. Fried studies high pressure chemistry and energetic materials using a combination of atomistic, statistical mechanical, and macroscopic simulation techniques. Dr. Fried began the Cheetah thermochemical code project in 1994. Dr. Fried is a Fellow of the American Physical Society.

Author Biographies for Volume "Detonation Dynamics"

Author of Chapter 2:

Andrew Higgins

McGill University, Department of Mechanical Engineering, 817 Sherbrooke St. W., Montreal, Quebec, H3A 2K6 Canada
andrew.higgins@mcgill.ca

Andrew Higgins has a bachelor of science in Aeronautical and Astronautical Engineering from the University of Illinois in Urbana/Champaign. His Master and Ph.D. degrees are from the University of Washington, Seattle, both in Aeronautics and Astronautics. He joined the Shock Wave Physics Group of McGill University in 1997 as a Postdoctoral Research Associate, and became an Assistant Professor in the Department of Mechanical Engineering of McGill University in 1999 and an Associate Professor in 2005. His research is in the fields of combustion, detonation, and shock wave physics. His work in gas-phase, condensed-phase, and heterogeneous explosives has included the application of detonation waves to aerospace propulsion, detonation as a means to reach extreme states of matter, an investigation of the existence of solid-solid detonation, and the development of an implosion-driven hypervelocity launcher. He teaches courses in thermodynamics, fluid dynamics, and combustion, and was recognized in 2004 by the Samuel and Ida Fromson Award for Outstanding Teaching (Faculty of Engineering, McGill University).

Authors of Chapter 3:

Hoi Dick Ng

Concordia University, Department of Mechanical and Industrial Engineering, 1455 de Maisonneuve Blvd. W., Montreal, Quebec, H3G 1M8 Canada
hoing@encs.concordia.ca

Dr. Hoi Dick Ng is an Associate Professor of Mechanical and Industrial Engineering at Concordia University, Montreal, Canada. He received his PhD in Mechanical Engineering from McGill University, Montreal, Canada in 2005. Prior to joining Concordia, he worked as a NSERC post-doctoral fellow at Princeton University, USA and at the University of Cambridge, UK. His research interests are in the general field of high-speed chemically reacting flows and compressible fluid dynamics, particularly in the areas of detonations and shock waves.

Author Biographies for Volume "Detonation Dynamics"

Fan Zhang

Defence R&D Canada – Suffield, PO Box 4000, Station Main, Medicine Hat, Alberta, T1A 8K6 Canada
fan.zhang@drdc-rddc.gc.ca

Fan Zhang is a Senior Scientist and the Head of the Advanced Energetics Group at Defence Research and Development Canada – Suffield and an adjunct Professor at the University of Waterloo in the Department of Mechanical Engineering. He specializes in shock waves, detonations and explosions, more specifically in multiphase reactive flow, heterogeneous explosives and high energy density systems. He obtained a Doctoral degree in Science in 1989 from the Aachen University of Technology (RWTH), Germany, and received a Borchers Medal, a Friedrich-Wilhelm Prize and several defence community awards. He is the author or co-author of more than 200 refereed journal and proceedings papers including book chapters and journal special issues. He has served for a number of international defence technical panels and academic committees.

Author of Chapter 4:

Anatoly A. Vasil'ev

Lavrentyev Institute of Hydrodynamics, SB RAS, 630090 Novosibirsk, Russia
gasdet@hydro.nsc.ru

Anatoly A. Vasil'ev is the Director of the Lavrentyev Institute of Hydrodynamics in Novosibirsk and a Professor in Physics at the Novosibirsk State University. He is a Russian State Prize Laureate. He received a D.Sc. at the Novosibirsk State University in physics, and is a leading scientist in physics of combustion and explosion for both experimental methods/instruments and modeling including chemical kinetics. He has published more than 175 scientific papers, reviews and patents. He is a member of the Editorial Board for scientific journals including Combustion, Explosions and Shock Waves, Journal of Engineering Physics and Thermophysics, and Combustion and Plasma-Chemistry. He also serves as a member of the Russian Section of the Combustion Institute and the Institute of Shock Wave, as well as an expert for the Russian Foundation of Basic Researches and INTAS.

Author Biographies for Volume "Detonation Dynamics"

Authors of Chapter 5:

Daniel Desbordes

Ecole Nationale Supérieure de Mécanique et d'Aérotechniques (ENSMA), Laboratoire de Combustion et de Détonique(LCD), 86961 Futuroscope-Chasseneuil, France
desbordes@lcd.ensma.fr

Dr. Daniel Desbordes is a Professor of Reactive Systems Engineering at ENSMA.
He has an extensive research background in the field of gaseous detonation and explosion, especially in detonation micro structures, detonability, detonation propulsion devices (standing oblique detonation, pulse and rotating detonation), deflagration to detonation transition and explosion hazards and related safety. He is the author or co-author of more than 130 publications and received the Numa Manson Medal.

Henri-Noël Presles

Ecole Nationale Supérieure de Mécanique et d'Aérotechniques (ENSMA), Laboratoire de Combustion et de Détonique(LCD), 86961 Futuroscope-Chasseneuil, France
presles@lcd.ensma.fr

Dr. Henri-Noël Presles was a senior scientist at Centre Nationale de la Recherche Scientifique (CNRS). His main areas of research were in three aspects: (i) Detonation in homogeneous and heterogeneous liquid-based explosives and also with hazardous solid mixtures based on lead azide or ammonium nitrate. He performed also some investigations on exotic gaseous reactive systems (e.g., gaseous nitromethane). (ii) Physico-chemical behaviors and properties of inert and reactive shocked liquids. (iii) Improvement of mechanical properties of metals by shock compression.

Author Biographies for Volume "Detonation Dynamics"

Author of Chapter 6:

Craig M. Tarver

Lawrence Livermore National Laboratory, Energetic Materials Center, Livermore, CA 94550 USA
tarver1@llnl.gov

Dr. Craig Tarver is a chemist at Lawrence Livermore National Laboratory. He obtained his Ph.D. degree at The Johns Hopkins University in 1973. He has worked on many aspects of the theory and computer modeling of high explosive and propellant reactions at SRI International (1973–1976) and Lawrence Livermore National Laboratory (1976–present). He is a Fellow of the American Physical Society and has published over 200 papers.

Authors of Chapter 7:

John B. Bdzil

Los Alamos National Laboratory, Los Alamos, NM USA (and) University of Illinois at Urbana-Champaign, Department of Mechanical Sciences & Engineering, Urbana, IL USA
jbbdzil@gmail.com

Dr. John Bdzil is a Fellow of the Los Alamos National Laboratory and Visiting Research Professor in the Department of Mechanical Sciences and Engineering at the University of Illinois, Urbana-Champaign. His research interests include continuum modeling in the areas of multidimensional reactive flows, multiphase flows, physical modeling, and front propagation models. He has worked on the theory and modeling of explosives and detonation since 1972; this has led to numerous publications. Along with Professor D. Scott Stewart, he co-developed both the theory and the software that implements Detonation Shock Dynamics, a model which is widely used to simulate detonation propagation in hydrodynamic codes.

Author Biographies for Volume "Detonation Dynamics"

D. Scott Stewart

University of Illinois at Urbana-Champaign, Department of Mechanical Sciences & Engineering, 1206 West Green Street, Urbana, IL 61801 USA
dss@illinois.edu

D. Scott Stewart is the Shao Lee Soo Professor of Mechanical Science and Engineering at the University of Illinois. He received his Ph.D. in 1981 from Cornell. He has published in the areas of combustion theory, theory for multi-dimensional detonation, detonation stability, continuum mechanics of multi-phase flows, phase transformations, modeling of condensed phase energetic materials, solid rocket motor combustion, advanced level set methods, computational modeling of reactive flow and miniaturization of explosive systems. With John Bdzil, Professor Stewart is the co-developer of detonation shock dynamics. His most recent interests include modeling and large-scale numerical simulation of multi-physics condensed phase reactive flow in meso-scale environments. Professor Stewart is a Fellow of the American Physical Society, (Division of Fluid Dynamics), Fellow of the Institute of Physics and Associate Fellow of AIAA, and won a National Academy of Science Senior Research Fellowship in 2008.

9783642229664